FUNDAMENTALS OF PLASMA PHYSICS

V. E. GOLANT

A. P. ZHILINSKY

I. E. SAKHAROV

A. F. JOFFE PHYSICOTECHNICAL INSTITUTE
LENINGRAD, USSR

SANBORN C. BROWN

EDITOR OF THE ENGLISH LANGUAGE EDITION
PROFESSOR EMERITUS, DEPARTMENT OF PHYSICS
MASSACHUSETTS INSTITUTE OF TECHNOLOGY

TRANSLATED FROM THE RUSSIAN BY
K. Z. VEDENEYEVA AND V. F. AGRANAT

JOHN WILEY & SONS, NEW YORK · CHICHESTER · BRISBANE · TORONTO

Library of Congress Cataloging in Publication Data

Golant, Viktor Evgen'evich.
Fundamentals of plasma physics.

(Wiley series in plasma physics)
Translation of Osnovy fiziki plazmy.
Bibliography: P.
Includes index.
1. Plasma (Ionized gases) I. Zhilinskiĭ, Alekseĭ Petrovich, joint author. II. Sakharov, Igor Evge'evich, joint author. III. Title.
QC718.G6213 530.4'4 79–19650
ISBN 0–471–04593–4

Printed in the United States of America

10 9 8 7 6 5 4 3 2 1

PREFACE

The book is based on a course of lectures in plasma physics, which were given by the authors to undergraduate and graduate physics students, and research workers at the Leningrad Kalinin Polytechnic Institute and the Joffe Physicotechnical Institute. It is an attempt to present systematically the fundamentals of plasma physics embracing a wide range of conditions, including weakly and highly ionized plasmas both in the absence of a magnetic field and under its considerable effect. We tried our best to combine sufficiently detailed and substantiated quantitative analysis of the processes determining the behavior of plasmas with a qualitative description of the physical pattern of these processes. The limited size of the book prevented the full coverage of the problems associated with the excitation and propagation of waves in plasmas. The reader can familiarize himself with these by using the monographs listed in the bibliography.

This text is translated from the Russian edition published in the USSR in 1977. In the course of translation we introduced some adjustments and corrected the detected errors. The bibliography includes the principal manuals and monographs on plasma physics published in English in recent years.

V. E. Golant
A. P. Zhilinsky
I. E. Sakharov

October 1979

CONTENTS

SYMBOLS

a	Plasma radius
$\mathbf{a}$	Acceleration
b_α	Mobility of α-type particles
c	Light velocity
D_α	Diffusion coefficient of α-type particles
D_A	Ambipolar diffusion coefficient
D_α	Thermal diffusion coefficient
D_V	Diffusion coefficient in velocity space
D_H	Magnetic field diffusion coefficient
d	Dipole moment
$\mathbf{E}$	Electric field strength
e	Electron charge
$\mathscr{E}$	Internal energy
$\mathscr{E}_{lm}$	l- to m-level transition energy
$\mathscr{E}_{0j} = \mathscr{E}_j$	j-Level excitation energy
$\mathscr{E}_i$	Ionization energy
F	Six-dimensional distribution function
$\mathbf{F}$	Force
$f(\vartheta)$	Scattering amplitude
$f_\alpha(v)$	Velocity distribution function of α-type particles
f_v	Total-velocity distribution function

f_k	Energy distribution function
f_0, f_1	Isotropic and directed components of distribution function
$\mathbf{G}$	Force per plasma unit volume
g	Statistical weight
$\mathbf{H}$	Magnetic field strength
$H^{(n)}_{k_1}, k_2 \dots k_n$	L'Hermite–Chebyshev polynomials
$\mathbf{h}$	Unit vector in $\mathbf{H}$ direction
$\hbar = h/2\pi$	Planck constant
J_s	First-kind, sth-order Bessel function
j	Current density
I	Current
i	$\sqrt{-1}$
K	Kinetic energy
k^q	Constant of q process
k	Wave number
$\mathcal{K}_\alpha$	Thermal conductivity coefficient of α-type particles
L	Characteristic plasma size
L_α	Coulomb logarithm at α-type particle collision cross section
l	Length
$\mathbf{M}$	Angular momentum
m_α	α-type particle mass
N	Particle number
n_α	α-Type particle concentration
n	Charged particle concentration
P	Power
P_E	Power of heating of plasma unit volume by electric field
$\mathbf{p}_\alpha$	α-Type particle momentum
P_α	Flux tensor of α-type particle momentum
p_α	α-Type particle pressure
p_H	Magnetic pressure
$Q_{\alpha\beta}$	Number of collisions of α- and β-type particles per unit time
$\mathbf{Q}_\alpha$	Energy flux density of α-type particles
q_α	Heat flux density of α-type particles
q	Stability margin of toroidal magnetic trap
$\mathbf{R}$	Six-dimensional radius vector
R	Curvature radius

$\mathbf{R}_{\alpha\beta}$	Friction force acting on α-type particles as a result of their collisions with β-type particles
$\mathbf{R}^T_{\alpha\beta}$	Thermal force due to collisions of α- and β-type particles
r	Distance
$\mathbf{r}$	Radius vector
r_a	Atomic radius
r_s	Radius of strong interaction
r_D	Debye radius
S	Area
$S^q_{\alpha\beta}$	Collision term of kinetic equation describing q-type collisions between α- and β-type particles
S_0, S_1	Collision terms of equations for components of electron distribution function
$s^q_{\alpha\beta}$	Total cross section of q-type collisions between α- and β-type particles
$S^t_{\alpha\beta}$	Transport cross section of elastic collisions
$s^n_{\alpha\beta}, S^l_{\alpha\beta}, s^h_{\alpha\beta}$	Summary inelastic collision cross section, cross section of inelastic collisions with low and high energy loss
T	Temperature
T_α	α-Type particle temperature
t	Time
U	Potential interaction energy
$\mathbf{u}$	Directed velocity (average velocity)
u_α	Directed velocity of α-type particles
$\dot{\mathbf{u}}_E, \mathbf{u}_n, \mathbf{u}_T, \mathbf{u}_P$	Directed velocity components associated with motion in the field under the effect of density, temperature, and pressure gradients
$\mathbf{u}_{d\alpha}$	Drift velocity of α-type particles
$\mathbf{u}_p, \mathbf{u}_g$	Wave phase and group velocities
V	Volume
$\mathbf{v}$	Velocity
$\mathbf{v}_\alpha$	α-Type particle velocity
$\mathbf{v}_0$	Center-of-inertia velocity
$\mathbf{v}, \mathbf{v}_{\alpha\beta}$	Relative velocity of α- and β-type particles
$\mathbf{v}'_\alpha, \mathbf{v}'_\beta, \mathbf{v}_{\alpha\beta}$	Postcollision velocity vectors
$v_{T_\alpha} = ((3T_\alpha)m_\alpha)^{1/2}$	Thermal velocity
W	Energy
$\mathbf{w}$	Random velocity component
Z_α	Charge number of α-type particles

α	Recombination coefficient
β	Kinetic-to-magnetic pressure ratio
Γ_α	Flux density of α-type particles
Γ_v	Flux density in velocity space
γ	Instability increment
γ	Adiabatic index
$\Delta q = q' - q$	Changes in q on collision
δ_{lk}	Kronecker delta
δp_x, δx	Quantum-mechanical uncertainty of momentum and coordinate
ε	Dielectric constant
η	Degree of ionization
η_α	Reflection index of α-type particles
ϑ	Scattering angle
$\kappa_{\alpha\beta}$	Coefficient of energy transfer on collisions of α- and β-type particles
Λ	Diffusion length
Λ	Disturbance wavelength (in Chapter 10)
$\lambda_{\alpha\beta}$	Mean free path of α-type particles up to their collision with β-type particles
λ_α	deBroglie wavelength of α-type particles
$\mu_{\alpha\beta}$	Reduced mass on collision of α- and β-type particles
μ	Magnetic moment
$\nu_{\alpha\beta}$	Collision frequency of α- and β-type particles
$\nu^t_{\alpha\beta}(v)$	Transport frequency of elastic collisions
$\nu^n_{\alpha\beta}(v), \nu^h_{\alpha\beta}(v), \nu^l_{\alpha\beta}(v)$	Frequencies of inelastic collisions (total summary frequency, and frequency of collisions with high and low energy losses, respectively)
$\bar{\nu}_{\alpha\beta}$	Collision frequency averaged with weight v^2
ξ	Impact parameter
ξ	Displacement of plasma volume element
π	Viscosity tensor
ρ_e	Charge density
ρ	Density of matter
$\rho_{H\alpha}$	Larmor radius of α-type particles
Σ	Statistical sum
σ	Conductivity
$\sigma_{\alpha\beta}(\vartheta)$	Differential collision cross section of α- and β-type particles
τ	Characteristic time

$\tau_{\alpha\beta}$	Average time between collisions of α- and β-type particles
$\tau_{\alpha\beta}^{T}$	Average time of energy exchange between α- and β-type particles
Φ	Magnetic flux
$\varphi(r)$	Electrostatic potential
φ	Meridional angle
χ_α	Temperature conductivity coefficient of α-type particles
Ψ	Wave function
ψ	Scattering angle (meridional)
Ω	Solid angle
ω	Frequency
$\omega_{H\alpha}$	Cyclotron frequency α-type particles
ω_p	Plasma frequency
$d^3r = dx\,dy\,dz$	Volume element
$d^3v = dv_x\,dv_y\,dv_z$	Volume element in velocity space
$d^3R = d^3r\,d^3v$	Volume element in six-dimensional phase space

OPERATORS

grad	Gradient
div	Divergence
∇^2	Laplace operator
grad_v	Gradient in velocity space
div_v	Divergence in velocity space
$\delta/\delta t$	Operator denoting rate of variation of a value on collisions
$\langle\ \rangle$	Averaging operator

SUBSCRIPTS

a	Neutral atom
A	Ambipolar characteristic
e	Electron
e	Region outside plasma
E	Value associated with electric field
g	Value of parameter at plasma boundary
H	Value associated with magnetic field
i	Region inside plasma
i	Ion

j	Value associated with current
t	Direction parallel to force acting in plane perpendicular to magnetic field
α, β	Particle type
$\parallel$	Direction parallel to magnetic field
$\perp$	Direction perpendicular to magnetic field

SUPERSCRIPTS

e	Elastic collisions
E	Value associated with electric field
h	Inelastic collisions with high energy losses
H	Value associated with magnetic field
i	Ionization
l	Inelastic collisions with low energy losses
n	Inelastic collisions
q	Type of collision

FUNDAMENTALS OF PLASMA PHYSICS

1

INTRODUCTION

1.1 IONIZED GASES AND PLASMAS

At a nonzero absolute temperature any gas has a certain number of ionized atoms; that is, some charged particles—electrons and ions—are present along with the neutrals. However, the charged particles substantially affect the properties of the gas only at concentrations at which the space charge formed by them restricts their motion. As the concentration increases, this restriction becomes more and more stringent, and at sufficiently high concentrations the interaction of positively and negatively charged particles results in persistent macroscopic neutrality in volumes commensurate with that of the gas. Then any disturbances of macroscopic neutrality induce strong electric fields, which quickly restore it. An ionized gas at such concentrations is called *a plasma*. This term was proposed in 1923 by the American physicist Langmuir.

Thus at sufficiently high charged particle concentrations an ionized gas turns into a plasma. The most natural method for obtaining a plasma is to heat the gas to temperatures at which the average energy of the particles is comparable with the ionization energy of the atoms or molecules. At temperatures much below the ionization energy the ratio between the concentration of ions and neutral atoms is small. It increases with temperature, and when the average particle energy approaches the ionization energy, the gas almost completely turns into an ionized plasma.

Because a plasma can be obtained by heating a substance in the gaseous state (the third state of aggregation), it is sometimes called *the fourth state of matter*. The state of an equilibrium plasma, as well as of any gas, depends on its composition, components concentration, and temperature. Let us denote the partial concentration of the plasma components by n_α, where the subscript α stands for a(neutral particles),

i (ions), or e (electrons), as the case may be. Generally speaking, a plasma may contain more than one species of atoms and ions. Unless otherwise specified, we refer here to so-called *simple plasma,* which consists of neutral particles of one species, single-charged ions of the same species, and electrons. Then the *degree of ionization* η can be defined as the ratio of the ion concentration to the total concentration of ions and neutral atoms:

$$\eta = \frac{n_i}{n_i + n_a}. \tag{1.1}$$

The plasma temperature T is expressed in energy units; it is related to the generally used expression for the temperature

$$T = kT_K \tag{1.2}$$

where T_K is the Kelvin temperature and k is the Boltzmann constant. The relationship between the average energy of thermal motion of particles W and the plasma temperature is given, as for any equilibrium gas, by the following equality:

$$W = \tfrac{3}{2}T \tag{1.3}$$

In an equilibrium plasma the prescribed concentrations and temperature completely characterize its state. The temperature of such a plasma determines not only the average energy, but also the particle velocity distribution (Maxwellian distribution). From the plasma concentration and temperature one can find the degree of ionization, the concentration of ions, excited atoms, photons, and so on. However, not always can a plasma be considered at equilibrium. In particular, a gas-discharge plasma, which is usually obtained in the laboratory, greatly deviates from equilibrium. One occasionally comes across so-called partial equilibrium, at which the velocity distribution of charged and neutral particles is Maxwellian, but the temperatures determining this distribution for the electrons and heavy particles are different. For such a nonisothermal plasma one can introduce the electron and ion temperatures T_e and T_i. In the general case of a nonequilibrium plasma, the velocity distribution of charged particles may be essentially non-Maxwellian. But here, too, we speak of the temperature of the plasma components, defining it as the measure of the average energy of random motion of particles in accordance with Eq. 1.3. Naturally, to obtain complete information on the behavior of a nonequilibrium plasma it is not sufficient to know the average energies (temperature) of the components; it is also necessary to know the particle velocity distribution function.

1.2 QUASI-NEUTRALITY OF PLASMAS

A characteristic feature of plasmas is their macroscopic neutrality, which is maintained because of the mutual compensation of the space charge of the positive ions and electrons. This compensation, however, takes place only in terms of averages, that is, in sufficiently large volumes and for sufficiently long time intervals. Therefore a plasma is said to be a quasi-neutral medium. The dimensions of the areas and time intervals within which the compensation of the volume charge may be disturbed are called the *space* and *time scales* of *charge separation.*

Let us now determine the space scale of charge separation. Imagine that the neutrality is disturbed in some volume of the plasma. We assume for simplicity that this disturbance is due to the displacement of a plane layer of electrons, which produces layers of a negative and a positive volume charge (Fig. 1.1*a*). The electric field between the layers is equivalent to that of a parallel-plate capacitor. The field strength E is

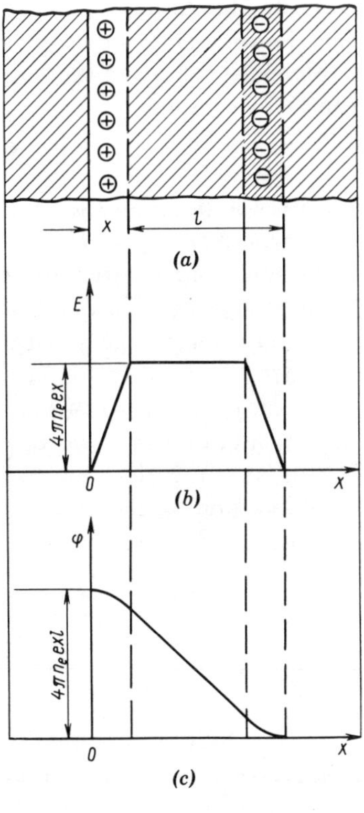

Fig. 1.1 *a*: Layers of negative and positive volume charge. *b*: Distribution of field E. *c*: Distribution of potential ϕ.

determined by the surface density of the charge σ on the plates:

$$E = 4\pi\sigma = 4\pi n_e e x \tag{1.4}$$

where e is the charge, n_e is the electron density, and x is the layer displacement. The distribution of the field and potential φ is shown in Figs. 1.1*b*, *c*. The total potential difference φ_l is equal to

$$\varphi_l \approx El = 4\pi n_e e x l \tag{1.5}$$

(l is the layer thickness). Obviously, the disturbance of neutrality owing to the displacement of the electron layer can be maintained only when the height of the potential barrier of the space charge field is less than the energy of random motion of the electrons and ions: $e\varphi_l < T_e, T_i$. Otherwise the particle motion under the effect of the electric field quickly restores the neutrality. Substituting $\Delta\varphi_e$ into this inequality and assuming $x \approx l$, we obtain $4\pi n_e e^2 l^2 < T$, or

$$l < \sqrt{T/4\pi n_e e^2} \tag{1.6}$$

where T is the smaller of the values T_e and T_i. The quantity on the right-hand side determines the maximum scale of charge separation in the plasma with an accuracy to a numerical factor. This quantity is called the *Debye radius* (or *length*) after the physicist who was the first to introduce it for electrolytes. We denote the Debye radius as

$$r_D = \sqrt{T/4\pi n_e e^2} \tag{1.7}$$

Since it is widespread in plasma physics and often appears in the text, we also write its numerical value $r_D \approx 500\sqrt{T/n}$, where r_D is given in cm, n in cm^{-3}, and T in eV. When the electron and ion temperatures differ substantially, sometimes the electron and ion Debye radii are introduced: $r_{D_e} \sim \sqrt{T_e}$ and $r_{D_i} \sim \sqrt{T_i}$. The scale of long-term charge separation is characterized by the smaller of these values.

Let us determine the time scale of charge separation. To do this we turn again to Fig. 1.1 and consider the electron motion after the disturbance of neutrality. At the site of the electron layer the electrons are acted on by the force of attraction from the ion side equal to $eE = 4\pi n_e e^2 x$ (see Eq. 1.4). The electron motion equation has the form

$$\frac{m_e d^2 x}{dt^2} = -4\pi n_e e^2 x \tag{1.8}$$

It describes harmonic oscillations with a frequency of

$$\omega_p = \sqrt{4\pi n_e e^2/m_e} \tag{1.9}$$

Since the quantity ω_p is often found in plasma physics, we also give its numerical value: $\omega_p = 5.6 \times 10^4 \sqrt{n_e}$ (where ω_p is given in $\sec^{-1}$ and n_e in cm^{-3}).

It is easy to understand the nature of these oscillations of the electron layer. The layer is attracted to the ion layer, passes by it by inertia, is attracted again, and so on. The oscillations do not damp out, because the thermal motion of the charged particles and the dissipation are not considered. Oscillations of a space charge upon disturbance of quasi-neutrality were first discovered by Langmuir. They are termed *plasma*, or *Langmuir*, oscillations. The frequency (Eq. 1.9) is accordingly called the *plasma*, or *Langmuir*, frequency.

Plasma oscillations determine the quasi-neutrality restoration mechanism. Obviously, on the average over many oscillation periods a plasma can be considered neutral. Therefore the time scale of charge separation in a plasma is determined by

$$t_D \approx \frac{1}{\omega_p} = \sqrt{m_e/4\pi n_e e^2} \tag{1.10}$$

Its relationship with the space scale of charge separation (Eq. 1.7) is very simple:

$$t_D = \frac{r_{D_e}}{v_{T_e}} \tag{1.11}$$

where $v_{T_e} \approx (3T_e/m_e)^{1/2}$ is the thermal velocity of the electrons.

We have estimated the scales of spontaneous disturbances of the macroscopic neutrality of a plasma owing to the thermal motion of charged particles. Let us now consider the disturbances of neutrality under the effect of external electric fields. If a charged body is introduced into a plasma or moves near its boundary, charge separation occurs in the vicinity of this body—the unlike charges are attracted to the body, and the like charges are repulsed from it. Plasma polarization results in the shielding of the external field. The characteristic space scale of this shielding is equal to the Debye radius.

Consider, for instance, the field of a fixed spherical charge in a plasma. The field potential φ outside the charge must satisfy the Poisson equation

$$\Delta\varphi = -4\pi\rho \tag{1.12}$$

For a plasma containing single-charged ions and electrons the space charge density is determined by the difference of their concentrations:

$$\rho = e(n_i - n_e) \tag{1.13}$$

If the shielding region contains a great number of charged particles the instantaneous values of the concentrations can be assumed to be practically equal to their average values. The relationship between these average concentrations and the potential with a Maxwellian particle velocity distribution is determined by the Boltzmann equation (see Section 4.1):

$$n_i = n\exp\left(\frac{-e\varphi}{T_i}\right); \qquad n_e = n\exp\left(\frac{e\varphi}{T_e}\right) \tag{1.14}$$

where n is the charged particle concentration in the unperturbed region (in which the plasma is neutral; i.e., $n_e = n_i = n$). Substituting the expressions for n_i and n_e into the Poisson equation, we obtain the self-consistent equation for the potential:

$$\Delta\varphi = -4\pi e n\left[\exp\left(\frac{-e\varphi}{T_i}\right) - \exp\left(\frac{e\varphi}{T_e}\right)\right] \tag{1.15}$$

It is easy to find the solution of Eq. 1.15 for sufficiently large distances from the charge, when $e\varphi \ll T_i, T_e$. At such distances we can expand the exponents in power series and restrict ourselves to the first two terms of the expansion:

$$\exp\left(\frac{-e\varphi}{T_i}\right) \approx \frac{1-e\varphi}{T_i}; \qquad \exp\left(\frac{e\varphi}{T_e}\right) \approx \frac{1+e\varphi}{T_e}$$

Substituting them into Eq. 1.15, we reduce it to the form

$$\Delta\varphi - \frac{1}{r_D^2}\varphi = 0 \tag{1.16}$$

where

$$r_D = \left[\frac{T_e T_i}{4\pi n e^2(T_e + T_i)}\right]^{1/2} \tag{1.17}$$

is the general expression for the Debye radius for an arbitrary ratio between T_e and T_i. For $T_e \gg T_i$ or $T_i \gg T_e$ this expression changes to Eq. 1.7. The spherically symmetric solution of the equation, which vanishes at infinity, has the form

$$\varphi = \frac{C}{r}\exp\left(\frac{-r}{r_D}\right)$$

This is easy to ascertain by substituting it into Eq. 1.16.

For this solution to turn into the usual *coulomb potential* of the charge q at small distances, when there is no shielding, we must put $C = q$. Then the expression for the potential can be written

$$\varphi = \frac{q}{r}\exp\left(\frac{-r}{r_D}\right) \tag{1.18}$$

It is seen that for $r \ll r_D$ the potential practically coincides with the coulomb, and for $r > r_D$ is much less than the coulomb because of the shielding effect of the plasma. Thus the characteristic scale of the shielding region is determined by the Debye radius. This result has been obtained for the field of a spherical charge under conditions where the potential energy $e\varphi \approx qe/r_D$ in the shielding region is less than the thermal energy. The nature of shielding when $e\varphi \leqslant T_e$ must evidently remain the same for any other method of creating fields, because it can be described by a set of point charges. At potentials exceeding this limit the characteristic scale of the region of quasi-neutrality disturbance is also often of the order of Debye radius.

The above consideration makes it possible to determine the conditions of quasi-neutrality more accurately. According to it interaction of space charges of electrons and ions maintains the electric neutrality of the plasma in volumes substantially exceeding r_D and for periods much longer than the reciprocal plasma frequency $t_D \approx 1/\omega_p$. For these conditions to be fulfilled in a plasma, the following inequalities must hold:

$$L \gg r_D; \qquad \tau \gg t_D \tag{1.19}$$

where L is the characteristic plasma dimension and τ is the characteristic time of variation in its parameters. In fact, these inequalities determine the charged particle concentration above which the ionized gas can be called a plasma. They are usually fulfilled with a large safety margin. Therefore even in relatively small volumes, much smaller than that of the plasma, the positive and negative charges offset each other. For a plasma in which the negative charge is produced by the electrons and the positive one by single-charged ions, their concentrations must be practically equal:

$$n_e \approx n_i \gg |n_i - n_e| \tag{1.20}$$

Bearing this in mind, we sometimes omit the subscripts e and i at the charged particle concentration.

1.3 CHARACTERISTICS OF CHARGED PARTICLE MOTION IN PLASMAS

The basic properties of a plasma are determined by the motion of charged particles in it. Here we only outline some of its peculiarities. In the absence of external fields the motion of charged particles in a weakly ionized plasma, where it is mainly affected by collisions with neutral particles, is similar to that of atoms in an ordinary gas. When the mean free path is much shorter than the plasma dimension, the trajectory consists of more or less extended line segments corresponding to intercollisional periods and of collision regions within which the particle direction and velocity change (Fig. 1.2*a*). In this case we actually deal with point collisions, since the effective radius of interaction between the charged particles and neutrals is much shorter than the mean free path.

The considerable difference between the properties of a plasma and those of a neutral particle gas is due primarily to the effect exerted by the electric and magnetic fields on the charged particle motion. The electric field (external field and space charge field) accelerates the charged particles in the intercollisional period. On the average, over a large number of periods such acceleration gives rise to directed motion of particles and to increased random motion velocity; that is, it results in the heating of the plasma. The electron and ion components are heated to a different extent. The electrons usually acquire a higher energy than the ions. The magnetic field twists the paths of the charged particles in a plane perpendicular to the field. If the gyroradius of the charged particles in the magnetic field (so-called *Larmor radius*) is much shorter than the mean free path, the effect of the field on the particle motion is very substantial. Their displacement in a direction perpendicular to the magnetic field over distances exceeding the Larmor radius may depend on charged particle collisions or drift associated with the electric field or with the inhomogeneity of the magnetic field.

Another important peculiarity of the charged particle motion shows up in a highly ionized plasma, when the collisions of charged particles with each other play the determining role. This peculiarity is due to a slow decrease of the coulomb potential (which determines interaction between the charged particles) with the distance. Therefore the interaction is considerable at distances greatly exceeding the size of the atoms. The limited range of the coulomb forces is associated only with the shielding of the fields of interaction in the plasma; the limiting radius may be assumed equal to the Debye shielding radius. Since a sphere with a radius equal to the Debye usually contains a great number of electrons

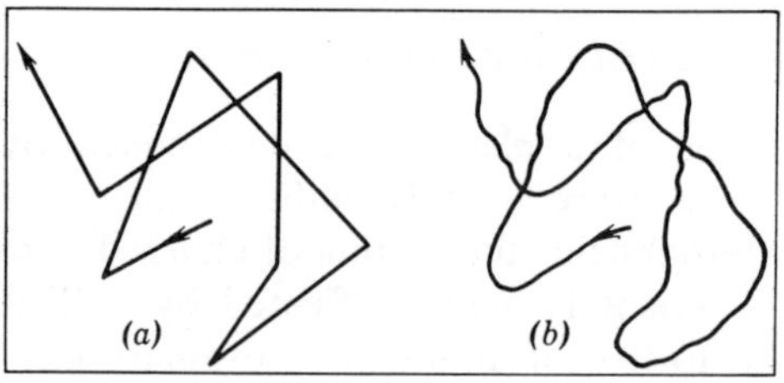

Fig. 1.2 Trajectories of particles on collision. (a) in a weakly ionized plasma; (b) in a highly ionized plasma.

and ions, charged particle interaction does not, strictly speaking, amount to pair collisions. Because of these collisions the particle path can no longer be represented as a polygonal line with clearly defined periods of collisions and intercollisional periods; it becomes more intricate (see Fig. 1.2*b*). In our analysis we have to consider the interaction of each particle with many others, which are within the Debye sphere. Since, however, the interaction with distant particles (at distances of the order of Debye radius) causes the particles to deviate by small angles, we can, by approximation, replace the collective interaction by a set of pair collisions. This cannot be done at high concentrations of charged particles, when the energy of interaction between them becomes comparable with thermal energy. The properties of a plasma in this case differ substantially from those of an ideal gas, for which the pair collision model is applicable.

Let us estimate the concentration determining this transition. The potential energy of interaction at distances of the order of the Debye radius is given by the relation (see Eq. 1.7)

$$U \approx \frac{e^2}{r_D} = e^3\sqrt{4\pi n/T} \tag{1.21}$$

It is small compared with thermal energy ($U \ll T$), provided that

$$n \ll \frac{T^3}{4\pi e^6} \tag{1.22}$$

or $n \ll 10^{19}T^3$ (here n is given in cm^{-3} and T in eV). It is the inequality 1.22 that determines the conditions of "weak" interaction, which are characteristic of an ideal gas. It usually holds good.

To conclude, we note that in considering the particle motion in a plasma during the intercollisional period we can practically always use the classical description. To demonstrate this, we examine the restrictions imposed by the quantum-mechanical uncertainty relation. At a concentration of charged particles equal to n the average distance between them is $\delta r \approx n^{-1/3}$. It determines the maximum permissible uncertainty in the coordinate when describing the particle motion. The uncertainty relation makes it possible to find the associated momentum uncertainty

$$\delta p \approx \frac{\hbar}{\delta r} \approx \hbar n^{1/3} \tag{1.23}$$

($\hbar$ is the Planck constant). If the classical description is to be used, this uncertainty must be much less than the average particle momentum:

$$\delta p \ll \bar{p} \approx \sqrt{mT} \tag{1.24}$$

Using Eq. 1.23, we obtain the condition for the applicability of the classical description of the electron motion:

$$n \ll \frac{(m_e T_e)^{3/2}}{\hbar^3} \tag{1.25}$$

or $n \ll 10^{23} T_e^{3/2}$ (n is in cm^{-3} and T_e in eV). This condition is satisfied for practically all kinds of gas plasma (see also Section 1.4, Fig. 1.3). Therefore we can say that a plasma is a classical object. Sometimes, we imply by the term quantum solid plasma the behavior of current carriers in solids—electrons in metals or electrons and holes in semiconductors. For such a plasma a condition of the type 1.24 breaks down, and its properties must be described in terms of quantum mechanics.

1.4 PLASMA PARAMETERS

We give the parameters of some typical kinds of plasma. We begin with plasma objects in outer space. Most of the stars, including our sun, are known to have temperatures at which matter is in the plasma state. The interstellar gas is ionized and, despite the relatively low concentration of charged particles, it can also be considered a plasma, because its characteristic space scale exceeds the Debye radius by many orders. Plasma is also widespread in the near space. It fills the earth's magnetosphere and forms ionospheric layers. Magnetosphere perturbations associated with particle fluxes from the sun (so-called solar wind) are also found to involve plasma. Table 1.1 lists the plasma parameters for some of these space objects.

Various kinds of gas discharge find wide application in technology and experimental investigations. These are primarily stationary or pulsed electrode discharges—so-called glow discharges with cold electrodes, which occur at relatively low currents, and arc discharges characterized by high currents and strong electrode heating. This type of discharge has been used for a long time in radio engineering, current commutation, and machining. Relatively recently it found use in the pumping of gas lasers. Gas-discharge plasma sources—plasmatrons—have lately acquired extensive application in many chemical engineering ventures, primarily for triggering high-temperature chemical reactions (in so-called plasma chemistry). Along with the traditional electrode discharges, use is made of a plasma induced by high-frequency fields (HF and SHF discharges) under the effect of laser radiation (laser discharges). Table 1.2 presents the typical parameters of some kinds of gas-discharge plasma.

Plasma is the object of many experiments. Investigations on plasma confinement by means of a magnetic field and on its heating have been

TABLE 1.1 Plasma parameters of some space objects

Objects	Gas	L (cm)	H (Oe)	n_a (cm^{-3})	n_e (cm^{-3})	η	T_e (eV)	λ_e (cm)	r_D (cm)	ρ_{He} (cm)
Ionosphere (E-layer)	Air	10^6	1	$10^{12}-10^{13}$	10^5	10^{-8}–10^{-7}	0.03	10^2–10^3	0.3	1
Ionosphere (F-layer)	Air	10^6–10^7	1	10^9–10^{12}	10^5–10^6	10^{-7}–10^{-4}	0.03–0.1	10^3–10^5	0.2	1
Solar photosphere	H	5×10^{10}	1(10^3)	—	10^{14}	1	1	5×10^{-2}	5×10^{-4}	3 (3×10^{-3})
Solar crown	H	10^{11}–10^{12}	1(10^2)	—	10^4–10^8	1	100	10^8	1–100	30 (0.3)
Interstellar space	H	10^{18}	—	—	10^{-3}–10	0.1–1	0.01–1	10^8-10^{12}	10^2–10^4	—

Notes: 1. The table is of an illustrative nature. Most of the parameters are given with an accuracy to one order of magnitude.
2. In the table, L is the characteristic plasma size, H is the magnetic field strength, λ_e is the electron mean free path, and ρ_{He} is the Larmor radius of electrons.
3. The figures in parentheses refer to solar spot fields.

TABLE 1.2 Parameters of gas-discharge plasma

Type of discharge	p (mm Hg)	a (cm)	I (A)	P (W/cm^3)	n_e (cm^{-3})	η	T_e (eV)	T_a (eV)	λ_e (cm)	r_D (cm)
Low-pressure discharge	10^{-2}	1	1	10	10^{11}	3×10^{-4}	3–7	3×10^{-2}	1	3×10^{-3}
Glow discharge	1	1	10^{-2}	10^{-1}	10^{10}	3×10^{-7}	1–3	3×10^{-2}	10^{-2}	5×10^{-2}
Arc discharge	1	1	10	10–10^2	10^{13}	3×10^{-4}	0.5–2	10^{-1}	10^{-2}	3×10^{-4}
Ultrahigh-frequency discharge	10	1	—	10	10^{12}	3×10^{-6}	1–3	5×10^{-2}	10^{-3}	5×10^{-3}
High-pressure discharge	10^3	1	1	10^2	10^{15}	3×10^{-5}	0.5–1	0.5	10^{-4}	10^{-5}
Superhigh-pressure discharge	10^5	10^{-1}	1	10^3–10^4	10^{17}	3×10^{-5}	0.5–1	0.5–1	10^{-6}	10^{-7}
Stationary laser plasma	10^3	10^{-1}	—	10^4	10^{17}	3×10^{-3}	1–3	1–3	10^{-4}	10^{-7}

Notes:
1. The table lists tentative characteristics of discharges in some typical regime. They may vary appreciably depending on the regime and gas filling. The parameter values are given with an accuracy to one order of magnitude.
2. In the table p is the gas pressure, a is the plasma radius, I is the discharge current, P is the power introduced into a unit volume of the plasma, T_e is the electron temperature, T_a is the gas temperature, and λ_e is the electron mean free path.

FUNDAMENTALS OF PLASMA PHYSICS

WILEY SERIES IN PLASMA PHYSICS

SANBORN C. BROWN ADVISORY EDITOR
RESEARCH LABORATORY OF ELECTRONICS
MASSACHUSETTS INSTITUTE OF TECHNOLOGY

BEKEFI · RADIATION PROCESSES IN PLASMAS
BROWN · ELECTRON-MOLECULE SCATTERING
BROWN · INTRODUCTION TO ELECTRICAL DISCHARGES IN GASES
GILARDINI · LOW ENERGY ELECTRON COLLISIONS IN GASES
GOLANT, ZHILINSKY, AND SAKHAROV · FUNDAMENTALS OF PLASMA PHYSICS
HEALD AND WHARTON · PLASMA DIAGNOSTICS WITH MICROWAVES
HUXLEY AND CROMPTON · THE DIFFUSION AND DRIFT OF ELECTRONS IN GASES
LICHTENBERG · PHASE-SPACE DYNAMICS OF PARTICLES
MACDONALD · MICROWAVE BREAKDOWN IN GASES
McDANIEL · COLLISION PHENOMENA IN IONIZED GASES
McDANIEL AND MASON · THE MOBILITY AND DIFFUSION OF IONS IN GASES
MITCHNER AND KRUGER · PARTIALLY IONIZED GASES
NASSER · FUNDAMENTALS OF GASEOUS IONIZATION IN PLASMA ELECTRONICS
TIDMAN AND KRALL · SHOCK WAVES IN COLLISIONLESS PLASMAS

TABLE 1.3 Plasma parameters in experiments on controlled fusion

Type of experiment	a (cm)	H(kOe)	n_e (cm^{-3})	T_e (eV)	T_i (eV)	τ_E (sec)	λ_e (cm)	r_D (cm)	ρ_{He} (cm)	ρ_{Hi} (cm)
Tokamak	20	40	5×10^{13}	3×10^{3}	10^{3}	10^{-2}	10^{6}	5×10^{-3}	5×10^{-3}	0.1
Theta-pinch	1	100	10^{17}	10^{3}	5×10^{3}	10^{-5}	100	5×10^{-5}	10^{-3}	0.03
Laser experiment	1	—	10^{22}	300	300	—	10^{-5}	5×10^{-8}	—	—
Quasi-stationary thermo-nuclear reactor	300	50	10^{14}	10^{4}	10^{4}	5	10^{5}	10^{-3}	10^{-3}	0.5
Laser thermonuclear reactor	0.01–1	—	10^{22}–10^{24}	10^{4}	10^{4}	10^{-8}–10^{-10}	10^{-4}–10^{-2}	10^{-6}–10^{-7}	—	—

Notes:
1. The data listed in the table have been obtained for a plasma set up in hydrogen or deuterium. The degree of plasma ionization is near unity.
2. The first two lines include tentative plasma parameters on typical modern experimental setups, and the last two lines, typical plasma parameters in the prospective thermonuclear reactors discussed in the literature.
3. In the table a is the plasma radius, H is the magnetic field strength, τ_E is the time of retention of the energy in the plasma, T_e, T_i are the electron and ion temperatures, respectively, ρ_{He}, ρ_{Hi} are the electron and ion Larmor radii, respectively, and λ_e is the electron mean free path.

conducted in recent years in connection with nuclear fusion. They embrace a wide range of conditions corresponding to different schemes, from quasi-stationary to single-action pulsed ones. Extensive plasma investigations are also being carried out in connection with magnetohydrodynamic and thermionic energy conversion, the development of plasma jet engines, and spacecraft propulsion through the atmosphere. Being unable to embrace the characteristics of all the types of plasma used in these investigations, we only give plasma parameters in typical current experiments on nuclear fusion and the parameters of prospective thermonuclear reactors (Table 1.3). Figure 1.3 gives typical temperatures and concentrations of the following plasma objects: solid plasma (1), ionosphere (2), solar corona (3), gas-discharge plasma (4), laser thermonuclear experiment (5), quasi-stationary thermonuclear experiment (6), laser thermonuclear reactor (projected) (7), and quasi-stationary thermonuclear reactor (projected) (8).

These data cover a very wide range of parameters of plasmas used in

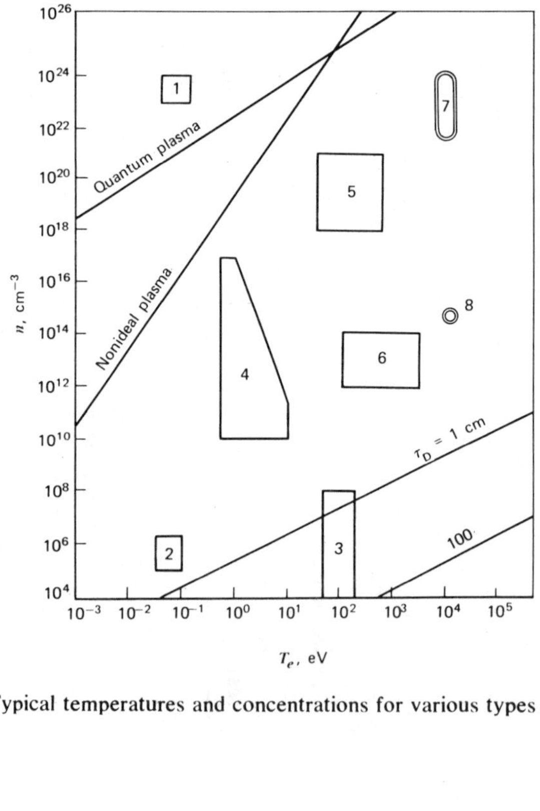

Fig. 1.3 Typical temperatures and concentrations for various types of plasmas.

laboratory investigations and in technology. One can, however, single out the most common ranges of the main parameters. For a low-temperature plasma it is the concentration range from 10^{10} to $10^{18}\,cm^{-3}$ and the electron temperature range from 1 to 10 eV. For a high-temperature plasma in fusion experiments it is the concentration range from 10^{12} to $10^{18}\,cm^{-3}$ and the temperature range of 100–10 keV.

2

COLLISIONS IN PLASMAS

2.1 APPLICATION OF CONSERVATION LAWS TO PARTICLE COLLISIONS

As noted in Chapter 1, a plasma is a multicomponent gas of weakly interacting particles. Therefore, in analyzing its behavior one usually applies the approach adopted in the kinetic theory of gases, which is based on dividing the particle trajectory into collision regions and intercollisional intervals. With this approach, in the particle interaction region the effect of external fields is neglected, and in the intercollisional intervals the particle interaction forces are not taken into account. Since the interaction radius is much shorter than the mean free path, when considering the kinetics of charged particle motion one need not determine their trajectories in the course of collisions; it is sufficient to know the consequences of the collisions, that is, the changes in the velocities and states of the colliding particles. Strictly speaking, the described approach is applicable only to collisions of charged particles with neutrals, and of neutrals with each other. Interaction of charged particles with each other, as already noted, involves larger distances and is collective in nature. As an approximation, however, it can also be regarded as a combination of independent pair binary collisions (see Section 2.4).

In this chapter we discuss different kinds of collision, which are essential for the description of plasma behavior. Some general characteristics of collisions can be expressed in terms of the energy and momentum conservation laws without looking into the specific type of interaction forces. The simplest method is to analyze binary collisions, in which the nature of the particles remains unchanged.

Consider collision of α- and β-type particles. We assume that in collision the particles are not affected by any external forces. The energy and momentum of the system of interacting particles are known to remain unchanged. The system momentum is the sum of the momenta of the colliding particles:

$$\mathbf{p} = \mathbf{p}_\alpha + \mathbf{p}_\beta = m_\alpha \mathbf{v}_\alpha + m_\beta \mathbf{v}_\beta \tag{2.1}$$

where m_α and m_β are the masses, and $\mathbf{v}_\alpha$ and $\mathbf{v}_\beta$ the velocity vectors of the particles. The momentum conservation law leads to the equality of the system's pre- and postcollision momenta:

$$\mathbf{p} = \mathbf{p}'$$

or

$$m_\alpha \mathbf{v}_\alpha + m_\beta \mathbf{v}_\beta = m_\alpha \mathbf{v}'_\alpha + m_\beta \mathbf{v}'_\beta \tag{2.2}$$

(hereafter the values characterizing the postcollision particle motion are primed). The energy conservation law makes it possible to determine the change in the net kinetic energy of the system in collision:

$$K = K_\alpha + K_\beta = \frac{m_\alpha v_\alpha^2}{2} + \frac{m_\beta v_\beta^2}{2} \tag{2.3}$$

which is determined by the equality $K = K' + \Delta\mathscr{E}$, or

$$\frac{m_\alpha v_\alpha^2}{2} + \frac{m_\beta v_\beta^2}{2} = \frac{m_\alpha v_\alpha'^2}{2} + \frac{m_\beta v_\beta'^2}{2} + \Delta\mathscr{E} \tag{2.4}$$

where $\Delta\mathscr{E}$ is the net change in the internal particle energy as a result of collision.

It is customary to distinguish *elastic* collisions, which do not affect the internal state of the particles, from *inelastic* collisions. In elastic collisions, $\Delta\mathscr{E}$ is clearly equal to zero. Among the inelastic collisions one distinguishes *collisions of the first kind*, in which $\Delta\mathscr{E} > 0$, from *collisions of the second kind*, with $\Delta\mathscr{E} < 0$. An example of the former is collisions causing the atoms to shift from the ground to the excited state. An example of the latter are collisions with opposite transfers. Thus for collisions with a prescribed change in the state of the particles the energy and momentum conservation laws yield four equations limiting the change in particle velocities (equality 2.4 and vector equality 2.2 equivalent to three equations for the velocity components). Therefore only two of the six components of the vectors of the postcollision particle velocity $\mathbf{v}'_\alpha$ and $\mathbf{v}'_\beta$ are independent; they are determined by the forces of interaction between the particles and by their relative positions.

For a more detailed consideration of the constraints imposed by the conservation laws, it is best to switch to the center-of-mass system. The coordinates of the center-of-mass system are known to be related to the coordinates of these particles by

$$\mathbf{r}_0 = \frac{m_\alpha \mathbf{r}_\alpha + m_\beta \mathbf{r}_\beta}{m_\alpha + m_\beta} \tag{2.5}$$

where $\mathbf{r}_\alpha$ and $\mathbf{r}_\beta$ are the radius vectors characterizing the position of α- and β-type particles. Accordingly, the center-of-mass velocity is

$$\mathbf{v}_0 = \frac{d\mathbf{r}_0}{dt} = \frac{m_\alpha \mathbf{v}_\alpha + m_\beta \mathbf{v}_\beta}{m_\alpha + m_\beta} \tag{2.6}$$

It remains constant in collision in conformity with the momentum conservation law. Therefore we can change to a reference system in which the center of inertia is at rest; that is, $\mathbf{v}_0 = 0$. In such a system the particle velocities are related to each other by $m_\alpha \mathbf{v}_{\alpha_0} + m_\beta \mathbf{v}_{\beta_0} = 0$, or

$$\mathbf{v}_{\beta_0} = -\left(\frac{m_\alpha}{m_\beta}\right) \mathbf{v}_{\alpha_0} \tag{2.7}$$

So we can express them via a single vector value, namely, the relative particle velocity:

$$\mathbf{v} = \mathbf{v}_\alpha - \mathbf{v}_\beta = \mathbf{v}_{\alpha_0} - \mathbf{v}_{\beta_0} \tag{2.8}$$

Substituting Eq. 2.7 into Eq. 2.8,

$$\mathbf{v}_{\alpha_0} = \frac{m_\beta \mathbf{v}}{m_\alpha + m_\beta}; \qquad \mathbf{v}_{\beta_0} = -\frac{m_\alpha \mathbf{v}}{m_\alpha + m_\beta} \tag{2.9}$$

Thus the motion of colliding particles is completely determined by the center-of-mass velocity $\mathbf{v}_0$ (Eq. 2.6) and the relative-motion velocity $\mathbf{v}$ (Eq. 2.8). It is easy to express the particle velocities in the laboratory system by using these values. Taking advantage of Eq. 2.9, we find

$$\begin{aligned} \mathbf{v}_\alpha &= \mathbf{v}_0 + \mathbf{v}_{\alpha_0} = \mathbf{v}_0 + \frac{m_\beta}{m_\alpha + m_\beta}\mathbf{v}; \\ \mathbf{v}_\beta &= \mathbf{v}_0 + \mathbf{v}_{\beta_0} = \mathbf{v}_0 - \frac{m_\alpha}{m_\alpha + m_\beta}\mathbf{v} \end{aligned} \tag{2.10}$$

With the aid of Eq. 2.10 we determine the net kinetic energy:

$$K = \frac{m_\alpha v_\alpha^2}{2} + \frac{m_\beta v_\beta^2}{2} = \left(\frac{m_\alpha + m_\beta}{2}\right) v_0^2 + \frac{\mu_{\alpha\beta} v^2}{2} \tag{2.11}$$

The first term on the right-hand side is usually called *the center-of-mass energy*, and the second, the *relative-motion energy*; the quantity

$$\mu_{\alpha\beta} = \frac{m_\alpha m_\beta}{m_\alpha + m_\beta} \tag{2.12}$$

is called *the reduced mass*. Using Eq. 2.11, we can write the energy conservation law for collision (Eq. 2.4):

$$\frac{(m_\alpha + m_\beta)v_0^2}{2} + \frac{\mu_{\alpha\beta}v^2}{2} = \frac{(m_\alpha + m_\beta)v_0'^2}{2} + \frac{\mu_{\alpha\beta}v'^2}{2} + \Delta\mathscr{E}$$

Since in collision the center-of-mass velocity and kinetic energy remain unchanged ($v_0' = v_0$), we get

$$\frac{\mu v^2}{2} = \frac{\mu v'^2}{2} + \Delta\mathscr{E} \tag{2.13}$$

By Eq. 2.13, only the part of the net kinetic energy K corresponding to the relative-motion energy can convert to internal energy.

The relations obtained enable one to determine the restraints imposed on collisions by the conservation laws. As seen from Eq. 2.9, the particle velocities in the center-of-mass system $\mathbf{v}_{\alpha 0}$ and $\mathbf{v}_{\beta 0}$ are proportional to the relative velocity $\mathbf{v}$. In accordance with Eq. 2.13 they are unambiguously determined by the energy conservation law. Therefore only the angles determining the rotation of the relative velocity vector $\mathbf{v}$ and hence of the vectors $\mathbf{v}_{\alpha 0}$ and $\mathbf{v}_{\beta 0}$ are the independent parameters in this system and thus are not limited by the conservation laws.

Let us discuss elastic collisions in more detail. As follows from Eq. 2.13, in elastic collisions, when $\Delta\mathscr{E} = 0$, the relative velocity vector does not change its value, but merely reverses its direction. We define the rotation of the vector by two angles ϑ and ψ corresponding to the spherical system of coordinates (Fig. 2.1). The angle ϑ, which is the angle between the vectors of pre- and postcollision velocities $\mathbf{v}$ and $\mathbf{v}'$, is called the scattering angle. It is clearly determined by the colliding-particles spacing and the forces of their interaction. The meridional

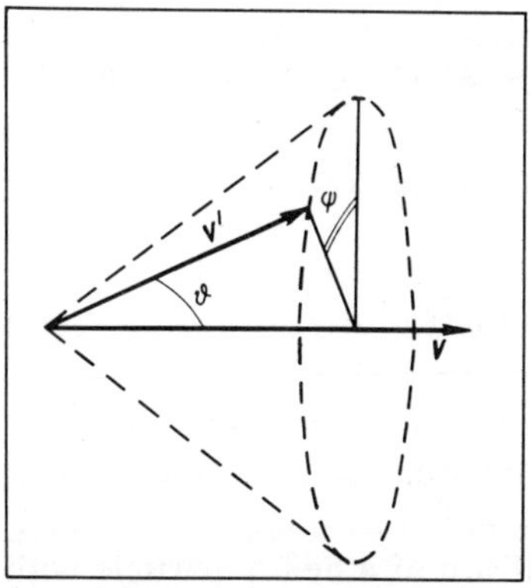

Fig. 2.1 Spherical system of coordinates.

angle ψ determines the position of the interaction plane (plane in which the vectors $\mathbf{v}$ and $\mathbf{v}'$ lie) relative to some fixed plane through $\mathbf{v}$. With centrally symmetric interaction forces this angle depends exclusively on the relative positions of the colliding particles.

Express the change in the momentum and kinetic energy of colliding particles in terms of angles ϑ and ψ. The change in the momentum of an α-type particle

$$\Delta \mathbf{p}_\alpha = m_\alpha \mathbf{v}'_\alpha - m_\alpha \mathbf{v}_\alpha = m_\alpha \mathbf{v}'_{\alpha_0} - m_\alpha \mathbf{v}_{\alpha_0} \tag{2.14}$$

can be related, by means of Eq. 2.9, to the change in relative velocity vector:

$$\Delta \mathbf{p}_\alpha = \frac{m_\alpha m_\beta}{m_\alpha + m_\beta}(\mathbf{v}' - \mathbf{v}) = \frac{m_\alpha m_\beta}{m_\alpha + m_\beta}\Delta \mathbf{v} \tag{2.15}$$

The quantity $\Delta \mathbf{v}$ can be expressed via angles ϑ and ψ. To do this, we represent it as a sum of three projections (see Fig. 2.1): a projection on the direction of $\mathbf{v}$, that is, $\Delta v_{(1)} = v \cos \vartheta - v = -v(1 - \cos \vartheta)$ and two projections on the perpendicular directions of $\Delta v_{(2)} = v \sin \vartheta \cos \psi$, and $\Delta v_{(3)} = v \sin \vartheta \sin \psi$. Since the angle ψ is determined only by the relative positions of the particles, in statistical consideration of collisions the values are averaged over the angle ψ. Then the components Δv_2 and Δv_3 vanish, because $\sin \psi = \cos \psi = 0$. Therefore the ψ-averaged change in vector $\mathbf{v}$ is equal to $\Delta \mathbf{v} = -(1 - \cos \vartheta)\mathbf{v} = -(1 - \cos \vartheta)(\mathbf{v}_\alpha - \mathbf{v}_\beta)$. Substituting it into Eq. 2.15, we obtain

$$\Delta \mathbf{p}_\alpha = -\frac{m_\alpha m_\beta}{m_\alpha + m_\beta}(1 - \cos \vartheta)(\mathbf{v}_\alpha - \mathbf{v}_\beta) = -\mu_{\alpha\beta}(1 - \cos \vartheta)(\mathbf{v}_\alpha - \mathbf{v}_\beta) \tag{2.16}$$

It is seen that the change in momentum is proportional to the relative velocity of the colliding particles. Its dependence on the scattering angle is determined by the factor $1 - \cos \vartheta$, which is maximum in head-on collisions ($\vartheta = \pi$, $\cos \vartheta = -1$) and small in far collisions ($\vartheta \to 0$, $\cos \vartheta \to 1$). 1).

For a collision with a slow particle ($\mathbf{v}_\beta \ll \mathbf{v}_\alpha$), Eq. 2.16 yields the relative loss of momentum in collision:

$$\frac{\Delta p_\alpha}{p_\alpha} = -\frac{m_\beta}{m_\alpha + m_\beta}(1 - \cos \vartheta) \tag{2.17}$$

Its maximum value is determined by the mass ratio. In collision of a light particle with a heavy one ($m_\alpha \ll m_\beta$) the momentum may reverse its sign: $|\Delta p_\alpha / p_\alpha|_{\max} = 2$; when the masses are comparable ($m_\alpha \approx m_\beta$) the momentum may be lost completely; finally, for collison of a heavy particle with a fixed light one ($m_\alpha \gg m_\beta$) the maximum loss of momentum is of the order of the small mass ratio m_β/m_α.

The change in the kinetic energy of a particle in the laboratory system as a result of collision can be related to the change in momentum:

$$\Delta K_\alpha = \frac{m_\alpha v_\alpha'^2}{2} - \frac{m_\alpha v_\alpha^2}{2} = \frac{m_\alpha}{2}[(\mathbf{v}_0 + \mathbf{v}'_{\alpha_0})^2 - (\mathbf{v}_0 + \mathbf{v}_{\alpha_0})^2]$$
$$= m_\alpha \mathbf{v}_0(\mathbf{v}'_{\alpha_0} - \mathbf{v}_{\alpha_0}) = \mathbf{v}_0 \Delta \mathbf{p}_\alpha \tag{2.18}$$

Using Eq. 2.16 and accounting for Eqs. 2.6 and 2.8, we obtain

$$\Delta K_\alpha = -\left(\frac{m_\alpha \mathbf{v}_\alpha + m_\beta \mathbf{v}_\beta}{m_\alpha + m_\beta}\right)\frac{m_\alpha m_\beta}{m_\alpha + m_\beta}(\mathbf{v}_\alpha - \mathbf{v}_\beta)(1 - \cos\vartheta)$$
$$= -\frac{m_\alpha m_\beta}{(m_\alpha + m_\beta)^2}(1 - \cos\vartheta)[m_\alpha v_\alpha^2 - m_\beta v_\beta^2 + (m_\beta - m_\alpha)\mathbf{v}_\beta \mathbf{v}_\alpha]$$

If the velocity distribution of the β-type particles is isotropic, then, upon averaging, the third term in the parentheses is reduced to zero. Thus

$$\Delta K_\alpha = -2\frac{m_\alpha m_\beta}{(m_\alpha + m_\beta)^2}(1 - \cos\vartheta)\left(\frac{m_\alpha v_\alpha^2}{2} - \frac{m_\beta v_\beta^2}{2}\right)$$
$$= -\kappa_{\alpha\beta}(1 - \cos\vartheta)(K_\alpha - K_\beta) \tag{2.19}$$

where the coefficient

$$\kappa_{\alpha\beta} = 2\frac{m_\alpha m_\beta}{(m_\alpha + m_\beta)^2} \tag{2.20}$$

characterizes the efficiency of energy exchange in collision (it is called *the energy transfer coefficient*). It can be seen that energy exchange between the colliding particles is proportional to the difference of their initial energies. The dependence of the efficiency of energy exchange on the scattering angle is determined, as for momentum transfer, by the factor $1 - \cos\vartheta$. The dependence on the mass ratio is determined by the coefficient κ, which is maximum at $m_\alpha = m_\beta$ ($\kappa_{max} = 0.5$) and is of the order of the small mass ratio at $m_\alpha \ll m_\beta$, or $m_\alpha \gg m_\beta$.

The expressions obtained can easily be simplified for the important case of collision of electrons with heavy particles (atoms or ions), when the masses of the colliding particles differ sharply: $m_\alpha = m_e \ll m_\beta = m_a$. At atom energies where $K_a \lesssim K_e$, their velocity $v_a = \sqrt{2K_a/m_a}$ is much less than the electron velocity: $v_\alpha = v_e \gg v_\beta = v_a$. The position of the center of mass in this case evidently coincides with that of the heavy particle; the relative velocity is practically equal to the electron velocity

$\mathbf{v} = \mathbf{v}_e - \mathbf{v}_a \approx \mathbf{v}_e$, and the reduced mass, to the mass of the electron:

$$\mu_{ea} = \frac{m_e m_a}{m_e + m_a} \approx m_e$$

Therefore the problem on collision in the center-of-mass system amounts to the problem on electron motion in the field of a fixed atom. The expressions 2.16 and 2.19 for the energy and momentum transport in elastic collisions, with due regard for the above inequalities, take the form

$$\Delta \mathbf{p}_e = -\mathbf{p}_e (1 - \cos \vartheta);$$
$$\Delta K_e = -(K_e - K_a) 2 \frac{m_e}{m_a} (1 - \cos \vartheta) \qquad (2.21)$$

Here ϑ is the electron scattering angle in collision with the fixed atom.

In inelastic collision Eq. 2.13, which follows from the energy conservation law, takes the form

$$\frac{m_e v_e'^2}{2} = \frac{m_e v_e^2}{2} - \Delta \mathscr{E} \qquad (2.22)$$

It means that the change in internal energy of the heavy particle is equal to the change in kinetic energy of the electron; the kinetic energy of the heavy particle remains unchanged in collision. This conclusion is valid with an accuracy to the electron/atom mass ratio for any electron collisions. It can be obtained directly from the momentum conservation law. For an inelastic collision of an electron with a fixed atom this law leads to the equality

$$m_e \mathbf{v}_e = m_e \mathbf{v}_e' + m_a \mathbf{v}_a' \qquad (2.23)$$

Transferring the terms associated with the electrons to the left-hand side and squaring the equation, we find $m_a^2 v_a'^2 = m_e^2 (\mathbf{v}_e - \mathbf{v}_e')^2$. Considering that $v_e' < v_e$, we get the inequality for the postcollision kinetic energy of the atom:

$$\frac{m_a v_a'^2}{2} < \frac{4m_e}{m_a} \frac{m_e v_e^2}{2}; \qquad K_a' < \frac{4m_e}{m_a} K_e$$

It shows that the maximum portion of the kinetic energy of the electron that can be transferred to the atom is equal to $4m_e/m_a$.

The same ratio can readily be obtained for more complex collisions. Thus for ionization in collision of an electron with atoms the momentum and energy of the newly produced electron must be included in the laws of conservation. The momentum conservation law then acquires the form

$$m_e \mathbf{v}_e = m_a \mathbf{v}_a' + m_e \mathbf{v}_e' + m_e \mathbf{v}_e''$$

The last term of the equality stands for the momentum of the newly produced electron. Transferring the terms associated with the electrons to the left-hand side and squaring the equality, we obtain, as before

$$\frac{m_a v_a'^2}{2} = \frac{m_e^2}{2m_a}(\mathbf{v}_e - \mathbf{v}_e' - \mathbf{v}_e'')^2 < \frac{6m_e}{m_a}\frac{m_e v_e^2}{2} \qquad (2.24)$$

that is, $K_a' < 6(m_e/m_a)K_e$. Thus in collisions accompanied by ionization the portion of kinetic energy transferred to the heavy particle again does not appreciably exceed the mass ratio m_e/m_a. With an accuracy to this small ratio the energy conservation law for ionization can be written as

$$K = K' + K'' + \mathscr{E}_i \qquad (2.25)$$

where K'' is the energy of the newly produced electron and $\mathscr{E}_i$ is the energy spent on detachment of the electron from the atom (ionization energy).

2.2 METHODS FOR DESCRIPTION OF COLLISIONS

A plasma is an assembly of a large number of randomly moving and colliding particles. In this assembly the effect of collisions on the macroscopic parameters of the plasma is averaged over a large number of collisions; that is, it is usually manifested statistically as the average result of a great number of collisions. Moreover, if it is impossible to neglect the quantum-mechanical effects, the result of each collision is related to the initial state by probabilistic equations. Therefore we must introduce the statistical characteristics of collisions.

As noted in the preceding section, in considering collisions of arbitrarily moving particles it is convenient to switch to the center-of-mass system. Then the particle velocities are related unambiguously to the relative-motion velocity **v**. Since the center-of-mass velocity remains unchanged in collision, the result of the collision amounts to a change in relative velocity vector **v**, and in the case of elastic collisions, to a change in its direction. The classical equation describing the change in vector **v** is readily obtained from the equations of the colliding-particles motion:

$$\frac{d\mathbf{v}_\alpha}{dt} = \frac{\mathbf{F}_{\alpha\beta}}{m_\alpha}; \qquad \frac{d\mathbf{v}_\beta}{dt} = \frac{\mathbf{F}_{\beta\alpha}}{m_\beta} \qquad (2.26)$$

where $\mathbf{F}_{\alpha\beta} = -\mathbf{F}_{\beta\alpha}$ is the particle interaction force. The difference of these equations $d\mathbf{v}_\alpha/dt - d\mathbf{v}_\beta/dt = (1/m_\alpha + 1/m_\beta)\mathbf{F}_{\alpha\beta}$ determines the relative velocity equation

$$\mu_{\alpha\beta} \frac{d\mathbf{v}}{dt} = \mathbf{F}_{\alpha\beta} \tag{2.27}$$

in which $\mu_{\alpha\beta}$ is the reduced mass (Eq. 2.12).

Particle interaction forces usually can be considered central. The potential $U_{\alpha\beta}$ (r), determining such interaction, depends only on the distance between the colliding particles $\mathbf{r} = |\mathbf{r}_\alpha - \mathbf{r}_\beta|$. Accordingly, the force $\mathbf{F}_{\alpha\beta} = -\text{grad } U(r)$ is directed parallel to the vector $\mathbf{r} = r_\alpha - r_\beta$. In this case Eq. 2.27 can be represented as

$$\mu \frac{d^2\mathbf{r}}{dt^2} = -\text{grad } U(r) \tag{2.28}$$

That is, it is equivalent to the equation of the motion of a particle of mass μ in the centrally symmetric field of a fixed center of force $U(r)$. We have precisely this equivalent problem below. Its classical solution consists in integrating the equation of motion (Eq. 2.28) for a given law of interaction and given initial conditions. The initial conditions include the precollision velocity $\mathbf{v}$ of the particle and its position.

The position of the particles in the classical collision problem is usually characterized by the *impact parameter* ξ (distance between the center of force and the undisturbed particle trajectory) and the position in space of the interaction plane through the center of force and the undisturbed trajectory. In the center-of-force field the position of the interaction plane naturally cannot change in the course of collision. Therefore the angle ψ, characterizing the position of the plane in which the postcollision velocity of the particle lies, is determined solely by the initial position of the particles. The shape of the particle trajectory and the scattering angle ϑ, with a given interaction law, are defined unambiguously by the impact parameter ξ. The meaning of this parameter is explained in Fig. 2.2, which depicts the particle trajectory in the field of

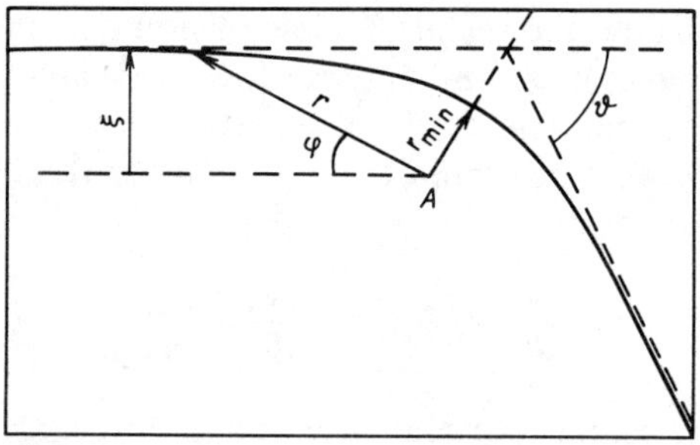

Fig. 2.2 Particle trajectory in the field of the central force A.

the center of force A. The value of ξ may vary from zero (head-on collision) to large values at which the interaction forces are small (far collisions). Accordingly, the scattering angle varies from π (reflection) to zero.

It is determination of the relationship between the scattering angle and the impact parameter that is actually the subject of the classical collision theory. The expression for this relationship can be obtained from the conservation laws which follow from the equation of motion (Eq. 2.28). The first one is the energy conservation law:

$$K = \frac{\mu v_\infty^2}{2} = \frac{\mu v^2(r)}{2} + U(r) = \frac{\mu v_r^2}{2} + \frac{\mu v_\varphi^2}{2} + U(r)$$

The second is the angular momentum conservation law:

$$M = \mu \xi v_\infty = \mu [\mathbf{r} \times \mathbf{v}]_z = \mu r v_\varphi$$

Here v_∞ is the precollision velocity; v_r and v_φ are the velocity components in the polar system of coordinates in the interaction plane. Determining these components from the above equalities, we obtain the equation for the trajectory:

$$\frac{d\varphi}{dr} = \frac{1}{r}\frac{v_\varphi}{v_r} = \frac{M}{r^2\sqrt{2\mu(K-U) - M^2/r^2}} = \frac{\xi}{r^2\sqrt{1 - \xi^2/r^2 - U/K}}$$

Integrating, we find for $\Delta\varphi = \pi - \vartheta$ (see Fig. 2.2),

$$\pi - \vartheta = 2\int_{r_{\min}}^{\infty} \frac{\xi\, dr}{r\sqrt{r^2 - \xi^2 - U(r)r^2/K}} \tag{2.29}$$

where $r_{\min}$ is the distance of maximum approach at which the denominator of the integral equations reduces to zero. With a given interaction potential $U(r)$ Eq. 2.29 establishes the relationship between the scattering angle ϑ and the impact parameter ξ. For small scattering angles, when $U \ll K$, it can be transformed to

$$\vartheta = -\frac{\xi}{K}\int_{\xi}^{\infty} \frac{dU}{dr}\frac{dr}{\sqrt{r^2 - \xi^2}} \tag{2.30}$$

Passing over to the statistical description, we isolate such collisions for which the relative velocities lie in the interval from $\mathbf{v}$ to $\mathbf{v} + d\mathbf{v}$. The particles of the isolated group differ from each other as regards their impact parameter ξ. Since the relative positions of the colliding particles are accidental, we must consider the scattering of a group of particles impinging on a fixed center of force in a wide flux. For electron–heavy particle collisions this problem fully corresponds to the actual situation, because a heavy particle is a practically immobile center of force. In the general case its solution enables one to determine the results of actual collisions of different particles after switching to the laboratory system of coordinates.

Thus let a particle flux with a velocity $\mathbf{v}$ and a density n impinge on the scattering center A (Fig. 2.3). We determine the average number of particles dN' scattering per unit time into an element of the solid angle $d\Omega$ near the direction given by the angles ϑ and ψ, that is, into $d\Omega = \sin\vartheta\, d\vartheta\, d\psi$. The scattering probability is clearly proportional to $d\Omega$. At the same time the number of scattered particles is proportional to the density of the impinging flux (to the number of particles crossing a unit area per unit time). Therefore the number of particles scattered per unit time into the solid angle $d\Omega$ is equal to

$$dN' = \sigma n v\, d\Omega \tag{2.31}$$

The proportionality factor σ, which has the dimension of area, characterizes the probability of particle scattering in a definite direction. It is called *the differential scattering cross section.* This term corresponds to the physical meaning of σ. Indeed, in accordance with Eq. 2.31 the quantity $\sigma\, d\Omega$ determines the area of a surface element perpendicular to the direction of the incident flux, where the arriving particles scatter into the solid angle $d\Omega$: $ds = \sigma\, d\Omega = \sigma \sin\vartheta\, d\vartheta\, d\psi$. It is easy to correlate this area with the corresponding change in impact parameter. Let us write the relationship between them (Fig. 2.4): $ds = \xi\, d\xi\, d\psi$. Here the change in impact parameter $d\xi$ corresponds to the change in scattering angle $d\vartheta$. The increment $d\vartheta$ is opposite in sign to $d\xi$ because the angle ϑ

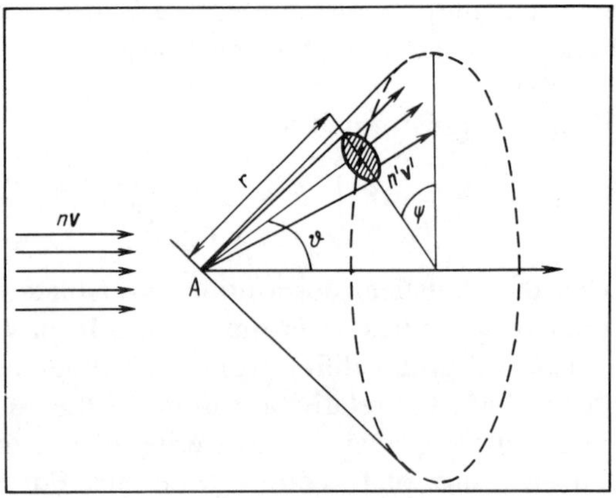

Fig. 2.3 Particle flux impinging on the scattering center A.

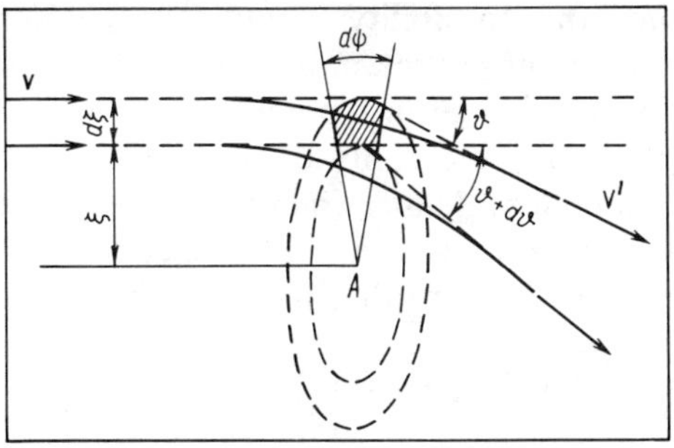

Fig. 2.4 Differential scattering cross section.

decreases with increasing ξ. Comparing this relation with the preceding, we obtain the equation of classical mechanics for the differential cross section:

$$\sigma = \frac{\xi}{\sin\vartheta}\left|\frac{d\xi}{d\vartheta}\right| \tag{2.32}$$

It determines the quantity σ as a function of the scattering angle ϑ, provided the relationship between ϑ and ξ is known (Eq. 2.29).

We have given the classical definition of the scattering cross section. But classical consideration is possible only within a limited range of conditions. It is inapplicable for analysis of inelastic processes accompanied by a change in the internal state of the particles. As regards elastic collisions, their classical analysis is possible if the concept of the trajectory is meaningful, that is, if the particle coordinates and momentum are determined accurately enough at each instant. This requires, in the first place, that the precollision de Broglie wavelength be much less than the characteristic scale of the interaction region: $\lambda\!\!\!^{-} = \hbar/\mu v \ll r_0$, $v \gg \hbar/\mu r_0$. If one of the particles is a neutral atom, the range of the forces is determined by its size; that is, it is of the order of the Bohr radius $r_a \ll \hbar^2/m_e e^2 \approx 10^{-8}$ cm. Then the condition reduces to the inequality

$$v \gg \frac{e^2}{\hbar}\frac{m_e}{\mu}; \qquad K \gg \frac{m_e^2 e^4}{\mu\hbar^2} \tag{2.33}$$

Hence we find that for electron–atom collisions $K_e \gg m_e e^4/\hbar^2 \approx 10$ eV, and for ion–atom collisions

$$K_i \gg \frac{m_e^2}{m_i}\frac{e^4}{\hbar^2} \approx 10^{-2}\frac{m_H}{m_i}\ \text{eV}$$

where m_H is the mass of the hydrogen atom.

At the same time the possibility of utilizing classical mechanics for determining the change in particle momentum on collision is limited by the constraint that the change in momentum Δp should greatly exceed the quantum-mechanical uncertainty of the momentum: $\Delta p \gg \delta p \approx \hbar/\delta r$. Since δr must be much less than r_m (minimum interparticle distance), the inequality $\Delta p \gg \hbar/r_m$ must hold true. The change in momentum Δp in the course of interaction is equal to the force momentum:

$$\Delta p = F\Delta t \approx \frac{F(r_m)r_m}{v} \approx \frac{U(r_m)}{v}$$

Using this relation, we find that $v \ll r_m U(r_m)/\hbar$. For charged–neutral particle collisions considerable scattering must be observed when they penetrate the electron shell. Therefore for the purpose of estimation we can assume the value of U to be equal to the coulomb potential at $r = r_m$. Thus we obtain the second condition for the applicability of the classical description of collisions, which limits the particle velocity from above:

$$v \ll \frac{e^2}{\hbar}, \qquad K \ll \frac{\mu e^4}{\hbar^2} \tag{2.34}$$

Substitution of numerical values yields for electron–atom collisions

$$K_e \ll \frac{m_e e^4}{\hbar^2} \approx 10 \text{ eV}$$

and for ion–atom collisions

$$K_i \ll \frac{m_i e^4}{\hbar^2} \approx 10^4 \frac{m_i}{m_H} \text{ eV}$$

A comparison of the inequalities 2.33 and 2.34 shows that the description of electron–atom collisions at any energy requires the application of quantum mechanics. For elastic ion–atom and atom–atom collisions there is a wide range of energies within which the classical description is possible. Note that it covers the energies of heavy plasma particles in nearly all cases of practical interest.

As we did above, we now determine the criteria of applicability of the classical description for collisions of charged particles whose interaction is due to the coulomb potential $U = e^2/r$. The radius of a strong interaction characterizing deviation of the particles by large angles corresponds to a potential interaction energy of the order of the kinetic energy: $\mu v^2/2 \approx e^2/r_s$, $r_s \approx e^2/\mu v^2$. Therefore the criterion of determinacy of the trajectory takes the form $\hbar/\mu v \ll e^2/\mu v^2$, or

$$v \ll \frac{e^2}{\hbar} \tag{2.35}$$

The second condition of the applicability of classical mechanics, the condition of the determinacy of the change in momentum, is the same as for charged–neutral particle collisions. It coincides with Eq. 2.35. Thus both conditions of the applicability of classical mechanics for charged particle collisions limit the energy only from above: for electron–electron and electron–ion collisions we find $K_e \ll 10\,\text{eV}$ and for ion–ion collisions $K_i \ll 10^4 m_i/m_H$ eV.

Let us now discuss the quantum-mechanical approach to the description of particle collisions. In quantum mechanics the trajectory of interacting particles is not defined, and only probabilistic characteristics can be established for each collision. To define them it is necessary to find the wave functions of the interacting particles. As in classical consideration, the problem is best solved in the center-of-mass system, where it is equivalent to the problem of scattering on a center of force. For elastic collisions the problem amounts to finding the stationary solution of the Schrödinger equation with a given interaction potential. It can be written thus

$$\Delta\Psi + k^2\Psi = \frac{2\mu}{\hbar^2} U\Psi \qquad (2.36)$$

In the absence of interaction the solution to the equation is a plane wave $\Psi_0 = \exp(i\mathbf{kr})$ where $\mathbf{k} = \mathbf{p}/\hbar$ is the wave vector and $\mathbf{p} = \mu\mathbf{v}$ is the particle momentum in the equivalent problem.

Considering the right-hand side of Eq. 2.34 as an inhomogeneity, we can formally write its solution in the integral form:

$$\Psi(\mathbf{r}) = \left[\exp(\mathbf{ikr}) - \frac{\mu}{2\pi\hbar^2}\int \frac{\exp(\mathrm{i}k|\mathbf{r}-\mathbf{r}'|)}{|\mathbf{r}-\mathbf{r}'|} U(r')\Psi(\mathbf{r}')d^3r'\right] \qquad (2.37)$$

Integration is done over the entire interaction region. The collision result is evidently determined by the asymptotic behavior of the wave function at large distances from the center of force (for $r \to \infty$). To determine it, we must put $r \gg r'$ in the integrand in Eq. 2.37. Then

$$|\mathbf{r}-\mathbf{r}'| = \sqrt{(\mathbf{r}-\mathbf{r}')^2} \approx r\left(1 - \frac{\mathbf{rr}'}{r^2}\right) = r - \mathbf{r}_1\mathbf{r}'$$

and the integral is transformed:

$$\int \frac{\exp(\mathrm{i}k|\mathbf{r}-\mathbf{r}'|)}{|\mathbf{r}-\mathbf{r}'|} U(r')\Psi(\mathbf{r}')d^3r' \approx \frac{\exp(\mathrm{i}kr)}{r}\int \exp(-\mathrm{i}k\mathbf{r}_1\mathbf{r}')U(r')\Psi(\mathbf{r}')d^3r'$$

where $\mathbf{r}_1$ is a unit vector in the direction of $\mathbf{r}$. The remaining integrand no longer contains r; the integral depends only on the direction of the vector r, that is, on the angle ϑ between $\mathbf{r}$ and the wave vector $\mathbf{k}$, which determines the direction of the precollision particle motion (this is the only isolated direction). Substituting the integral into Eq. 2.37, we obtain the general expression for the asymptotic representation of the wave function

$$\Psi = \exp(i\mathbf{kr}) + \frac{f(\vartheta)}{r}\exp(ikr) \tag{2.38}$$

where

$$f(\vartheta) = -\frac{\mu}{2\pi\hbar^2}\int \exp(-i\mathbf{k}\mathbf{r}_1\mathbf{r}')U(r')\Psi(\mathbf{r}')d^3r'$$

The first term of Eq. 2.38 describes the particle flux incident on the center of force; the second term is the flux of scattered particles. The quantity $f(\vartheta)$ is called the *scattering amplitude.* It is easy to see that it actually determines the cross section. Indeed, according to Eq. 2.31 the cross section is determined by the ratio of the number of particles scattered into an element of a solid angle $dN' = n'v'r^2d\Omega$ to the particle flux incident on the center of force nv:

$$\sigma = \frac{1}{nv}\frac{dN'}{d\Omega} = \frac{n'v'}{nv}r^2 \tag{2.39}$$

where n' is the density of the scattered particles and v' is their velocity.

The particle densities in the incident and scattered fluxes are determined by the square of the moduli of the respective terms of the asymptotic expression 2.38:

$$\frac{n'}{n} = \frac{|\Psi_s|^2}{|\Psi_0|^2} = \frac{|f(\vartheta)|^2}{r^2}$$

Since in an elastic collision $v' = v$, we get

$$\sigma = |f(\vartheta)|^2 \tag{2.40}$$

In analyzing inelastic scattering in the asymptotic representation of the wave function we must keep in mind the possibility of a change in the internal state of the atom. For the wave function of the electron–atom system it can be written as

$$\Psi(r, \rho) = \exp(i\mathbf{kr})\Phi_v(\rho) + \sum_j f_{0j}(\vartheta)\Phi_j(\rho)\frac{\exp(ik'_jr)}{r} \tag{2.41}$$

where ρ is the set of internal coordinates of the atom, j is the number of the atomic state; the subscript '0' denotes the initial state, and Φ_0 and Φ_j are the wave functions of the initial and final states. The quantity $f_{0j}(\vartheta)$ is the amplitude of scattering accompanied by transition of the atom from the 0 state to the j state. It determines the scattering cross section.

In conformity with Eq. 2.39,

$$\sigma_{0j}(\vartheta) = \frac{v'}{v}|f_{0j}(\vartheta)|^2 \tag{2.42}$$

Thus the solution of the quantum-mechanical problem on scattering amounts to finding, with the aid of Schrödinger's asymptotic representation, the wave function 2.41 and the scattering amplitudes $f_{0j}(\vartheta)$ contained in it. In calculating the wave function it is customary to represent it as an expansion in the eigenfunctions of the remaining values. In the problem on the particle motion in the center-of-force field this quantity is the angular momentum $M = \mu[\mathbf{r} \times \mathbf{v}]_z = \mu\xi v$, which is determined by the initial value of the velocity and by the impact parameter. In classical consideration both these parameters are fixed. In quantum mechanics it is impossible to assign the velocity and the impact parameter simultaneously because of the uncertainty relation. Therefore when solving the problem with a fixed initial velocity we have to bear in mind the possibility of a wide spectrum of values of the impact parameter and angular momentum. The possible momentum values are determined by the orbital quantum number $M = l\hbar$, and the eigenfunctions corresponding to these values are proportional to the Legendre polynomials P_l (cos ϑ). Therefore the wave function is represented as an expansion in Legendre polynomials. With this expansion its asymptotic form corresponds to the sum of converging and diverging spherical waves:

$$\Psi = \frac{\exp(ikr)}{r} \sum_l a_l P_l(\cos\vartheta) + \frac{\exp(-ikr)}{r} \sum_l b_l P_l(\cos\vartheta)$$

In the absence of scattering the coefficients a_l and b_l are chosen so that their superposition yields a plane wave exp(i**kr**). Inclusion of scattering changes the coefficients a_l, which can be found by substituting the expansion into the Schrödinger equation. In elastic scattering an additional phase shift occurs in each coefficient; in inelastic scattering the phase amplitudes also change.

The change in coefficients determines the scattering amplitude $f(\vartheta) = \sum_l f_l P_l(\cos\vartheta)$ and, in accordance with Eq. 2.42, the differential cross section.

The described general approach makes it possible, in principle, to solve the particle collision problem. But the solution is in the form of infinite series. At high energies of the colliding particles it can often be assumed that the interaction affects the initial state but slightly. This enables one to solve the collision problem by the perturbation method. Such treatment is called the *Born approximation.* It can be applied where the change in momentum in the course of collision is much less than the quantum–mechanical uncertainty of the momentum in the unperturbed state, $\Delta p \ll \delta p$. If the inequality is fulfilled, we can, as a zero approximation, neglect the particle interaction and consider it as perturbation.

It can be seen that this criterion is the opposite of one of the conditions of applicability of the classical consideration and can be represented as an inequality opposite to Eq. 2.34:

$$v \gg \frac{U(r_m)r_m}{\hbar} \approx \frac{e^2}{\hbar} \tag{2.43}$$

For electron–atom collisions the inequality 2.43 leads to the condition $K_e \gg m_e^4/\hbar^2 \approx 10\,\text{eV}$. If this condition is fulfilled, the application of the perturbation theory yields comparatively simple integral expressions for the amplitude of scattering of electrons by atoms. For elastic collisions, for instance, they can be found from Eq. 2.38. Substituting the undisturbed wave function $\Psi_0 = \exp(i\mathbf{kr})$ into the integrand of Eq. 2.38, we obtain

$$f(\vartheta) = -\frac{m_e}{2\pi\hbar^2}\int \exp(-i\Delta\mathbf{kr})U(r)\,d^3r \tag{2.44}$$

where the vector $\Delta\mathbf{k} = \mathbf{k}' - \mathbf{k}$ is the change in wave vector in collision; if $k' = k$, then $\Delta k = 2k \sin(\vartheta/2) = 2(mv/\hbar)\sin(\vartheta/2)$.

The amplitude of inelastic scattering of electrons by an atom is determined similarly in the Born approximation:

$$f_{0j} = -\frac{m_e}{2\pi\hbar^2}\int \exp(-i\Delta\mathbf{kr})U(\mathbf{r}, \boldsymbol{\rho})\Phi_j(\rho)\Phi_0(\rho)\,d^3r\,d^{3s}\rho \tag{2.45}$$

where, as before, $\Delta\mathbf{k} = \mathbf{k}' - \mathbf{k} = m_e\mathbf{v}'/\hbar - m_e\mathbf{v}/\hbar$; $U(\mathbf{r}, \rho)$ is the potential of interaction between a free electron and the atomic electrons and nucleus; Φ_0 and Φ_j are the wave functions of the initial and final states of the atom; and integration is done over all the coordinates d^3r of the free electron and the coordinates $d^{3s}\rho$ of the atomic electrons. By substituting the expressions for the scattering amplitudes into Eq. 2.40 and 2.42 we determine the elastic and inelastic scattering cross sections in the Born approximation.

2.3 INTEGRAL CHARACTERISTICS OF COLLISIONS

In Section 2.2 we introduced the differential collision cross section, which characterizes scattering in the equivalent problem on particle motion in the field of a fixed center of force. Using σ, we can determine the characteristics of real collisions.

First of all we write the expression for the number of α- and β-type particle collisions in unit volume per unit time. To do this we take advantage of the determination of a cross section (Eq. 2.31). The density of the particles incident on an equivalent center of force, which appears in this equation, must evidently be replaced by the number of pairs of colliding particles in unit volume. The number of such pairs with a relative velocity in the range from v to $v+dv$ can be written as $n_\alpha n_\beta f_{\alpha\beta} dv$, where $f_{\alpha\beta}(v)dv$ is the fraction of pairs with relative velocities in the selected range from v to $v+dv$. Thus the number of collisions in unit volume per unit time accompanied by scattering into the solid angle $d\Omega$ is determined by the relation

$$dq_{\alpha\beta} = n_\alpha n_\beta \sigma_{\alpha\beta} v f_{\alpha\beta}(v) dv\, d\Omega \tag{2.46}$$

When studying the effect of collisions on the plasma behavior we usually need not consider separately the collisions leading to scattering at different angles. This effect is determined by the integral characteristics of collisions summed over the scattering angles. It is possible, for instance, to calculate the total number of collisions undergone by particles of a given velocity. To this end we integrate Eq. 2.46 over the solid angle:

$$dQ_{\alpha\beta} = n_\alpha n_\beta v s_{\alpha\beta} f_{\alpha\beta}(v)\, dv \tag{2.47}$$

the quantity

$$s_{\alpha\beta} = \int \sigma_{\alpha\beta}(\vartheta)\, d\Omega = 2\pi \int_0^\pi \sigma_{\alpha\beta}(\vartheta) \sin\vartheta\, d\vartheta \tag{2.48}$$

is called the *total scattering cross section.*

In the classical approach the total cross section determines the area of the surface perpendicular to the relative velocity in which particles collide with scattering at an arbitrary angle.* Besides the total cross section, other associated quantities are introduced. The quantity

$$\nu_{\alpha\beta}(v) = n_\beta v s_{\alpha\beta}(v) \tag{2.49}$$

which appears in Eq. 2.47, is called the *α,β-type particle collision frequency* (or the number of collisions per unit time). In accordance with the classical concept of the cross section it determines the number of times that an α-type particle impinges on targets formed by β-type particles. Indeed, the number of such hits per unit time is equal to the product of the velocity by the collision cross section and by the target density. The reciprocal of ν yields the average intercollisional time

$$\tau_{\alpha\beta}(v) = \frac{1}{\nu_{\alpha\beta}(v)} \tag{2.50}$$

*Note that the term "cross section" characterizes primarily hard-sphere collisions, which were previously regarded as a model of atomic collisions.

It also allows us to determine *the mean free path*, the average distance between collisions:

$$\lambda_{\alpha\beta}(v) = v\tau_{\alpha\beta} = \frac{1}{n_\beta s_{\alpha\beta}(v)} \tag{2.51}$$

The total cross section and the related characteristics have often been used for describing collisions. It should be borne in mind, however, that they are seldom applicable. For elastic collisions the total cross section often is devoid of any physical meaning. The reason is that the integral 2.48 determining the total cross section contains, with the same weight, close collisions, which drastically change the trajectory of the colliding particles, and far collisions, which hardly change it at all. The classical total cross section can be obtained only for collisions with a strictly limited interaction radius, for instance, for neutral particle collisions. In collisions of electrons and ions with atoms or with each other the interaction radius is not limited; interaction (even a very small one) takes place at any distance. Therefore the integral 2.48, which determines the total cross section, diverges. In the quantum-mechanical approach the total cross section is finite for a rapidly decreasing potential ($U < A/r^3$). But even when a total cross section of elastic collisions does exist, it cannot be used directly for the kinetics of plasma particle motion, because it neglects the difference in the effect of close and far collisions.

To describe the effect of elastic collisions on the particle motion, one usually introduces the integral cross section, which characterizes the change in momentum and energy collision. In accordance with Eqs. 2.16 and 2.19 the scattering-angle dependence of these changes is determined by the factor $1-\cos\vartheta$. Therefore one should introduce this weighting factor into the integral cross section of elastic collisions, which defines the change in momentum and kinetic energy. Then the expression for the cross section takes the form

$$s^t_{\alpha\beta} = \int_{(\Omega)} (1-\cos\vartheta)\sigma_{\alpha\beta}(\vartheta)\, d\Omega = 2\pi \int_0^\pi \sigma_{\alpha\beta}(\vartheta)(1-\cos\vartheta)\sin\vartheta\, d\vartheta \tag{2.52}$$

The weighting factor $1-\cos\vartheta$ is close to zero for far collisions (at $\vartheta \to 0$); therefore the integral converges for all kinds of electron–atom and ion–atom interactions. It is only for the coulomb law, which is used for charged particle interaction, that the integral 2.52 diverges logarithmically. In this case it can be determined with due allowance for the screening effect of the plasma (see Section 2.4). The integral cross section s^t, which defines the change in energy and momentum in elastic collisions, is called the *transport cross section*. The last term is used because this cross section appears in the basic equations of the plasma kinetics describing the transport of particles and energy (see Chapters 5 and 6). Along with the transport cross section one can introduce other related integral collision characteristics, as was done above in determining the total cross section (see Eqs. 2.49–2.51). These are the effective collision frequency, the intercollisional time, and the mean free path:

$$\begin{aligned} \nu^{t}_{\alpha\beta}(v) &= n_\beta s^{t}_{\alpha\beta}(v)\, v; \\ \tau^{t}_{\alpha\beta}(v) &= \frac{1}{n_\beta s^{t}_{\alpha\beta}(v)\, v}; \\ \lambda^{t}_{\alpha\beta}(v) &= \frac{1}{n_\beta s^{t}_{\alpha\beta}(v)} \end{aligned} \tag{2.53}$$

Generally speaking all the integral characteristics introduced depend on the relative particle velocity. For calculations and estimates they must be averaged over the velocities of the colliding particles, the averaging method depending on the particular problem.

For instance, let us determine, by using the transport cross section (Eq. 2.52), the change in energy during elastic collisions. The change in the kinetic energy of an α-type particle colliding with a β-type particle is determined by the relation 2.19. Multiplying it by the number of collisions in unit volume per unit time (see Eq. 2.46), we obtain the change in the energy of α-type particles in unit volume as the result of such collisions:

$$d(n_\alpha K_\alpha) = -\kappa_{\alpha\beta} n_\alpha n_\beta v \sigma_{\alpha\beta}(1-\cos\vartheta)(K_\alpha - K_\beta) f_{\alpha\beta}(v)\, dv\, d\Omega$$

Integrating over the solid angle and the velocities, we find the change in the average α-type particle energy per unit time:

$$\begin{aligned} \frac{dK_\alpha}{dt} &= -\kappa_{\alpha\beta} \int_{(v)} n_\beta s^{t}_{\alpha\beta} v (K_\alpha - K_\beta) f_{\alpha\beta}(v)\, dv \\ &= -\kappa_{\alpha\beta} \int_{(v)} \nu^{t}_{\alpha\beta}(v)(K_\alpha - K_\beta) f_{\alpha\beta}(v)\, dv \end{aligned} \tag{2.54}$$

We obtain similarly the expression for the change in momentum. By using Eq. 2.16 we find the change in the average momentum of the α-type particles as a result of their elastic collisions with β-type particles:

$$\frac{d\mathbf{p}_\alpha}{dt} = -\mu_{\alpha\beta} \int_{(v)} \nu^{t}_{\alpha\beta}(v)(\mathbf{v}_\alpha - \mathbf{v}_\beta) f_{\alpha\beta}(v)\, dv \tag{2.55}$$

Let us now discuss the determination of integral cross sections for inelastic collision. For inelastic processes the total collision cross section, which is determined by an equation of the type 2.48,

$$s_{\alpha\beta}^{(j)} = 2\pi \int_0^{\pi} \sigma_{\alpha\beta}^{(j)} \sin \vartheta \, d\vartheta \tag{2.56}$$

has a clear physical meaning: it yields the number of inelastic acts of a given type (j). Thus for excitation of a given level this cross section determines the total number of excitation acts; for ionization, the total number of ionization acts; and so on. Naturally, the integral contained in the expression for the cross section converges in all cases. As for elastic collisions, one can introduce frequencies of inelastic collisions leading to some process $\nu_{\alpha\beta}^{(j)} = n_\beta v s_{\alpha\beta}^{(j)}$ and the corresponding mean free paths.

As distinct from elastic collisions, the change in the energy of the electrons elastically colliding with atoms is determined by the total cross section (Eq. 2.56) and the corresponding collision frequency. Indeed, in each such collision the change in electron energy is practically equal to that in the internal energy of the atom $\Delta K_e = -\mathscr{E}_{0j}$, since in conformity with the conservation laws the change in kinetic atom energy is of the order of the small mass ratio m_e/m_a. Multiplying ΔK_e by the number of inelastic collisions in unit volume per unit time and averaging over the velocities, we obtain the change in average electron energy caused by collisions of the given type:

$$\left(\frac{d\bar{K}_e}{dt}\right)_{ea}^{(j)} = -\mathscr{E}_{0j} \int n_a v s_{ea}^{(j)} f_e(v) \, dv = -\bar{\nu}_{ea}^{(j)} \mathscr{E}_{0j} \tag{2.57}$$

where $s_{ea}^{(j)}$ is a total cross section of the type of 2.56, and the collision frequency $\nu_{ea}^{(j)}$ is averaged over the electron velocities (relative velocity in electron–atom collisions is practically equal to the electron velocity). The total electron energy losses in inelastic collisions are obtained by summing the expressions 2.57 over all the possible processes of atom excitation and ionization:

$$\left(\frac{d\bar{K}_e}{dt}\right)_{ea}^{n} \approx -\sum_j \bar{\nu}_{ea}^{(j)} \mathscr{E}_{0j} \tag{2.58}$$

When the average electron energy is much lower than the energy of excitation of the lowest atomic level $\mathscr{E}_{01}$, inelastic processes are caused by electrons from the "tail" of the distribution function, whose number rapidly diminishes with increasing energy. Here each level is excited by electrons with an energy close to that of excitation, and one can assume approximately that inelastic collisions lead to a complete loss of energy by the electron. Then the expression for the total electron energy losses takes the form

$$\left(\frac{dK_e}{dt}\right)^n_{ea} \approx -\sum_j \nu^{(j)}_{ea} K_e = -\nu^n_{ea} K_e \tag{2.59}$$

where $\nu^n_{ea} = n_a v s^n_{ea} = n_a v_j \Sigma_j s^j_{ea}$ is the summary frequency of inelastic collisions.

2.4 ELASTIC COLLISIONS BETWEEN CHARGED PARTICLES

We now survey the available data on charged particle collisions; the survey is brief and certainly not complete.* Its purpose is merely to give a general idea of the characteristics of collisions that substantially affect the plasma properties. In this section we consider elastic collisions of charged particles with each other.

The main forces responsible for charged particle interaction are the long-range coulomb forces. The internal structure of particles is usually insignificant. It manifests itself only in close ion collisions, which usually make only a small contribution to the integral characteristics of the collisions.

The solution of the problem on scattering in a coulomb field with a potential

$$U = \frac{e^2}{r} \tag{2.60}$$

is well known. In the classical approach it can be obtained according to the scheme described in Section 2.2. Substitution of the coulomb potential into Eq. 2.29 makes it possible to determine the relationship between the scattering angle and the impact parameter. Integration leads to the following expression:

$$\xi = \frac{e^2}{\mu v^2} \cot \frac{\vartheta}{2} = r_s \cot \frac{\vartheta}{2} \tag{2.61}$$

We introduce here the parameter $r_s = e^2/\mu v^2$, which is called *the strong-interaction radius*. It is equal to the distance over which the potential interaction energy $U(r_s)$ is twice as high as the kinetic energy. The scattering angle corresponding to $\xi = r_s$ is equal to $\pi/2$. Substituting Eq. 2.61 into the cross section expression 2.32, we obtain the Rutherford equation:

$$\sigma = \frac{(e^2/2\mu v^2)^2}{\sin^4(\vartheta/2)} \tag{2.62}$$

*A systematized presentation of the physics of charged particle collisions can be found in the monographs listed in the bibliography.

It determines the rapid increase in cross section with decreasing scattering angle (i.e., with increasing impact parameter) which is associated with the relatively slow field decrease over large distances ($U \sim 1/r$). The slow decrease in field strength also leads to a strong energy dependence of the effective-interaction radius ($r_s \sim 1/K$) and the scattering cross section ($\sigma \sim 1/K^2$).

As to the scattering of different types of particles in the coulomb field, the quantum-mechanical solution of this problem also leads to the Rutherford equation. There is a certain difference in the case of collision of identical particles. Here the quantum-mechanical approach requires an allowance for exchange interaction because the colliding particles cannot be distinguished from each other. The superposition of their wave functions introduces an interference term into the collision cross-section expression. With this term taken into account the differential cross section of electron–electron collision acquires the form

$$\sigma(\vartheta)+\sigma(\pi-\vartheta)=\left(\frac{e^2}{m_e v^2}\right)^2\left[\frac{1}{\sin^4(\varphi/2)}+\frac{1}{\cos^4(\vartheta/2)}\right.$$
$$\left.-\frac{1}{\sin^2(\vartheta/2)\cos^2(\vartheta/2)}\cos\left(\frac{e^2}{\hbar v}\ln\tan^2\frac{\vartheta}{2}\right)\right] \qquad (2.63)$$

(the equation is written for the sum of cross sections with angles ϑ and $\pi-\vartheta$, because replacement of ϑ by $\pi-\vartheta$ is equivalent to electron permutation and therefore is indistinguishable). This expression is illustrated in Fig. 2.5, in which the curves correspond to the following values of the parameter: 1, $e^2/\hbar v=1$; 2, $e^2/\hbar v=10$; and 3, $e^2/\hbar v=100$. In the range from $\pi/4$ to $3\pi/4$ the value of σ is magnified 100-fold. At low velocities $v \ll e^2/\hbar$ the interference term in Eq. 2.63 rapidly oscillates with the variation in ϑ. It reduces to zero on averaging even over a narrow range of angles; therefore it can be neglected (see the dashed curve). Otherwise this term exerts an appreciable effect on scattering at large angles. At small scattering angles $\vartheta \ll 1$ the cross section is practically identical with the Rutherford cross section in all cases.

Let us now find the transport cross section (Eq. 2.52). Using Eq. 2.62, we write the integral appearing in the cross-section expression, as follows:

$$\begin{aligned} s_\vartheta^t &= 2\pi\int_{\vartheta_{\min}}^{\pi}\sigma(1-\cos\vartheta)\sin\vartheta\,d\vartheta \\ &= \frac{\pi}{2}\left(\frac{e^2}{\mu v^2}\right)^2\int_{\vartheta_{\min}}^{\pi}\frac{(1-\cos\vartheta)\sin\vartheta}{\sin^4(\vartheta/2)}\,d\vartheta \\ &= 4\pi\left(\frac{e^2}{\mu v^2}\right)^2\ln\frac{1}{\sin(\vartheta_{\min}/2)} \end{aligned} \qquad (2.64)$$

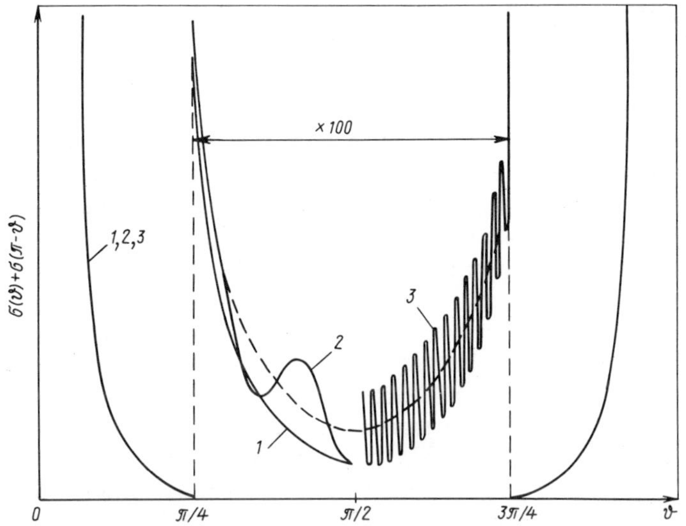

Fig. 2.5 Differential cross section of electron-electron collision. The different curves correspond to: (1) $e^2/\hbar v = 1$ (2) $e^2/\hbar v = 10$ (3) $e^2/\hbar v = 100$.

Taking into account the relationship between the scattering angle and the impact parameter, which is obtained from Eq. 2.61, that is, assuming that $\sin(\vartheta/2) = 1/\sqrt{1+(\xi/r_s)^2}$, we can write Eq. 2.64 for $\xi_{max} \gg r_s$ in the form

$$s^t = 4\pi \left(\frac{e^2}{\mu v^2}\right)^2 \ln \frac{\xi_{max}}{r_s}$$

In integrating over the range from $\vartheta_{min} = 0$ to $\vartheta_{max} = \pi/2$ or, respectively, from $\xi_{max} = \infty$ to $\xi_{min} = 0$, the integral in Eq. 2.64 diverges logarithmetically. This divergence is caused by the large contribution of far collisions leading to small-angle scattering (divergence appears as $\vartheta_{min} \to 0$, $\xi \to \infty$). To eliminate the divergence one should limit the coulomb interaction radius. In a plasma such a limitation is due to the shielding of the field of each interacting particle by other charged particles. The size of the screening area, as shown in Section 1.2, is determined by the Debye radius. We can therefore assume $\xi_{max} = r_D = (T/4\pi ne^2)^{1/2}$; for collisions involving electrons one can usually put $T = T_e$, because the ions are unable to take part in screening within the collision time. For ion–ion collisions it is necessary to take into account the screening associated with ions and substitute the smaller of the values T_e and T_i into r_D (see Eq. 1.17). The substitution of ξ_{max} makes it possible to represent the logarithmic factor appearing in Eq. 2.64 (so-called *coulomb logarithm*) in the form

$$L = \ln \frac{r_D}{r_s} \approx \ln \left[\frac{T^{3/2}}{e^3 n^{1/2}} \right] \tag{2.65}$$

where the relative velocity is replaced by the root-mean-square velocity of particle motion in the plasma, $\bar{v} = \sqrt{3T/\mu}$. If the average electron and ion energy are different, then the coulomb logarithm, which determines the cross sections of electron–electron and electron–ion collisions, will evidently contain the electron temperature T_e (it determines the screening radius and the relative velocity in collisions). At $T_e > T_i$ the coulomb logarithm appearing in the cross section of ion collisions is determined by the ion temperature, and at $T_i > T_e$ by the electron temperature. In the quantum-mechanical approach the minimum angle of scattering on a sphere with a radius equal to the Debye length is limited because of diffraction: $\vartheta_{\min} \approx \lambda\!\!\!^{-}/r_D = \hbar/\mu v r_D$. If this quantity is greater than the angle $\vartheta_{\min} = r_s/\xi_{\max} = r_s/r_D$, then it actually determines the coulomb logarithm in conformity with Eq. 2.64. At $\lambda\!\!\!^{-} > r_s$

$$L \approx \ln \frac{2r_D}{\lambda\!\!\!^{-}} \approx \frac{1}{2} \ln \left[\frac{\mu T^2}{e^2 \hbar^2 n} \right] \tag{2.66}$$

Note that inaccuracies in determination of the maximum impact parameter (approximate allowance for the screening conditions and a crude estimate of the quantum-mechanical limitation) affect the coulomb logarithm only slightly, since it is sufficiently large at typical plasma parameters. The equation obtained becomes invalid at very high magnetic fields, when the Larmor radius of the electrons is less than the Debye length. Such fields substantially change the trajectory of colliding electrons motion at large impact parameters. An allowance for this effect changes the expression for the coulomb logarithm.

Thus the transport scattering cross section (Eq. 2.64) has the form

$$s^t = 4\pi \left(\frac{e^2}{\mu v^2} \right)^2 L = 4\pi r_s^2 L \tag{2.67}$$

Note that this expression can also be used for describing electron–electron collisions when the effect of exchange interaction is substantial. As noted above, this effect changes the differential cross section at not-too-small scattering angles (see Eq. 2.63); such a change, however, only slightly affects the transport cross section.

In accordance with Eq. 2.67 the transport cross section of coulomb scattering exceeds by $4L$ times the value of πr_s^2, which determines the large-angle scattering cross section (as mentioned above, in the classical model at $\xi < r_s$ and $\vartheta > \pi/2$). Since L usually greatly exceeds unity, this points to the predominant role of far interactions, for which $\vartheta \ll \pi/2$. At far interactions the impact parameter is commensurate with the Debye radius, which greatly exceeds the average distance between the plasma particles (see Section 1.2). Under these conditions each particle interacts with many electrons and ions simultaneously. In such interaction the direction of the particle motion changes continuously, and its trajectory cannot be divided into alternating sections of free motion and interaction with other particles (see Fig. 1.2). Strictly speaking, it is only with this division that the concept of pair collision has a definite physical meaning. With coulomb interaction a consistent analysis of scattering requires consideration of the problem on simultaneous interaction of many bodies. The effect of each electron or ion on the test particle changes, in principle, its trajectory and hence affects its interaction with the other particles. This was not taken into account when the coulomb scattering cross sections in a plasma were calculated. Therefore direct utilization of the expressions obtained for s^t in computing the associated kinetic values in a plasma must be substantiated.

The legitimacy of describing charged particle interaction as a set of binary collisions can be substantiated quantitatively by comparing the results of such description with the rigorous solution of the problem. For some cases, for which a rigorous solution was obtained,* such a comparison leads to very good agreement. Qualitatively, this agreement can be explained as follows. Interaction of charged particles can be divided into close collisions, at distances of the order of r_s, and far collisions, at distances much larger than r_s (but less than r_D). Close collisions can be considered binary collisions without reservations, because under actual conditions r_s is much less than the interparticle distance ($r_s \ll n^{-1/3}$). When considering individual far interactions one should keep in mind that they are characterized by a small scattering angle. Therefore the effect of each plasma particle results in an insignificant change of the test particle trajectories. In analyzing scattering, in the field of the other electrons and ions affecting the test particle, this change is small and can be neglected to a first approximation. Then the total result of scattering on all electrons and ions can be represented as a sum of deviations on interaction with each of them. Hence it follows that far charged particle interactions can also be reduced, without an appreciable error, to a sequence of binary collisions. Thus it is possible, despite the collective nature of interaction between charged particles, to use the concept of

*The self-consistent problem on interaction of a large group of charged particles was solved for some simple cases by using the methods of the modern field theory (so-called *graphical calculation technique*).

binary collisions for describing their behavior. This approach greatly simplifies the plasma problems.

In accordance with the foregoing we make use of the transport cross sections (see Eq. 2.67) to describe charged particle collisions. For different types of collision (of electrons with each other, of electrons with ions, or of the same type of ions) these cross sections are determined as follows:

$$s_{ee}^{t} = 4\pi\left(\frac{e^2}{\mu_{ee}v^2}\right)^2 L_e = 16\pi\left(\frac{e^2}{m_e v^2}\right)^2 L_e;$$

$$s_{ei}^{t} = 4\pi\left(\frac{e^2}{\mu_{ei}v^2}\right)^2 L_e = 4\pi\left(\frac{e^2}{m_e v^2}\right)^2 L_e; \qquad (2.68)$$

$$s_{ii}^{t} = 4\pi\left(\frac{e^2}{\mu_{ii}v^2}\right)^2 L_i = 16\pi\left(\frac{e^2}{m_i v^2}\right)^2 L_i$$

The coulomb logarithm appearing in Eq. 2.68 can be obtained from Eqs. 2.65 and 2.66, which can be written approximately in the following form:

(*a*) *For electron–electron and electron–ion collisions*

at $T < m_e e^2/3\hbar^2 \approx 10$ eV,

$$L_e \approx 23 + \tfrac{3}{2}\ln T_e - \tfrac{1}{2}\ln n$$

at $T_e > m_e e^2/3\hbar^2 \approx 10$ eV,

$$L_e \approx 24 + \ln T_e - \tfrac{1}{2}\ln n$$

(*b*) *For ion–ion collisions*

at $T_i < m_i e^2/3\hbar^2 \approx 10^4 m_i/m_H$ eV,

$$L_i \approx 23 + \tfrac{3}{2}\ln T_{ie} - \tfrac{1}{2}\ln n$$

where n is expressed in cm^{-3} and T in eV; T_{ie} is the lower of the temperatures T_i and T_e. It can be seen that the dependence of L_e and L_i on the plasma parameters is usually weak. Therefore, when the concentration and temperatures vary within a narrow range, one can assume $L = \text{const}$. Here the cross sections are determined exclusively by the energy of the relative motion in collisions. Putting $L \approx 10$, we obtain the estimate of the cross sections: $s^t \approx 10^{-12}/K^2$, where s^t is measured in cm^2 and $K = \mu v^2/2$, in eV.

The collision cross sections (Eq. 2.68) make it possible to obtain the other integral characteristics as well, namely, the collision frequency, the intercollisional time, and the mean free path. Here we write only those collision frequency equations that are used below:

$$\nu_{ee}^{t} = \frac{16\pi e^4 n}{m_e^2 v^3} L_e;$$
$$\nu_{ei}^{t} = \frac{4\pi e^4 n}{m_e^2 v^3} L_e; \qquad (2.69)$$
$$\nu_{ii}^{t} = \frac{16\pi e^4 n}{m_i^2 v^3} L_i$$

At equal relative motion energies the ratio of these frequencies is equal to

$$\frac{\nu_{ee}}{\nu_{ei}} \approx \sqrt{2}; \qquad \frac{\nu_{ii}}{\nu_{ei}} \approx \sqrt{2}\left(\frac{m_e}{m_i}\right)^{1/2} \frac{L_i}{L_e}$$

2.5 ELASTIC COLLISIONS OF ELECTRONS WITH ATOMS

Let us consider elastic electron–neutral atom collisions, which usually determine the kinetics of electron motion in a weakly ionized plasma. As noted in Section 2.2, elastic electron–atom collisions must be described quantum mechanically. The full solution of the quantum-mechanical problem can be obtained only for the simple atoms (hydrogen, helium, hydrogenlike atoms, etc.). Three types of electron–atom interaction can be distinguished: interaction with an' undisturbed atom field, and polarization and exchange interaction.

At distances commensurate with the atomic radius or less the principal role is played by the first type of interaction. It is described by the potential of the coulomb field of the atomic nucleus, which is shielded by electron shells. For hydrogenlike atoms in the ground state the law of variation in potential energy is given by the relation (see Fig. 2.6, curve 1)

$$e\varphi_1 = U_1 = e^2\left(\frac{1}{r} + \frac{1}{r_a}\right)\exp\left(-2\frac{r}{r_a}\right) \qquad (2.70)$$

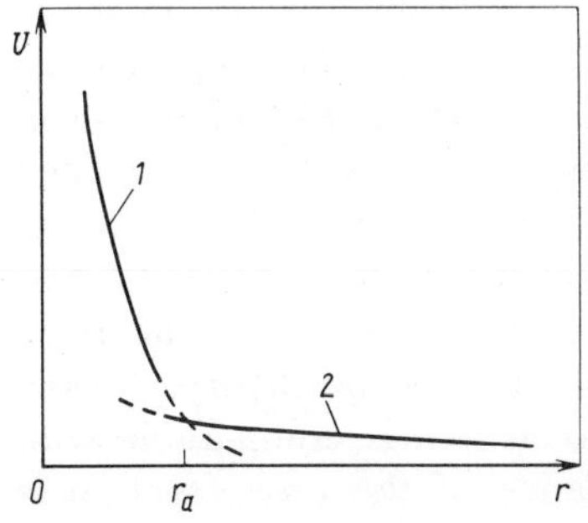

Fig. 2.6 Variation of potential energy as a function of radius.

where $r_a = \hbar^2/m_e e^2$ is the Bohr radius of the atom. The solution of the problem on electron scattering in such a field yields a differential cross section which decreases monotonically with the scattering angle. On rapid reduction in interaction potential at $r > r_a$ in the region of small scattering angles corresponding to far collisions, the cross section remains constant.

Polarization interaction takes place because an atom acquires an electrical dipole moment under the effect of the coulomb field of the free electron, which attracts the nucleus and repulses the electron shells of the atom. The value of the dipole moment d is proportional to the electric field induced by the electron near the atom, $d = \alpha_d e/r^2$. The proportionality factor $\alpha_d = \chi r_a^3$ represents the so-called polarizability of the atom, and χ is a numerical coefficient (for a hydrogen atom at low electron velocities $\chi \approx 4.5$). Accordingly, the potential energy of interaction of an induced dipole momentum with an electron at $r > r_a$ is equal to

$$e\varphi_2 = U_2 = 2\frac{\alpha_d e^2}{r^4} = 2\frac{\chi r_a^3 e^2}{r^4} \tag{2.71}$$

At large distances this potential decreases much slower than that of the undisturbed atom (Eq. 2.70). Therefore far collisions corresponding to small-angle scattering are determined by polarization interaction, which sharply increases the cross section at small angles.

At low velocities of the scattered electron, when its time of residence near the atom is large compared with the period of revolution of the atomic electron, the effect of exchange interaction between the electrons is substantial. This interaction is due to the possibility of exchange of a free and an atomic electron, or, more precisely, to the superposition of their wave functions. Since the probability of ejection of the atomic electron depends only slightly on the direction, inclusion of the exchange interaction results in an increase of the cross section at large scattering angles.

Taking the indicated factors into consideration results in good agreement of experiment with the theory of elastic scattering of electrons on hydrogen and helium atoms. This is illustrated for helium in Fig. 2.7. Curve 1 corresponds to scattering in the field of a shielded nucleus, curve 2 accounts for the exchange interaction as well, and curve 3 accounts for the exchange and polarization interactions. For heavier atoms a detailed theoretical analysis involves great difficulties, but the general shape of the curves is explained by the theory. At low electron energies, when the de Broglie wavelength is comparable with the size of the atom, the angular dependence of the cross section is

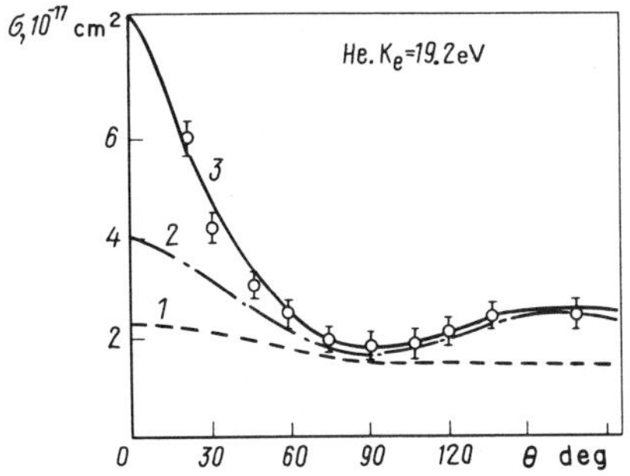

Figure 2.7 Elastic scattering of electrons in helium.

nonmonotonic, the number of maxima and minima on the curve increasing with the atomic number (Fig. 2.8). This diffraction effect is due to the fulfilment of conditions similar to the condition of resonance for one of the harmonics of the wave function that are proportional to the Legendre polynomials (see p. 31). The resonance condition corresponds to a definite ratio between the size of the potential well and the effective wavelength of the electron in the interaction region where the partial scattering amplitude increases sharply. An increase in electron velocity widens the range of values of the angular momentum within which the scattering is considerable ($0 < M < m_e v r_a$). The number of harmonics making a considerable contribution to the cross section ($l_{max} \approx M_{max}/\hbar$) increases accordingly. Therefore, as the velocity increases, the nonmonotonic angular dependence of the cross section smoothes out, and the cross section becomes a monotonically decreasing function of the scattering angle.

At high electron velocities the scattering cross section can be derived from the Born approximation. It is valid at $v \gg e^2/\hbar$ (see Eq. 2.43) and is usually applied from electron energies of 50–100 eV and higher. Sub-

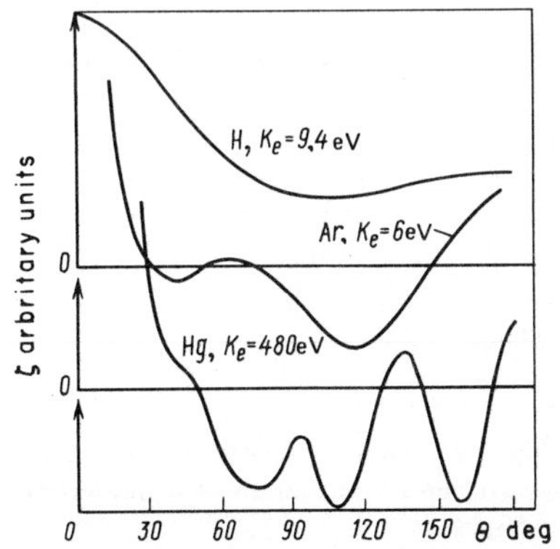

Fig. 2.8 Angular dependence of elastic scattering cross sections.

stitution of the potential of scattered electron–atom interaction into Eq. 2.44 yields the following general expression for the elastic scattering amplitude:

$$f(\vartheta) = \frac{e^2}{2m_e v^2 \sin^2(\vartheta/2)} [Z - F(q)] \tag{2.72}$$

where the first term is determined by interaction of the electron with the nucleus, and the second, with the atomic electrons; $F(q)$ is the so-called *atomic form factor*:

$$F(q) = \int g(r) \exp(i\mathbf{qr})\, d^3r, \qquad q = \frac{mv}{\hbar} \sin\frac{\vartheta}{2}$$

where $g(r) = |\psi(r)|^2$ is the distribution of electron concentration in the atom. The scattering cross section is determined by $|f(\vartheta)|^2$ (see Eq. 2.40). Its scattering-angle dependence is illustrated by Fig. 2.9. In the small-angle range, at

$$qr_a = \frac{m_e v r_a}{\hbar} \sin\frac{\vartheta}{2} \ll 1, \qquad \vartheta \ll \frac{\hbar}{m_e v r_a} \approx \frac{e^2}{\hbar v} \ll 1$$

the scattering amplitude (Eq. 2.72) and the cross section are independent of the scattering angle and the electron velocity:

$$\sigma = \left(\frac{m_e e^2}{3\hbar^2} \int g r^2\, d^3r\right)^2 = \left(\frac{Z\bar{r}^2}{3r_a}\right)^2 \tag{2.73}$$

It can be seen that σ is determined by the root-mean-square distance of the atomic electrons from the nucleus, $\bar{r}^2 = (1/Z) \int g r^2\, d^3r$. Otherwise, at

$$qr_a = \frac{m_e v r_a}{\hbar} \sin\frac{\vartheta}{2} \gg 1, \qquad \vartheta \gg \frac{e^2}{\hbar v}$$

$F(q) \ll Z$, and in the scattering amplitude the only essential term is the first, which determines the coulomb scattering of the electron in the nuclear field Ze/r. The weak effect of field shielding by atomic electrons

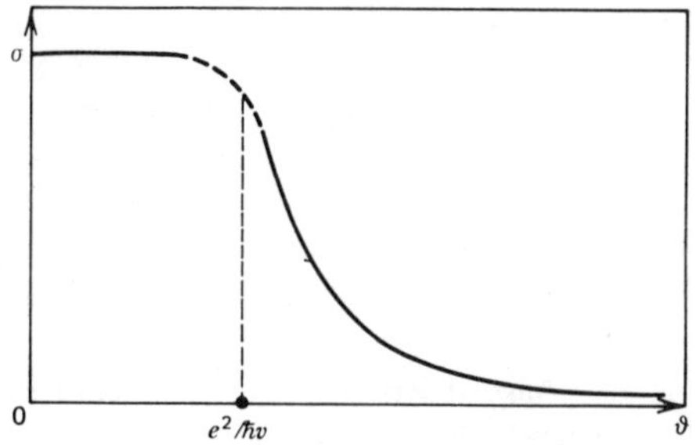

Fig. 2.9 Angular dependence of elastic scattering cross section at high electron velocities.

is explained by the deep penetration of the scattered electron into the atom. The scattering cross section for this case is obtained from the relation 2.62 in accordance with Eq. 2.72; it can be written as

$$\sigma = \left(\frac{Ze^2}{2m_e v^2}\right)^2 \frac{1}{\sin^4(\vartheta/2)} \tag{2.74}$$

That is, it is a rapidly decreasing function of the scattering angle.

On the basis of the differential cross section $\sigma(\vartheta)$ one can obtain the integral characteristics, in particular the transport cross section s^t (Eq. 2.52), which determines the losses of electron momentum and energy in collisions. At high electron velocities, when the Born approximation ($K_e > 50$–100 eV) is applicable, it is possible to obtain the general expression for the transport cross section from Eqs. 2.72–2.74. It is easy to ascertain that the principal contribution to it is made by the range of $\vartheta > e^2/\hbar v$ in which scattering is determined by the coulomb field of the nucleus. Bearing this in mind, we obtain, similarly to Eq. 2.64,

$$\begin{aligned} s^t &= 2\pi \int_{\vartheta_0}^{\pi} \sigma(1-\cos\vartheta)\sin\vartheta\, d\vartheta \\ &= 4\pi \left(\frac{Ze^2}{m_e v^2}\right)^2 \ln \frac{1}{\sin(\vartheta_0/2)} \end{aligned}$$

and then, since $\vartheta_0 \approx e^2/\hbar v$,

$$s^t = 4\pi \left(\frac{Ze^2}{m_e v^2}\right)^2 \ln \gamma \frac{\hbar v}{e^2} \tag{2.75}$$

where γ is a coefficient of about unity (its accurate determination requires taking into account small angles $\vartheta \leqslant \vartheta_0$ and cannot be carried out in the general form). The quantity s^t, however, only slightly depends on γ. In conformity with Eq. 2.75, at high electron energies the transport cross section decreases inversely as the square of the energy.

As noted above, in the range of low and intermediate energies (at $K_e < 50$–100 eV) reliable theoretical data on cross sections are available only for the simplest atoms, hydrogen and helium. For more complex atoms one usually utilizes experimental data. They have been obtained in two runs of experiments. The greatest amount of information was obtained for monokinetic electron beams. However, the experimenters measured not the transport cross section, but the total cross section, limited on the small-angle side:

$$s_\vartheta = 2\pi \int_{\vartheta_{\min}}^{\pi} \sigma \sin\vartheta\, d\vartheta$$

The value of the minimum angle $\vartheta_{\min}$ could not be determined accurately. The cross section s_ϑ differs from the transport cross section;

judging by the angular distribution, this difference may be as great as 30–50%.

The other run of cross section determinations involve measuring the coefficients for the transport processes in a plasma (plasma mobility, diffusion, and conductivity coefficients). These coefficients are directly related to the transport cross section. The accuracy of their determination, however, is not high either, since the results of measurements in a plasma are averaged over a wide distribution of electron velocities. Therefore the available experimental data cannot be considered accurate; they render correctly the velocity dependence of the cross sections, but the absolute values have been determined only with an accuracy to a coefficient of about unity. The experimental data on the cross sections of collisions of electrons with atoms of some gases are given in Figs. 2.10–2.12. The largest cross sections for different gases are within the range from 10^{-14} to $10^{-16}\,cm^2$. They are especially large for atoms of alkali metals. The common feature of the curves of the function $s(v)$ for all gases at not-too-low energies is a decrease in cross section with increasing electron velocity. This is attributed to the reduced particle interaction time $\Delta t \sim r_a/v$.

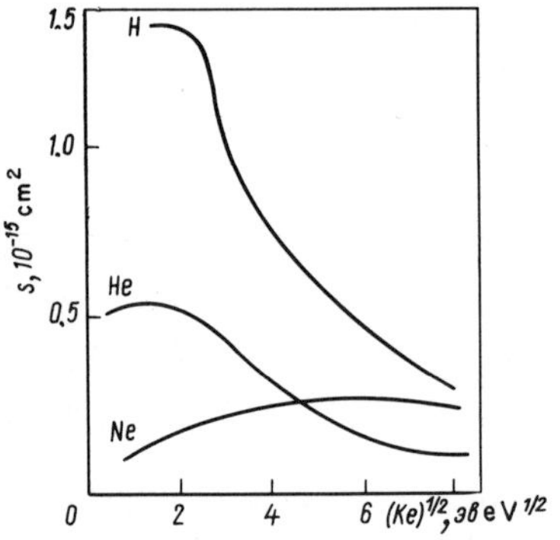

Fig. 2.10 Experimental elastic collision cross sections for H, He, and Ne.

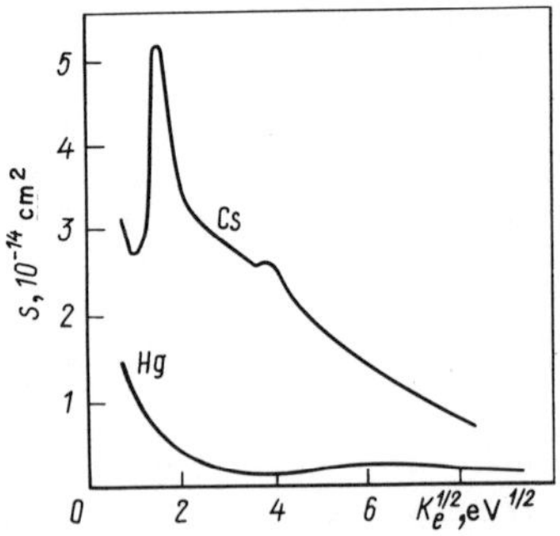

Fig. 2.11 Experimental elastic collision cross sections for Cs and Hg.

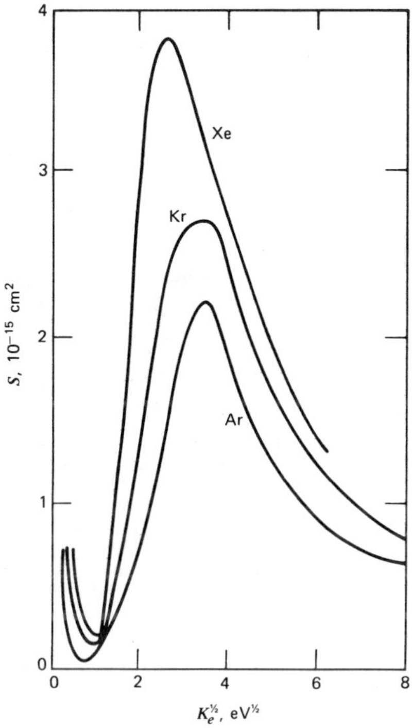

Fig. 2.12 Ramsauer effect in Ar, Kr, and Xe.

There is a sharp decrease in scattering cross section at low electron energies ($K_e < 1$ eV) for some of the heavy atoms, including those of heavy inert gases. This phenomenon is called the ***Ramsauer effect.*** Like the nonmonotonic nature of the dependence of σ on the angle ϑ, it is caused by diffraction of electrons on atoms. At low velocities, when this effect is observed, it is usually only the first term with $l = 0$, determining the isotropic scattering, that is significant in the expansion of the wave function in Legendre polynomials. Therefore the decrease in cross section is ascribed to the condition under which the phase shift of the zero harmonic in the interaction region differs by π from the phase shift in the absence of interaction. Far away from the atom the wave function is the same as in the absence of interaction; hence the scattering amplitude f_0, corresponding to the zero harmonic, vanishes. Harmonics with $l > 0$ contribute little to the scattering cross section. This leads to a deep minimum of the cross section. A similar effect must, in principle, be observed at higher velocities, at which the phase shift in the interaction region is equal to a whole number of π. The corresponding minima smooth out, however, because an increase in velocity enhances the effect of harmonics with $l > 0$.

For a kinetic description of the electron motion in a plasma it is convenient to approximate the dependence of the cross sections on the electron velocity by simple relations. As seen from Figs. 2.10–2.12, approximations may differ greatly for different atoms and molecules and for different electron energy regions. In the low-energy region ($K_e < 1$–3 eV) the cross sections for many atoms can be assumed approximately constant (for instance, for hydrogen, helium, and nitrogen). At high energies, the cross section for some of them is a decreasing function of the energy. Thus for hydrogen and helium, in the energy region of $3\ \text{eV} < K_e < 50\ \text{eV}$ it decreases approximately as $1/v$; in this case the collision frequency $\nu_{ea} = n_a s_{ea} v$ is velocity independent (for hydrogen $\nu_{ea}/n_a \approx 1.6 \times 10^{-7}\ \text{cm}^3/\text{sec}$, for helium $\nu_{ea}/n_a \approx 7 \times 10^{-8}\ \text{cm}^3/\text{sec}$). For some other atoms, such as nitrogen and neon, the cross sections in this energy region can be assumed constant. For heavy atoms, under a considerable influence of the Ramsauer effect, the shape of the curve $s_{ea}(v)$ varies with the energy—it has a minimum at low energies ($K_e < 1$ eV), rises at $1\ \text{eV} < K_e < 5$–15 eV, and drops at high energies. Accordingly, for such atoms the use of simple approximations of the cross section is possible only in a narrow energy range.

2.6 ELASTIC COLLISIONS OF IONS WITH ATOMS

We now consider elastic collisions of ions with atoms, which usually determine the kinetics of their motion in a plasma. The ion–atom interaction forces can be divided into two types. At small impact parameters, "overlapping" of the electron shells results in strong interaction and large-angle scattering. The forces produced by the overlapping effect decrease practically to zero at distances comparable with the size of the atom. Therefore the scattering cross section associated with this interaction must be of the order of πr_a^2. At distances exceeding the atomic radius the interaction is due to polarization; it determines small-angle scattering.

The mechanism of atom polarization by an ion does not differ, in principle, from the polarization mechanism in electron–atom collisions. A free ion induces a dipole moment $d = \alpha_d(e/r^2)$. The potential of its interaction with an ion is equal to $U = 2\alpha_d e^2/r^4$ (see Eq. 2.71). An analysis of ion scattering under the effect of this potential can easily be carried out within the framework of the classical model (as shown in Section 2.2, one can use classical mechanics in describing elastic ion scattering over a wide energy range). By substituting the polarization interaction potential into Eq. 2.29 we can obtain the expression for the differential scattering cross section. At small scattering angles ($\vartheta \ll 1$) it takes the form

$$\sigma \approx \frac{0.4}{\vartheta^{2.5}} \left(\frac{\alpha_d e^2}{\mu_{ia} v^2} \right)^{1/2} \tag{2.76}$$

From the dependence $\sigma(\vartheta)$ we can determine the transport cross section:

$$s^t \approx 10 \left(\frac{\alpha_d e^2}{\mu_{ia} v^2} \right)^{1/2} \tag{2.77}$$

It is inversely proportional to the square root of the kinetic energy of the relative motion and, with the energy fixed, depends exclusively on the polarizability of the atom $\alpha_d = \chi r_a^3$. In its order of magnitude the polarization transport cross section is equal to

$$s^t \sim r_a^2 \left(\frac{\chi \mathscr{E}_a}{K} \right)^{1/2}$$

where $\mathscr{E}_a = m_e e^4/\hbar \approx 10\,\text{eV}$. Since the values of χ lie between 1 and 300, at $K \ll \mathscr{E}_a$ the contribution of polarization interaction to the transport cross section greatly exceeds the effect of interaction associated with the overlapping of the electron shells of the ion and atom. This conclusion is confirmed by the fact that for different ions the cross sections of collisions with the same atoms are approximately the same. In Fig. 2.13 the region of experimental cross sections for different ions is shaded, and it is seen that they differ by no more than a factor of 2. These data were obtained in experiments with monokinetic ion beams. Here, as in electron beam scattering experiments, the quantity $s_\vartheta = 2\pi \int_{\vartheta_{\min}}^{\pi} \sigma \sin \vartheta \, d\vartheta$ was determined.

Investigation of ion scattering in the background gas yielded essentially different results (see Fig. 2.13). The cross sections for this scatter-

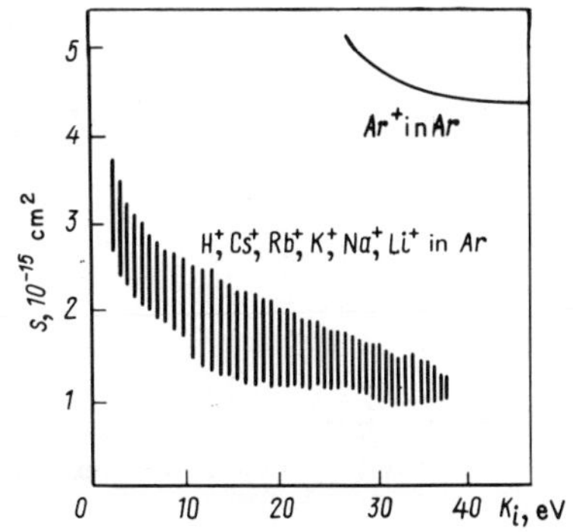

Fig. 2.13 Ion scattering in argon.

ing proved to be much larger than those for the other ions in the same gas. This difference is due to the *charge exchange effect*, arising from electron exchange between the ion and the atom $A_1^+ + A_2 \rightarrow A_1 + A_2^+$.

Ion charge exchange in the background gas can evidently occur without any change in the sum of the internal energies of the interacting particles—the electron moves from the ground level of one atom to that of another identical atom. This process with a zero "energy effect" is called *resonance charge exchange*. It can be classified as an elastic process, since the sum of the kinetic energies remains unchanged after such a collision.

The resonance charge exchange mechanism can be explained qualitatively in the following way. When an atom and an ion approach each other, the height and width of the potential barrier between them gradually decrease for the electron. Figure 2.14 shows schematically the radial dependences of the coulomb potential energy of the atomic residues of particles at a distance r from each other (the dot-and-dash lines refer to each particle; the solid line denotes the summary dependence), and the potition of the energy level $\mathscr{E}_0$ of the valency electron ground state. As can be seen from Fig. 2.14, the potential barrier reaches its maximum height at a distance $r/2$ from each atomic residue. Therefore the maximum potential is equal to $4e/r$. As r decreases, the probability of tunnel transition of the electron from the atom to the ion sharply increases because of the reduced width and height of the potential barrier. In the stationary state the probability that the electron is in any one of the atomic residues is equal to $\frac{1}{2}$. The time to stationary state rapidly reduces with decreasing barrier height $(\mathscr{E}_0 - 4e^2/r)$. As $r \rightarrow r_0 = 4e^2/\mathscr{E}_0$, the barrier height tends to zero, and the time to stationary state is of the order of the time of atomic electron revolution, that is, much less than the ion–atom interaction time at low energies

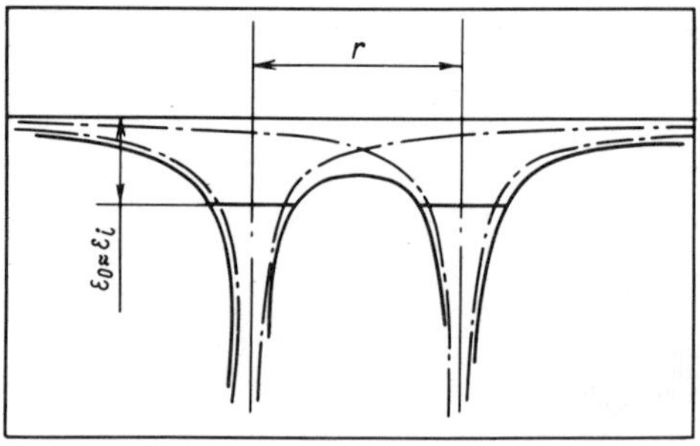

Fig. 2.14 Radial dependence of the coulomb potential energy as a function of particle distance.

($K < 1$ keV). If we neglect the displacement of the energy levels in the atomic residues because of their interaction, the energy $\mathscr{E}_0$ can be equated to the ionization energy $\mathscr{E}_i$. This yields a crude estimate of the charge exchange cross section:

$$s^c \approx \tfrac{1}{2}\pi r_0^2 = 8\,\frac{\pi e^4}{\mathscr{E}_i} \tag{2.78}$$

Taking into account the tunnel transition at distances exceeding r_0 results in some increase of the resonance charge exchange cross section, but the estimate (Eq. 2.78) correctly reflects the order of the cross section. At $K > 0.1$ eV it greatly exceeds the polarization cross section (Eq. 2.77). At high energies ($K \gtrsim 1$ keV) the charge exchange cross section falls off because of the reduced particle interaction time. Some data on the cross sections for ion scattering by atoms in the background gas (these cross sections are determined by resonance charge exchange) are presented in Fig. 2.15. Here the points 1 were obtained by theoretical calculations for $K = 0.1$ eV; the points 2 and 3 were found experimentally at a velocity of $v = 10^5$ and 10^6 cm/sec, respectively; and the dashed curve was calculated by the approximate equation 2.78.

The charge exchange effect on the positive ion motion in a plasma is strong. With no charge exchange fast ion–slow atom collisions generally result in small scattering at which the velocity and direction of the ions change only slightly. On charge exchange the fast particle (ion) also scatters mainly at small angles. But as a result of charge exchange this particle converts into an atom, whereas the slow particle becomes an ion. Therefore the charge exchange is an effective mechanism of energy exchange between the atoms and ions of the plasma. It leads to a "relay" mechanism of positive-charge transfer from particle to particle in the course of directed ion motion induced by the electric field. The

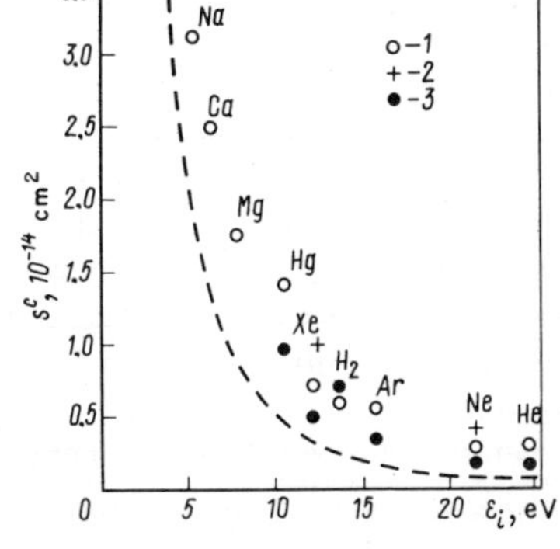

Fig. 2.15 Ion scattering cross sections in their background gas.

slow ion resulting from charge exchange is accelerated by the electric field until the next collision. After a charge exchange type of collision it turns into an atom, whereas the slow ion produced is again accelerated by the field. Therefore in each intercollisional period corresponding to the "relay" stage the charge transfer is accomplished by new particles.

2.7 INELASTIC COLLISIONS OF ELECTRONS WITH ATOMS

Let us consider inelastic electron–atom collisions leading to atom excitation. Such interaction is often the main channel of electron energy loss in a plasma. The excitation of atoms or atomic ions is caused by the rearrangement of their electron shells. The minimum energy necessary for excitation of electron levels is of the order of 10 eV (Table 2.1). Quantum-mechanical calculation of the excitation cross sections for electron states is most difficult in the region of colliding-electron energies commensurate with the excitation energy. In this energy range reliable calculations have been made only for simple cases, in particular for atoms of hydrogen, helium, and alkali metals. Fortunately, however, for many atoms experimental data on total excitation cross sections are available. The dependence of the excitation cross section of the level $3s3p\,^3P$ in helium on the electron kinetic energy is shown in Fig. 2.16 (s^{0j}). The cross section increases rapidly from the threshold, which is equal to the excitation energy $\mathscr{E}_{0j}$. For optically allowed transitions (when the corresponding matrix element of the atom dipole momentum is different from zero) the cross section maximum is usually achieved at energies $K = (1.5\text{–}3)\mathscr{E}_{0j}$, and the largest cross sections are usually from 10^{-19} to 10^{-17} cm^2. The decrease in cross section at high energies is due to the reduced time of electron–atom interaction, as in elastic collisions. An electron–atom collision may also excite metastable levels; transitions to these levels from the ground state on optical excitation are forbidden.

TABLE 2.1 Excitation and ionization energy of some atoms

Energy	H	H_2	He	Ne	Ar	Kr	Xe	Cs	Hg
$\mathscr{E}_1$	10.2	11.2	21.2	16.7	11.6	10.0	8.5	1.5	4.9
$\mathscr{E}_{1M}$	—	—	19.8	16.5	11.5	9.9	8.3	—	4.6
$\mathscr{E}_i$	13.6	15.4	24.5	21.6	15.8	14.0	12.1	3.9	10.4

Note: $\mathscr{E}_1$ is the excitation energy of the lowest level, transition to which from the ground state is permissible (for H_2—the lowest electron level); $\mathscr{E}_{1M}$ is the excitation energy of the lowest metastable level; $\mathscr{E}_i$ is the energy of ionization from the ground state.

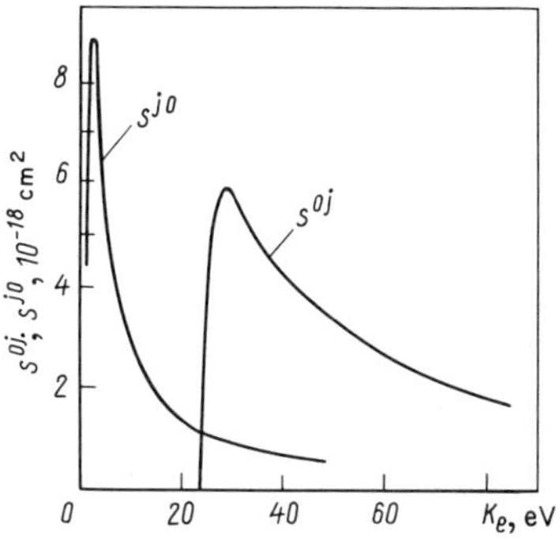

Fig. 2.16 Dependence of the excitation cross section on electron kinetic energy.

The excitation is associated with exchange interaction between the colliding and atomic electrons. Therefore the cross section for such a transition is appreciable only near the excitation threshold, when the time the electron remains near the atom is relatively long, and it rapidly reduces with increasing electron velocity. A typical curve of helium metastable level ion is presented in Fig. 2.17 (curve 1, for allowed level $1s\ 4s\,^1S_0$; and curve 2, for metastable level $1s\ 2s\,^2S_1$).

These curves show that the maximum of the excitation cross section lies close to the excitation threshold. When this is taken into account in approximations and estimations it is often assumed that inelastic collisions lead to complete losses of electron energy. As noted in Section 2.3, in such calculations the efficiency of losses is determined by the summary cross section of inelastic processes $s^n = \Sigma_j s^{(j)}$. For some gases such cross sections were found experimentally (Fig. 2.18) from the electron energy losses. The dependence of s^n on the electron energy near the threshold can be approximated in the following way with an error not exceeding 10–20%:

$$s^n = \beta(K - \mathscr{E}_{01})$$

At electron energies greatly exceeding that of excitation we can use the Born approximation, which yields the following expression for the level excitation energy:

$$s^{0j}_{ea} = 8\pi \left(\frac{e^2}{\hbar v}\right)^2 |d_{0j}|^2 \ln \frac{\gamma_j v \hbar}{e^2} \tag{2.79}$$

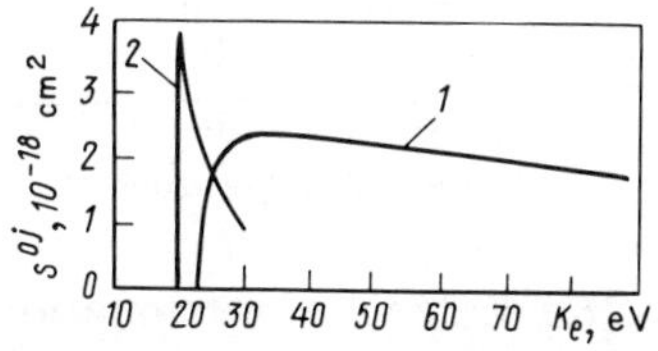

Fig. 2.17 Excitation cross section near the helium metastable level.

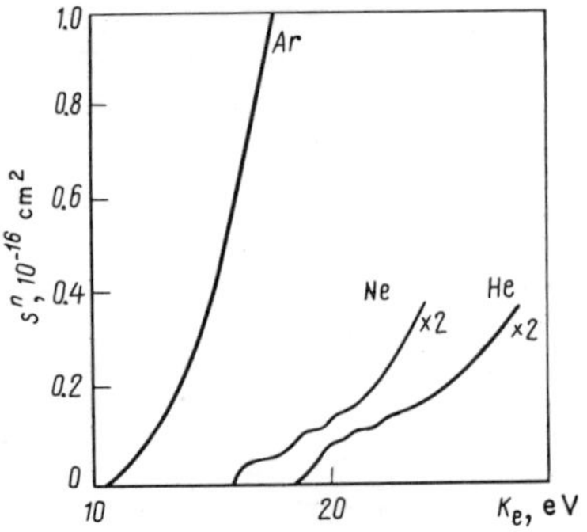

Fig. 2.18 Experimental inelastic collision for electrons in Ar, Ne, and He.

where d_{0j} is the matrix element of the transition dipole momentum and γ_j is a coefficient of the order of unity. Summing the cross sections over all the possible transitions, that is, over all the j states, and over the collisions resulting in ionization, we obtain the expression for the summary cross section of inelastic processes:

$$s_{ea}^{n} = 8\pi \left(\frac{e^2}{\hbar v}\right)^2 \bar{d}^2 \ln \frac{\gamma \bar{v} \hbar}{e^2} \tag{2.80}$$

where $\bar{d}^2$ is the average value of the square of the dipole momentum in the ground state ($d^2 = \Sigma_j |d_{0j}|^2$).

As seen from Eqs. 2.79 and 2.80, the inelastic collision cross section in the area of applicability of the Born approximation is inversely proportional to the energy (with an accuracy to a slowly varying logarithmic factor); that is, it reduces with increasing energy more slowly than the transport cross section of elastic collisions (Eq. 2.75). It is easy to see that the ratio of these cross sections is of the following order:

$$\frac{s_{ea}^{n}}{s_{ea}^{t}} = \frac{(\hbar^2 v^2/e^4)\bar{d}^2}{Z^2 e^2 r_a^2}$$

Thus at $v \gg e^2/\hbar$ the ratio greatly exceeds unity; that is, the role of elastic electron–atom collisions is insignificant.

In collisions of electrons in molecular gases, inelastic collisions may excite oscillation and rotation levels of molecules, as well as electron levels. The energy gap between the oscillation levels is 10^{-2}–1 eV, and between rotation levels, 10^{-3}–10^{-1} eV. For slow electrons the energy losses owing to the excitation of these levels may exceed the energy losses in elastic collisions $\Delta K = 2(m_e/m_a)K$. Therefore such inelastic collisions play an important role in the energy balance of slow electrons in a plasma.

The energy transferred by the electron to the molecule on excitation of rotational degrees of freedom can be estimated using the laws of conservation of energy, momentum, and angular momentum. Their joint consideration indicates that the maximum energy transfer at $R < \xi < R\sqrt{m_a/m_e}$ is of the order of

$$\frac{\Delta K}{K} \approx \frac{m_e}{m_a} \frac{\xi^2}{R^2}$$

where m_a is the mass of the molecule, R is its effective size, and ξ is the impact parameter. It is seen that rotational level excitation at low electron energies is caused by far collisions $\xi > R$, in which the initial angular momentum of the electron $M \sim m_e v \xi$ is sufficiently large. The excitation probability in such collisions is low, however, and therefore the integral rotational level excitation cross section is substantially (usually one or one and a half orders) lower than the elastic scattering cross section.

Collisions due to excitation of oscillatory degrees of freedom of molecules begin to play an appreciable role at electron energies of 0.5–1 eV. They are highly efficient because electron–molecule interaction may produce unstable intermediate bound states, called *autoionization states*. The integral cross sections are largest at energies from 1 to 3 eV.

To characterize the effect of excitation of low-lying levels of molecules on the electron energy balance it is customary to introduce the effective *coefficient of electron energy transfer* in collision κ^s_{ea}, which is determined, similarly to Eq. 2.54, from the inequality

$$\left(\frac{dK}{dt}\right)_{ea} = -\kappa^s_{ea} \nu^s_{ea} K \tag{2.81}$$

where ν^s_{ea} is the summary collision frequency, which depends on the sum of the transport elastic collision cross sections and the summary cross section of inelastic electron collisions. Measurements of the relaxation rate of the slow electron energy in molecular gases yield experimental data on the value of κ_{ea} averaged over the electron velocity distribution. Such data for hydrogen (1) and nitrogen (2) are presented in Fig. 2.19 (dashed line—κ^e_{ea} for H_2). It is seen that with increasing average electron energy the quantity κ^s_{ea} increases as compared with the elastic transfer coefficient $\kappa^e_{ea} = 2m_e/m_a$. For energies from 10^{-2} to 10^{-1} eV this increase is caused by the excitation of rotational levels, and at $10^{-1} - 1$ eV, by the excitation of the oscillation levels of molecules.

The above inelastic electron collisions lead to the appearance of excited atoms in the plasma. The atom concentration in each excited state is determined by the frequency of collisions ν^{0j} producing a state with a lifetime τ_j. From the equality of the rates of excited state

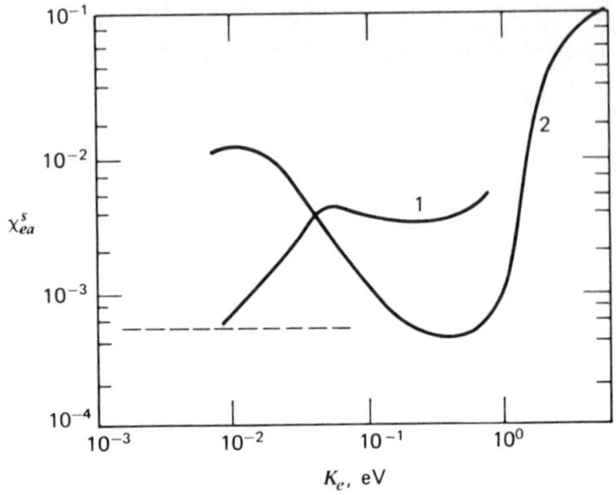

Fig. 2.19 Coefficient of electron energy transfer in (1) hydrogen and (2) nitrogen.

formation and decay it is easy to find the ratio between concentration of atoms in the ground and an excited state n_0 and n_j:

$$\nu^{0j} n_0 = \frac{n_j}{\tau_j}, \qquad \frac{n_j}{n_0} = \nu^{0j} \tau_j \tag{2.82}$$

The lifetime of an excited state relative to radiation is usually of the order of 10^{-8} sec. But it can be five to eight orders higher for the first excited level, when its transition to the ground state is optically forbidden. Therefore the atom concentration in such a metastable state is much higher than in the other excited states. At a high density of the neutral gas the increase in atom concentration in the lowest excited state may also be due to the effect called the *diffusion of resonance radiation.* Since the energy of the emitted quantum is equal to the excitation energy, the quantum may be absorbed by an atom in the ground state, and this will result in transition of the atom to the excited state: the cross section of such photoexcitation usually ranges from 10^{-18} to 10^{-16} cm^2. Therefore, if the system greatly exceeds in size the mean free path of the radiation ($L \gg 1/n_0 s_\nu$), the excited state will be restored repeatedly as a result of emitted-quantum absorption.

At considerable concentrations of excited atoms, along with inelastic collisions leading to the excitation of the atoms a certain role is played by second-order inelastic collisions, which cause transition of the atoms from the excited to the ground state and hence increase the electron energy. The relationship between the cross section of such collisions and that of the excitation processes can be found with the aid of the *detailed balancing principle.*

In quantum mechanics the detailed balancing principle stems from the symmetry of scattering with respect to the time inversion (so-called reciprocity theorem). This symmetry leads to the equality of the scattering amplitude modulus for forward and reverse transitions between two states of a colliding-particles system. For an inelastic collision between an electron of a given energy K and an atom, which results in excitation of a definite state of the atom $e + A^{(0)} \leftrightarrows A^{(j)} + e$ the reverse process seems to consist in deexcitation on collision of this atom with an electron that has a correspondingly lower energy ($K' = K - \mathscr{E}_{0j}$). The equality of scattering amplitudes for such collisions leads to the following relationship between the cross sections of the forward and reverse processes:

$$g_0 v_{e0}^2 s^{0j}(v_{e0}) = g_j v_{ej}^2 s^{j0}(v_{ej}) \tag{2.83}$$

where s^{0j} is the excitation cross section of the j level, s^{j0} is the cross section of deexcitation on collision, g_0 and g_j are the statistical weights of the levels, which determine the number of degenerate states in each state, and the velocities v_{e0} and v_{ej} are related by the energy conservation law: $v_{ej}^2 - v_{e0}^2 = 2\mathscr{E}_{0j}/m_e$. The expression 2.83 is actually the detailed balancing principle for the forward and reverse processes.*

The detailed balancing principle makes it possible to find cross sections of second-order inelastic collisions accompanied by deexcitation, provided the cross sections of the excitation processes are known. From this principle it follows that at $K_e \gg \mathscr{E}_{0j}$ the cross sections of the two processes are of the same order; therefore the second-order collisions can play an appreciable role only if the density of the excited atoms is close to that of the atoms in the ground state. The cross sections of second-order collisions accompanied by deexcitation are maximum at energies lower than the excitation energy (see Fig. 2.16, curve s^{j0}). No collisions accompanied by excitation are possible, and second-order collisions may play a definite role even at relatively low densities of the excited atoms.

2.8 IONIZATION ON COLLISIONS OF ELECTRONS WITH ATOMS

Ionization of atoms in their collisions with electrons is usually the main source of charged particle production in a plasma. The ionization process can be recorded thus: $e + A \rightarrow A^+ + e + e'$. For ionization, the

*Sometimes the detailed balancing principle is taken to mean relations of the type 2.83 averaged over the Maxwellian particle velocity distribution. Such relations can be obtained from consideration of forward and reverse transitions under conditions of thermodynamic equilibrium.

electron energy must exceed the binding energy of one of the outer (valence) electrons in the atom, that is, the ionization energy. The values of the ionization energy $\mathscr{E}_i$ for some atoms can be found in Table 2.1. The rate of charged particle production as a result of ionization in a plasma is determined by the total cross section of the process $s^i(v)$ or by the ionization frequency $\nu^i(v)$. In conformity with Eq. 2.47 we have for $(\partial n/\partial t)_i$

$$\left(\frac{\partial n}{\partial t}\right)_i = n_e n_a \langle v s^i_{ea}(v)\rangle = k^i_{ea} n_e n_a = \langle \nu^i_{ea}\rangle n_e \tag{2.84}$$

where the so-called *process constant* $k^i = \langle v s^i(v)\rangle$ is introduced and the averaging is done over the velocities of the electrons whose energy is sufficient for ionization ($mv^2/2 > \mathscr{E}_i$).

The quantum-mechanical solution of the problem on ionization in electron–atom collision makes it possible, conceptually, to determine the ionization cross section and find the distribution of the ionization-produced electrons over the velocities and scattering angles. As for the other types of collision, a complete solution has been found only for the simplest atoms. A simple estimate of the ionization cross section can be obtained from Thomson's classical theory. It is based on a model in which the atomic electron is assumed free in consideration of energy exchange between an ionizing and an atomic electron. This assumption enables one to use Rutherford's equation (Eq. 2.62) for the differential scattering cross section $\sigma(\vartheta)$. The relationship between the energy transferred and the scattering angle can be written as follows, with due allowance for Eq. 2.19:

$$\Delta K = \tfrac{1}{2}(1 - \cos\vartheta) \quad K = K \sin^2\left(\frac{\vartheta}{2}\right)$$

Assuming that atom ionization, accompanied by electron detachment from the outer shell, occurs as a result of transfer to it of an energy exceeding the ionization energy, that is, at $\Delta K > \mathscr{E}_i$, $\sin^2(\vartheta/2) > \mathscr{E}_i/K$, we obtain the ionization cross section in Thomson's model with the aid of Eq. 2.62:

$$s^i = 2\pi Z_w \int_{\vartheta_{\min}}^{\pi} \sigma \sin\vartheta\, d\vartheta = 2\pi Z_w \left(\frac{e^2}{m_e v_0}\right)^2 \int_{\vartheta_{\min}}^{\pi} \frac{\sin\vartheta\, d\vartheta}{\sin^4(\vartheta/2)}$$

and after integration

$$s^i = Z_w \frac{\pi e^4}{K}\left(\frac{1}{\mathscr{E}_i} - \frac{1}{K}\right) \tag{2.85}$$

and at $K \gtrsim \mathscr{E}_i$ $s^i \approx (Z_w \pi e^4/\mathscr{E}_i^3)(K - \mathscr{E}_i)$, where Z_w is the number of electrons in the outer shell. This equation yields the correct order of the ionization cross section.

Experimental data on the ionization cross section have been obtained for many gases. Typical experimental dependences $s^i(K)$ are presented in Fig. 2.20. Near the threshold the dependence $s^i(K)$ can usually be considered linear:

$$s^i = b_i(K - \mathscr{E}_i) = \beta_i \frac{\pi e^4}{\mathscr{E}_i^3}(K - \mathscr{E}_i) \tag{2.86}$$

s^i reaches the maximum value at $K = (3\text{–}5)\mathscr{E}_i$; it is of the following order: $(s^i)_{\max} \approx \beta_i \pi e^4 / \mathscr{E}_i^2 = 10^{-17}$ to $5 \times 10^{-16}\ \text{cm}^2$.

Within a rather wide energy range the ionization cross section leading to the release of an electron from the outer shell of the atom can be approximated by the formula

$$s^i = Z_w \frac{10\mathscr{E}_i(K - \mathscr{E}_i)}{K(K + 8\mathscr{E}_i)} \tag{2.87}$$

The error of this approximation for atoms and ions in which all the electrons are in the same state does not exceed 20% in the range of energies less than 1 keV.

At high energies $K \gg \mathscr{E}_i$ the ionization cross section can be found using the Born approximation. In this approximation we obtain the following expression for the ionization cross section:

$$s^{i(k)} = \frac{2\pi e^4}{m_e v^2} \frac{\zeta_i Z_w}{\mathscr{E}_i} \ln \gamma_i \frac{m_e v^2}{2\mathscr{E}_i} = \frac{\pi e^4}{K} \frac{\zeta_i Z_w}{\mathscr{E}_i} \ln \gamma_i \frac{K}{\mathscr{E}_i} \tag{2.88}$$

where ζ_i and γ_i are numerical factors (for hydrogen, for instance, $\zeta_i = 0.28$ and $\gamma_i = 0.2$). At high energies the difference between this expression and the Thomson equation (Eq. 2.85) is determined by a logarithmic factor, which slightly changes the dependence of s_i on K. It is easy to see that in the range of applicability of the Born approximation the expressions for the ionization cross section (Eq. 2.88)

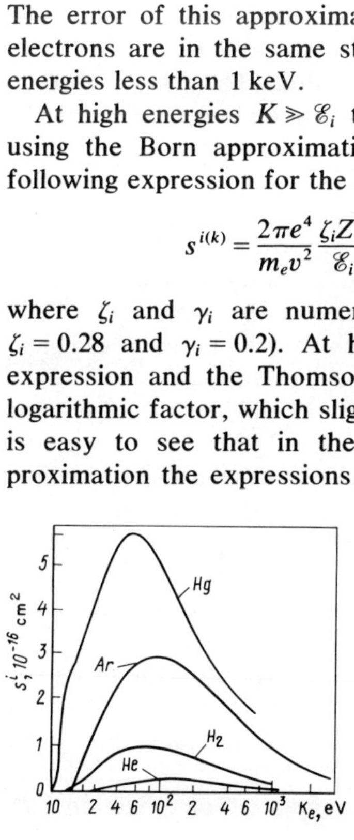

Fig. 2.20 Experimental data on the ionization cross sections for electrons in Hg, Ar, H_2, and He.

and for the total cross section of inelastic collisions (Eq. 2.80) have practically the same energy dependence and the same order of magnitude. Their ratio, which determines the fraction of the ionization processes among the inelastic collisions, usually varies between 0.2 and 0.5 (for hydrogen, for instance, $s^i/s^n = 0.31$).

Note that in electron–atom collision, in addition to ionization with detachment of one electron, so-called *multiple ionization* is possible, which leads to simultaneous detachment of two or several electrons and the production of a multicharge ion. At electron energies greatly exceeding the ionization energy the probability of such a process can be estimated by assuming independent interaction of the ionizing electron with the atomic ones; at $K > \mathscr{E}_i$ it is comparable with the probability of single ionization.

Along with direct ionization of atoms on collisions with electrons considerable *stepwise ionization* may take place, which occurs in two stages, the first being excitation of the atom, and the second, ionization of the excited atom. Let us write the scheme of this process: $e + A \rightarrow A^* + e$; $e + A^* \rightarrow A^+ + e + e'$ (here A^* denotes the excited atom). The efficiency of stepwise ionization depends on the cross sections of both reactions and on the lifetime of the excited atom. The rate of charged particle production as a result of ionization of the atom that was in the excited state j is equal to

$$\left(\frac{\partial n}{\partial t}\right)_{i(j)} = n_e n_a^{(j)} \langle v s^{i(j)}(v) \rangle \tag{2.89}$$

where $n_a^{(j)}$ is the concentration of excited atoms in the j state; $s^{i(j)}$ is the cross section of ionization from this state; and the quantity $\langle v s^{i(j)} \rangle$ is averaged over the velocities of the electrons whose energy is sufficient for ionization. In the stationary case the concentration of the excited atoms produced in collision of electrons with atoms in the normal state satisfies the relation 2.82. Substituting it into Eq. 2.89 and summing over the j states, we obtain the expression for the rate of stepwise ionization:

$$\left(\frac{\partial n}{\partial t}\right)_{is} = n_a n_e^2 \sum_j \frac{\langle v s^{0j} \rangle}{\tau j} \langle v s^{i(j)} \rangle = k^{is} n_a n_e^2 \tag{2.90}$$

The quantity k^{is}, which determines this rate, is called *the stepwise-ionization constant.*

When the lifetime of the excited states is independent of n_e and n_a (when it is determined by radiation) the constant k^{is} depends exclusively on the electron temperature. It evidently greatly increases in the presence of metastable states with a large life time, or of states maintained as a result of diffusion or resonance radiation (see Section 2.7).

The ratio between the rate of stepwise ionization and that of direct ionization (Eq. 2.84) varies with the concentration and temperature of the electrons:

$$\frac{(\partial n/\partial t)_{is}}{(\partial n/\partial t)_i} = \frac{k^{is}}{k^i} n_e$$

At low electron temperatures ($T_e \ll \mathscr{E}_i$) the number of electrons possessing sufficient energy for direct ionization is much less than the number of electrons whose energy is sufficient for ionization from the excited state; with a Maxwellian distribution their ratio is of the order of $\exp[-(\mathscr{E}_i - \mathscr{E}_i^{(j)})/T_e]$. Accordingly, the k^{is}/k^i ratio rapidly increases with decreasing electron temperature (at $T_e \ll \mathscr{E}_i$), and at sufficiently small T_e and large n_e the efficiency of stepwise ionization exceeds that of direct ionization.

2.9 INELASTIC COLLISIONS OF IONS WITH ATOMS

In ion–atom collisions various inelastic processes may take place owing to a change in the internal state of the colliding particles and to their ionization. As follows from the conservation laws (see Section 2.1), the change in internal energy is offset in such collisions by a change in the kinetic energy of the relative motion. In particular, excitation of some state or other is possible if the energy of relative motion exceeds that of excitation. With an equal mass of the ion and atom this condition leads to the inequality $\mu_{ia}v^2/2 = m_i v^2/4 > \mathscr{E}_{0j}$, or $K_i > 2\mathscr{E}_{0j}$ for $v_i \gg v_a$.

Near the threshold, however, the excitation cross section of the atom is very small. It is comparable with the cross section of excitation by electrons only at energies exceeding the threshold energy by three or four orders. The cause for the low efficiency of inelastic collisions of ions at low energies is that their velocity is much lower than that of the atomic electrons (near the threshold $v_i/v_{ae} \approx \sqrt{m_e/m_i}$). At such energies the effect of the electric field of the ion on the atom is nearly quasi-stationary. The slow mutual approach of the particles leads to adiabatic displacement and splitting of electron energy levels in the atom under the influence of the ionic field, as in the Stark effect. As the ion recedes, the levels return to the unperturbed state. The probability of a change in the internal state of the atom, associated with adiabatic disturbance, is low with such a slow motion of the ion. In analyzing collisions of complex ions with atoms one must naturally take into account not only the effect of the ionic field, but also the interaction due to the overlapping of electron shells. Its influence on the inelastic processes, however, also decreases at low velocities.

A quantitative evaluation of the conditions under which the excitation is efficient can be obtained from the following reasoning. The transition of the atom from the state with energy $\mathscr{E}_0$ to the state with energy $\mathscr{E}_j$ can be regarded as excitation of some oscillator with a frequency $\omega_{0j} = (\mathscr{E}_j - \mathscr{E}_0)/\hbar = \mathscr{E}_{0j}/\hbar$. The interaction potential of particles moving at a relative velocity v can be represented as the Fourier integral:

$$U(t) = \frac{1}{2\pi}\int_{-\infty}^{\infty} g(\omega)\exp(i\omega t)\,d\omega$$

The oscillator excitation probability is proportional to the spectral density near $\omega = \omega_{0j}$. The value of $g(\omega_{0j})$ is close to the maximum if the interaction time $\tau \sim r_a/v$ is of the order of $1/\omega_{0j}$; therefore the cross section is maximum at velocities

$$v_0 \sim r_a\omega_{0j} = \left(\frac{\mathscr{E}_{0j}}{\hbar}\right) r_a \tag{2.91}$$

or at energies $K_0 \approx \mu v_0^2/2 \approx \mu r_a^2 \mathscr{E}_{0j}^2/2\hbar^2$ (so-called *Massey criterion*). Note that this estimate corresponds to the energy uncertainty relation. The probability of transition from the 0 to the j state is high if the uncertainty of the energy state of the interacting particles $\delta\mathscr{E} \approx \hbar/\tau$ is of the order of $\mathscr{E}_{0j}$ because of the finite interaction time.

The equality 2.91 thus determines the velocity range in which the cross section of inelastic excitation of the atom is maximum. For an interaction radius of the order of the atomic size $r_a \approx \hbar^2/m_e e^2$ it corresponds to atomic electron velocities $v_{ea} \approx e^2/\hbar \approx 10^8$ cm/sec. Therefore in electron–atom collisions the criterion of effective excitation is already fulfilled near the threshold (at $K_e \approx 10$ eV). In ion–atom collisions the same velocities are associated with much higher energies, $K_i \approx m_i e^4/\hbar^2$. Even for the lightest hydrogen atoms (protons) they are of the order of 10 keV.

The optimum energy for other types of inelastic ion–atom collisions is estimated in a similar way. This refers, in particular, to ionization and *nonresonance charge exchange* in ion-atom collisions. The optimum energies for these processes can be determined with the aid of the Massey criterion by substituting the ionization energy $\mathscr{E}_i$ and the energy defect in charge exchange $\Delta\mathscr{E}$, respectively, into Eq. 2.91. The dependence of hydrogen and helium ionization cross sections on experimental velocity is presented in Fig. 2.21. The kinetic energies of electrons and ions are laid off on the x axis, along with the velocity. As seen from the curves, the maxima of the cross sections of ionization by ions are in the range of velocities of $(3–4) \times 10^8$ cm/sec, which are close to those of atomic electrons. At high velocities the ionization cross sections in

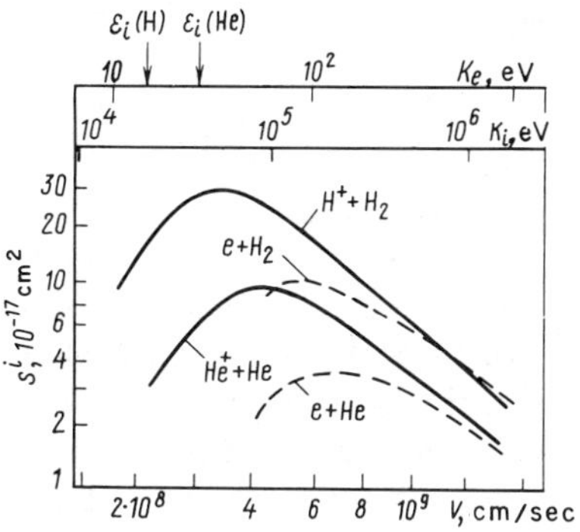

Fig. 2.21 Experimental velocity dependence of ionization cross sections for hydrogen and helium.

electron and ion collisions are practically the same. The difference in the curves for resonance and nonresonance charge exchange in hydrogen is illustrated in Fig. 2.22. It is seen that the resonance charge exchange cross section decreases monotonically with increasing electron energy, whereas the nonresonance charge exchange cross section has a maximum at $K_i \approx 10\,\text{keV}$.

It is worth noting that collisions of high-energy complex ions with atoms may give rise to more complicated inelastic processes. These include multiple ionization,

$$L + M^+ \rightarrow L^{(k+)} + M^+ + ke$$

further ionization of ions (so-called stripping),

$$L + M^+ \rightarrow L + M^{(n+)} + (n-1)e$$

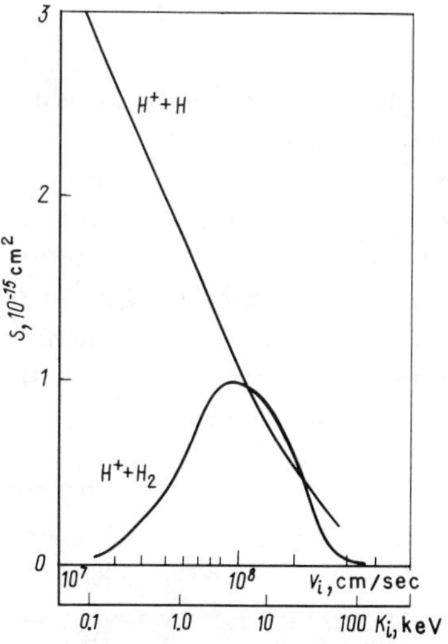

Fig. 2.22 Resonance and nonresonance charge exchange in hydrogen.

and combined ionization of atoms and ions,

$$L + M^+ \rightarrow L^{(k+)} + M^{(n+)} + (n + k - 1)e$$

At ion energies of 10–100 keV and higher, the cross sections for these processes are comparable with those for single ionization.

Thus the basic inelastic processes in collisions of atomic particles (ions and atoms) are significant only at energies of the order of 1 keV or more. These energies are characteristic of a high-temperature thermonuclear plasma. In a gas-discharge plasma the average ion and atom energies usually do not exceed 10 eV. At such energies inelastic collisions of the above type cannot play any appreciable role. Only elastic collisions of ions and atoms and inelastic collisions with a very small energy defect ($\Delta\mathscr{E} < 1$ eV) can be significant in such a plasma. These resonance (or, more precisely, near-resonance) processes are usually possible in gas mixtures. An example is the *Penning effect* discovered in an argon–neon mixture. The metastable excitation level of neon has an excitation potential of $\mathscr{E}_{0i} = 16.5$ eV, which is close to the ionization potential of argon in the ground state ($\mathscr{E}_i = 15.8$ eV). Therefore collision of a neon atom in the metastable state with an argon atom may be accompanied by near-resonance argon ionization $\mathrm{Ar} + \mathrm{Ne}^* \rightarrow \mathrm{Ne} + \mathrm{Ar}^+ + e$. The energy defect of such a process is only 0.7 eV, and its cross section is of the order of the atomic one, even at low energies of the atoms. Therefore, even a small addition of argon to neon sharply increases the ionization efficiency and facilitates the discharge. A similar effect can be observed in other mixtures.

2.10 ELECTRON–ION COLLISIONAL RECOMBINATION

Electron–ion collisions may lead to recombination, that is, the production of a neutral atom as a result of the capture of an electron by an ion. The efficiency of processes responsible for recombination is considerable at low electron energies at which the electron–ion interaction time is sufficiently large. Accordingly, at low electron temperatures (much less than the ionization energy) these processes strongly affect the balance of the charged plasma particles. The rate of charged particle removal due to recombination in the volume can be obtained from the total recombination cross section s_{ei}^r. In conformity with the general definition of the cross section (see Eq. 2.47) we write

$$\left(\frac{\partial n}{\partial t}\right)_r = -n_e n_i \langle s_{ei}^r(v) v \rangle = -n^2 \langle v s^r(v) \rangle \tag{2.92}$$

where we put $n_e = n_i = n$; the relative velocity v in electron–ion collisions can usually be assumed equal to the electron velocity; and $\langle vs^r(v)\rangle$ is velocity averaged. As a rule, to characterize recombination in a plasma, one introduces ***the recombination coefficient*** α, taking that

$$\left(\frac{\partial n}{\partial t}\right)_r = -\alpha n^2, \qquad \alpha = \langle vs^r(v)\rangle \tag{2.93}$$

The coefficient is determined by the recombination cross section and depends on the electron velocity distribution, and with a prescribed distribution, on the electron temperature.

The recombination process can be written in the form of a reaction, $e + L^+ \to L$. It is easy to see that it fails to occur in the absence of third bodies. Indeed, in reactions with the production of a single particle the energy and momentum conservation laws cannot be satisfied simultaneously. These laws are represented by four equations (three projections of the momentum conservation law, and the energy conservation law), whereas the particle produced has three velocity components. Therefore in order to single out the "extra" momentum or "extra" energy, a third body must take part in the recombination. Such a third body may take the form of electrons, ions, neutral particles, or recombination-produced photons. Accordingly, the following processes of electron and ion recombination may take place:

1. $e + L^+ + e \to L + e$.
2. $e + L^+ + M^+ \to L + M^+$.
3. $e + L^+ + M \to L + M$.
4. $e + L^+ \to L + \hbar\nu$.

In estimating the probability of these processes one must bear in mind that each recombination reaction is the reverse of the corresponding ionization. Therefore it is possible to use the detailed balancing principle, according to which the probability of the forward process is proportional to the probability of the reverse process (see Section 2.7). Accordingly, we can compare the first three processes. As shown in Section 2.9, for the energies of ions and atoms usually found in a plasma (K_i, $K_a \ll 10$ keV) the probability of ionization under the effect of heavy particles is much lower than under the effect of electrons. Therefore the probability of recombination (2) with an ion as a third body is always negligibly small compared with two-electron recombination (1). As for recombination with a neutral third body (3), it is appreciable only at very low degrees of plasma ionization.

Recombination involving two electrons and an ion (it is sometimes called *impact recombination*) plays the determining role at high charged particle concentrations. Such a recombination is efficient for slow electrons, which remain in the ion field for a long time. The recombination process here consists of two stages. First, the capture of the electron into a distant orbit in the ion field. Such a capture is due to the energy exchange between the electrons, as a result of which the trajectory of one of them changes from hyperbolic to elliptical (Fig. 2.23). The electron is captured into an orbit near which the potential electron–ion interaction energy is of the order of the energy transferred upon electron–electron interaction, that is, of the order of their average energy $e^2/r_s \approx T_e$, $r_s \approx e^2/T_e$ (the corresponding distance r_s is the average radius of a strong interaction (see Eq. 2.61)). At $T_e \ll \mathscr{E}_i$ this orbit corresponds to a highly excited state of the atom. The second stage of recombination is transition of the electron from the highly excited state to lower-lying levels as a result of collisions or radiation. The recombination coefficient in collision of two electrons and an ion can be estimated from the cross section determining the probability of such a collision in a region of the order of r_s. It is obviously equal to the two-particle collision cross section (of the order of πr_s^2) multiplied by the probability of existence of a third particle (of the order of $\pi r_s^3 n$) in this region:

$$s^r \approx \pi^2 n r_s^5 = \frac{\pi^2 n e^{10}}{T_e^5}$$

The order of magnitude of the recombination coefficient is defined by the product of this cross section times the average electron velocity $v_{T_e} \approx \sqrt{T_e/m_e}$. Accordingly, the expression for this coefficient can be written as

$$\alpha = a \frac{e^{10}}{\sqrt{m_e}} \frac{n}{T_e^{4,5}} \approx 10^{-25} \frac{n}{T_e^{4.5}} \tag{2.94}$$

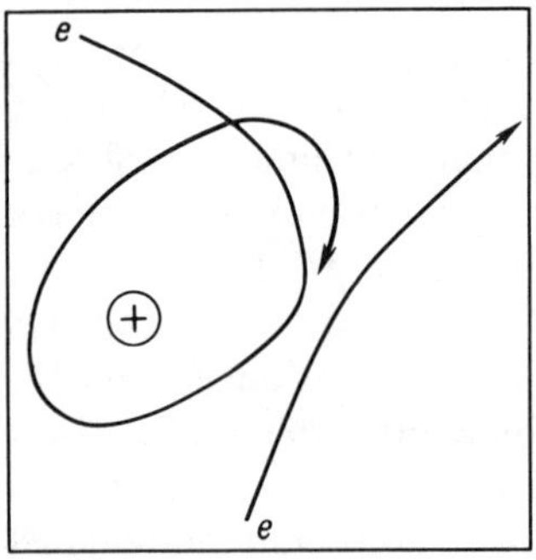

Fig. 2.23 Capture of an electron by an ion.

where a is a coefficient of the order of unity. In numerical estimation α is given in cm^3/sec, T_e in eV, and n in cm^3. The coefficient a includes, in particular, the efficiency of the second recombination stage, which depends on the ratio between the probability of transition of the atom from the highly excited state to the ground state upon electron capture, and the probability of its ionization.

As the electron concentration reduces and the electron temperature increases, the coefficient of impact recombination with the participation of two electrons drops off. The radiation recombination then plays an ever-increasing role. To estimate its efficiency one can use the results of quantum-mechanical consideration of the photoionization process, which is the reverse of radiation recombination. According to this consideration the photionization cross section rapidly increases near the threshold (at $\hbar v \gtrsim \mathscr{E}_i$) to a value of $s^i_{\nu n} \approx (e^2/\hbar c) r_a^2$, which differs from the atomic cross section by a coefficient of the order of the fine-structure constant ($\alpha_R = e^2/\hbar c = 1/137$). With a further increase in photon energy the photoionization cross section falls off. The relationship between the cross sections of radiation recombination and photoionization is found using the detailed balancing principle from a relation similar to Eq. 2.83:

$$m_e^2 v^2 s^r_{ei} \approx p_\nu^2 s^i_{\nu n} \approx \left(\frac{\mathscr{E}_\nu^2}{c^2}\right) s^i_{\nu n}$$

where $p_\nu = \mathscr{E}_\nu / c$ is the photon momentum. At electron energies much below the ionization potential, when $\mathscr{E}_\nu \approx \mathscr{E}_i$, this relationship leads to the following estimate of the average cross section:

$$\overline{s^r} \approx \frac{e^2}{\hbar c^3} \frac{r_a^2 \mathscr{E}_i^2}{m_e^2 \bar{v}^2} \approx \left(\frac{e^2}{\hbar c}\right)^3 r_a^2 \frac{\mathscr{E}_i}{T_e}$$

whence we find the order of magnitude of the recombination coefficient

$$\alpha \approx a^1 \left(\frac{e^2}{\hbar c}\right)^3 r_a^2 \frac{\mathscr{E}_i}{\sqrt{m_e T_e}} \approx \frac{10^{-13}}{\sqrt{T_e}} \tag{2.95}$$

(α is given in cm^3/sec and T_e in eV). At $T_e \gtrsim \mathscr{E}_i$ the recombination coefficient reduces more rapidly with increasing electron temperature.

Impact recombination with the participation of two electrons and radiation recombination can be regarded as the extreme cases of the overall process, which is sometimes called *impact-radiation recombination*. For a detailed consideration one must bear in mind that the participation of one or several additional electrons is possible and that radiation may occur on capture of an electron by an ion and on transition of the capture-produced excited atom to the ground state. The results of this consideration are shown in Fig. 2.24, which presents the

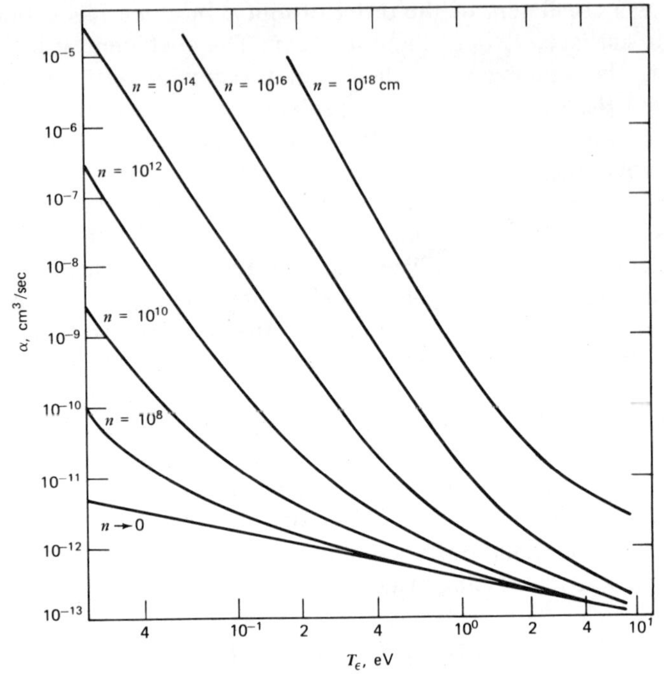

Fig. 2.24 Dependence of the coefficient of impact-radiation recombination in hydrogen on electron concentration and temperature.

dependence of the coefficient of impact-radiation recombination in hydrogen on the electron concentration and temperature.

In the presence of molecular ions in the plasma a recombination accompanied by ion dissociation $e + (ML)^+ \to M + L$ may prove most efficient; it is called *dissociative recombination*. This mechanism does not require triple collision, since the third body is produced in the course of recombination. Therefore the probability of such a recombination is much higher than that of recombination on triple collisions. At an electron energy of the order of the dissociation energy the dissociative recombination cross section is close to the effective cross section of a molecular ion. With a decrease in energy the cross section increases. The typical values of the dissociative recombination coefficient at $T_e <$ 1 eV lie in the range of $\alpha \approx (10^{-8}\text{–}10^{-10})/T_e^{3/2}$ (α is given in cm^3/sec and T_e in eV).

In estimating the efficiency of dissociative recombination one must take into account that molecular ions are produced not only in molecular gases. If the pressure is not too low they are also formed in atomic gases (even in inert gases) as a result of collisions of atomic ions with atoms of the neutral gas $L^+ + 2L \rightarrow L_2^+ + L$. Since the dissociative recombination in atomic gases is usually more efficient than the production of molecular ions, the recombination rate is controlled by the efficiency of the latter.

In so-called *electronegative gases*, whose atoms can form stable negative ions, two-stage recombination is possible in addition to the processes discussed. In the first stage the electron is captured by a neutral atom or a molecule to produce a negative ion. The second stage consists in recombination of the positive and negative ions. Most of the chemically active molecules (for instance O_2, Cl_2, and H_2O) and many organic compounds capture electrons very efficiently. An electron can be captured in several different ways, as follows.

With radiation;

$$e + M \rightarrow M^- + h\nu$$

In triple collision;

$$e + M + L \rightarrow M^- + L^*$$

And through molecular dissociation;

$$e + ML \rightarrow M^- + L^*$$

The last reaction, which does not require the participation of a third particle, is most efficient.

The recombination of positive and negative ions $L^+ + M^- \rightarrow L + M$ also does not require the presence of a third particle. Because of the large size of negative ions and low velocities of their relative motion the recombination cross sections of positive and negative ions may exceed the atomic cross section by one or two orders. The efficiency of ion–ion recombination increases considerably with the pressure of the neutral gas. Then, as a result of collision of ions with a neutral atom one of the ions may be captured into the orbit in the field of the other. In such a quasi-molecule there is a considerable probability of tunnel transition of the electron from the negative ion to the positive one, which completes the recombination process. The ion–ion recombination cross section usually considerably exceeds that of electron capture by neutral atoms. Therefore the efficiency of two-stage recombination is governed by the slower process of electron capture.

2.11 INTERACTION OF CHARGED PARTICLES WITH SOLID SURFACES

Under actual conditions plasmas are usually in contact with various solids: a plasma is surrounded by the chamber walls, it encloses the electrodes, and so on. Therefore it is important to know what happens to the charged particles when they find themselves on the surface of these bodies (we call them walls for brevity). Interactions of particles with solid surfaces are extremely diverse. Below, we mention only those that may substantially affect the state of plasmas.

In the first place, the wall plays a considerable role in the balance of the plasma particles. Solid surfaces may be sources of charged particles (*thermionic*, *secondary*, *autoemission*, and *photoemission*), but the recombination efficiency of the electrons and ions that get onto the wall from the plasma volume is high. This is because the wall is an ideal third body, since it possesses a practically infinite mass and a continuous spectrum of variations of internal energy (temperature). Besides, ion neutralization and electron absorption can be separated in time there, and also in space—in the case of a metal wall.

Bombardment of the walls by particles also results in a change of plasma composition owing to knocking out of particles of the wall material in the form of atoms and ions (so-called *cathode sputtering*). Moreover, incorporation of ions into the wall "hardens" the gas in the course of discharge (decreases the total pressure of the heavy particles). Finally, for the plasma energy balance to be maintained it is important to know to what degree the energy of the particles changes on their reflection from the walls, and also how much energy the particles emitted by the wall carry to the plasma.

Let us consider the processes involved in the incidence of the electrons on the wall. At low energies elastic reflection and absorption are of major importance. In the high-energy region the predominant process is so-called *secondary electron emission*, that is, the knocking of electrons out of the wall by the incident electrons. These secondary electrons possess an energy of several electron volts. Inelastic reflection of primary electrons is also observed.

The efficiency of reflection depends on the wall material, the state (purity) of the surface, and the energy of the incident electrons. The coefficient of elastic reflection (the ratio of the number of reflected particles to that of the incident ones) reaches its maximum (0.1–0.4 for metals and 0.5–0.8 for dielectrics) at energies from 3 to 20 eV. With an increase in energy its value drops. The coefficient of inelastic reflection does not exceed 0.5 for any materials; it is considerable at energies of hundreds of electron volts.

Electron energy losses through reflection can be related to interzone transitions of electrons of the substance and the excitation of natural collective vibrations of the electron gas as a whole; all this yields *discrete* or so-called *characteristic losses*. In addition, the primary electron slows down in interaction with conduction electrons (coulomb scattering). If the momentum of such an electron reverses, it may return to the plasma, having lost part of its energy.

The coefficient of secondary electron emission exhibits a non-monotonic dependence on the energy (Fig. 2.25). For practically all substances (with the exception of some metals such as Te, Al, and Be) there is a range of electron energy $K_e' - K_e''$, where $\sigma > 1$. For metals $K_e \lesssim 100$–200 eV and $K_e'' \approx 1$ keV. For dielectrics $K_e' \approx 10$–30 eV and $K_e'' \approx 2$–10 keV. For metals σ_{max} does not exceed 2, whereas for dielectrics it may be as large as 20–30. A further mechanism leading to knocking electrons out of the walls is the photoeffect, which is caused by plasma radiation.

The above processes may exert a strong influence on the state of the plasma either in the absence of a magnetic field or if an electron exchange is possible between the wall and the plasma along the lines of force of the field. Twisting the trajectories of the charged particles, the magnetic field prevents the secondary and reflected electrons from moving into the plasma volume by returning them to the wall.

Incidence of ions onto a solid surface, as well as incidence of electrons, may be accompanied by elastic and inelastic reflection, absorption (or incorporation, which is the same), and *ion–electron* emission. Moreover, neutralization (reflection of a particle with acquisition of the lacking electron) is usually efficient. Cathode sputtering may also play an important role. Since it occurs in the form of atoms and ions this phenomenon could also be called *ion–atom* and *ion–ion* emission by analogy with the electron processes.

Electron emission from a substance under ion bombardment may be caused by two factors. If the ionization energy of the impinging ion exceeds the doubled work function, so-called *potential pickup* is possible. The mechanism of this phenomenon is as follows. The system "ion at the surface–solid body" can be assumed in an excited state. Transfer to the ground state results from the capture of one of the electrons of the solid by an ion. The excess energy released on neutralization can be

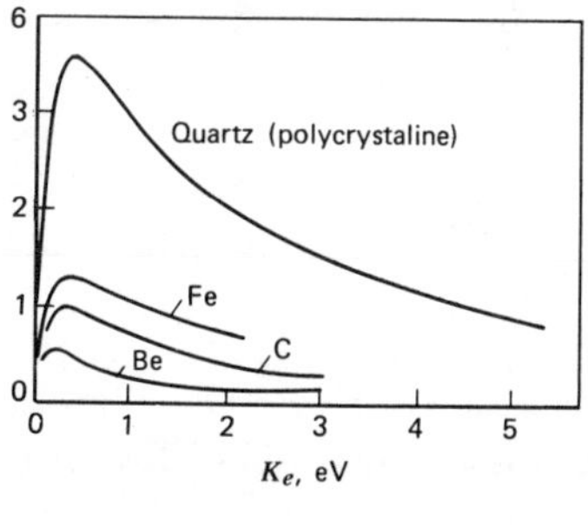

Fig. 2.25 Coefficient of secondary electron emission as a function of energy.

transferred to another electron (*Auger process*), which, having acquired an energy exceeding the work function, may leave the solid. The probability of potential electron pickup by single-charged ions is almost always less than unity. It is only under bombardment by helium ions that the coefficient of such secondary ion–electron emission σ_i may reach 0.3–0.5. In the case of multicharge ions the efficiency of potential pickup increases, and σ_i may exceed unity. In the high-energy range (more than several hundred electron volts) kinetic ion–electron emission predominates. Its mechanism, roughly speaking, is similar to impact ionization of atoms by ions (see Section 2.9). Kinetic ion–electron emission is also inefficient at low energies, when the perturbation of the electron states in a crystal can be considered adiabatically slow. There is usually a threshold beginning with which the kinetic emission is comparable with the potential emission. The threshold velocity is of about the same order $(0.5–1)\times 10^7$ cm/sec for different ions. Kinetic ion–electron emission reaches a maximum at energies of tens of kiloelectron volts for light ions and hundreds of kiloelectron volts for heavy ones. The maximum value of the secondary emission coefficient may be a few units for metals and more than 10 for dielectrics.

Let us consider the behavior of atomic particles in bombardment of a solid surface by ions. Principal attention should be given to the comparatively low values of the reflection coefficient of the primary particles incident on the surface. In the case of light ions (H^+, He^+, and Li^+) it does not exceed 20% for most targets. With increasing mass of the impinging ion the reflection coefficient increases, reaching 40–60%. But if the ion mass exceeds that of the target atoms, it drops sharply to a few percent for heavy ions. In the range from several hundred to tens of kiloelectron volts the reflection coefficient is generally independent of the ion energy. At high energies the reflection efficiency decreases monotonically. In the low-energy range (below 100 eV) the reflection coefficient is slightly higher. At energies below 10 eV practically no experimental data are available. If the energy of the impinging ions greatly exceeds the binding energy of the atoms in the target the ion–atom interaction can be regarded, to a first approximation, as a succession of independent collisions. Here, as in binary collisions in a gas, the energy losses are determined by the transfer coefficient (the ratio of the atom and ion masses) and by the possibility of ionization. As a rule, reflection occurs even on the first collision. This explains the energy spectrum of the reflected particles.

The efficiency of ion neutralization is very high in most cases. For the ions of inert gases the coefficient of reflection from metals without neutralization is 10^{-1}–10^{-2}%, and from dielectrics, less than 1%. For protons it does not exceed a few percent. The mechanism of ion neutralization on reflection is associated with the possibility of transfer of one of the electrons of the solid to the free level of the ion. If this transfer from the filled states of a crystal is energetically advantageous the neutralization efficiency is high. If the ion energy is not too high the system "ion–crystal lattice" for the electrons may be considered quasi-static (adiabatically variable). Under these conditions the neutralization probability can be estimated, to a first approximation, from the probability of filling the corresponding states of the ion and the lattice. Therefore the degree of neutralization is high if the energy of the electron in the atom at the surface is less than the Fermi energy. The situation reverses only for atoms of alkali metals reaching the metal surface with a high work function. This leads to surface ionization of the atoms. Upon reflection the electron leaves the atom, entering the solid wall. For a stationary flux the process of surface ionization is efficient only at a high temperature of the metal surface (over 2000°C). An elevated temperature is necessary for efficient evaporation of the atoms of the alkali metal, which penetrates into the surface layers as a result of bombardment. Otherwise the work function decreases, and surface ionization practically ceases.

It is clear from the foregoing that a hot wall whose material has a high work function (for instance, tungsten) can serve as a plasma source in the presence of vapors of alkaline earth metals. Then ions are produced by surface ionization, and electrons by thermionic emission. With a sufficient density of the electron and ion fluxes the space charge near the wall automatically creates a potential difference, which regulates the fluxes so that quasi-neutrality in the plasma is ensured.

Cathode sputtering of solid particles begins with certain threshold values of the ion energy. This threshold energy is equal to the ratio of the sublimation energy of the atoms of the given solid to the coefficient of energy transfer from the ion to the atom in the solid. The order of magnitude of the latter is close to that of the energy transfer coefficient

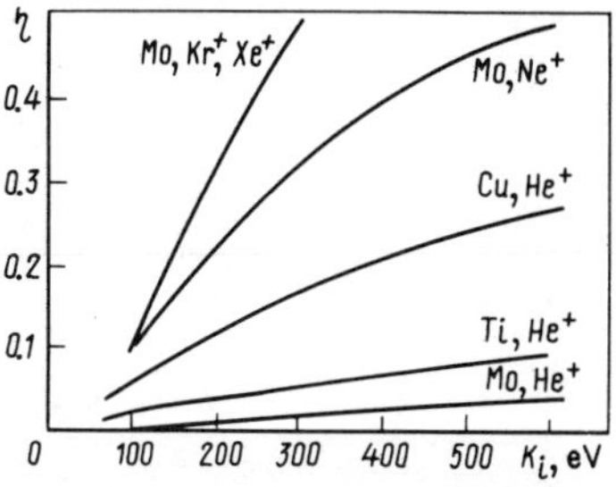

Fig. 2.26 Sputtering coefficient as a function of energy.

κ in atomic collisions. The sublimation energy lies in the range from 1.5 to 9 eV. Therefore the threshold energy is commonly from 3 to 20 eV. With increasing primary beam energy the sputtering coefficient η (ratio of the number of sputtered particles to that of the incident ones) increases (Fig. 2.26). At an energy of about 1000 eV it becomes comparable with unity. On the whole, the dependence of η on K_i represents a smooth curve with a maximum. For light ions this maximum is rather sharp. It is observed at energies of around 10 keV. For heavy ions (Ar^+, Kr^+) a broad maximum is observed in the energy range from 30 to 100 keV. As in reflection, cathode sputtering mainly produces neutral particles (atoms and sometimes whole complexes). Their average energy grows with the atomic number of the sputtered ions, but it does not exceed 10 eV. The number of atoms emitted per ion depends on the atomic number and the structure of the electron shells of the ion and the target atoms. The value of η increases with the mass and with the degree of filling of the outer shells with electrons. Hydrogen and helium atoms have the lowest sputtering efficiency. Heavy elements such as tantalum and tungsten are most stable to sputtering under the effect of these atoms. For these, the atomization coefficient at the maximum is less than 10^{-2}.

3

KINETIC EQUATIONS FOR CHARGED PARTICLES

3.1 DISTRIBUTION FUNCTIONS

As has already been noted, a plasma is an assembly of a great number of moving and interacting particles. An accurate description of the behavior of the assembly, based on an analysis of the trajectories of all the constituent particles, is virtually impossible.* Therefore plasma problems are solved by statistical methods of physical kinetics. The main statistical characteristic of an assembly is the distribution function. To find this function the whole configuration space and the entire range of particle velocities must be divided into small intervals such that the particle density variation within each interval could be neglected. If the number of particles per interval is sufficiently large the fluctuations associated with the random motion of the particles will be small. It would be legitimate to use statistically averaged values of the number of particles in each interval to describe the behavior of the assembly. The averages of the ratio of the number of particles in each interval to the interval itself are called *distribution functions.*

We now express the distribution function quantitatively. To do this we introduce designations of the intervals in the configuration and velocity spaces. Let us denote a volume element in the configuration space by $\mathbf{r}\, d\mathbf{r}$. Its position is determined by the vector $\mathbf{r}$ with coordinates x, y, z, and its linear dimensions by the coordinate increments dx, dy, and dz. The volume is denoted by $d^3r(d^3r = dx\,dy\,dz)$. The velocity interval with

*Such an approach is now used in so-called computer experiments. The trajectories of some assembly of particles at preassigned initial conditions are calculated with the aid of an electronic computer. Contemporary computers, however, are capable of simulating only the simplest problems for a comparatively small number of particles in the assembly.

a variation in the values of projection in the range from v_x to $v_x + dv_x$, from v_y to $v_y + dv_y$, and from v_z to $v_z + dv_z$ is denoted by $\mathbf{v}\, d\mathbf{v}$. This interval, by analogy with the space one, can be regarded as a volume element in the velocity space whose position is defined by the vector $\mathbf{v}$, and the size, by the increments dv_x, dv_y, and dv_z; accordingly, the volume is equal to $d^3v = dv_x\, dv_y\, dv_z$. Because finding a complete definition of the parameter interval requires simultaneous assignment of the intervals of the coordinates and of the velocity components it is customary to use the concept of a six-dimensional phase space, whose three coordinate axes correspond to those of the configuration space, and the other three to those of the velocity space. The radius vector in the six-dimensional space $\mathbf{R}$ then has projections x, y, z, v_x, v_y, v_z; the volume element corresponds to the selected intervals

$$d^6R = d^3r d^3v = dx\, dy\, dz\, dv_x\, dv_y\, dv_z$$

Assuming that the chosen coordinate and velocity intervals satisfy the above-formulated conditions, we can write the expression for the average number of particles dN_{rv} in the six-dimensional interval:

$$dN_{rv} = F(\mathbf{v}, \mathbf{r}, t) d^6R = F(\mathbf{v}, \mathbf{r}, t) d^3r\, d^3v \tag{3.1}$$

Here $F(\mathbf{v}, \mathbf{r}, t)$ is the distribution function, which determines the number of particles in a given interval of coordinates and velocity components at a given instant. It clearly has the meaning of concentration (density) of particles in six-dimensional space. Distribution functions can be introduced for each type of particles (in plasmas, for electrons, ions, and atoms). If there are particles in different internal states (for instance, atoms in different excited states) it is possible to introduce distribution functions characterizing the particle distribution not only over the space coordinates and velocities, but also over the various internal states.

Let us integrate Eq. 3.1 over the velocities. This integration will obviously yield the total number of particles per volume element in configuration space:

$$dN_r = d^3r \int_{(\mathbf{v})} F(\mathbf{v}, \mathbf{r}, t) d^3v = n(\mathbf{r}, t) d^3r$$

This number is proportional to the volume element, and the proportionality factor is equal, by definition, to the usual concentration (density) of particles in the configuration space

$$n(\mathbf{r}, t) = \int_{(\mathbf{v})} F(\mathbf{v}, \mathbf{r}, t) d^3v \tag{3.2}$$

The distribution function $F(\mathbf{v}, \mathbf{r}, t)$ can be represented as the product

$$F(\mathbf{v}, \mathbf{r}, t) = n(\mathbf{r}, t) f(\mathbf{v}, \mathbf{r}, t) \tag{3.3}$$

where the function $f(\mathbf{v}, \mathbf{r}, t)$ is called the *velocity distribution function.* The condition of its normalization can be obtained by integrating Eq. 3.3 over the velocities. Taking into account Eq. 3.2, we find

$$\int_{(\mathbf{v})} f(\mathbf{v}, \mathbf{r}, t) d^3v = 1 \tag{3.4}$$

The velocity distribution function $f(\mathbf{v}, \mathbf{r}, t)$ defines the relative number of particles with velocities lying in the interval $\mathbf{v}\, d\mathbf{v}$ (at a point $\mathbf{r}$ at an instant t). With respect to a single particle it determines the probability that under the given conditions the particle will possess a velocity lying in the interval $\mathbf{v}\, d\mathbf{v}$. The condition 3.4 corresponds to the natural requirement that the probability of defection of a particle velocity in the interval covering the entire velocity space be equal to unity.

Besides $f(\mathbf{v}, \mathbf{r}, t)$ use is often made of some "derivative" distribution functions. For some problems the *unidimensional distribution function* is useful, since it gives the particle distribution with respect to one of the velocity components. The unidimensional distribution function can be obtained from $f(\mathbf{v}, \mathbf{r}, t)$ by integrating over the remaining velocity components. Thus the function

$$f_x(v_x, \mathbf{r}, t) = \int_{(v_y)} \int_{(v_z)} f(\mathbf{v}, \mathbf{r}, t)\, dv_y\, dv_z \tag{3.5}$$

determines the relative number of particles with an x-velocity component within the range from v_x to $v_x + dv_x$. The normalization condition for f_x naturally has a form similar to Eq. 3.4:

$$\int_{-\infty}^{\infty} f_x(v_x, \mathbf{r}, t)\, dv_x = 1$$

It is also useful to introduce the distribution function with respect to the absolute value of the velocity (*distribution function for complete velocities*). It is obtained from $f(\mathbf{v}, \mathbf{r}, t)$ if we change to the spherical system of coordinates in velocity space and integrate over all the angular coordinates. The volume element in the spherical system is equal to $d^3v = v^2 \sin\vartheta\, d\vartheta\, d\varphi\, dv$, and $f_v(v, \mathbf{r}, t)$ is, accordingly,

$$f_v(v, \mathbf{r}, t) = \int_{(\vartheta)} \int_{(\varphi)} f(\mathbf{v}, \mathbf{r}, t) v^2 \sin\vartheta\, d\vartheta\, d\varphi \tag{3.6}$$

When $f(\mathbf{v}, \mathbf{r}, t)$ is isotropic, that is, depends solely on the value of the velocity, but not on the angles ϑ and φ, the relation 3.6 takes the form $f_v(v, \mathbf{r}, t) = 4\pi v^2 f(v, \mathbf{r}, t)$.

The *kinetic energy distribution function* can be obtained from Eq. 3.6 by replacing the velocity interval dv with the corresponding energy interval, that is, $f_v\,dv = f_k(K, \mathbf{r}, t)dK$ or, since $dK = mv\,dv$,

$$f_K = \frac{f_v}{mv} = \frac{f_v}{\sqrt{2mK}} \tag{3.7}$$

A knowledge of the concentrations and the velocity distribution functions of the plasma particles yields comprehensive information on the macroscopic values characterizing the properties of the plasma. The distribution function makes it possible, in particular, to average out any value of interest over the particle velocities. In conformity with the definition the averaging amounts to integration by the equation

$$\langle g(\mathbf{v}_\alpha, \mathbf{r}, t)\rangle = \int_{(\mathbf{v}_\alpha)} g(\mathbf{v}_\alpha, \mathbf{r}, t) f_\alpha(\mathbf{v}_\alpha, \mathbf{r}, t) d^3\, v_\alpha \tag{3.8}$$

where g is a quantity depending on thc particle velocity α and f_α is the velocity distribution function for these particles (it is taken into account that the function f_α is normalized to unity). By using Eq. 3.8 we can find the average velocity of particles of each type, the average energy, the average momentum and energy fluxes, and the other related values (see Section 6.1).

The distribution functions also determine the rate of the processes involved in particle collisions: momentum and energy exchange, excitation, ionization, and recombination (see Section 3.3). Thus the number of binary collisions between α-type particles with velocities in the range of $\mathbf{v}_\alpha\, d\mathbf{v}_\alpha$ and β-type particles with velocities $\mathbf{v}_\beta\, d\mathbf{v}_\beta$ in unit volume per unit time can be written as follows in accordance with Eq. 2.47:

$$dQ^{(p)}_{\alpha\beta}(\mathbf{v}_\alpha, \mathbf{v}_\beta, \vartheta) = (n_\alpha f_\alpha d^3 v_\alpha)(n_\beta f_\beta d^3 v_\beta) v_{\alpha\beta}\sigma^{(p)}_{\alpha\beta}\, d\Omega \tag{3.9}$$

where the first two co-factors define the density of the α- and β-type particles in the indicated velocity ranges, $\mathbf{v}_{\alpha\beta} = \mathbf{v}_\alpha - \mathbf{v}_\beta$ is the relative velocity, and $\sigma^{(p)}_{\alpha\beta}$ is the differential cross section of collisions of the given type (p). The total number of such collisions in unit volume per unit time is obtained by integrating over the angles and over the velocities of the α- and β-type particles:

$$\begin{aligned} Q^{(p)}_{\alpha\beta} &= n_\alpha n_\beta \int_{(\mathbf{v}_\alpha)} \int_{(\mathbf{v}_\beta)} v_{\alpha\beta} f_\alpha f_\beta d^3 v_\alpha d^3 v_\beta \int_{(\Omega)} \sigma^{(p)}_{\alpha\beta}\, d\Omega \\ &= n_\alpha n_\beta \langle v_{\alpha\beta} s^{(p)}_{\alpha\beta}(v_{\alpha\beta})\rangle \end{aligned} \tag{3.10}$$

It can also be expressed in terms of collision frequencies (see Eq. 2.49):

$$Q^{(p)}_{\alpha\beta} = n_\alpha \langle \nu^{(p)}_{\alpha\beta}\rangle = n_\beta \langle \nu^{(p)}_{\beta\alpha}\rangle \tag{3.11}$$

where $\nu^{(p)}_{\alpha\beta} = n_\beta v_{\alpha\beta} s^{(p)}_{\alpha\beta}(v_{\alpha\beta})$. Similar expressions can be obtained for energy and momentum transfer in collisions (see Section 3.3).

3.2 KINETIC EQUATIONS

We now obtain the equation for the distribution function of charged plasma particles $F = nf$. The distribution function, which represents the particle concentration in six-dimensional space, gives the number of particles in each of its elements. Therefore, to determine the changes in distribution function owing to various factors one must consider their effect on the number of particles in an element of a six-dimensional volume. A change in the number of each type of particles in an element $\mathbf{r}\,d\mathbf{r}, \mathbf{v}\,d\mathbf{v}$ may occur, firstly because of the appearance of new and disappearance of existing particles (in the case of charged particles, for instance, as a result of ionization and recombination); and secondly because the fluxes "into" the element of the six-dimensional volume under consideration and "out" of it may not offset each other.* These fluxes are attributed to the motion of the particles and to the change in their velocity under the action of various forces.

Let us first determine the change in the number of particles in the element $\mathbf{r}\,d\mathbf{r}$, $\mathbf{v}\,d\mathbf{v}$ owing to their motion. Consider, for instance, the "flux" of the particles into this element in the direction of the x axis through an area $dy\,dz$ (Fig. 3.1). Within the time dt all the particles that are less than $v_x\,dt$ distant from this area will have passed through it. The number of particles is equal to the product of the density of the particles associated with the velocity range under study $nf\,d^3v$ and the volume bounded by the area $dy\,dz$ and the height $v_x\,dt$, that is, $(nfv_x)_x dt\,d^3v\,dy\,dz$; the values of n, f, v_x must be determined at the point x. The number of particles flowing out of the volume through the area $dy\,dz$ is obtained in the same way, but it must be determined at the point $x + dx$, that is, $(nfv_x)_{x+dx} dt\,d^3v\,dy\,dz$. The difference of these fluxes yields the change in the number of particles in the volume because of the

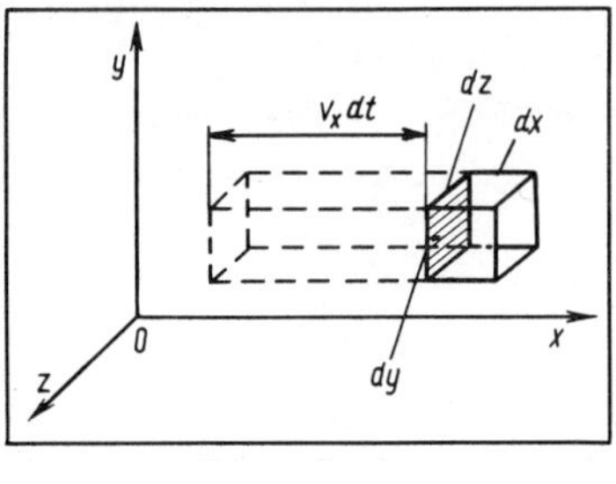

Fig. 3.1 Flux of particles in the x direction through an area $dy\,dz$.

*Note that the change in distribution function due to a particle flux in six-dimensional space is similar to a change in an ordinary three-dimensional density as defined by the flux divergence.

motion along the x axis:

$$\left[\frac{\partial(nf)}{\partial t}\right]_x dt\, d^3r\, d^3v = [(nfv_x)_x - (nfv_x)_{x+dx}]dt\, d^3v\, dy\, dz$$
$$= -\left[\frac{\partial(nfv_x)}{\partial x}\right] dt\, d^3r\, d^3v$$

where the difference of the values of nfv_x at two closely spaced points is obtained from the partial derivative with respect to x, which is taken at fixed values of the other six variables (t, y, z, v_x, v_y, v_z). Canceling out dt and the space elements d^3r and d^3v, we find the rate of variation of the distribution function due to the particle motion along the x axis:

$$\left[\frac{\partial(nf)}{\partial t}\right]_x = -\frac{\partial(nfv_x)}{\partial x} = -v_x\frac{\partial(nf)}{\partial x}$$

Similar expressions are obtained for the components $(\partial nf/\partial t)_y$ and $(\partial nf/\partial t)_z$. Their sum determines the total change in distribution function due to the particle motion:

$$\left[\frac{\partial(nf)}{\partial t}\right]_r = \frac{v_x\partial(nf)}{\partial x} - \frac{v_y\partial(nf)}{dy} - \frac{v_z\partial(nf)}{\partial z}$$
$$= \mathbf{v}\,\mathrm{grad}(nf) \tag{3.12}$$

Let us now find the change in the number of particles in the element $\mathbf{r}\, d\mathbf{r}, \mathbf{v}\, d\mathbf{v}$ caused by the change in their velocity. Consider the flux of particles into an element of a volume in the velocity space across the "area" $dv_y\, dv_z$ (Fig. 3.2). It is associated with the x component of acceleration a_x. Within the time dt the volume is penetrated by particles whose x component of velocity was within the range from v_x to $v_x - a_x\, dt$. The total number of particles in the volume element d^3r is equal to $n\, d^3r$. The portion of the particles with such an x component is within the range $\Delta v_x = a_x\, dt$ and the components v_y and v_z within the range of dv_y, dv_z, is equal to $fa_x\, dt\, dv_y\, dv_z$. Therefore the number of particles entering the volume element by way of the area dv_y, dv_z is

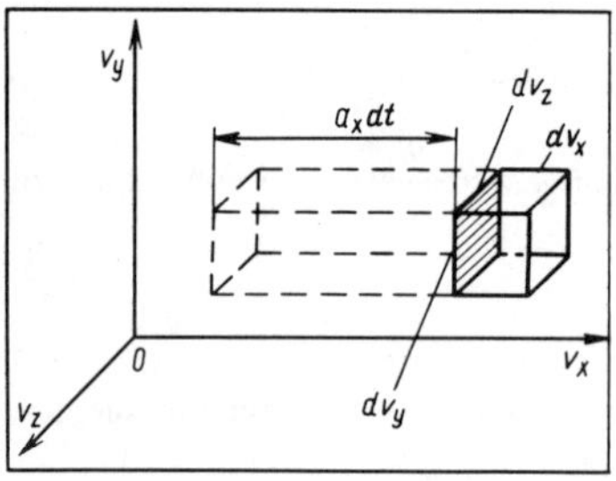

Fig. 3.2 Flux of particles in velocity space.

equal to $(nfa_x)_{v_x} dt\, d^3r\, dv_y\, dv_z$, and the values of n, f, a_x must be determined for the x component of the velocity v_x. The number of particles leaving the element under review because of the change in their velocity is obtained in the same way, but for the velocity $v_x + dv_x$, that is, $(nfa_x)_{v_x+dv_x}\, dt\, d^3r\, dv_y\, dv_z$. The difference of these fluxes gives the change in the number of particles because of their acceleration in the direction of the x axis:

$$\left[\frac{\partial(nf)}{\partial t}\right]_{v_x} dt\, d^3r\, d^3v = -\left[\frac{\partial(nfa_x)}{\partial v_x}\right] dt\, d^3r\, d^3v$$

where the partial derivative $\partial/\partial v_x$ is taken at fixed values of all the other variables (t, x, y, z, v_y, v_z). Similar relations are obtained for changes in the number of particles under the effect of acceleration in the y and z directions. Their sum determines the total change in distribution function as a result of change in particle velocity:

$$\left[\frac{\partial(nf)}{\partial t}\right]_v = -\frac{\partial}{\partial v_x}(nfa_x) - \frac{\partial}{\partial v_y}(nfa_y) - \frac{\partial}{\partial v_z}(nfa_z) \tag{3.13}$$

The acceleration of the particles that appears in Eq. 3.13 is affected by the force acting on them: $\mathbf{a} = \mathbf{F}/m$. The charged plasma particles are subjected to the electric and magnetic fields.* The acceleration due to these forces has the form

$$\mathbf{a} = \frac{1}{m}\left(Ze\mathbf{E} + \frac{Ze}{c}[\mathbf{v} \times \mathbf{H}]\right) \tag{3.14}$$

where $\mathbf{E}$ and $\mathbf{H}$ are the strengths of the electric and magnetic field, respectively, and Ze is the particle charge. When substituting Eq. 3.14 into Eq. 3.13 one should bear in mind that each of the acceleration components is independent of the parallel velocity component. The force $Ze\mathbf{E}$ is independent of $\mathbf{v}$, and the Lorentz force $(Ze/c)[\mathbf{v} \times \mathbf{H}]$ depends solely on the velocity component perpendicular to it. Therefore in Eq. 3.13 the acceleration components can be taken out of the derivative operators. Then we get

$$\left[\frac{\partial(nf)}{\partial t}\right]_v = -a_x \frac{\partial(nf)}{dv_x} - a_y \frac{\partial(nf)}{\partial v_y} - a_z \frac{\partial(nf)}{\partial v_z} = -\mathbf{a}\, \mathrm{grad}_v(nf)$$

where grad_v is the operator of the gradient in velocity space with components $\partial/\partial v_x$, $\partial/\partial v_y$, and $\partial/\partial v_z$. Taking into account Eq. 3.14, we find the change in distribution function due to the acceleration in the electric

*We neglect the effect of gravitational forces, which are appreciable for the plasma of some space objects.

and magnetic fields:

$$\left[\frac{\partial(nf)}{\partial t}\right]_{E,H} = -\frac{Ze}{m}\left[\mathbf{E}+\frac{\mathbf{v}\times\mathbf{H}}{c}\right]\mathrm{grad}_v(nf) \tag{3.15}$$

The electric and magnetic fields acting on each particle consist of fields external to the plasma and fields induced by all the other particles. The latter fields are actually particle interaction forces. Interaction forces can usually be divided into two classes. The first is associated with the collective motion of relatively large particle volumes. These forces may result from charge separation and the passage of currents, which induce macroscopic electric and magnetic fields. The space scales of variation in these fields greatly exceed the average distance between the particles. Therefore the fields related to collective motions can be considered external to the groups of particles contained in each of the elements $\mathbf{r}\,d\mathbf{r}$, $\mathbf{v}\,d\mathbf{v}$. They can usually be determined by averaging the space charge and current, neglecting the fluctuations caused by the random particle motion.

The second type of interaction—close interaction of particles—can be reduced to collisions. As noted in Chapter 2, elastic and inelastic collisions of charged particles with neutrals can be regarded as point collisions in plasma kinetics. This means that their effect on the trajectory of a charged particle can be reduced to an instantaneous change in particle velocity vector. Although a coulomb charged particle interaction lasts a comparatively long time, it can also be (as noted in Section 2.4) represented as a succession of point collisions. Each point collision is obviously equivalent to a dual act: disappearance of a charged particle with the precollision velocity and appearance of a particle with a new velocity in the same place. Therefore all elastic and inelastic collisions of charged particles can together with ionization and recombination be regarded as leading to a change in the number of particles inside a volume element in six-dimensional space. Such a change in element $\mathbf{r}\,d\mathbf{r}, \mathbf{v}\,d\mathbf{v}$ due to collisions will be written in the form $\delta(nf)/\delta t\, d^3r\, d^3v$. Since different collisions occur independently of each other the rate of change in distribution function because of various types of collision can be summed. We do this, for instance, for α-type particles:

$$\frac{\delta(n_\alpha f_a)}{\delta t} = \sum_\beta \sum_p S_{\alpha\beta}^{(p)}(n_\alpha f_a) \tag{3.16}$$

Here the summation is extended to all species of β-type particles and all types of p collisions. The explicit form of the terms $S_{\alpha\beta}^{(p)}$ for some types of collision is given in Section 3.3.

Adding the expressions obtained for the change in distribution function due to the particle motion $[\partial(nf)/\partial t]_r$ (Eq. 3.12) with the change in velocity in the electric and magnetic fields $\partial(nf)/\partial t_{E,H}$ (Eq. 3.15) and with the collisions $\delta(nf)/\delta t$, we find the equation for the distribution function:

$$\frac{\partial(nf)}{\partial t} = -\mathbf{v}\,\text{grad}(nf) - \left(\frac{Ze}{m}\right)\left\{\mathbf{E} + \frac{1}{c}[\mathbf{v}\times\mathbf{H}]\right\}\text{grad}_v(nf) + \frac{\partial(nf)}{\delta t}$$

or, transferring the second and third terms to the left-hand side,

$$\frac{\partial(nf)}{\partial t} + \mathbf{v}\,\text{grad}(nf) + \left(\frac{Ze}{m}\right)\left\{\mathbf{E} + \frac{1}{c}[\mathbf{v}\times\mathbf{H}]\right\}\text{grad}_v(nf = \frac{\delta(nf)}{\delta t} \quad (3.17)$$

This equation is called the *kinetic, or Boltzmann, equation.* It can be written for each type of particle (electrons, ions, and neutral particles). To establish the particular form of the equation it is necessary to determine the fields **E** and **H** and the collision terms $\delta(nf)/\delta t$ in it. As has been mentioned above, the fields **E** and **H** may be induced not only by external sources, but also by space charges and currents created by the charged plasma particles. Their self-consistent determination generally requires the use of the field equations, which include among the sources the charge and current densities of the charged particles as determined by their distribution functions. The kinetic and field equations must be considered jointly.

3.3 COLLISION TERM OF KINETIC EQUATION

As noted above, collisions of different types enter additively into the collision term of the kinetic equation (Eq. 3.16). Therefore their effect on the distribution function can be considered independently. Let us first determine the change in distribution function attributed to elastic collisions, for instance, the change for α-type particles as a result of their elastic collisions with β-type particles. To do this we find the collision-produced change in the number of α-type particles in the volume element $\mathbf{r}\,d\mathbf{r}$ and in the velocity range $\mathbf{v}_\alpha\,d\mathbf{v}_\alpha$. This number is obtained from the distribution function:

$$dN_\alpha = n_\alpha(\mathbf{r})f_\alpha(\mathbf{r}, \mathbf{v}_\alpha)d^3\,r\,d^3\,v_\alpha$$

Assuming the collisions to be of the point type, we presume that they drastically change the velocities of the colliding particles and do not directly affect their position in the configuration space. In particular, each elastic collision in the isolated velocity range changes the velocity of an α-type particle and knocks it out of this range. Let us determine the number of collisions of the α-type particles under consideration with β-type particles moving with velocities in the range of $\mathbf{v}_\beta\,d\mathbf{v}_\beta$ their density is equal to $dn_\beta = n_\beta(\mathbf{r})f_\beta(\mathbf{r}, \mathbf{v}_\beta)d^3v_\beta$. As shown in Sections 2.1 and 2.2, in elastic collision the conservation laws unambiguously determine the relationship between the initial velocities $\mathbf{v}_\alpha$ and $\mathbf{v}_\beta$ and the final velocities of the particles $\mathbf{v}'_\alpha$ and $\mathbf{v}'_\beta$ if the scattering angles ϑ and ψ are given. Therefore a collision of particles from the ranges $\mathbf{v}_\alpha\, d\mathbf{v}_\alpha$ and $\mathbf{v}_\beta\, d\mathbf{v}_\beta$, accompanied by scattering into the solid angle $d\Omega$, results in their transition to quite definite ranges of the velocity space $\mathbf{v}'\alpha\, d\mathbf{v}'_\alpha$ and $\mathbf{v}'_\beta\, d\mathbf{v}'_\beta$ (Fig. 3.3). The number of such collisions per unit time is determined by the differential cross section (see Eq. 2.46):

$$\begin{aligned} d^3r\, dQ(\mathbf{v}_\alpha, \mathbf{v}_\beta | \mathbf{v}'_\alpha, \mathbf{v}'_\beta) &= dN_\alpha\, dn_\beta\, v_{\alpha\beta}\, \sigma_{\alpha\beta}(v_{\alpha\beta}, \vartheta)d\Omega \\ &= n_\alpha n_\beta f_\alpha(\mathbf{v}_\alpha)f_\beta(\mathbf{v}_\beta)v_{\alpha\beta}\sigma_{\alpha\beta}(v_{\alpha\beta}, \vartheta)d\Omega\, d^3v_\alpha\, d^3v_\beta\, d^3r \end{aligned} \tag{3.18}$$

where $\mathbf{v}_{\alpha\beta} = \mathbf{v}_\alpha - \mathbf{v}_\beta$ is the relative velocity. The total removal of α-type particles from the volume element under discussion due to their elastic collisions with β-type particles is obtained by summing Eq. 3.18 over the scattering angles and over the velocities of the β-type particles:

$$dQ^-_{\alpha\beta} = n_\alpha n_\beta \int_{(\mathbf{v}_\beta)} \int_{(\Omega)} f_\alpha(\mathbf{v}_\alpha)f_\beta(\mathbf{v}_\beta)v_{\alpha\beta}\sigma_{\alpha\beta}(v_{\alpha\beta}, \vartheta)d\Omega\, d^3v_\alpha\, d^3v_\beta \tag{3.19}$$

Because of collisions α-type particles may move into the element $\mathbf{v}_\alpha\, d\mathbf{v}_\alpha$ from other elements. In particular, this occurs in collisions opposite to those considered above (see Fig. 3.3)—collisions of particles from the elements $\mathbf{v}'_\alpha\, d\mathbf{v}'_\alpha$ and $\mathbf{v}'_\beta\, d\mathbf{v}'_\beta$ that are determined by the rotation angles of the vector of the relative velocity $\vartheta' = -\vartheta$ and $\psi' = -\psi$. The number of such collisions per unit time can be written as in Eq. 3.18:

$$dQ^+_{\alpha\beta}(\mathbf{v}'_\alpha, \mathbf{v}'_\beta | \mathbf{v}_\alpha, \mathbf{v}_\beta) = n_\alpha n_\beta f_\alpha(\mathbf{v}'_\alpha)f_\beta(\mathbf{v}'_\beta)v_{\alpha\beta}\sigma_{\alpha\beta'}(v_{\alpha\beta}, \vartheta)d\Omega\, d^3v'_\alpha\, d^3v'_\beta \tag{3.20}$$

This expression takes into account that the relative velocity remains unchanged in elastic collision ($v'_{\alpha\beta} = v_{\alpha\beta}$). The total increment of α-type particles in the element $\mathbf{v}_\alpha\, d\mathbf{v}_\alpha$ can be obtained by integrating Eq. 3.20

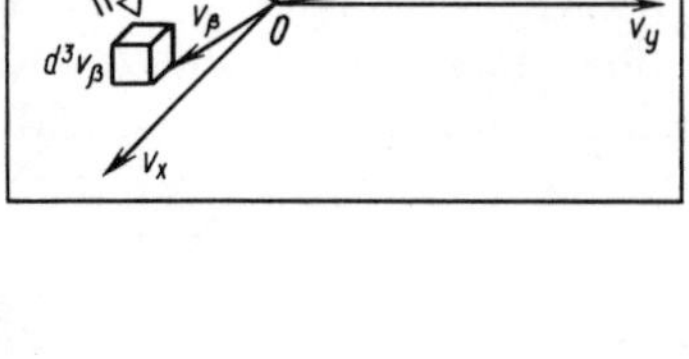

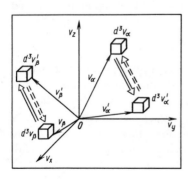

Fig. 3.3 Collisions of particles determined by the rotation angles of the vector of their relative velocities.

over all the combinations of the angles Ω and the velocities $\mathbf{v}'_\alpha$ and $\mathbf{v}_\beta$ that lead to the preassigned value of the velocity $\mathbf{v}_\alpha$:

$$dQ^+_{\alpha\beta} = n_\alpha n_\beta \int_{(\Omega)} \int_{(\mathbf{v}'_\alpha)} \int_{(\mathbf{v}'_\beta)} f_\alpha(\mathbf{v}'_\alpha) f_\beta(\mathbf{v}'_\beta) v_{\alpha\beta} \sigma_{\alpha\beta}(v_{\alpha\beta}, \vartheta)\, d\Omega\, d^3v'_\alpha\, d^3v'_\beta \tag{3.21}$$

It is best to take into account the condition $\mathbf{v}_\alpha = \text{const}$ by changing over from integration with respect to the variables $\mathbf{v}'_\alpha$ and $\mathbf{v}'_\beta$ (postcollision velocities) to integration with respect to $\mathbf{v}_\alpha$ and $\mathbf{v}_\beta$ (precollision velocities). As has already been noted, the relationship between these velocities at the given scattering angles is unambiguous. It is easy to see that the Jacobian of this transition is equal to unity; that is, we can put $d^3v'_\alpha\, d^3v'_\beta = d^3v_\alpha\, d^3v_\beta$ in the integrand.* Therefore after the changeover to $\mathbf{v}_\alpha$ and $\mathbf{v}_\beta$ the integration with respect to $\mathbf{v}'_\alpha$ and $\mathbf{v}'_\beta$ for $\mathbf{v}_\alpha = \text{const}$ in Eq. 3.21 can be replaced by integration with respect to $\mathbf{v}_\beta$. Then we get

$$dQ^+_{\alpha\beta} = n_\alpha n_\beta \int_{(\mathbf{v}_\beta)} \int_{(\Omega)} f_\alpha(\mathbf{v}'_\alpha) f_\beta(\mathbf{v}'_\beta) v_{\alpha\beta} \sigma(v_{\alpha\beta}, \vartheta) d\Omega\, d^3v_\beta\, d^3v_\alpha \tag{3.22}$$

where the velocities $\mathbf{v}'_\alpha$ and $\mathbf{v}'_\beta$ must be expressed via $\mathbf{v}_\alpha$ and $\mathbf{v}_\beta$ in terms of the conservation laws.

The relations 3.19 and 3.22 make it possible to calculate the change in the number of α-type particles in the element $\mathbf{r}\, d\mathbf{r}, \mathbf{v}_\alpha\, d\mathbf{v}_\alpha$ due to elastic collisions with β-type particles:

$$S^e_{\alpha\beta}\, d^3v_\alpha = dQ^+_{\alpha\beta} - dQ^-_{\alpha\beta} = n_\alpha n_\beta \left[\int_{(\mathbf{v}\beta)} \int_{(\Omega)} (f'_\alpha f'_\beta - f_\alpha f_\beta) v_{\alpha\beta} \sigma_{\alpha\beta}\, d\Omega\, d^3v_\beta \right] d^3v_\alpha$$

where the notation $f_\alpha(\mathbf{v}'_\alpha) = f'_\alpha, f_\beta(\mathbf{v}'_\beta) = f'_\beta$ is introduced for brevity. Finally, we obtain for the rate of change in distribution function caused by elastic collisions:

$$S^e_{\alpha\beta} = n_\alpha n_\beta \int_{(\mathbf{v}_\beta)} \int_{(\Omega)} (f'_\alpha f'_\beta - f_\alpha f_\beta) v_{\alpha\beta} \sigma_{\alpha\beta}\, d\Omega\, d^3v_\beta \tag{3.23}$$

We can similarly find the change in distribution function because of inelastic collisions. Let us find it for inelastic collisions associated with a change in the internal state of the particles involved in the collisions (ions or atoms). When considering such collisions one must use the

*This can be ascertained by switching from integration with respect to $\mathbf{v}_\alpha$ and $\mathbf{v}_\beta$ to integration with respect to $\mathbf{v}_0$ (center-of-mass velocity) and $\mathbf{v}_{\alpha\beta}$ (relative velocity) by using the equations from Section 2.1. Since elastic collisions do not affect the velocity $\mathbf{v}_0$ at all and changes in velocity $\mathbf{v}_{\alpha\beta}$ amount to its rotation, it is obvious that in this system $d^3v'_0\, d^3v'_{\alpha\beta} = d^3v_0\, d^3v_{\alpha\beta}$.

distribution function that assigns the particle distribution not only over the space coordinates and velocities, but also over their internal states. We accordingly assume the distribution function to depend on the set of the quantum numbers characterizing the internal state; we denote this set by j, which enumerates all the discrete states (including the various degenerate states with an identical internal energy). We now determine, as before, the change in the number of α-type particles in the volume element $\mathbf{r}\, d\mathbf{r}$ and in the velocity range $\mathbf{v}\, d\mathbf{v}$ due to their collisions with β-type particles resulting in the preassigned change in the internal state of the particles. The collisions of the isolated group of α-type particles lead, as before, to the removal of particles from the element. The number of collisions of this type, summed over all the scattering angles and all the velocities of β-type particles, is obtained from an equation similar to Eq. 3.19:

$$dQ^{-}\binom{j_\alpha j_\beta}{j'_\alpha j'_\beta} = n_\alpha n_\beta\, d^3v_\alpha \int_{(\mathbf{v}_\beta)}\int_{(\Omega)} f_\alpha(\mathbf{v}_\alpha, j_\alpha) f_\beta(\mathbf{v}_\beta, j_\beta) v_{\alpha\beta}\sigma_{j_\alpha j_\beta}^{j'_\alpha j'_\beta}(v_{\alpha\beta}, \vartheta)\, d\Omega\, d^3v_\beta \tag{3.24}$$

in which $\sigma_{j_\alpha j_\beta}^{j'_\alpha j'_\beta}$ is the differential cross section of the process in question.

As above, we also determine the number of reverse inelastic processes leading to the appearance of α-type particles in the velocity range $\mathbf{v}_\alpha\, d\mathbf{v}_\alpha$:

$$dQ^{+}\binom{j'_\alpha j'_\beta}{j_\alpha j_\beta} = n_\alpha n_\beta \int_{(\mathbf{v}'_\alpha)}\int_{(\mathbf{v}'_\beta)}\int_{(\Omega)} f_\alpha(\mathbf{v}'_\alpha, j'_\alpha) f_\beta(\mathbf{v}'_\beta j'_\beta) \times v'_{\alpha\beta}\sigma_{j'_\alpha j'_\beta}^{j_\alpha j_\beta}(v'_{\alpha\beta}, \vartheta)\, d\Omega\, d^3v'_\alpha\, d^3v'_\beta\Big|_{\mathbf{v}_\alpha=\text{const}} \tag{3.25}$$

Since integration must be carried out with $\mathbf{v}_\alpha = \text{const}$, here too it is convenient to switch from the variables $\mathbf{v}'_\alpha$ and $\mathbf{v}'_\beta$ to variables $\mathbf{v}_\alpha$ and $\mathbf{v}_\beta$. For this replacement we can use the detailed balancing principle, according to which the probabilities of forward and reverse transitions between the various states must be the same. For the transitions under study we have

$$v_{\alpha\beta}\sigma_{j_\alpha j_\beta}^{j'_\alpha j'_\beta}(v_{\alpha\beta}, \vartheta)\, d^3v_\alpha\, d^3v_\beta\, d\Omega = v'_{\alpha\beta}\sigma_{j'_\alpha j'_\beta}^{j_\alpha j_\beta}(v'_{\alpha\beta}, \vartheta)\, d^3v'_\alpha\, d^3v'_\beta\, d\Omega'$$

Allowing for this equality, we obtain, with the aid of Eqs. 3.24 and 3.25, the change in the number of α-type particles and the associated change in distribution function:

$$S_{j_\alpha j_\beta}^{j'_\alpha j'_\beta} = \frac{dQ^{+}_{\alpha\beta} - dQ^{-}_{\alpha\beta}}{d^3v_\alpha} = n_\alpha n_\beta \int_{(\mathbf{v}\beta)}\int_{(\Omega)} [f_\alpha(\mathbf{v}'_\alpha, j'_\alpha) f_\beta(\mathbf{v}'_\beta, j'_\beta) - f_\alpha(\mathbf{v}_\alpha, j_\alpha) f_\beta(\mathbf{v}_\beta, j_\beta)] v_{\alpha\beta}\sigma_{j_\alpha j_\beta}^{j'_\alpha j'_\beta}(v_{\alpha\beta}, \vartheta)\, d\Omega\, d^3v_\beta \tag{3.26}$$

Summing Eq. 3.26 over all the initial states of the β-type particles (with respect to j_β) and over all the possible transitions (over all the possible values of j'_α and j'_β), we get the total change in distribution function of the α-type particles following their collisions with the β-type particles:

$$S_{\alpha\beta} = n_\alpha n_\beta \sum_{j_\beta, j'_\alpha, j'_\beta} \iint [f'_\alpha f'_\beta - f_\alpha f_\beta] v_{\alpha\beta} \sigma_{j_\alpha j_\beta}^{j'_\alpha j'_\beta}(v_{\alpha\beta}, \vartheta)\, d\Omega\, d^3v_\beta \qquad (3.27)$$

The expression 3.27 includes all the kinds of binary collisions of α- and β-type particles. The collision term of the kinetic equation for the distribution function of the α-type particles (Eq. 3.16) can be obtained by summing Eq. 3.27 over all the "species" of β-type particles. The sum also contains an integral which makes allowance for collisions of α-type particles with each other at $\beta = \alpha$; it must be included with a weight of $\frac{1}{2}$, because at $\beta = \alpha$ each collision is included twice in the integral. If necessary, one can also include in the collision term collisions involving three particles (for instance, ionization and recombination in collision).

Thus the collision term of the kinetic equation is a complicated integral expression. It is therefore called the *collision integral.* If the kinetic equation 3.17 includes the collision term, it becomes an integro-differential equation that contains partial derivatives with respect to seven independent variables, and integral operators. Besides, as seen from Eq. 3.27, the kinetic equations for the different plasma components are generally not independent. They are interrelated because the integrand for the collision terms for binary interactions includes distribution functions of two particle species. This interrelation is natural because in collisions particles exchange energy and momentum, which leads to mutual influence of their velocity distributions. The solution of a set of coupled integro-differential equations is usually extremely intricate, but in many specific situations it can be simplified.

One of the possibilities is the solution of the kinetic equation for electrons, since the velocity of electrons is practically always much higher than that of heavy particles because of the low electron mass. In many cases one may also assume that the average kinetic energy of the electrons greatly exceeds that of the heavy particles. The integral of collisions of electrons with heavy particles is practically independent of their distribution functions, and its expression can be obtained from the general equations. It is, however, simpler to obtain it directly, assuming the atoms to be fixed. To do this, let us determine the change in the density of electrons whose velocity is in the range $\mathbf{v}_e\, d\mathbf{v}_e$. This density is equal to $n_e f_e\, d^3v_e$. The rate of removal of electrons (in unit volume) from the velocity space element under consideration as a result of collisions with atoms is equal to

$$dQ^- = n_a n_e \int_\Omega v_e \sigma(v_e, \vartheta)\, d\Omega f_e(\mathbf{v}_e)\, d^3 v_e \tag{3.28}$$

The appearance of electrons in the element under review is due to reverse collisions. The electron arrival rate has the form

$$dQ^+ = n_a n_e \int_\Omega v'_e \sigma(v'_e, \vartheta)\, d\Omega f_e(v'_e)\, d^3 v'_e \tag{3.29}$$

where, with given scattering angles (with a fixed $d\Omega$), the postcollision electron velocity $\mathbf{v}'_e$ is uniquely related to their precollision velocity. The difference between dQ^+ and dQ^- yields the change in electron distribution function:

$$\begin{aligned} S_{ea}\, d^3 v_e = dQ^+ - dQ^- = n_e n_a \int_\Omega \Big[v'_e \sigma(v'_e, \vartheta) f_e(\mathbf{v}'_e) \\ \times \frac{d^3 v'_e}{d^3 v_e} - v_e \sigma(v_e, \vartheta) f_e(\mathbf{v}_e) \Big]\, d\Omega\, d^3 v_e \end{aligned} \tag{3.30}$$

The ratio of the elementary velocity space volumes for elastic collisions is determined by the change in velocity:

$$\frac{d^3 v'_e}{d^3 v_e} = \frac{(v'_e)^2\, dv'_e}{v_e^2\, dv_e} = \frac{v_e'^3}{v_e^3}$$

Taking this into account, we obtain the integral of electron–heavy particle elastic collisions:

$$S^e_{ea} = \frac{n_a n_e}{v_e^3} \int_{(\Omega)} [v_e'^4 \sigma_{ea}(v'_e, \vartheta) f_e(\mathbf{v}'_e) - v_e^4 \sigma_{ea}(v_e, \vartheta) f_e(\mathbf{v}_e)]\, d\Omega \tag{3.31}$$

Expression 3.31 is valid for both elastic and inelastic collisions of electrons when their energy greatly exceeds that of the atoms. Since the expression is independent of the atom distribution function the kinetic equation for the electron distribution function is not related to the kinetic equations of the heavy particles.

Another possibility for simplifying the collision integral has to do with elastic collisions of charged particles with each other. As shown in Section 2.4 the determining role in such collisions belongs to far interactions leading to a slight change in velocity. It is easy to show that in this case the collision integral can be represented as a divergence of the flux density in velocity space:

$$S(nf) = -\text{div}_v\, \mathbf{\Gamma}_v = -\left(\frac{\partial \Gamma_{vx}}{\partial v_x} + \frac{\partial \Gamma_{vy}}{\partial v_y} + \frac{\partial \Gamma_{vz}}{\partial v_z}\right) \tag{3.32}$$

where the components of the flux density vector are equal to

$$\Gamma_{vk} = nf\langle \Delta v_k \rangle_1 - \frac{1}{2}\sum_l \frac{\partial}{\partial v_l}[nf\langle \Delta v_l \Delta v_k \rangle_1] \tag{3.33}$$

and the values $\langle \Delta v_k \rangle_1$ and $\langle \Delta v_l \Delta v_k \rangle_1$ are summed over the particle collisions per unit time.

Calculation of $\langle \Delta v_k \rangle_1$ and $\langle \Delta v_l \Delta v_k \rangle_1$ for collisions with a slight velocity change leads to the following expression for the components of the density of α-type particle flux caused by β-type particle collisions:

$$\begin{aligned}\Gamma_{vk} = &\frac{1}{2}\frac{m_\alpha m_\beta}{(m_\alpha + m_\beta)^2} n_\alpha n_\beta \sum_l \int_{(\mathbf{v}_\beta)} \nu^t_{\alpha\beta}(v_{\alpha\beta})(v^2\delta_{kl} - v_l v_k) \\ &\times \left[f_\alpha \frac{\partial f_\beta}{\partial v_{\beta k}} - \frac{m_\beta}{m_\alpha} f_\beta \frac{\partial f_\alpha}{\partial v_{\alpha k}}\right] d^3 v_\beta\end{aligned} \tag{3.34}$$

This expression was obtained by Landau in investigating the effect of Coulomb collisions on the distribution function and is often used when considering processes in a highly ionized plasma. The collision term obtained by substituting Eq. 3.34 into Eq. 3.32 is called the *Landau collision integral.*

The relation 3.33 can be written in the form of the vector equality

$$\mathbf{\Gamma}_v = \mathbf{g}nf - \check{\mathbf{D}}_v\, \text{grad}_v(nf) \tag{3.35}$$

The first term determines the directed variation in the velocity of particles of a given species as a result of their collision with other particles, which can be called the friction effect; accordingly, the vector $\mathbf{g}$ is called the *dynamic friction coefficient.* The second term describes the flux due to the derivatives of the velocity component distribution function. It is present because it tends to smooth the decrease in distribution function. By analogy with the diffusion flux in configuration space (which is proportional to the concentration gradient) this flux is called the *diffusion in velocity space*, and the tensor $\check{\mathbf{D}}_v$, the *diffusion coefficient.*

Substituting the collision term of Eq. 3.32 into Eq. 3.17, we reduce the kinetic equation to

$$\frac{\partial(nf)}{\partial t} + \text{div}(\mathbf{v}nf) + \text{div}_v\left(\frac{\mathbf{F}}{m}nf + \mathbf{\Gamma}_v\right) = 0 \tag{3.36}$$

The second term in the equation determines the divergence of the flux in coordinate space, which is associated with the particle motion, and the third term determines the divergence of the flux in the velocity space associated with the external force **F** and with collisions. The sum of these terms defines the divergence in six-dimensional space. Therefore the kinetic equation in the form 3.36 is the continuity equation for the distribution function in six-dimensional space. It is called the *Fokker–Planck equation.* The use of this equation sometimes substantially simplifies the solution of kinetic problems.

4

EQUILIBRIUM PLASMAS

4.1 DISTRIBUTION FUNCTION IN EQUILIBRIUM PLASMAS

In this chapter we consider the characteristics of a plasma in the state of statistical or thermodynamic equilibrium. First we find the particle distribution functions in an equilibrium plasma. In a closed system, in the absence of external forces, the equilibrium state is spatially homogeneous. The kinetic equation 3.17 for particles of a homogeneous plasma in the absence of external forces has the form

$$\frac{\partial(n_\alpha f_\alpha)}{\partial t} = \frac{\delta(n_\alpha f_\alpha)}{\delta t} \tag{4.1}$$

For the stationary state, the equation amounts to reducing the collision term to zero. When only elastic collisions are substantial it can be written as follows:

$$\sum_\beta S^e_{\alpha\beta} = n_\alpha \sum_\beta n_\beta \int_{(\mathbf{v}_\beta)} \int_{(\Omega)} (f'_\alpha f'_\beta - f_\alpha f_\beta) v_{\alpha\beta} \sigma_{\alpha\beta}\, d\Omega\, d^3 v_\beta = 0 \tag{4.2}$$

Equation 4.2 is obviously satisfied if the following relation holds true for distribution functions:

$$f_\alpha(\mathbf{v}'_\alpha) f_\beta(\mathbf{v}'_\beta) = f_\alpha(\mathbf{v}_\alpha) f_\beta(\mathbf{v}_\beta) \tag{4.3}$$

for any pairs of particles.

As will be shown in Section 4.2, under arbitrary initial conditions distribution functions tend, with time, to a stationary state, which is determined by Eq. 4.3. Taking the logarithm of the relation, we get

$$\ln f_\alpha(\mathbf{v}'_\alpha) + \ln f_\beta(\mathbf{v}'_\beta) = \ln f_\alpha(\mathbf{v}_\alpha) + \ln f_\beta(\mathbf{v}_\beta) \tag{4.4}$$

This equality means that the sum of the logarithms of the distribution function remains unchanged in collisions. As has been established, there are four independent combinations of particle velocities whose sum remains unchanged in elastic collisions. These are the three components of the momentum $p_k = mv_k$ and the kinetic energy $K = mv^2/2$ (see Section 2.1). Obviously, $\ln f(\mathbf{v})$ must be their linear combination:

$$\ln f_\alpha = a + b_x m_\alpha v_{\alpha x} + b_y m_\alpha v_{\alpha y} + b_z m_\alpha v_{\alpha z} - \frac{c m_\alpha v_\alpha^2}{2} \tag{4.5}$$
$$= a' - c\,\frac{m_\alpha}{2}\left[\left(\frac{v_{\alpha x} - b_x}{c}\right)^2 + \left(\frac{v_{\alpha y} - b_y}{c}\right)^2 + \left(\frac{v_{\alpha z} - b_z}{c}\right)^2\right]$$

It is easy to see that here the sum of $\ln f$ for any pairs of particles remains unchanged if the constant coefficients a, b_k, and c are the same for all the particles.

Denoting $b_x/c = u_x$, $b_y/c = u_y$, $b_z/c = u_z$, $c = 1/T$ we find

$$f_\alpha = A \exp\left\{-\frac{m_\alpha}{2T}\left[(v_{\alpha x} - u_x)^2 + (v_{\alpha y} - u_y)^2 + (v_{\alpha z} - u_z)^2\right]\right\} \tag{4.6}$$
$$= A \exp\left[-\frac{m_\alpha}{2T}(\mathbf{v} - \mathbf{u})^2\right]$$

The constant A can be found from the normalization condition 3.4:

$$A \int_{(\mathbf{v})} \exp\left[-\frac{m_\alpha}{2T}(\mathbf{v} - \mathbf{u})^2\right] d^3v$$
$$= A \int_{(v_x)} \int_{(v_y)} \int_{(v_z)} \exp\left[-\frac{m_\alpha}{2T}(\mathbf{v} - \mathbf{u})^2\right] dv_x\, dv_y\, dv_z = 1 \tag{4.7}$$

The triple integral here is the product of three integrals of the same type, each of which is equal to

$$\int_{-\infty}^{\infty} \exp\left[-\frac{m_\alpha}{2T}(v_k - u_k)^2\right] dv_k = \sqrt{\frac{2T}{m_\alpha}} \int_{-\infty}^{\infty} \exp(-\xi^2)\, d\xi = \sqrt{\pi}\,\sqrt{\frac{2T}{m_a}}$$

Substituting this result into Eq. 4.7, we obtain $A = (m_a/2\pi T)^{3/2}$. Thus the distribution function takes the form

$$f_\alpha(\mathbf{v}_\alpha) = \left(\frac{m_\alpha}{2\pi T}\right)^{3/2} \exp\left[-\left(\frac{m_\alpha}{2T}\right)(\mathbf{v}_\alpha - \mathbf{u})^2\right] \tag{4.8}$$

The velocity distribution described by this function is known as the *Maxwellian distribution.*

The distribution function 4.8 can be represented as the product of one-dimensional functions $f_\alpha(\mathbf{v}_\alpha) = f_x(v_{\alpha x}) f_y(v_{\alpha y}) f_z(v_{\alpha z})$ where

$$f_k(v_{\alpha k}) = \left(\frac{m_\alpha}{2\pi T}\right)^{1/2} \exp\left[-\frac{m_\alpha}{2T}(v_{\alpha k} - u_k)^2\right] \tag{4.9}$$

As follows from Eq. 3.5 each of them yields the distribution over one of the velocity components.

The parameters **u** and T appearing in the distribution function 4.8 define the average velocity and the average energy of the particles. We now demonstrate this. The average value of any one of the velocity components can be found by using the unidimensional distribution functions 4.9:

$$\begin{aligned}\langle v_{\alpha k}\rangle &= \int_{-\infty}^{\infty} v_{\alpha k} f_k(v_{\alpha k})\, dv_{\alpha k} \\ &= \left(\frac{m_\alpha}{2\pi T}\right)^{1/2} \int_{-\infty}^{\infty} v_{\alpha k} \exp\left[-\frac{m_\alpha}{2T}(v_{\alpha k}-u_k)^2\right] dv_{\alpha k} \\ &= \frac{u_k}{\sqrt{\pi}} \int_{-\infty}^{\infty} \exp(-\xi^2)\, d\xi = u_k\end{aligned} \tag{4.10}$$

The average velocity **u** is also called the *directed velocity.* Since the vector **u** is the same for particles of all species it determines the motion of the entire plasma as a whole. Therefore in considering the physical characteristics it is convenient to change to a reference system in which the directed velocity is zero. Let us find the average kinetic energy of the particles in this system, that is, the average energy of the random motion. The components of the average energy associated with each of the velocity components are equal to

$$\begin{aligned}\left\langle \frac{m_\alpha (v_{\alpha k}-u_k)^2}{2} \right\rangle &= \frac{m_\alpha}{2} \int_{-\infty}^{\infty} (v_k - u_k)^2 f_k(v_k)\, dv_k \\ &= \left(\frac{m_\alpha}{2\pi T}\right)^{1/2} \int_{-\infty}^{\infty} \frac{m_\alpha (v_k-u_k)^2}{2} \exp\left[-\frac{m_\alpha (v_k-u_k)^2}{2T}\right] dv_k \\ &= \frac{T}{\sqrt{\pi}} \int_{-\infty}^{\infty} \xi^2 \exp(-\xi^2)\, d\xi = \frac{T}{2}\end{aligned}$$

Their sum yields an average energy

$$\left\langle \frac{m_\alpha (\mathbf{v}_\alpha - \mathbf{u})^2}{2} \right\rangle = \sum_k \left\langle \frac{m_\alpha (u_{\alpha k} - u_k)^2}{2} \right\rangle = \frac{3T}{2} \tag{4.11}$$

This equation shows that the parameter T, which determines the average energy of the random motion, is the temperature of the system. Thus elastic collisions lead to the Maxwellian distribution of the velocities of particles of all species (see Eq. 4.8) with the same average energy and the same directed velocity.

Let us now determine the effect of inelastic collisions on the distribution function of the particles of an equilibrium plasma. To do this we take advantage of Eq. 3.27 for the collision term, which makes allowance for inelastic collisions. Its integrand, as in Eq. 4.2, includes the difference of the products of the pre- and postcollision distribution functions $(f'_\alpha f'_\beta - f_\alpha f_\beta)$. Therefore, for inelastic collisions the condition for reducing the collision term to zero can also be written in the form 4.3 or 4.4; however, one must take into consideration the possibility of a change in the internal state of the particles; here the energy conservation law includes the total energies of the particles $E_{\alpha j} = m_\alpha v_{\alpha j}^2/2 + \mathscr{E}_{\alpha j}$ ($\mathscr{E}_{\alpha j}$ is the internal energy in the j state), rather than the kinetic energies. Accordingly, $\ln f$ for particles of each species must be a linear combination of the components of the momentum and total energy (cf. Eq. 4.5):

$$\ln f_\alpha(\mathbf{v}_\alpha, j_\alpha) = a + \mathbf{b} m_\alpha \mathbf{v}_\alpha - cE_{\alpha j} = a + \mathbf{b} m_\alpha \mathbf{v}_\alpha - \frac{c m_\alpha v_\alpha^2}{2} - c\mathscr{E}_{\alpha j} \quad (4.12)$$

With the aid of Eq. 4.12 we obtain the following distribution function, as before:

$$f_\alpha(\mathbf{v}_\alpha, j_\alpha) = A \exp\left[-m_\alpha \frac{(\mathbf{v}_\alpha - \mathbf{u})^2}{2T}\right] \exp\left(-\frac{\mathscr{E}_{\alpha j}}{T}\right) \quad (4.13)$$

it yields the distribution both over the velocities and the internal states. The normalization constant A is found by integrating Eq. 4.13 over the velocities and summing over all the states of the discrete spectrum:

$$A \sum_j \exp\left(-\frac{\mathscr{E}_j}{T}\right) \int_{(\mathbf{v})} \exp\left[-m_\alpha \frac{(\mathbf{v} - \mathbf{u})^2}{2T}\right] d^3v = 1$$

$$A = \left(\frac{m_\alpha}{2\pi T}\right)^{3/2} \left[\sum_j \exp\left(-\frac{\mathscr{E}_j}{T}\right)\right]^{-1} = \left(\frac{m_\alpha}{2\pi T}\right)^{3/2} \left[\sum_j{}^* g_j \exp\left(-\frac{\mathscr{E}_j}{T}\right)\right]^{-1}$$

Note that in summing over j, Σ_j takes into account all the states irrespective of degeneration; in Σ_j^* the degenerate states are covered by one term, which includes the statistical weight g_i.

Substituting A into Eq. 4.13, we find the normalized distribution function. It can be represented as

$$f(\mathbf{v}_\alpha, j) = f_\mathbf{v}(\mathbf{v}_\alpha) f_j(j) \quad (4.14)$$

where the first cofactor is the Maxwellian distribution over the velocities (Eq. 4.8), and the second determines the distribution over the internal states,

$$f_j = \frac{\exp(-\mathscr{E}_j/T)}{\Sigma_l \exp(-\mathscr{E}_l/T)} = \frac{\exp(-\mathscr{E}_j/T)}{\Sigma_l{}^* g_l \exp(-\mathscr{E}_l/T)} \quad (4.15)$$

which is called the *Boltzmann distribution.* Thus in inelastic collisions, too, the equilibrium particle velocity distribution remains Maxwellian. At the same time inelastic collisions lead to equilibrium distribution among the various internal states of the atoms and ions, which is determined by their energy.

The equilibrium state with a Maxwellian velocity distribution function may occur in the presence of external constant potential fields as well, provided the directed velocity is zero ($\mathbf{u} = 0$). Naturally, in this case the plasma can no longer be considered homogeneous. The kinetic equation 3.17 for the stationary distribution function at $\partial(nf)/\partial t = 0$ takes the form

$$\mathbf{v}\,\text{grad}(nf) + \left(\frac{\mathbf{F}}{m}\right)\text{grad}_v(nf) = \frac{\delta(nf)}{\delta t} \tag{4.16}$$

for particles of any species.

Assume that the velocity distribution is determined by Eq. 4.8 for $\mathbf{u} = 0$:

$$f(\mathbf{v}) = \left(\frac{m}{2\pi T}\right)^{3/2} \exp\left(\frac{-mv^2}{2T}\right) \tag{4.17}$$

We substitute this into Eq. 4.16. Recalling that for this distribution the collision term $\delta(n, f)/\delta t$ vanishes, we obtain $f\mathbf{v}\,\text{grad}\,n - f(n/T)\mathbf{v}\mathbf{F} = 0$. Then, using the expression of the force via the potential $\mathbf{F} = -\text{grad}\,U(\mathbf{r})$, we reduce this equation to $(1/n)\,\text{grad}\,n + (1/T)\,\text{grad}\,U = 0$.

Integrating, we obtain the dependence of the concentration on the coordinates:

$$n = n_0 \exp\left(\frac{-U(\mathbf{r})}{T}\right) \tag{4.18}$$

where n_0 is the concentration at the point $U = 0$. This equation is called the *Boltzmann equation*, and the distribution nf, which is determined by Eqs. 4.17 and 4.18, the *Maxwell-Boltzmann distribution*.

4.2 FINDING EQUILIBRIUM DISTRIBUTION FUNCTION

In the preceding section we obtained distribution functions satisfying the stationary kinetic equation for equilibrium conditions. We now show that any spatially homogeneous distribution in the absence of external fields tends to equilibrium with time. We introduce the function

$$H(t) = -\sum_{\alpha} \int_{(\mathbf{v}_\alpha)} n_\alpha f_\alpha \ln(n_\alpha f_\alpha)\, d^3 v_\alpha \tag{4.19}$$

In accordance with the thermodynamic definition it is equal to the entropy of a unit volume. Let us find the nature of the time variation of the function H. To this end we calculate its derivative

$$\frac{dH}{dt} = -\sum_{\alpha} \int_{(\mathbf{v}_\alpha)} [1 + \ln(n_\alpha f_\alpha)]\, \frac{\partial(n_\alpha f_\alpha)}{\partial t}\, d^3 v_\alpha \tag{4.20}$$

Under these conditions the time variation of the distribution function is determined by the collision term (see Eq. 4.1).

Assuming that only elastic collisions are substantial, we substitute the collision term in the form 4.2 into Eq. 4.20. Then

$$\frac{dH}{dt}=\sum_{\alpha,\beta} n_\alpha n_\beta \iiint [1+\ln(n_\alpha f_\alpha)][f_\alpha f_\beta - f'_\alpha f'_\beta] v_{\alpha\beta}\sigma_{\alpha\beta}\, d\Omega\, d^3v_\alpha\, d^3v_\beta \tag{4.21}$$

where the summation must be extended to the collisions of all the possible species of α- and β-type particles.

Let us consider one of the terms of this double sum:

$$J_{\alpha\beta}=n_\alpha n_\beta \iiint [1+\ln(n_\alpha f_\alpha)][f_\alpha f_\beta - f'_\alpha f'_\beta] v_{\alpha\beta}\sigma_{\alpha\beta}\, d\Omega\, d^3v_\alpha\, d^3v_\beta \tag{4.22}$$

We denote formally the velocities under the integral from $\mathbf{v}_\alpha$ and $\mathbf{v}_\beta$ (unprimed) to $\mathbf{v}'_\alpha$ and $\mathbf{v}'_\beta$ (primed), and vice versa. Since collisions are reversible (to each collision with a velocity change $\mathbf{v}'_\alpha, \mathbf{v}'_\beta \rightarrow \mathbf{v}_\alpha, \mathbf{v}_\beta$ there corresponds a reverse one), this merely changes the order of summation of collisions, but does not affect the integral itself. With the new notation it takes the form

$$J_{\alpha\beta}=n_\alpha n_\beta \iiint [1+\ln n_\alpha f'_\alpha][f'_\alpha f'_\beta - f_\alpha f_\beta] v_{\alpha\beta}\sigma_{\alpha\beta}\, d\Omega\, d^3v'_\alpha\, d^3v'_\beta \tag{4.23}$$

In elastic collisions the relative velocity $v_{\alpha\beta}$ and the cross section $\sigma_{\alpha\beta}(v_{\alpha\beta})$ remain unchanged. Moreover, the equality $d^3v'_\alpha\, d^3v'_\beta = d^3v_\alpha\, d^3v_\beta$ holds good, since the Jacobian of transition from the variables $\mathbf{v}_\alpha, \mathbf{v}_\beta$ to $\mathbf{v}'_\alpha, \mathbf{v}'_\beta$ is equal to unity (see p. 87). Taking this into account and adding Eqs. 4.22 and 4.23, we get

$$J_{\alpha\beta}=\frac{1}{2} n_\alpha n_\beta \iiint \left(\ln\frac{f'_\alpha}{f_\alpha}\right)(f'_\alpha f'_\beta - f_\alpha f_\beta) v_{\alpha\beta}\sigma_{\alpha\beta}\, d\Omega\, d^3v_\alpha\, d^3v_\beta \tag{4.24}$$

A similar expression can be found for $J_{\beta\alpha}$ (it is obtained from Eq. 4.24 by permuting the subscripts α and β). Their sum is equal to

$$J_{\alpha\beta}+J_{\beta\alpha}=\frac{1}{2} n_\alpha n_\beta \iiint (f'_\alpha f'_\beta - f_\alpha f_\beta)\left(\ln\frac{f'_\alpha f'_\beta}{f_\alpha f_\beta}\right) v_{\alpha\beta}\sigma_{\alpha\beta}\, d\Omega\, d^3v_\alpha\, d^3v_\beta$$

Substituting it into Eq. 4.21, we find

$$\frac{dH}{dt}=\frac{1}{4}\sum_{\alpha,\beta} n_\alpha n_\beta \iiint (f'_\alpha f'_\beta - f_\alpha f_\beta)\left(\ln\frac{f'_\alpha f'_\beta}{f_\alpha f_\beta}\right) v_{\alpha\beta}\sigma_{\alpha\beta}\, d\Omega\, d^3v_\alpha\, d^3v_\beta \tag{4.25}$$

A similar expression for dH/dt can easily be found by making allowance for inelastic collisions. For this, we must substitute the collision integral in the form 3.27 into Eq. 4.20 and use the detailed balancing principle (Eq. 2.83) in transformations. As a result we obtain an equation differing from Eq. 4.25 only in that the right-hand side is summed over all the states of the α- and β-type particles before and after the collision.

The integrand of each term in Eq. 4.25 includes the function $(f'_\alpha f'_\beta - f_\alpha f_\beta)\ln(f'_\alpha f'_\beta / f_\alpha f_\beta)$. If $f'_\alpha f'_\beta > f_\alpha f_\beta$, then both cofactors in it are positive; if $f'_\alpha f'_\beta < f_\alpha f_\beta$, both cofactors are negative. Therefore the function cannot be negative irrespective of the velocities. Since this conclusion is true for any one of the terms appearing in Eq. 4.25 (and in a similar expression taking into account inelastic collisions), it can be inferred that

$$\frac{dH}{dt} \geqslant 0 \tag{4.26}$$

That is, the function H can only increase with time (this result is called *Boltzmann's H theorem*). At the same time the function H cannot become infinitely large. Indeed, in the integrals (Eq. 4.19) that determine it the integrand expressions are finite at finite velocities, and for $\mathbf{v} \to \infty$ the distribution functions $f(\mathbf{v}) \to 0$, and the products $nf \ln(nf)$ appearing in the integrals tend to zero. Thus at any initial particle velocity distribution a stationary distribution at which $dH/dt = 0$ must set in after a sufficiently long period. Since all the terms appearing in Eq. 4.25 are nonnegative the sum can vanish only if Eq. 4.3 or

$$f_\alpha(\mathbf{v}'_\alpha)f_\beta(\mathbf{v}'_\beta) = f_\alpha(\mathbf{v}_\alpha)f_\beta(\mathbf{v}_\beta) \tag{4.27}$$

is satisfied for any collision. As established in the preceding section, these equalities hold for all collisions at equilibrium (Maxwellian) velocity distribution functions and at equilibrium distributions over the internal states.

Thus collisions between particles in a homogeneous plasma result in an equilibrium particle distribution; collisions between particles of each species lead to "maxwellization" of the velocity distribution; collisions of particles of different species equalize the average velocities and temperatures; inelastic collisions promote the attainment of an equilibrium distribution over the internal states of the particles.

4.3 IONIZATION EQUILIBRIUM

As has been established, the particle velocity distribution in an equilibrium plasma is Maxwellian, and the temperature and directed velocity that determine it are the same for all the components. Since the directed velocity characterizes the plasma motion as a whole it cannot affect the processes in the plasma. Therefore the characteristics of an equilibrium plasma are defined unambiguously by its temperature and the concentration of its components. A knowledge of the distribution function allows one to trace the course of any processes with known cross sections in a plasma. In particular, it helps determine the ionization and recombination efficiency and, from the condition for the balance of these processes (ionization equilibrium), the relationship between the concentrations of the neutral and charged particles. However, in a closed equilibrium plasma this relationship can be found without analyzing the kinetics of the ionization and recombination processes with the aid of the cross sections. To determine it, one can use the general thermodynamic relations.

Let us consider in more detail the conditions for ionization equilibrium in a closed plasma. As indicated in Sections 2.8 and 2.10, various ionization and recombination processes may occur in a plasma, and to each ionization process there corresponds the reverse process of recombination. An analysis of each of them enables one to establish the relationship between the concentrations of the neutral and charged particles. Consider two processes by way of example: (1) impact ionization $L + e \rightleftarrows L^+ + 2e$ (the reverse process being triple recombination); and (2) photoionization $L + h\nu \rightleftarrows L^+ + e$ (the reverse process being recombination with radiation).

For the first process the rate of the forward reaction (impact ionization) is proportional to the concentrations of the neutral atoms and electrons $Q_{i1} = k_{i1} n_a n_e$. Accordingly, the rate of the reverse reaction is proportional to the square of the electron concentration and to ion concentration $Q_{r1} = k_{r1} n_e^2 n_i$. In a closed equilibrium system the rates of the forward and reverse processes must be the same. Equating Q_i to Q_r, we obtain the relationship between the particle concentrations

$$\frac{n_e n_i}{n_a} = \frac{k_{i1}}{k_{r1}} = \mathcal{K}_1 \tag{4.28}$$

For the second process (photoionization) the rates of the forward and reverse reactions are found from the equations $Q_{i2} = k_{i2} I n_a$ and $Q_{r2} = k_{r2} n_e n_i$, where I is the equilibrium radiation intensity. Equating them, we get

$$\frac{n_e n_i}{n_a} = \frac{I k_{i2}}{k_{r2}} = \mathcal{K}_2 \tag{4.29}$$

$\mathcal{K}_1$ and $\mathcal{K}_2$ are the equilibrium constants of the processes at hand. Since both processes occur simultaneously in the same system, it follows from Eqs. 4.28 and 2.29 that $\mathcal{K}_1 = \mathcal{K}_2 = \mathcal{K}$. This means that in a thermodynamically equilibrium plasma the relationship between the neutral and charged particles is not governed by a particular type of ionization and recombination reactions. The equilibrium constant $\mathcal{K}$ depends exclusively on the temperature and the internal structure of the particles.

To find this constant we take advantage of the thermodynamic relations for reversible reactions. In the general form, a reversible reaction can be written:

$$aA + bB + \cdots \leftrightarrows mM + lL + \cdots + \Delta\mathscr{E}$$

where $A, B, \ldots$ are the initial components, and $M, L, \ldots$ the reaction products; the lowercase letters denote the stoichiometric coefficients determining the ratio between the number of particles of different species involved in the reaction; $\Delta\mathscr{E}$ is the energy yield of the reaction, which is equal to the difference of the sums of the internal energies of the initial components and the reaction products.

In the state of equilibrium the total system entropy must be maximum. This means that the partial derivatives of the entropy with respect to the number of particles of any species must be zero. Since the total entropy is additively composed of the entropy of the separate parts of the entire system

$$S = S_A + S_B + \cdots + S_L + S_M + \cdots$$

The entropy-maximum condition has the form

$$\frac{\partial S}{\partial N_A} = \frac{\partial S_A}{\partial N_A} + \frac{\partial S_B}{\partial N_B}\frac{\partial N_B}{\partial N_A} + \cdots + \frac{\partial S_L}{\partial N_L}\frac{\partial N_L}{\partial N_A} + \frac{\partial S_M}{\partial N_M}\frac{\partial N_M}{\partial N_A} + \cdots = 0$$

where clearly $\partial N_B/\partial N_A = b/a, \ldots;$ $\partial N_L/\partial N_A = -l/a;$ $\partial N_M/\partial N_A = -m/a, \ldots.$ For each component the quantity $\partial S_k/\partial N_k$ is equal to the ratio of the chemical potential (μ_k) to the temperature. Therefore the reaction equilibrium condition can be written as follows:

$$a\mu_a + b\mu_b + \cdots = l\mu_l + m\mu_m + \cdots \tag{4.30}$$

The condition 4.30 defines the so-called *mass action law.*

Let us consider the reversible reaction of ionization. In the general form it can be written as the equality $q + A \leftrightarrows e + A^+ + q$, where q is the particle inducing ionization (it may be an electron, a photon, an ion, or a neutral atom). For such a reaction the equilibrium condition (Eq. 4.30) takes the form:

$$\mu_a = \mu_e + \mu_i \tag{4.31}$$

μ_q appears on both the right- and left-hand sides and can be canceled out. The chemical potential of the plasma components can be obtained from the ideal gas equation

$$\mu_\alpha = -T \ln\left(\frac{\Sigma_\alpha}{N_\alpha}\right) \tag{4.32}$$

where $N_\alpha = n_\alpha V$ is the number of particles of a given species in the plasma volume and Σ_α is the so-called statistical sum

$$\Sigma_\alpha = \sum_j g_{\alpha j} \exp\left(-\frac{E_{\alpha j}}{T}\right) \tag{4.33}$$

Here the summation is extended to all the states of the α-type particles; the factor $\exp(-E_{\alpha j}/T)$ determines the relative probability of a state with an energy $E_{\alpha j}$ ($E_{\alpha j}$ are reckoned from the general level of the group of particles involved in the reaction), and $g_{\alpha j}$ is the statistical weight. Substituting Eq. 4.32 into Eq. 4.31, we obtain the equilibrium condition

$$T \ln\left(\frac{\Sigma_a}{N_a}\right) = T \ln\left(\frac{\Sigma_e}{N_e}\right) + T \ln\left(\frac{\Sigma_i}{N_i}\right)$$

or

$$\frac{N_e N_i}{N_a} = \frac{\Sigma_e \Sigma_i}{\Sigma_a} \tag{4.34}$$

Let us now adjust the expression for the statistical sum Σ (we omit the subscript α for simplicity). The total particle energy appearing in it is composed of the energy associated with the internal degrees of freedom $\mathscr{E}_j$ and the energy of the translational motion K. Accordingly, the quantity Σ can be written thus:

$$\begin{aligned}\Sigma &= \sum_{j,v} g_j g_v \exp\left[-\frac{\mathscr{E}_j + K}{T}\right] \\ &= \sum_j g_j \exp\left(-\frac{\mathscr{E}_j}{T}\right) \sum_v g_v \exp\left(-\frac{K}{T}\right)\end{aligned} \tag{4.35}$$

where Σ_j denotes summation over the internal states, and Σ_v over the velocities. The first sum is conveniently represented as

$$\begin{aligned}\sum_j g_j \exp\left(-\frac{\mathscr{E}_j}{T}\right) &= \exp\left(-\frac{\mathscr{E}_0}{T}\right) \sum_j g_j \exp\left(-\frac{\Delta\mathscr{E}_j}{T}\right) \\ &= \exp\left(-\frac{\mathscr{E}_0}{T}\right) G\end{aligned} \tag{4.36}$$

where the energy of the ground (lowest) state of the particle $\mathscr{E}_0$, is singled out, and the "internal" statistical sum G is introduced. As pointed out above, the energy $\mathscr{E}_0$ must be reckoned from the general level of the system. The difference of the energy of the electron–ion system before and after ionization is, of course, equal to the ionization energy

$$\mathscr{E}_{0a} - \mathscr{E}_{0i} - \mathscr{E}_{0e} = -\mathscr{E}_i \tag{4.37}$$

Precisely this defect of the ionization reaction energies appears in the expression for the statistical sum ratio (Eq. 4.34).

The internal statistical sums of the atoms and ions can be found from the equation

$$G = \sum_j (2l+1)(2s+1)\exp\left(-\frac{\Delta\mathscr{E}_j}{T}\right) \tag{4.38}$$

where the quantum numbers l and s determine the orbital angular momentum and the spin. At $T < \Delta\mathscr{E}_1$ the terms of the sum 4.38 rapidly fall off. In making calculations it is usually possible to restrict oneself to two terms in this sum for the atoms and to one for the ions. The electrons have no internal structure, and their internal statistical weight is equal to $G = 2$; it corresponds to two directions of the spin.

Let us now determine the statistical sum related to the translational degrees of freedom. Let us find the number of the possible states, proceeding from the concepts of the quasi-classical approximation of quantum mechanics. The uncertainty relation interconnects the intervals of coordinates and momenta of each particle which are indistinguishable: $\delta x\, \delta p_x \approx \delta y\, \delta p_y \approx \delta z\, \delta p_z \approx h$. Referring them to the same state, we find the volume of the corresponding cell of six-dimensional space: $\delta^{(6)} R = \delta x\, \delta y\, \delta z\, \delta v_x\, \delta v_y\, \delta v_z \approx h^3/m^3$.

The number of states per velocity interval $\Delta^{(3)}v$ in the entire plasma volume V is equal to

$$g_v = \frac{\Delta^{(6)}R}{\delta^{(6)}R} = \frac{V\Delta^{(3)}v}{\delta^{(6)}R} = \frac{m^3V}{h^3}\Delta^{(3)}v$$

Substituting it into the statistical sum, we obtain

$$\Sigma_v = \sum_{\mathbf{v}} g_v \exp\left(-\frac{mv^2}{2T}\right) = \frac{m^3}{h^3} V \sum_{\mathbf{v}} \exp\left(-\frac{mv^2}{2T}\right)\Delta^{(3)}v$$

Replacing summation over the velocities by integration, we find

$$\Sigma_v = \frac{m^3}{h^3} V \int_{(\mathbf{v})} \exp\left(-\frac{mv^2}{2T}\right) d^{(3)}v = \frac{V}{h^3}(2\pi mT)^{3/2} \tag{4.39}$$

(see the calculation of this integral on p. 94). With the aid of Eqs. 4.35, 4.36, and 4.39 we obtain the total statistical sum

$$\Sigma = \frac{V(2\pi mT)^{3/2}}{h^3} G \exp\left(-\frac{\mathscr{E}_0}{T}\right) \tag{4.40}$$

Using Eq. 4.40 for particles of all species and allowing for the relation 4.37, we reduce Eq. 4.34 to

$$\frac{n_e n_i}{n_a} = \frac{(2\pi m_e)^{3/2}}{h^3}\frac{2G_i}{G_a} T^{3/2} \exp\left(-\frac{\mathscr{E}_i}{T}\right) \tag{4.41}$$

This equation, which gives the ionization equilibrium constant, is called *Saha's equation.* It enables one to calculate the ratio between the concentrations of charged and neutral particles. Bearing in mind that in a quasineutral plasma $n_e = n_i = n$, we get from Eq. 4.41

$$\rho = \frac{n}{n_a} = \frac{(2\pi m_e)^{3/4}}{h^{3/2} n_a^{1/2}} \left(\frac{2G_i}{G_a}\right)^{1/2} T^{3/4} \exp\left(-\frac{\mathscr{E}_i}{2T}\right) \tag{4.42}$$

Substituting the numerical values of the constants into Eq. 4.42 and assuming $G_i \approx G_a$, we can write Eq. 4.42 approximately as

$$\rho \approx 10^{11} n_a^{-1/2} T^{3/4} \exp\left(-\frac{\mathscr{E}_i}{2T}\right)$$

Here T is expressed in eV and n in cm^{-3}. The degree of gas ionization is usually determined as the ratio of the number of ions (n_i) to the total number of heavy particles—atoms and ions ($n_0 = n_i + n_a$). According to Eq. 4.42 it is equal to $\eta = n_i/n_0 = \rho/(1+\rho)$.

Figure 4.1 presents the temperature dependences of the degree of ionization for hydrogen, helium, and cesium calculated using Eq. 4.42. It is seen that at sufficiently high temperatures the degree of ionization sharply increases, reaching values close to unity at $T \approx 0.05$–$0.2\mathscr{E}_i$. The curves of Fig. 4.1 characterize the transition from the weakly ionized state of the gas to that of an ionized plasma.

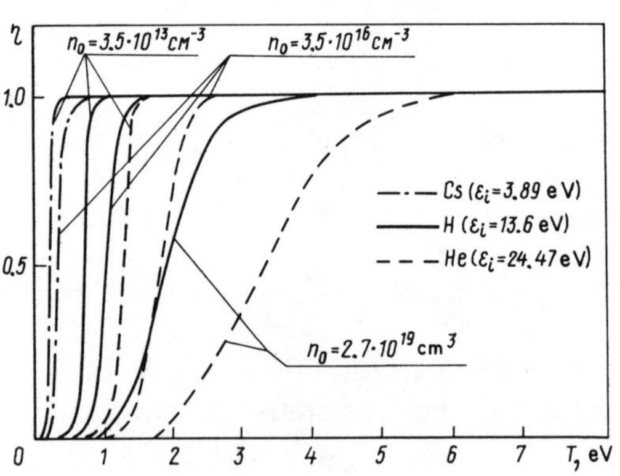

Fig. 4.1 Temperature dependence of the degree of ionization for H, He, and Cs.

In deriving Saha's equation only single ionization was taken into account. For sufficiently high temperatures one should also make allowance for processes of multiple ionization of atoms with $Z > 1$. Let us write the corresponding ionization reaction:

$$q + A^{k+} \rightarrow A^{(k+1)+} + e + q$$

where A^{k+} is an ion with charge number k. This reaction is similar to the one for single ionization (the difference lies in replacing the atom A by the ion A^{k+}). Therefore it leads to Saha's equation similar to Eq. 4.41:

$$\frac{n_e n_{i(k)}}{n_{i(k+1)}} = \frac{(2\pi m_e)^{3/2}}{h^3} \frac{2G_{i(k+1)}}{G_{i(k)}} T^{3/2} \exp\left(-\frac{\mathscr{E}_{i(k)}}{T}\right) \qquad (4.43)$$

The subscript $i(k)$ refers to the ion A^{k+}, and the subscript $i(k+1)$ to the ion $A^{(k+1)+}$; $\mathscr{E}_{i(k)}$ is the energy of "detachment" of the $(k+1)$th electron. Equations 4.43 for different k make it possible to determine the equilibrium concentrations of ions with any charge.

4.4 PARTIAL EQUILIBRIUM IN PLASMAS

A laboratory plasma is, as a rule, far from being at thermodynamic equilibrium since it cannot, even crudely, be considered a closed system. In contact with the environment, the plasma continuously loses the accumulated energy along various channels, primarily through electromagnetic radiation and energy transport by charged and neutral particles to the chamber walls. For the plasma state to be maintained, the energy losses must be replenished from external sources. When the total fluxes of the input and output energy are the same, a stationary state of the plasma is attained, which, however, is not thermodynamically balanced, since the energy input and output usually follow different channels. In the absence of thermodynamic equilibrium the average energies of the electrons, ions, and atoms may be different, and the particle velocity distribution function may be non-Maxwellian.

In some cases so-called *partial equilibrium* with respect to separate processes may take place. As shown below, at low particle densities the plasma may have a near-Maxwellian electron and ion velocity distribution, for which $T_i \approx T_a$, but $T_e \neq T_i$. This kind of plasma is usually called *nonisothermal*. The distribution for particles of the given species is practically Maxwellian, provided the rate of energy exchange with particles of other species and with the environment is low compared with the rate of energy exchange between them. The energy exchange between the electrons and the heavy particles proceeds slowly because of the great difference in masses. Therefore one can speak separately of partial equilibria of the electron component and of the heavy particles. All processes involving only one component must proceed under these conditions in the same way as in a thermodynamically equilibrium plasma. Ionization and recombination, generally speaking, involve all three plasma components. Therefore in deciding whether the thermodynamical relations are applicable in calculating the balance of the number of charged particles in a plasma one must proceed from the existing conditions.

As a rule, only one of the possible ionization reactions and one of the recombination reactions play the determining role. If they follow the same route, the process is mutually reversible, and the relations obtained from thermodynamical considerations can be used to calculate the ionization equilibrium constant. Indeed, the probabilities of ionization and recombination depend on the relative velocity of the interacting particles. If the particle masses differ appreciably, it can be assumed approximately that the relative velocity is equal to the velocity of a light particle. Therefore the conditions of applicability of Saha's equation amount to the requirement that the forward and reverse reactions ensuring the balance of the charged particles follow the same route and that the notion of partial equilibrium be applicable to the light components of the plasma. In particular, if ionization and recombination are due to inelastic collisions of heavy particles with electrons it is the electron gas that must be in the state of partial equilibrium. If the photoionization and radiative recombination are the determining processes, then the radiation must also be in equilibrium with the electron gas.

To evaluate the conditions of equilibrium with radiation, it is customary to use the optical width concept. The intensity of a directed radiation flux passing through a homogeneous plasma (as well as through any medium) is known to reduce according to the exponential law $I = I_0 \exp(-\gamma x)$, the damping coefficient γ characterizing both the absorption and scattering of the radiation. If the product of γ by the characteristic plasma dimension L (termed the *optical width*) is small, then the plasma is an *open system* with respect to the radiation. Accordingly, the probability of interaction of both the external radiation and the one induced inside the plasma with particles is negligibly small. With an optically dense plasma, radiation escape into the environment is possible only through repeated reradiation and scattering. The radiation is then trapped in the plasma volume. The number of the quanta of the electromagnetic field emitted from the plasma surface per unit time is much less than the number of the quanta absorbed and emitted by the particles inside the volume, and one can say that an equilibrium radiation density exists in the volume.

In actual conditions it often happens that a plasma is not a closed system with respect to the radiation and the radiation resulting from recombination freely leaves the plasma. Then if the concentration of the charged particles is not high, ionization and recombination follow different routes. Ionization is accomplished by an electron impact, and the removal of the charged particles in the volume by recombination with radiation. If the efficiency of this removal mechanism greatly exceeds that of the other mechanisms (in particular, those associated with the departure of particles from the volume), the steady state is achieved when the rates of the indicated processes $Q_i = k_{i1}n_a n_e$, $Q_r = k_{r2}n_i n_e$ are equal, and this leads to the relation

$$\frac{n_i}{n_a} = \frac{k_{i1}}{k_{r2}} \tag{4.44}$$

This relation (which is called the *Elvert equation*) shows that under the conditions discussed the degree of ionization is determined exclusively by the temperature and is independent of the particle concentration. It is clear that other relations describing the ionization equilibrium in a plasma are also possible, depending on the particular mechanisms of charged particle ionization and removal.

The question of the applicability of some relations to actual systems reduces to an analysis of the concrete state of the plasma. This analysis requires establishing the type of the distribution function for different species of particles and the nature of interaction between the particles and radiation.

5

DISTRIBUTION FUNCTION OF CHARGED PARTICLES IN ELECTRIC FIELD

5.1 EFFECT OF ELECTRIC FIELD ON CHARGED PARTICLE VELOCITY DISTRIBUTION

Under the effect of external forces particle distribution functions deviate from equilibrium. Not only can the average energy and directed velocity of the particles change in such cases, but also the type of the distribution function. The effect of the forces on the distribution function is different for different species of particles. In the electric field charged particles are accelerated, changing their energy. The magnetic field changes their trajectory. Neutral particles remain completely unaffected by these forces. The neutral particles of a plasma are usually in close contact with the environment (for instance, with the chamber walls). Therefore the average energy of the neutral particles is lower than that of the other components, and their velocity distribution is closer to equilibrium.

Let us determine the conditions under which an electric field substantially affects the charged particle velocity distribution in a weakly ionized plasma. Accelerating in the electric field, the charged particles acquire an additional velocity within the intercollisional time τ:

$$\Delta \mathbf{v}_E \approx \frac{e\mathbf{E}}{m}\tau \approx \frac{e\mathbf{E}}{m\nu} \tag{5.1}$$

The corresponding energy increment is equal to

$$\Delta K_E = \frac{m(\mathbf{v}+\Delta\mathbf{v}_E)^2}{2} - \frac{m\mathbf{v}^2}{2} = m\mathbf{v}\Delta\mathbf{v}_E + \frac{m(\Delta\mathbf{v}_E)^2}{2}$$

or, after averaging (at $\langle \mathbf{v} \rangle = 0$),

$$\langle \Delta K_E \rangle = \left\langle \frac{m(\Delta \mathbf{v}_E)^2}{2} \right\rangle = \left\langle \frac{e^2 E^2}{2} m\nu^2 \right\rangle \tag{5.2}$$

At the same time the energy losses of charged particles in collisions with neutral ones are proportional to the difference of their energies (see Section 2.1):

$$\Delta K_\nu = \kappa (K - K_a) \tag{5.3}$$

In a weakly ionized plasma these losses usually predominate. Therefore in the stationary state the average energy acquired in the field must be equal to that lost in collisions $\langle \Delta K_E \rangle = \langle \Delta K_\nu \rangle$. The energy balance determines the difference between the average energies of the charged and neutral particles:

$$\langle K - K_a \rangle \approx \frac{e^2 E^2}{m \kappa \nu^2} \tag{5.4}$$

where ν and κ are the averaged values of the collision frequency and of the energy transfer coefficient, respectively.

With the aid of Eq. 5.4 we obtain the condition for a small electric field effect on the average energy $\langle K - K_a \rangle \ll \langle K \rangle \approx T$ or

$$E \ll E_p = \frac{\nu}{e} \sqrt{\kappa m T} = \frac{\sqrt{\kappa} T}{e \lambda} \tag{5.5}$$

where $\lambda \approx v_T/\nu \approx (1/\nu)(\sqrt{T/m})$ is the mean free path. If this condition fails, that is, if $E \gtrsim E_p$, the average energy of the charged particles considerably exceeds that of the neutral particles in accordance with Eq. 5.4. At $E \gg E_p$ the average random velocity can be obtained from the relation

$$v_T = \sqrt{2\langle K \rangle / m} \approx \frac{(1/\sqrt{\kappa}) e E}{m \nu} \tag{5.6}$$

The average directed velocity u_E is approximately equal to the averaged velocity increment within the intercollisional time (Eq. 5.1), since collisions result in a substantial change in velocity direction. From this we find

$$u_E = \langle \Delta v_E \rangle \approx \frac{eE}{m\nu} = v_T \sqrt{\kappa} \tag{5.7}$$

For ions the energy transfer coefficient in elastic collisions is close to unity ($\kappa \approx m_i/m_a \approx 1$). Here the condition 5.5 becomes $E \ll E_{pi} \approx T_i/e\lambda_{ia}$ or, in numerical form, E (V/cm) $\ll 10^2 T_i$ (eV) p (mm Hg). (Here we used the average value $\lambda_{ia} \approx 10^{-2}/p$; the values of λ_{ia} for various gases do not differ greatly from it). If the condition is not fulfilled, the average energy and the distribution function deviate considerably from equilibrium. Then, for ions, $u_E \sim v_T$ in accordance with Eq. 5.7; that is, the distribution function is essentially anisotropic.

For electrons the energy transfer coefficient is usually much less than unity; in elastic collisions $\kappa \approx 2m_e/m_a$. Therefore the criterion of a weak effect of the electric field on the distribution function 5.5 is much more rigid than for the ions $E \ll E_{pe} = \sqrt{\kappa_{ea}}\,T/e\lambda_{ea}$. For electrons in hydrogen, for instance, $E_p \approx T_e p$ at $K < 2\,\text{eV}$ and $E_p \approx 1.5T_e^{1/2}p$ at $K > 2\,\text{eV}$, whereas in neon $E_p \approx 5 \times 10^{-2} T_e p$ (here E_p is given in V/cm, T_e in eV, and p in mm Hg). Hence the average electron energy depends on the electric field, even if it is comparatively weak. But irrespective of the field value the directed velocity 5.7 is much less than the random velocity $u_E/v_T \sim \sqrt{\kappa}$; that is, the anisotropy of the distribution function is low. As has already been noted, the directed velocity is relatively low because in each collision an electron sharply changes its direction, whereas the velocity and energy change vary little ($\Delta K \approx \kappa K$); that is, the electron "accumulates" its energy during many intercollisional periods. In other words, the momentum relaxation time is determined by the intercollisional time ($\tau_p \approx 1/\nu$), whereas the energy relaxation time greatly exceeds it ($\tau_K \approx 1/\kappa\nu \gg \tau_p$).

In a highly ionized plasma, when considering the energy balance of the electrons one must take into account their collisions not only with neutral particles, but also with ions. The condition for a weak effect of the electric field on the average electron energy can, as before, be represented in the form of the inequality 5.5 by using the total frequency of collisions of electrons with atoms and ions $\nu_e = \nu_{ea} + \nu_{ei}$. Note that as the electric field strength approaches its critical value (Eq. 5.5) the heating of the electrons reduces the frequency of their collisions with ions ($\nu_{ei} \sim 1/T_e^{3/2}$). The reduction of ν, in turn, increases the energy received by the electrons from the field (Eq. 5.2) and decreases the energy losses (Eq. 5.3); that is, it results in a further rise of the electron temperature. This process is limited either by collisions of electrons with neutral particles or by other types of losses, such as those due to radiation or thermal conduction. In the first case the conclusion about the weak anisotropy of the distribution function remains valid.

5.2 METHOD FOR SOLVING KINETIC EQUATION

When the deviation of the distribution function from equilibrium is small, the solution of the kinetic equation can be found by the method of successive approximations. When using this method the distribution function is represented as a series in the powers of the parameters determining its deviation from equilibrium (forces affecting the particles and the concentration and temperature gradients):

$$f = f_{(0)} + f_{(1)} + f_{(2)} + \cdots \tag{5.8}$$

The first term of the series $f_{(0)}$ is the equilibrium (Maxwellian) distribution, the second term $f_{(1)}$ includes a linear combination of the parameters, the third $f_{(2)}$ a quadratic combination, and so on. Substituting the series into the kinetic equation yields a sequence of equations of different orders of smallness. The first-approximation equation includes functions $f_{(0)}$ and $f_{(1)}$ and the second-approximation equation $f_{(0)}, f_{(1)}, f_{(2)}$, and so on. Solving them successively, we can find corrections of different orders to the distribution function. The set of these corrections describes both the deviation of the spherically symmetric part of the distribution function from the Maxwellian distribution and the anisotropy of the distribution function owing to deviations from equilibrium. Determination of the anisotropic component is particularly important in this case of small deviations from the equilibrium distribution, since it enables one to calculate such important macroscopic characteristics as the directed velocity, the energy flux, the momentum flux, and so on. We do not dwell on these calculations; in Chapter 6 we describe a method for obtaining anisotropic characteristics from the moments equation.

When the deviation of the distribution function from equilibrium is large, the method for solving the kinetic equation that uses expansion in the powers of perturbation is inapplicable. But for electrons in an electric field one can use another method of expansion based on the low anisotropy of the distribution function. As demonstrated in Section 5.1, the directed velocity is usually much lower than the random because of the small losses of electron energy in collision with heavy particles. Therefore, even in a strong electric field, with deviations from equilibrium the anisotropy of the electron velocity distribution remains low. Owing to this, when solving the kinetic equation one can use an expansion of the distribution function in parameters characterizing its anisotropy. The rapid convergence of the series permits us to restrict ourselves to a small number of terms and find rather easily both the anisotropy and the symmetric part of the distribution function.

Let us consider in more detail the case where a homogeneous electric field **E** is the source of disequilibrium. Here the field determines the only isolated direction (we assume that the $0z$ axis is parallel to it). Accordingly, the electron velocity distribution function may depend only on the velocity v and on the angle Θ between the directions of velocity $\mathbf{v}$ and field **E**. The dependence on the angle Θ, which is due to the anisotropy of the distribution function, must be low. Therefore it is natural to represent this dependence as an expansion in orthogonal Legendre polynomials $P_n(\cos\Theta)$:

$$f(\mathbf{v}) = \sum_{n=0}^{\infty} f_n(v) P_n(\cos\Theta) \tag{5.9}$$

where the functions f_n depend only on the velocity. By substituting the expansion 5.9 into the kinetic equation it is easy to get a system of coupled equations for the functions $f_n(v)$. Each of them is obtained by multiplying the kinetic equation by one of the polynomials $P_1(\cos\Theta)$ and integrating over all the values of $\cos\Theta$. The first two terms of the sum in Eq. 5.9 contain the polynomials $P_0 = 1$ and $P_1 = \cos\Theta$. With small anisotropy we can restrict ourselves to these terms in solving many problems. Then we have

$$f(\mathbf{v}) = f_0(v) + \cos\Theta\, f_1(v) = f_0(v) + \frac{v_z}{v} f_1(v) \tag{5.10}$$

It is easy to see that the function $f_0(v)$ defines the average electron energy and the average value of any other energy-dependent quantity. Indeed, by averaging $g(v)$ with the aid of Eq. 3.8 we obtain

$$\begin{aligned}\langle g(v)\rangle &= \int_{(\mathbf{v})} g(v) f(\mathbf{v})\, d^3v = \iiint g(v)[f_0(v) + f_1(v)\cos\Theta] v^2 \sin\Theta\, dv\, d\Theta\, d\varphi \\ &= 4\pi \int_0^{\infty} g(v) f_0(v) v^2\, dv \end{aligned} \tag{5.11}$$

(here we changed to spherical coordinates v, Θ, φ in the integrand).

From Eq. 5.11 we can see the relationship between f_0 and the distribution functions with respect to the total velocities f_v(Eq. 3.6) and energies f_K (Eq. 3.7):

$$f_v(v) = 4\pi v^2 f_0(v); \qquad f_K(K) = \frac{4\sqrt{2}\pi}{m^{3/2}} \sqrt{K} f_0\left(\sqrt{\frac{2K}{m}}\right)$$

This relationship also determines the normalization condition

$$\int_0^{\infty} 4\pi v^2 f_0(v)\, dv = 1 \tag{5.12}$$

In a similar manner we can ascertain that the average electron velocity is defined by the function $f_1(v)$. For the velocity component parallel to the electric field we get

$$\begin{aligned} u_E = \langle v\cos\Theta\rangle &= \iiint v\cos\Theta[f_0(v) + f_1(v)\cos\Theta] v^2 \sin\Theta\, dv\, d\Theta\, d\varphi \\ &= \frac{4\pi}{3}\int_0^{\infty} v^3 f_1(v)\, dv \end{aligned} \tag{5.13}$$

Accordingly, the function $f_0(v)$ is called the *isotropic component*, and the function $f_1(v)$ *the directed component of the distribution function.*

In order to obtain the equation for $f_0(v)$ and $f_1(v)$ we must substitute the sum of Eq. 5.10 into the kinetic equation. For a homogeneous plasma in a homogeneous electric field the kinetic equation 3.17 can be written as

$$\frac{\partial(nf)}{\partial t} - \frac{en}{m_e}\mathbf{E}\,\mathrm{grad}_v(nf) = \frac{\partial(nf)}{\delta t} \tag{5.14}$$

Substituting Eq. 5.10, we transform the derivative $\delta f/\delta v_z$ appearing in the second term of the equation:

$$\begin{aligned}\frac{\partial f}{\partial v_z} &= \frac{\partial}{\partial v_z}\left[f_0(v) + f_1(v)\frac{v_z}{v}\right]\\ &= \frac{v_z}{v}\frac{\partial f_0}{\partial v} + \frac{f_1}{v} + \frac{v_z^2}{v}\frac{\partial}{\partial v}\left(\frac{f_1}{v}\right)\\ &= \cos\Theta\,\frac{\partial f_0}{\partial v} + \frac{f_1}{v} + \cos^2\Theta\, v\,\frac{\partial}{\partial v}\left(\frac{f_1}{v}\right)\end{aligned}$$

Then

$$\frac{\partial(nf_0)}{\partial t} + \cos\Theta\,\frac{\partial(nf_1)}{\partial t} - \frac{enE}{m_e}\left[\frac{f_1}{v} + \cos\Theta\,\frac{\partial f_0}{\partial v} + \cos^2\Theta\, v\,\frac{\partial}{\partial v}\left(\frac{f_1}{v}\right)\right] = \frac{\delta(nf)}{\delta t} \tag{5.15}$$

whence it is easy to obtain the equations for the functions f_0 and f_1. The first equation is found by multiplying Eq. 5.15 by $d(\cos\Theta)$ and integrating term by term over all the values of $\cos\Theta$ from -1 to 1. As a result of integration the terms proportional to $\cos\theta$ vanish in Eq. 5.15, and the equation takes the form

$$\frac{\partial(nf_0)}{\partial t} - \frac{enE}{m_e}\left[\frac{f_1}{v} + \frac{1}{3}v\frac{\partial}{\partial v}\left(\frac{f_1}{v}\right)\right] = S_0$$

or

$$\frac{\partial(nf_0)}{\partial t} - \frac{enE}{3m_e v^2}\frac{\partial}{\partial v}(v^2 f_1) = S_0 \tag{5.16}$$

where the following notation is introduced:

$$S_0 = \frac{1}{2}\int_{-1}^{+1}\frac{\delta(nf)}{\delta t}\,d(\cos\Theta) \tag{5.17}$$

The second equation is obtained by multiplying Eq. 5.15 by $\cos\Theta\, d(\cos\Theta)$ and integrating again over all the values of $\cos\Theta$. In this case the terms independent of θ and proportional to $\cos^2\Theta$ vanish. Then

$$\frac{\partial(nf_1)}{\partial t} - \frac{enE}{m_e}\frac{\partial f_0}{\partial v} = S_1 \tag{5.18}$$

where

$$S_1 = \frac{3}{2}\int_{-1}^{+1} \frac{\delta(nf)}{\delta t} \cos\Theta \, d(\cos\Theta) \tag{5.19}$$

Simultaneous solution of Eqs. 5.16 and 5.18 makes it possible to find both components of the distribution function $f_0(v)$ and $f_1(v)$.

A similar method for expanding the electron distribution function can be applied to a more general case, where a magnetic field is present as well as the electric one and the plasma cannot be considered homogeneous. Then the distribution function may depend on all the velocity components. With a weak anisotropy it can be sought, as in Eq. 5.10, in the form of the sum of the isotropic component $f_0(v)$ and three directed components $f_{1k}(v)$, which define the components of the average electron velocity:

$$\begin{aligned} f(\mathbf{v}) &= f_0(v) + \frac{v_x}{v} f_{1x}(v) + \frac{v_y}{v} f_{1y}(v) + \frac{v_z}{v} f_{1z}(v) \\ &= f_0(v) + \frac{\mathbf{v}}{v}\mathbf{f}_1(v) \end{aligned} \tag{5.20}$$

where a vector function $\mathbf{f}_1(v)$ with components f_{1x}, f_{1y}, and f_{1z} is introduced. It is easy to see that these functions define the components of the directed velocity. Making use of the representation 5.20, we obtain

$$u_k = \int_{\mathbf{v}} v_k f(\mathbf{v})\, d^3v = \int_{\mathbf{v}} \left[v_k f_0(v) + \sum_l \frac{v_k v_l}{v} f_{1l}(v) \right] v^2\, dv\, d\omega$$

where we switched to spherical coordinates in the integrand ($d\omega = \sin\Theta\, d\Theta\, d\varphi$). After integration over the angles the sum retains only the term proportional to $(v_k^2/v)f_{1k}$, and the others vanish. Then we have

$$u_k = \frac{4\pi}{3}\int_0^\infty v^3 f_{1k}(v)\, dv \tag{5.21}$$

By substituting the representation 5.20 into the kinetic equation 3.17 we can, as before, find the associated equations for the functions f_0, f_{1x}, f_{1y}, and f_{1z}. The first one is obtained by averaging the kinetic equation over all the velocity directions (by integrating over the solid angle), and the other three by averaging after multiplying by the direction cosines of the angles between the velocity vector and the coordinate axes. Without dwelling on the calculations we give the resulting equations:

$$\begin{aligned} &\frac{\partial(nf_0)}{\partial t} + \frac{v}{3}\operatorname{div}(n\mathbf{f}_1) - \frac{en}{3m_e v^2}\frac{\partial}{\partial v}(v^2\mathbf{E}\mathbf{f}_1) = S_0; \\ &\frac{\partial(n\mathbf{f}_1)}{\partial t} + v\,\operatorname{grad}(nf_0) - \frac{en}{m_e}\mathbf{E}\frac{\partial f_0}{\partial v} - \frac{en}{m_e c}[\mathbf{H}\times\mathbf{f}_1] = \mathbf{S}_1 \end{aligned} \tag{5.22}$$

The latter equation is the vector form of the three equations for the components $f_1(f_{1x}, f_{1y}, f_{1z})$.

The collision terms S_0 and $\mathbf{S}_1$ in Eq. 5.22 are obtained from the equalities

$$S_0 = \frac{1}{4\pi}\int_{(\omega)} \frac{\delta(nf)}{\delta t}\, d\omega;$$
$$S_{1k} = \frac{3}{4\pi}\int_{(\omega)} \frac{\delta(nf)}{\delta t} \cos\Theta_k\, d\omega \tag{5.23}$$

where Θ_k is the angle between the k axis and the velocity vector, and integration is performed over the solid angles covering all the velocity directions.

5.3 COLLISION INTEGRALS FOR ELECTRONS

Let us now determine the collision terms S_0 and $\mathbf{S}_1$ appearing in Eqs. 5.16 and 5.18. Each of them can be represented as a sum of integral expressions related to different types of electron collisions:

$$S_n = \sum_\beta S_{n\beta}^{(q)} \tag{5.24}$$

where the summing is generally done over all the species of β-type particles and embraces elastic (e) and inelastic (n) collisions with excitation of different j levels.

We first consider the terms due to elastic collisions of electrons with heavy particles (atoms or ions). For a plasma whose stationary state is maintained by an electric field the average energy of electrons usually greatly exceeds that of the heavy particles. Bearing this in mind, we can use the collision integral for elastic collisions as in Eq. 3.31:

$$S_{ea}^{(e)} = \frac{n_e n_a}{v^3}\int_{(\Omega)} [v'^4\sigma(v', \vartheta)f(\mathbf{v}') - v^4\sigma(v, \vartheta)f(\mathbf{v})]\, d\Omega \tag{5.25}$$

Recall that here $\mathbf{v}'$ is the velocity of an electron before collision, which places it in the velocity range $\mathbf{v}\, d\mathbf{v}$ under review. Substituting Eq. 5.25 into Eqs. 5.17 and 5.19 and using the representation of $f(v)$ as the sum of Eq. 5.10, we obtain the expression for S_{0a} and S_{1a}:

$$S_{0a} = \frac{n_e n_a}{2v^3}\int_{-1}^{+1} d(\cos\Theta)\int_{(\Omega)} d\Omega[v'^4\sigma'(f_0' + f_1'\cos\Theta' - v^4\sigma(f_0 + f_1\cos\Theta)];$$
$$S_{1a} = \frac{3}{2}\frac{n_e n_a}{v^3}\int_{-1}^{+1} \cos\Theta\, d(\cos\Theta)\int_{(\Omega)} d\Omega[v'^4\sigma'(f_0' + f_1'\cos\Theta')$$
$$- v^4\sigma(f_0 + f_1\cos\Theta)] \tag{5.26}$$

in which the primed quantities σ, f_0', f_1', and Θ are determined by the electron velocity $\mathbf{v}'$. The relationship between the angles Θ' (between the vectors $\mathbf{v}'$ and $\mathbf{E}$) and Θ (between $\mathbf{v}$ and $\mathbf{E}$) can be established using a spherical triangle formed by the vectors $\mathbf{v}$, $\mathbf{v}'$, and $\mathbf{E}$ (Fig. 5.1). From the well-known formula of spherical trigonometry we find $\cos\Theta' = \cos\Theta\cos\vartheta - \sin\Theta\sin\vartheta\cos\psi$, where ϑ and ψ are the angles determining the solid scattering angle. Substituting this relation into Eq. 5.26 and integrating with respect to Θ, we get

$$S_0 = \frac{n_e n_a}{v^3}\int_{(\Omega)} (v'^4\sigma' f_0' - v^4\sigma f_0)\, d\Omega;$$
$$S_1 = \frac{n_e n_a}{v^3}\int_{(\Omega)} (v'^4\sigma' f_1' \cos\vartheta - v^4\sigma f_1)\, d\Omega \tag{5.27}$$

where it is taken into account that on integrating with respect to $d\Omega = \sin\vartheta\, d\vartheta\, d\psi$ the terms proportional to $\cos\psi$ reduce to zero.

In elastic collisions with heavy particles the electron velocity change is very small (of the order of the mass ratio). Neglecting it, that is, assuming $v' = v$ and $\sigma' = \sigma$, $f_0' = f_0$, $f_1' = f_1$, we obtain the following expression for S_{1a}:

$$S_{1a}^{(e)} = -n_e n_a f_1 \int_{(\Omega)} v\sigma(1-\cos\vartheta)\, d\Omega = -n_e \nu_{ea}^t f_1 \tag{5.28}$$

where the collision frequency ν_{ea}^t is found through the transport cross section $s^t = \int \sigma(1-\cos\vartheta)\, d\Omega$ in accordance with Eq. 2.52. In this approximation the integral $S_{0a}^{(e)}$ vanishes. The expression 5.28 yields the change in the directed distribution function component f_1 as a result of

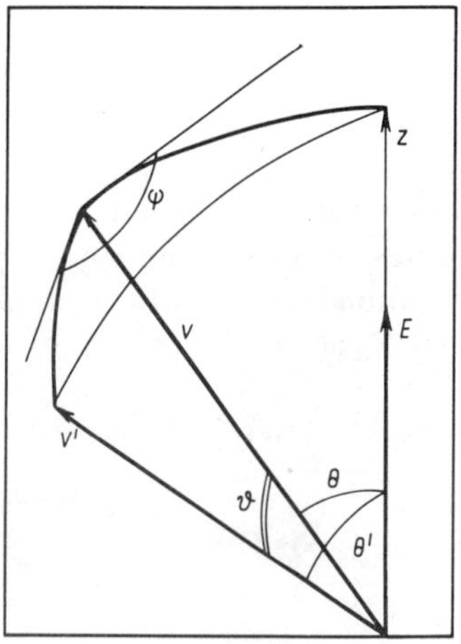

Fig. 5.1 Definition of terms in Eq. 5.26.

collisions. Its meaning is easy to understand, bearing in mind that f_1 characterizes the fraction of electrons moving in the isolated direction (in the $0z$ direction). Since collisions lead to a substantial change in the direction of electron motion, the characteristic time of variation in f_1 must be of the order of the time between collisions $\tau \approx 1/\nu$ and its variation rate $\delta f_1/\delta t$ must depend on the product of f_1 by the collision frequency ν. This conclusion agrees with Eq. 5.28.

In calculating the integral S_0, which determines the change in the isotropic distribution function component, one must take into account the change in the value of electron velocity in collisions. This change is given by the relations obtained in Chapter 2. In elastic collisions of an electron with an atom whose kinetic energy is much less than that of the electron ($K_a \ll K_e$), the change is equal to

$$v - v' = \left(-\frac{m_e}{m_a}\right) v(1 - \cos\vartheta) \tag{5.29}$$

(in distinction to Eq. 2.21, here v' is the electron velocity before collision, and v, after it). Since $|v - v'| \ll v$, the integrand of Eq. 5.27 can be represented as

$$\begin{aligned} v'^4\sigma' f_0' - v^4\sigma f_0 &\approx (v' - v)\frac{\partial}{\partial v}(v^4\sigma f_0) \\ &= \frac{m_e}{m_a}(1 - \cos\vartheta)v\frac{\partial}{\partial v}(v^4\sigma f_0) \end{aligned} \tag{5.30}$$

Substituting Eq. 5.30 into Eq. 5.27, we obtain

$$\begin{aligned} S_0 &= \frac{n_e n_a}{v^2}\frac{m_e}{m_a}\frac{\partial}{\partial v}\left[v^4 f_0 \int_{(\Omega)} \sigma(1 - \cos\vartheta)\, d\Omega\right] \\ &= \frac{n_e}{v^2}\frac{m_e}{m_a}\frac{\partial}{\partial v}(v^3 \nu_{ea}^t f_0) \end{aligned} \tag{5.31}$$

This expression defines the change in the isotropic distribution function component owing to the velocity reduction (because of energy losses) in elastic conditions.

When the average electron energy is comparable with that of the heavy particles the use of the approximate collision integral 5.25 is illegitimate, and to determine S_0 one must substitute the overall expression for the collision integral (Eq. 3.23) into Eq. 5.17. The integral formula obtained can be simplified, inasmuch as the electron velocity greatly exceeds that of the heavy particles and the electron velocity change in collisions is small. Calculations yield the following expression for S_0:

$$S_0 = \frac{n_e}{v^2} \frac{\partial}{\partial v} \left[\frac{m_e}{m_a} \nu_{ea}^t v^3 f_0 + \frac{T_a}{m_a} \nu_{ea}^t v^2 \frac{\partial f_0}{\partial v} \right] \tag{5.32}$$

where $T_a = \frac{2}{3}\langle K_a \rangle$ is the temperature of the atoms, which characterizes their average energy. When it is much below the electron energy $T_a \ll m_e v^2/2$, the second term is also much less than the first, and Eq. 5.32 becomes Eq. 5.31.

The expression 5.32 can be represented as the divergence of a spherically symmetric flux in velocity space. In accordance with the well-known expression for divergence in spherical coordinates we write

$$S_0 = -\text{div}_v \, \Gamma_v = -\frac{1}{v^2} \frac{\partial}{\partial v} (v^2 \Gamma_v) \tag{5.33}$$

where the density of the flux in the direction of increasing velocity Γ_v is equal to

$$\Gamma_v = -g_v(nf_0) - D_v \frac{\partial (nf_0)}{\partial v} \tag{5.34}$$

Here $g_v = (m_e/m_a) v \nu_{ea}^t$; $D_v = (T_a/m_a) \nu_{ea}^t$. The first term describes the reduction in velocity (friction) and the second, the diffusion in velocity space. Substitution of Eqs. 5.33 and 5.34 into the equation for the isotropic distribution function component changes it to the Fokker–Planck equation (see Section 3.3).

To establish the physical meaning of the expression 5.34 we estimate the particle flux across the spherical surface element of the velocity space dG associated with the collisions (Fig. 5.2*a*). The number of

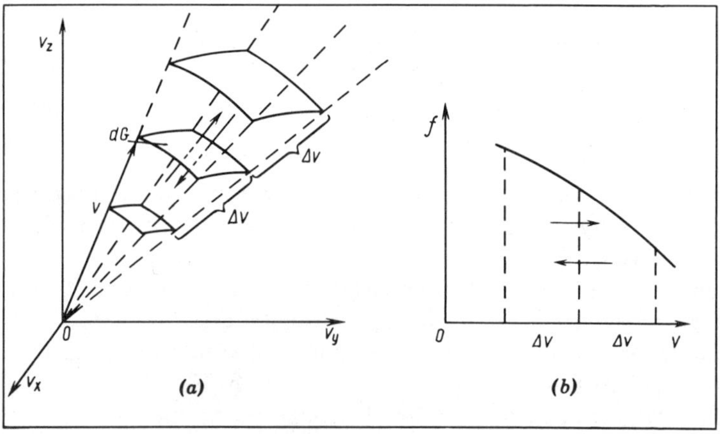

Fig. 5.2 *a* Particle flux across a spherical surface element in velocity space. *b* Change of the distribution functions with velocity.

particles in unit volume of configuration space crossing this element at a given velocity increment Δv can be found from the equality

$$dQ = nf_0\left(v - \frac{1}{2}\Delta v\right)\Delta v\, dG \approx nf_0(v)\Delta v\, dG - \frac{1}{2}n\frac{\partial f_0}{\partial v}(\Delta v)^2\, dG$$

where we use the average value of f_0 within a volume of height Δv and take into account the smallness of the velocity increment on elastic collisions. By correlating this number with the mean time between the collisions $\tau = 1/\nu$ and by averaging over the collisions we obtain an approximate expression for the flux density*:

$$\Gamma_v = \frac{1}{\tau}\frac{dQ}{dG} \approx n\nu f_0\langle\Delta v\rangle - \frac{1}{2}n\nu\langle(\Delta v)^2\rangle\frac{\partial f_0}{\partial v} \tag{5.35}$$

In accordance with this expression the friction coefficient is proportional to the average velocity decrease,

$$g_v = -\nu\langle\Delta v\rangle \tag{5.36}$$

and the diffusion coefficient to the mean-square increment,

$$D_v = \tfrac{1}{2}\nu\langle(\Delta v)^2\rangle \tag{5.37}$$

If the change in velocity is random, then $\langle\Delta \mathrm{v}\rangle = 0$, and Eq. 5.35 retains only the second term determining the diffusion flux. It is directed toward decreasing $f_0(v)$, since with equal probability of positive and negative changes in velocity the flux from the velocity region containing more particles is stronger (see Fig. 5.2*b*). Therefore diffusion tends to "straighten" the distribution function f_0. The whole picture is quite similar to that of diffusion in ordinary configuration space, which results from averaging the random motion of the particles (see Section 7.3).

Let us estimate the coefficients g_v and D_v for elastic electron–atom collisions. The change in electron velocity on such collisions can be obtained by using the general equation 2.18. Considering that $m_e \ll m_a$ and $v_e \gg v_a$, we obtain from it

$$\Delta(v_e^2) = 2\mathbf{v}_0\Delta\mathbf{v}_e \approx -2\mathbf{v}_0\mathbf{v}_e = -2\frac{m_e}{m_a}v_e^2 - 2\mathbf{v}_e\mathbf{v}_a$$

whence

$$\Delta v_e \approx -\frac{m_e}{m_a}v_e - v_a\cos\varphi \tag{5.38}$$

where φ is the angle between $\mathbf{v}_e$ and $\mathbf{v}_a$. The first term in Eq. 5.38 depends on the energy losses that are independent of the atom motion.

*For a more rigorous determination of Γ_v one must include the dependence of ν and s on v.

As seen from Eq. 5.36, the velocity decrease due to these losses $[\Delta v_1 = -(m_e/m_a)v_e]$ results in Eq. 5.34 for the friction coefficient. The second term in Eq. 5.38 is defined by the energy exchange between the electron and atom caused by the motion of the atom. The velocity change on such an exchange is $\Delta v_2 = v_a \cos\varphi$. With an isotropic atom velocity distribution the average value of this change is $\langle \Delta v_2 \rangle = 0$. Therefore the corresponding flux (Eq. 5.35) is associated with a mean-square displacement and represents diffusion in velocity space. Substituting $\Delta v_2 = v_a \cos\varphi$ into Eq. 5.37 and averaging over φ and v_a, we obtain the expression for the diffusion coefficient:

$$D \approx \tfrac{1}{2}\nu v_a^2 \langle \cos^2\varphi \rangle = \tfrac{1}{2}\frac{\nu T_a}{m_a}$$

which differs from Eq. 5.34 only by the coefficient $\frac{1}{2}$ (the correct numerical coefficient is obtained by a more accurate averaging over the collisions).

Let us now consider the collision terms S_0 and S_1 due to inelastic collisions. The electron distribution function may be affected by various inelastic collisions—excitation of different levels, ionization, and recombination. We find here the collision terms for excitation processes in the two extreme cases: when the electron energy greatly exceeds the excitation energy and when these energies are similar. These conditions are often observed in a stationary-gas-discharge plasma, when the average electron energy greatly exceeds the excitation energy of the vibration and rotation levels of the molecule, and is much less than the excitation energy of the electron (atom) levels.

Electron energy losses in inelastic collisions accompanied by excitation are equal to the excitation energy with an accuracy to the small mass ratio m_e/m_a (see Section 2.1):

$$\frac{mv^2}{2} - \frac{mv'^2}{2} = \mathscr{E}_j \tag{5.39}$$

Therefore in the first case, at an excitation energy much less than the electron energy, the velocity change is small:

$$v - v' \approx \frac{\mathscr{E}_j}{m_e v} \ll v \tag{5.40}$$

and the collision terms can be obtained in the same way as for elastic collisions. One should use the expression for the collision integral in the form 3.30, taking into account the relation between the volumes of velocity space before and after the collisions, which follows from Eq.

5.39, namely $v'\,dv' = v\,dv$. Then we obtain, with the aid of Eqs. 5.17 and 5.29,

$$S_{1a}^{(j)} = -nf_1\nu_{ea}^{t(j)} \qquad (5.41)$$

where ν_{ea}^{t} is the transport frequency of collisions accompanied by excitation of the j level, and

$$S_{0a}^{(j)} = \frac{n_e}{v^2}\frac{\partial}{\partial v}\left(\frac{\mathscr{E}_j}{m_e}v\nu^{t(j)}f_0\right) = \frac{1}{2}\frac{n_e}{v^2}\frac{\partial}{\partial v}(\kappa^{(j)}\nu^{tj}v^3 f_0) \qquad (5.42)$$

where $\kappa^j = 2\mathscr{E}_j/m_e v^2$ is the energy transfer coefficient, which determines the fraction of energy lost by the electron in inelastic collision.

The sum of the collision terms $S_{0a}^{(j)}$ and $S_{1a}^{(j)}$ attributable to inelastic collisions with a low energy loss can be written in a form similar to Eqs. 5.41 and 5.42:

$$S_{1a}^{l} = -nf_1\nu_{ea}^{l};$$
$$S_{0a}^{l} = \frac{1}{2}\frac{n_e}{v^2}\frac{\partial}{\partial v}(\kappa_{ea}^{l}\nu_{ea}^{l}v^3 f_0) \qquad (5.43)$$

Here ν_{ea}^{l} is the summary frequency of inelastic collisions with a low energy loss, and κ_{ea}^{l} is the averaged energy transfer coefficient:

$$\nu_{ea}^{l} = \sum_j \nu_{ea}^{t(j)}; \qquad \kappa_{ea}^{l} = \frac{\sum_j \kappa_{ea}^{(j)}\nu_{ea}^{t(j)}}{\sum_j \nu_{ea}^{t(j)}} \qquad (5.44)$$

The summary collision frequency ν_{ea}^{l} is usually much less than the frequency of elastic collisions ν_{ea}^{e}, but the energy transfer coefficient κ_{ea}^{l} in molecular gases may be much larger, $\kappa_{ea}^{l} = 2m_e/m_a$.

In the second case, when the electron energy exceeds the excitation energy only slightly, one can assume that an inelastic collision results in a total loss of electron energy. With this assumption the collision integral retains only the term defining particle removal from the preassigned velocity range. It takes the form

$$S_{-}^{(j)} = -\int_{(\Omega)} n_e n_a f v\sigma^{(j)}\,d\Omega = -n_e f\nu^{(j)} \qquad (5.45)$$

where $\nu^{(j)} = n_a v \int \sigma^{(j)}\,d\Omega$ is the frequency of inelastic collisions of the given type. The arrival of particles in the low-energy region can be covered by adding the term

$$S_{+}^{(j)} = \frac{\delta(v)}{4\pi v^2}Q^{(j)} \qquad (5.46)$$

where

$$Q^{(j)} = -\int_{(\mathbf{v})} \left(\frac{\delta nf}{\delta t}\right)_{-}^{(j)} d^3v = n_e \int_{(\mathbf{v})} f\nu^{(j)}\, d^3v$$

is the total number of inelastic collisions of the given type, and $\delta(v)$ is the delta function, which differs from zero only when $v = 0$. The integral over the velocity space volume from Eq. 5.46 is equal to

$$\int_{(\mathbf{v})} S_{+}^{(j)}\, d^3v = 4\pi \int_0^\infty S_{+}^{(j)} v^2\, dv = Q^{(j)}$$

Substituting the collision term 5.45 into Eqs. 5.17 and 5.19, we find

$$S_0^{(j)} = -nf_0\nu^{(j)}; \qquad S_1^{(j)} = -nf_1\nu^{(j)} \tag{5.47}$$

Similar equations are obviously obtained for collision terms describing ionization at an electron energy close to that of ionization. Then the increased number of electrons can be included by doubling the term 5.46, which reflects the appearance of particles in the low-velocity region. Summing Eq. 5.47 over all the inelastic collisions with a large energy loss results in replacement of the frequency ν^j by the summary frequency of inelastic collisions:

$$S_{0a}^h = -n\nu_{ea}^h f_0; \qquad S_{1a}^h = -n\nu_{ea}^h f_0 \tag{5.48}$$

where $\nu_{ea}^h = \Sigma \nu_{ea}^j$. In many cases $\nu_{ea}^h < \nu_{ea}^{te}$, and the contribution of inelastic collisions to the collision term S_{1a} is insignificant ($|S_{1a}^h| \ll |S_{1a}^e|$). At the same time they may determine, almost completely, the isotropic collision term S_{0a} since $S_{0a}^e \approx (m_e/m_a)\nu_{ea}^t nf_0 \ll \nu_{ea}^h nf_0$ in accordance with Eq. 5.31.

The expressions obtained define the collision terms related to electron–atom collisions. The terms determined by electron–ion collisions S_{0i} and S_{1i} have the same form. The expressions for the collision terms due to electron–electron collisions can be represented only in integral form. We do not give them here, but note that the equation for S_{0e} is derived from the general elastic collision integral 3.23 by replacing the total distribution function $f(\mathbf{v})$ with the isotropic component $f_0(\mathbf{v})$.

5.4 DISTRIBUTION FUNCTION OF ELECTRONS IN ELECTRIC FIELD WITH DETERMINING EFFECT OF ELASTIC ELECTRON–ATOM COLLISIONS

Under stationary conditions (when the distribution function is time invariant) the equations for the components of the distribution functions in a constant electric field (Eqs. 5.16 and 5.18) take the form

$$-\frac{eEn}{3m_e v^2}\frac{d}{dv}(v^2 f_1) = S_0; \qquad -\frac{eEn}{m_e}\frac{df_0}{dv} = S_1 \tag{5.49}$$

The collision terms S (Eq. 5.24) constitute the sum of the terms for collisions of electrons with atoms, ions, and each other, $S_n = S_{na} + S_{ni} + S_{ne}$.

To estimate the components $S_{0\beta}$ and $S_{1\beta}$ we can use the order equalities

$$S_{1\beta} \approx \nu_{e\beta} n f_1; \qquad S_{0\beta} \approx \kappa_{e\beta}\nu_{e\beta} n f_0 \tag{5.50}$$

These follow from the determination of the quantity $S_1 = \delta n f_1/\delta t$, $S_0 = \delta n f_0/\delta t$. The former characterizes the change of electron velocity direction in collision and therefore depends on the collision frequency ν, and the latter characterizes the change in velocity and is determined by the product of the collision frequency by the energy transfer coefficient $\kappa\nu$. For electron–heavy particle collisions the relations 5.50 are obtained directly from the equations of Section 5.3. With their aid one can evaluate the relative role of the different types of collision.

Let us consider the solution of Eq. 5.49 for a plasma with a low degree of ionization, when only electron–atom collisions are substantial. The conditions under which the effect of electron–electron and electron–ion collisions on the distribution function can be neglected are found from Eq. 5.50:

$$\nu_{ee} \approx \nu_{ei} \ll \nu_{ea}; \qquad \nu_{ee} \ll \kappa_{ea}\nu_{ea} \tag{5.51}$$

(It is taken into account here that $\kappa_{ei} \approx \kappa_{ea}$ since $m_i = m_a$ and $\kappa_{ee} \approx 1$). The second inequality is more rigid than the first. For elastic collisions it takes the form $\nu_{ee} \ll (m_e/m_a)\nu_{ea}$, or

$$\eta = \frac{n_e}{n_a} \ll \frac{m_e}{m_a}\frac{s_{ea}}{s_{ee}} \tag{5.52}$$

It can be seen that only at very small values of η ($\eta \lessdot m_e/m_a$) is it possible to neglect the effect of electron–electron collisions on the distribution function.

We first find the solution of Eq. 5.49 for the case where only elastic electron–atom collisions are substantial. Substituting Eqs. 5.28 and 5.32 for S_0 and S_1 into Eqs. 5.49, we obtain

$$\frac{eEn}{m_e}\frac{df_0}{dv} = n\nu' f_1 \tag{5.53}$$

$$-\frac{eEn}{3m_e v^2}\frac{d}{dv}(v^2 f_1) = \frac{1}{2}\frac{n}{v^2}\frac{d}{dv}\left[\kappa\nu' v^2\left(v f_0 + \frac{T_a}{m_e}\frac{df_0}{dv}\right)\right] \tag{5.54}$$

Equation 5.53 describes the stationary balance between the acquisition of a directed velocity in an electric field and its losses on collisions. It enables one to find the relationship between the functions f_1 and f_0:

$$f_1 = \frac{eE}{m_e \nu^t} \frac{df_0}{dv} \tag{5.55}$$

Using Eq. 5.55, we obtain from Eq. 5.13 the general expression for the directed electron velocity:

$$\begin{aligned} u_E &= \frac{4\pi}{3} \int_0^\infty f_1(v) v^3 \, dv = \frac{4\pi}{e} \frac{eE}{m_e} \int_0^\infty \frac{v^3}{\nu^t} \frac{df_0}{dv} \, dv \\ &= -\frac{4\pi}{3} \frac{eE}{m_e} \int_0^\infty f_0(v) \frac{d}{dv} \left(\frac{v^3}{\nu^t} \right) dv \end{aligned} \tag{5.56}$$

(the last equality results from integration by parts). If the collision frequency is velocity independent, that is, if $\nu = \text{const}$, expression 5.56 leads to an equation independent of the form of $f_0(v)$:

$$u = -\frac{eE}{m_e \nu^t} 4\pi \int_0^\infty f_0(v) v^2 \, dv = -\frac{eE}{m_e \nu^t} \tag{5.57}$$

The equation for the isotropic distribution function component is found by substituting Eq. 5.55 into Eq. 5.54:

$$-\frac{n}{v^2} \frac{d}{dv} \left(\frac{e^2 E^2 v^2}{3 m_e^2 \nu^t} \frac{df_0}{dv} \right) = \frac{1}{2} \frac{n}{v^2} \frac{d}{dv} \left[\kappa \nu^t v^2 \left(v f_0 + \frac{T_a}{m_e} \frac{df_0}{dv} \right) \right] \tag{5.58}$$

The left-hand side of the equation, which defines the electron velocity increase under the effect of the electric field, can, as well as the right-hand ("collision") side, be represented as a divergence of a spherically symmetric flux. The flux

$$\Gamma_{vE} = -n \frac{e^2 E^2 v^2}{3 m_e^2 \nu^t} \frac{df_0}{dv} = -D_E n \frac{df_0}{dv} \tag{5.59}$$

is proportional to the derivative df_0/dv, and since $f_0(v)$ is a decreasing function the flux is directed toward the higher velocities. The expression 5.59 describes the diffusion in velocity space. The energy accumulation by electrons in the electric field is of a diffusion nature because it continues for many periods between collisions. As a result, the acceleration during each period is added vectorially with an arbitrarily directed random velocity, and hence the change in velocity is accidental. It is easy to estimate directly the diffusion coefficient characterizing this process. The change in electron velocity vector within the intercollisional time τ is due to the electron acceleration in the electric field:

$$\Delta \mathbf{v}_E = \mathbf{v}^* - \mathbf{v} = -\left(\frac{e\mathbf{E}}{m_e} \right) \tau = -\frac{e\mathbf{E}}{m_e \nu}$$

The corresponding velocity change Δv is given by the relations

$$v^{*2} \approx v^2 + 2\mathbf{v}\Delta\mathbf{v}_E; \qquad \Delta v = v^* - v \approx \frac{\mathbf{v}\Delta\mathbf{v}_E}{v}$$

which take into account that $\Delta v_E \ll v$. Since the velocity $\mathbf{v}$ can be directed arbitrarily, the average value of $\Delta\mathbf{v}$ reduces to zero. The average value of $(\Delta v)^2$ is equal to

$$\langle(\Delta v)^2\rangle = \frac{1}{3}(\Delta v_E)^2 = \frac{1}{3}\frac{e^2E^2}{m_e^2\nu^2}$$

whence we find the diffusion coefficient (see Eq. 5.37):

$$D_E = \frac{\langle(\Delta v)^2\rangle}{2\tau} = \frac{1}{6}\frac{e^2E^2}{m_e^2\nu} \tag{5.60}$$

which agrees with Eq. 5.59 to within a numerical factor.

Thus it follows from Eq. 5.58 that the divergence of the density of the summary flux in velocity space—the flux determining energy acquisition in the field Γ_{vE} (Eq. 5.59) and the flux determining the energy losses on collisions $\Gamma_{v\nu}$ (Eq. 5.34)—reduces to zero; that is,

$$\frac{1}{v^2}\frac{d}{dv}[v^2(\Gamma_{vE} + \Gamma_{v\nu})] = 0 \tag{5.61}$$

Integrating the equation, we arrive at the constancy of the summary flux across the spherical surface of velocity space $4\pi v^2(\Gamma_{vE} + \Gamma_{v\nu}) = C$

Since the quantities Γ_E and Γ_ν are finite as $\mathbf{v} \to 0$, the integration constant C is zero. This means that the summary flux is zero as well. By using the expressions for its components Γ_{vE} (Eq. 5.59) and $\Gamma_{v\nu}$ (Eq. 5.34), we obtain the equation for f_0, which stems from Eq. 5.58:

$$\frac{2}{3}\frac{e^2E^2}{m_e^2(\nu_{ea}^t)^2}\frac{df_0}{dv} + \kappa_{ea}\left(vf_0 + \frac{T_a}{m_e}\frac{df_0}{dv}\right) = 0 \tag{5.62}$$

The solution of Eq. 5.62 has the form

$$f_0 = A\exp\left[-\int_0^v \frac{m_e v\,dv}{T_a + 2e^2E^2/3\kappa_{ea}(\nu_{ea}^t)^2 m_e}\right] \tag{5.63}$$

where the constant A is found from the normalization condition (Eq. 5.12):

$$4\pi\int_0^\infty f_0 v^2\,dv = 1$$

Equation 5.63 indicates the nature of the deviations of the electron velocity distribution from equilibrium. It can be seen that these deviations are determined by the ratio of the field strength E to the critical strength (Eq. 5.5) $E_p = \nu_{ea}\sqrt{\kappa_{ea} m_e T_a}/e$. At $E \ll E_p$ the second term in the denominator of the integrand in Eq. 5.63 is small, and the velocity distribution becomes Maxwellian with a temperature T_a. In strong fields, at $E \gg E_p$, the first term in the denominator can be neglected. Then Eq. 5.63 takes the form

$$f_0(v) = A \exp\left[-\frac{3}{2}\frac{m_e^2 \kappa_{ea}}{e^2 E^2}\int_0^v v(\nu_{ea}^t)^2\, dv\right] \qquad (5.64)$$

The expressions 5.63 and 5.64 show that the form of the velocity distribution function is determined by the dependence of the electron collision transport frequency on the velocity $\nu_{ea}^t\ (v)$. At a constant collision frequency $\nu^t = \text{const}$ the distribution function is Maxwellian irrespective of the field strength. From Eq. 5.63 we find

$$f_0 = A \exp\left(-\frac{m_e v^2}{2T_e}\right) = \left(\frac{m_e}{2\pi T_e}\right)^{3/2} \exp\left(-\frac{m_e v^2}{2T_e}\right) \qquad (5.65)$$

where

$$T_e = T_a + \frac{2}{3}\frac{e^2E^2}{\kappa_{ea} m_e (\nu_{ea}^t)^2} = T_a + \frac{1}{3}\frac{m_a e^2 E^2}{m_e^2(\nu_{ea}^t)^2} \qquad (5.66)$$

It is seen that the electron temperature, which characterizes the average electron energy, exceeds the atom temperature, and at $E \gg E_p$ is proportional to the square of the field.

If ν_{ea} is a function of the velocity, the distribution function may be substantially non-Maxwellian. Thus for a constant collision cross section, when $\nu_{ea}^t = n_a v s_{ea}^t = v/\lambda_{ea}^t$ is proportional to the velocity, the distribution function in a strong field (Eq. 5.64) has the form

$$f_0 = A \exp\left[-\frac{3m_e^2 v^4 \kappa_{ea}}{8e^2E^2\lambda_{ea}^2}\right] \qquad (5.67)$$

where, from the normalization condition,

$$A = \frac{0.37}{\pi}\left(\frac{m_e\sqrt{\kappa_{ea}}}{e\lambda_{ea}E}\right)^{3/2}$$

With the aid of Eq. 5.11 we find the average electron energy for this distribution:

$$\langle K_e \rangle = \frac{0.57 eE\lambda_{ea}^t}{\sqrt{\kappa_{ea}}} = 0.4\sqrt{m_a/m_e}\, eE\lambda_{ea}^t \qquad (5.68)$$

The directed velocity can be found by substituting Eq. 5.67 into Eq. 5.56. Integrating, we obtain

$$u_E = 0.69\kappa_{ea}^{1/4}\left(\frac{eE\lambda_{ea}^t}{m_e}\right)^{1/2} \qquad (5.69)$$

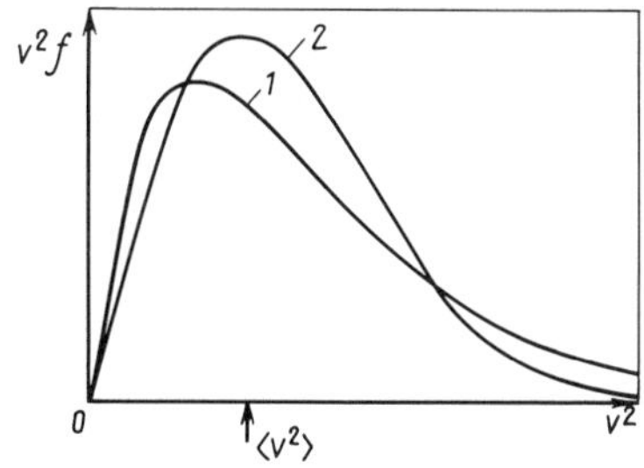

Fig. 5.3 Comparison of (1) Maxwellian and (2) Druyvesteyn distribution functions.

The distribution described by Eq. 5.67 is called the ***Druyvesteyn distribution.*** It is characterized by a much stronger velocity dependence than the Maxwellian; in Fig. 5.3: 1, Maxwellian distribution; 2, Druyvesteyn distribution. This is easy to explain. An increase in collision frequency at high velocities increases the flux $\Gamma_{v\nu}$ toward decreasing velocities. To offset it, the distribution function gradient, which determines the reverse flux, must increase compared with the Maxwellian.

5.5 EFFECT OF INELASTIC COLLISIONS ON ELECTRON DISTRIBUTION FUNCTION

As mentioned earlier, in a gas-discharge plasma steadily maintained by an electric field the average electron energy is usually much less than the excitation energy of the lower electron level. In atomic gases the distribution function is largely determined by elastic collisions. In molecular gases, however, a substantial role may be played by excitation of rotation and vibration levels of molecules whose energy is much less than the average electron energy. It was shown in Section 3.3 that inclusion of such collisions in the collision terms S_0 and S_1 amounts to a replacement of collision frequency and the energy transfer coefficient, which appear in these terms, by summary values defined by the relations 5.44. If the velocity dependence of these values is known, the distribution function, which takes into account inelastic processes with small energy losses, can be calculated using Eqs. 5.63 and 5.55.

Let us dwell in more detail on the effect of inelastic collisions with large energy losses. In a gas-discharge plasma they include excitation of electron levels and ionization. Changes in distribution function due to inelastic collisions are illustrated in Fig. 5.4, which depicts the energy distribution determined by the function $f_v = 4\pi v^2 f_0$ of Eq. 5.12. The distribution function without an allowance for inelastic collisions is denoted by f_{00}. We assume, for simplicity, that only one inelastic process with a threshold $\mathscr{E}_1$ is substantial. Because of inelastic collisions the

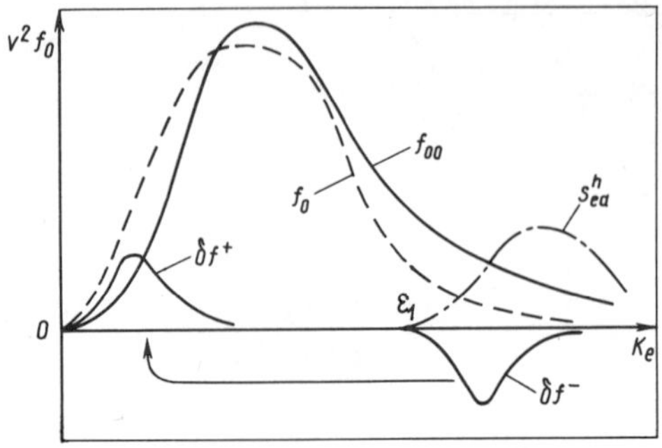

Fig. 5.4 Changes in distribution functions due to inelastic collisions.

electrons lose an energy approximately equal to $\mathscr{E}_1$ and move from the region of above-threshold energies to the low-energy region. Figure 5.4 shows a curve characterizing the reduction rate of distribution function as a result of this process $\delta f^- = -\nu^{(1)} f_0 \delta t$ in the high-energy region, and the reverse curve $\delta f^+ = -\delta f^-$ defining the arrival of electrons in the low-velocity region. The width of these curves depends on the rate of decrease in f_0 in the inelastic region (at $K_e > \mathscr{E}_1$). At an excitation energy greatly exceeding the average electron energy, f_0 falls off very fast, and the curve δf^- becomes narrow, "pressed" against the excitation threshold; accordingly, the curve δf^+ is "pressed" against the origin.

To find the change in function f_0 due to inelastic collisions, one must take into account not only transitions of δf^+ and δf^-, but also the diffusion in velocity space due to elastic collisions. The diffusion smoothes the curve $f_0(v)$. In particular, it increases the gradient of f in the region of $v \leqslant v_1$ and partly replenishes the loss of electrons from the high-velocity region (see Fig. 5.4). In actual conditions several inelastic processes with different excitation energies are usually substantial. The curves δf associated with such processes are added together, and the drop of the function in the inelastic collision region is increased still further.

Let us consider quantitatively the effect of inelastic collisions on the distribution function. We begin with finding the distribution function in the region of velocities exceeding the inelastic-process threshold $v > v_1 = (2\mathscr{E}_1/m_e)^{1/2}$, with the aid of Eqs. 5.49:

$$-\left(\frac{eEn}{3m_e v^2}\right)\left(\frac{d}{dv}\right)(v^2 f_1) = S_0; \qquad -\left(\frac{eEn}{m_e}\right)\left(\frac{df_0}{dv}\right) = S_1 \tag{5.70}$$

As noted before, when the distribution function falls off rapidly inelastic collisions of each type are experienced mainly by electrons whose energy is close to the excitation energy of the corresponding levels. Therefore we can use the expression for inelastic collision terms (Eq. 5.47) obtained by assuming the total electron energy loss. The overall equation for S_0 and S_1 then takes the form

$$S_0 = \frac{nm_e}{m_a v^2}\frac{d}{dv}(\nu_{ea}^t v^2 f_0) - \nu_{ea}^h n f_0;$$
$$S_1 = -(\nu_{ea}^t + \nu_{ea}^h) n f_1 \tag{5.71}$$

where ν_{ea}^h is the summary frequency of inelastic collisions with a large energy loss (Eq. 5.48).

In the inelastic collisions region the summary frequency usually lies largely within the following limits (see Chapter 2):

$$\left(\frac{m_e}{m_a}\right)\nu_{ea}^t \ll \nu_{ea}^h \ll \nu_{ea}^t \tag{5.72}$$

and owing to the smallness of m_e/m_a the inequality 5.72 is valid beginning with velocities close to the threshold of inelastic processes. With this in mind we simplify Eqs. 5.71. In the equation for S_0 we neglect the term representing elastic scattering (it is of the order of $\nu^t(m_e/m_a)nf_0$), and in the equation for S_1, the term for inelastic collisions. Then

$$S_0 = -\nu_{ea}^h n f_0; \qquad S_1 = -\nu_{ea}^t n f_1 \tag{5.73}$$

Substituting S_1 into the equation for f_1 yields the former relationship 5.55 between f_1 and f_0. Using it, we obtain from Eq. 5.70 the equation for f_0 in the inelastic collision region:

$$\frac{e^2E^2}{3m_e^2v^2}\frac{d}{dv}\left(\frac{v^2}{\nu_{ea}^t}\frac{df_0}{dv}\right) - \nu_{ea}^h f_0 = 0 \tag{5.74}$$

It can be written differently:

$$\frac{d^2f_0}{dv^2} + \left(\frac{2}{v} - \frac{1}{\nu_{ea}^t}\frac{d\nu_{ea}^t}{dv}\right)\frac{df_0}{dv} - \frac{q^2}{v_1^2}f_0 = 0 \tag{5.75}$$

where

$$q^2 = \frac{3m_e^2\nu_{ea}^t\nu_{ea}^h}{e^2E^2}v_1^2$$

The quantity q determines the decrease of the distribution function at $v > v_1$. We assume

$$q^2 \approx \frac{m_a}{m_e}\frac{\nu_{ea}^h}{\nu_{ea}^t}\frac{v_1^2}{v_{Te}^2} \approx \frac{m_a}{m_e}\frac{\nu_{ea}^h}{\nu_{ea}^t}\frac{\mathscr{E}_1}{T_0} \gg 1 \tag{5.76}$$

where $T_0 \approx e^2\mathscr{E}^2/m_e\kappa_{ea}(\nu_{ea}^t)^2$ is the average electron energy in an electric field in the absence of inelastic collisions. It is seen that this inequality follows from Eq. 5.72 at $T_0 \ll \mathscr{E}_1$. In solving Eq. 5.75 we can use, with the aid of Eq. 5.76, an approximation based on the fact that f_0 is a rapidly decreasing function of v. Therefore, in Eq. 5.75 the second term can be neglected compared with the first $|(1/v)\, df_0/dv| \ll d^2f_0/dv^2$, and q can be assumed a slowly varying function of the velocity $|(1/q)\, dq/dv| \ll |(1/f_0)\, df_0/dv|$. Then we obtain an approximate solution of the equation, which reduces to zero as $v \to \infty$, in the form*

$$f_0 = C \exp\left(-\int_{v_1}^{v} \frac{q}{v_1}\, dv\right) = C \exp\left[-\frac{\sqrt{3}m_e}{eE}\int_{v_1}^{v} \sqrt{\nu_{ea}^t \nu_{ea}^h}\, dv\right] \quad (5.77)$$

This solution holds in the velocity region bounded by the inequality $\nu^h \gg m_e\nu^t/m_a$, when the electron energy losses are due to inelastic collisions.

When the opposite inequality is true, one can use f_0 obtained without allowance for inelastic collisions (Eq. 5.64). The coefficient C is determined approximately by combining Eq. 5.77 with Eq. 5.64 at the region boundary for $\nu^h = m_e\nu^t/m_a$. This boundary is usually very close to the threshold of inelastic processes, and the combination may be done at the threshold velocity $v = v_1$.

The particular form of f_0 (Eq. 5.77) is determined by the velocity dependence of the collision frequencies ν^t and ν^h. Let us find it for $\nu^t = \text{const}$.

The dependence $\nu^h(v)$ near the threshold of inelastic processes can be approximated by the equation

$$\nu^h(v) = n_a S^h v_1 \approx \nu_0^h \frac{v}{v_1 - 1} \quad (5.78)$$

Using this approximation, we can compute the power index in Eq. 5.77:

$$\frac{\sqrt{3}m_e}{eE}\int_{v_1}^{v} \sqrt{\nu^t\nu^h}\, dv = \frac{\sqrt{3}m_e\sqrt{\nu^t\nu_0^h}}{eE}\int_{v_1}^{v}\sqrt{\frac{v}{v_1} - 1}\, dv = q_0\left(\frac{v}{v_1} - 1\right)^{3/2}$$

where (5.79)

$$q_0 = \frac{2}{\sqrt{3}}\frac{m_e\sqrt{\nu^t\nu_0^h}}{eE}v_1 = \frac{4}{3}\sqrt{\frac{1}{\kappa}\frac{\nu_0^h}{\nu^t}\frac{\mathscr{E}_1}{T_e}}$$

Substituting Eq. 5.79 into Eq. 5.77, we obtain the expression for f_0 at $v > v_1$:

$$f_0 = C \exp\left[-q_0\left(\frac{v}{v_1 - 1}\right)^{3/2}\right] \quad (5.80)$$

*This approximation is sometimes called quasi-classical by analogy with the quasi-classical approximation in quantum mechanics.

Using Eq. 5.80, we can find the efficiency of the processes determined by the electrons whose energy exceeds the inelastic-collision threshold. We first calculate the average number of inelastic collisions per unit time (see Eq. 5.46):

$$Q_n = n_e \int f \nu^h \, d^3 v = 4\pi n_e \int \nu^h f_0(v) v^{-2} \, dv$$

Substituting the approximation 5.78 and the function 5.80 and taking into account the fast decrease of f_0 and $v > v_1$, we find

$$\begin{aligned} Q_n &= 4\pi C n_e \nu_0^h \int v^2 \left(\frac{v}{v_1} - 1\right) \exp\left[-q_0 \left(\frac{v}{v_1} - 1\right)^{3/2}\right] dv \\ &\approx 4\pi C n_e \nu_0^h v_1^2 \int \left(\frac{v}{v_1} - 1\right) \exp\left[-q_0 \left(\frac{v}{v_1} - 1\right)^{3/2}\right] dv \end{aligned} \tag{5.81}$$

and further

$$Q_n \approx 4\pi C n_e \nu_0^h q_0^{-4/3} v_1^3 \int x \exp(-x^{3/2}) \, dx \approx 2.5\pi C n_e \nu_0^h v_1^3 q_0^{-4/3}$$

Let us now find the average ionization frequency characterizing the rate of appearance of new electrons in the plasma. To this end we use the approximation of the frequency of collisions leading to ionization (see Eq. 2.86):

$$\nu^i = n_a s^i v \approx \nu_0^i \frac{v^2 - v_i^2}{v_i^2} \tag{5.82}$$

where $v_i = (2e\mathscr{E}_i/m_e)^{1/2}$ is the threshold velocity for ionization. The average ionization frequency is obtained by averaging Eq. 5.82 with the aid of the function f_0 (see Eq. 5.80). Using Eq. 5.11, we obtain

$$\begin{aligned} \nu^i &= 4\pi \int_{v_i}^{\infty} \nu^i(v) f_0(v) v^2 \, dv \\ &= 4\pi C \nu_0^i \int_{v_i}^{\infty} \exp\left[-q_0 \left(\frac{v}{v_1} - 1\right)^{3/2}\right] \left(\frac{v^2}{v_i^2} - 1\right) v^2 \, dv \end{aligned} \tag{5.83}$$

When calculating the integral of Eq. 5.83 we take into account that $f_0(v)$ falls off rapidly, and the integrand function $\nu^i f_0$ approaches the ionization threshold. Therefore it can be assumed that under the integral $v - v_i = w \ll v_i$, and then the exponent contained in it has the form

$$\exp\left[-q_0 \left(\frac{v}{v_1} - 1\right)^{3/2}\right] = \exp\left[-q_0 \left(\frac{v_i}{v_1} - 1\right)^{3/2}\right] \exp(-\zeta w)$$

where $\zeta = \frac{3}{2}(q_0/v_1)(v_i/v_1 - 1)^{1/2}$. Assuming also that under the integral

$v^2/v_i^2 - 1 \approx 2w/v_i$ and $v^2 \approx v_i^2$, we get

$$\int_{v_i}^{\infty} \exp\left[-q_0\left(\frac{v}{v_1}-1\right)^{3/2}\right]\left(\frac{v^2}{v_i^2}-1\right) v^2\, dv$$

$$\approx 2v_i \exp\left[-q_0\left(\frac{v_i}{v_1}-1\right)^{3/2}\right]\int_0^{\infty} w \exp(-\zeta w)\, dw$$

$$= \frac{2v_i}{\zeta^2} \exp\left[-q_0\left(\frac{v_i}{v_1}-1\right)^{3/2}\right]$$

Substituting this expression into Eq. 5.83, we find

$$\frac{\langle \nu^i \rangle}{\nu_0^i} = \frac{32\pi c}{9q_0^2}\left(\frac{2\mathscr{E}_i}{m_e}\right)^{3/2}\left(1-\sqrt{\frac{\mathscr{E}_1}{\mathscr{E}_i}}\right)^{-1} \exp\left[-q_0\left(\sqrt{\frac{\mathscr{E}_i}{\mathscr{E}_1}}-1\right)^{3/2}\right] \qquad (5.84)$$

It can be seen that the quantity $\langle \nu^i \rangle$ contains a small exponential factor defining the decrease of f_0 in the region of $v > v_1$, which is associated with inelastic collisions.

We now determine the distribution function in the velocity range $v < v_1$. For this we use Eq. 5.58, adding to it the collision term 5.46, which describes the arrival of electrons as a result of inelastic collisions. Assuming that the average electron energy greatly exceeds the atom temperature, we write the equation thus:

$$-\frac{1}{3v^2}\frac{d}{dv}\left(\frac{e^2E^2}{m_e^2\nu^t} v^2 \frac{df_0}{dv}\right) = \frac{1}{2v^2}\frac{d}{dv}(\kappa \nu^t v^3 f_0) + \frac{Q_n}{4\pi v^2 n}\,\delta(v) \qquad (5.85)$$

Multiplying Eq. 5.85 by $4\pi v^2\, dv$ and integrating it from zero to v, we obtain

$$\frac{4\pi}{3}\frac{e^2E^2}{m_e^2\nu^t} v^2 \frac{df_0}{dv} - 2\pi\kappa\nu^t v^3 t_0 = Q_n \qquad (5.86)$$

The left-hand side of the equation represents the flux in velocity space across a spherical surface $4\pi v^2$; the first term is due to the energy received by the electrons in the electric field, and the second to the energy losses on elastic collisions (see Eqs. 5.34 and 5.59). The sum of these fluxes is determined by the rate of appearance of slow electrons as a result of elastic collisions.

Equation 5.86 is a linear inhomogeneous equation of the first order relative to f_0. Its solution can easily be found by the integrating-factor method. At $\nu^t = \text{const}$ it has the form

$$f_0(v) = A \exp\left(-\frac{m_e v^2}{2T_0}\right)\left[\int_v^{v_1} \frac{1}{v^2} \exp\left(\frac{m_e v^2}{2T_0}\right) dv + B\right] \qquad (5.87)$$

where $T_0 = \frac{2}{3}e^2E^2/m_e\kappa(\nu^t)^2$ is the temperature determining the average electron energy in the field E without inelastic collisions and $A = 3Qm_e^2 \nu_{ea}^t / 4\pi n e^2 E^2 \approx 1.4 C v_1 q_0^{2/3}$. For $v = v_1$ we must combine this solution with Eq. 5.77. Since $f(v_1) = 0$, we find

$$B \approx C/A \exp\left(\frac{\mathscr{E}_1}{T_0}\right) \approx \frac{0.71}{v_1 q_0^{2/3}} \exp\left(\frac{\mathscr{E}_1}{T_0}\right) \tag{5.88}$$

From Eq. 5.87 it follows that for $v \to 0$ the value of f_0 tends to infinity as $1/v$. This conclusion is obviously associated with the assumption of a total loss of energy by the electrons on inelastic collisions, as a result of which we introduced an electron source $Q_n \delta(v)$, infinite with respect to the amplitude, for $v = 0$. Note that although f_0 has a singularity as $v \to 0$, in this case the distribution function for total velocities $f_v = 4\pi v^2 f_0$ remains small, since $f_v \sim v^2$, $f_0 \sim v$; therefore this singularity does not cause any difficulty in averaging.

Let us first consider the case $T_0 \ll \mathscr{E}_1$. When analyzing Eq. 5.87 we can clearly see that the integral appearing in it depends on the lower limit only in the low-velocity range $v \lesssim v_1\ (\mathscr{E}_1/T_0) \exp(-\mathscr{E}_1/T_0)$ and in the near-threshold velocity range $(v_1 - v) < v_1 T_0/\mathscr{E}_1$. Outside these ranges it is practically independent of the lower limit and can be found approximately by integrating by parts:

$$\begin{aligned}
\int_v^{v_1} \frac{1}{v^2} \exp\left(\frac{m_e v^2}{2T_0}\right) dv &= \left(\frac{m_e}{2T_0}\right)^{1/2} \int_{v/v_{T_0}}^{v_1/v_{T_0}} \frac{1}{t^2} \exp t^2\, dt \\
&= \left(\frac{m_e}{2T_0}\right)^{1/2} \left[\frac{1}{2t^3} \exp t^2 - \frac{3}{2} \int \frac{1}{t^4} \exp t^2\, dt\right]_{v/v_{T_0}}^{v_1/v_{T_0}} \\
&\approx \frac{1}{2} \left(\frac{m_e}{2T_0}\right)^{1/2} \left(\frac{T_0}{\mathscr{E}_1}\right)^{3/2} \exp\left(\frac{\mathscr{E}_1}{T_0}\right)
\end{aligned}$$

Substituting this expression and the relation 5.88 into Eq. 5.87, we get

$$\begin{aligned}
f_0(v) &= D \exp\left(-\frac{m_e v^2}{2T_0}\right) \\
D &= C \exp\left(\frac{\mathscr{E}_1}{T_0}\right) \left[1 + 0.70 \frac{T_0}{\mathscr{E}_1} q_0^{2/3}\right]
\end{aligned} \tag{5.89}$$

To obtain the coefficients D and C we can use the approximate normalization condition, which allows for the smallness of the deviation of $f_0(v)$ from Eq. 5.64 at $v < v_1$ and its fast drop at $v > v_1$. Therefore the normalized coefficients can be found approximately from the condition

$$4\pi_0 \int_0^{v_1} f_0(v) v^2\, dv \approx 4\pi_0 \int_0^\infty f_0(v) v^2\, dv = 1$$

according to which $D = (m_e/2\pi T_0)^{3/2}$;

$$C = f_0(v_2) = \frac{m_e}{2\pi T_0}\exp\left(\frac{\mathscr{E}_1}{T_0}\right)\left[1 + 0.70\frac{T_o}{\mathscr{E}_1}q_0^{2/3}\right]^{-1} \tag{5.90}$$

Thus at $T_0 \ll \mathscr{E}_1$, in most of the region $v < v_1$ the distribution function is found to be the same as with elastic collisions alone (for $\nu^t = \text{const}$, it is Maxwellian). The deviations are substantial only at low velocities $[v \lessapprox v_1(\mathscr{E}_1/T_0)\exp(-\mathscr{E}_1/T_o)]$ and at velocities close to v_1 $[v_1 - v < v_1(T_0/\mathscr{E}_1)]$.

We have determined the function $f_0(v)$ for $T_0 \ll \mathscr{E}_1$. An increase in electric field strength and T_0 increases the number of electrons with energies exceeding the inelastic threshold. The effect of inelastic collisions on the form of the distribution function at $v < v_1$ increases correspondingly. Let us consider the case*

$$T_0 = \frac{2}{3}\frac{e^2E^2}{m_e\kappa(\nu^t)^2} \gg \mathscr{E}_1$$

assuming, however, that $T_0 \ll \mathscr{E}_1(\nu_0^h/\kappa\nu^t)$. Then the parameter q_0 of Eq. 5.79 remains large, and the solution of Eq. 5.80 for the inelastic-collision range $(v > v_1)$ holds true. In the range of $v < v_1$ the equation for the distribution function (Eq. 5.86) is simplified because the second term, which describes the energy losses on elastic collisions, is negligibly small as compared with the first one. The mechanism of energy distribution formation is as follows: at $v < v_1$ the electrons accumulate energy as a result of velocity diffusion in the electric field (see p. 124) and lose it at $v \gtrsim v_1$ owing to inelastic collisions. For $\mathscr{E}_1 \ll T_0$ we can assume the exponents in Eq. 5.89 to be equal to unity. The function f_0 then acquires the simple form

$$f_0 = A\left[\frac{1}{v} - \frac{1}{v_1}(1 - 0.71q_0^{-2/3})\right] \tag{5.91}$$

The coefficient A can be found from the approximate normalization condition:

$$\begin{aligned} 1 &= 4\pi\int_0^{v_1} f_0 v^2\,dv = \frac{2}{3}\pi A v_1^2(1 - 1.42q_0^{-2/3}) \\ A &\approx \frac{3}{2\pi v_1^2}(1 - 1.42q_0^{-2/3})^{-1} \approx \frac{3}{2\pi v_1^2} \end{aligned} \tag{5.92}$$

For the distribution obtained the average electron energy is equal approximately to

$$\langle K_e\rangle \approx \int_0^{v_1}\frac{m_e v^2}{2}4\pi v^2 f_0(v)\,dv \approx \frac{3}{10}\mathscr{E}_1 \tag{5.93}$$

*Note that both conditions can be correlated for a heavy-atom gas, since the coefficient $\kappa_{ea} = 2m_e/m_a$ is very small.

(we made an allowance for the fact that $f_0(v)$ drops fast at $v > v_1$ and $q_0 \gg 1$). The distribution function at the boundary of the inelastic region is given by the relation

$$C = f_0(v_1) \approx \frac{0.34}{v_1^3 q_0^{2/3}}$$

Using this, we can find from Eq. 5.11 the average ionization frequency for the case under consideration:

$$\nu^i/\nu_0^i = \frac{3.8}{q_0^{8/3}}\left(1 - \sqrt{\frac{\mathscr{E}_1}{\mathscr{E}_i}}\right)^{-1} \exp\left[-q_0\left(\sqrt{\frac{\mathscr{E}_i}{\mathscr{E}_1}} - 1\right)^{3/2}\right] \tag{5.94}$$

where $q_0 = (2/\sqrt{3})(m_e\sqrt{\nu^t\nu_0^h}/eE)v_1$.

We have considered the effect on the distribution function of inelastic collisions resulting in the excitation of different levels and the loss of electron energy. In principle, reverse processes are also possible, namely, electron-excited atom collisions, which increase the electron energy (second-type collisions). One can frequently assume that they are insignificant, since the excited atoms lose their energy mainly through radiation and collisions with neutral particles and the walls. Sometimes, however, such as at high neutral gas pressures, when the plasma is not completely transparent to radiation, one has to take into account second-type collisions. They lead to some compensation for the energy losses caused by excitation. In the limiting case, on changeover to a closed system, forward and reverse inelastic collisions ensure such energy exchange between electrons and atoms, which leads to equilibrium Maxwellian distribution (see Section 4.1).

5.6 EFFECT OF ELECTRON–ELECTRON COLLISIONS ON ELECTRON DISTRIBUTION FUNCTION

So far we have neglected the effect of electron–electron collisions on the distribution function. As shown in Section 5.4, the conditions under which this neglect (see Eq. 5.52) is admissible are fulfilled only for very low degrees of ionization. Let us now consider the opposite, when energy exchange of electrons as a result of electron–electron collisions is much more efficient than the energy losses in electron–atom collisions. This occurs when

$$\kappa_{ea}\nu_{ea} \ll \nu_{ee} \tag{5.95}$$

If the electron energy losses are caused by elastic collisions and $\kappa_{ea} = 2m_e/m_a$, the condition 5.95 takes a form opposite to Eq. 5.52:

$$\eta \gg \frac{m_e}{m_a}\frac{s_{ea}}{s_{ee}} \tag{5.96}$$

It can be seen that this inequality holds true even at comparatively low degrees of ionization, which are lower, the smaller the electron energy.

Bearing in mind the condition 5.95, let us discuss the solution of the equation for the isotropic distribution function component. In accordance with Eq. 5.49 this equation has the form

$$-\frac{eEn}{3m_e v^2}\frac{d}{dv}(v^2 f_1) = S_{0a} + S_{0i} + S_{0e} \tag{5.97}$$

In the equation, the stationary balance of the average electron energies is determined by the first three terms—the acquisition of energy from the electric field by electrons (described by the first term) is offset by electron energy losses in collisions with atoms and ions (described by the second and third terms). Accordingly, the first term must be of the same order of magnitude as the sum $S_{0a} + S_{0i}$. The collision term S_{0e}, which is due to electron–electron collisions, clearly cannot affect the variation in the average energy of the electrons; it only results in redistribution of energy among them. In our case the term S_{0e} is of the order of $\nu_{ee}nf_0$; it greatly exceeds the second ($S_{0a} \approx \kappa_{ea}\nu_{ea}nf_0$) and the third ($S_{0i} \approx \kappa_{ei}\nu_{ei}nf_0$) terms. Therefore, to a first approximation the form of the distribution functions is described by the equation

$$S_{0e} = \frac{1}{2}n_e^2 \int_{(v_1)}\int_{(\Omega)} [f_0(v')f_0(v_1') - f_0(v)f_0(v_1)]|\mathbf{v} - \mathbf{v}_1|\sigma\, d\Omega\, d^3v_1 = 0 \tag{5.98}$$

As noted in Section 3.3, we can write S_{0e} in the form of the collision integral in Eq. 3.23, replacing $f(v)$ by $f_0(v)$. As shown in Section 4.2, this equation leads to the Maxwellian velocity distribution:

$$f_0 = \left(\frac{m_e}{2\pi T_e}\right)^{3/2} \exp\left(\frac{-m_e v^2}{2T_e}\right) \tag{5.99}$$

The electron temperature T_e, however, cannot be found from Eq. 5.98. Since this equation includes only interelectron collisions, it is satisfied by the distribution 5.99 with any value of T_e.

To find T_e one can use the energy balance equation, which is derived from the initial equation 5.97 for f_0. In order to obtain it we multiply Eq. 5.97 term by term by $m_e v^2/2$ and by the weighting factor $4\pi v^2\, dv$ and integrate over the velocities from zero to infinity. The integral of the first term is calculated by integrating by parts:

$$\int_0^\infty \frac{m_e v^2}{2} 4\pi v^2\, dv \left[-\frac{eEn}{3m_e v^2}\frac{d}{dv}(v^2 f_1)\right] = -\frac{2\pi}{3} eEn \int_0^\infty v^2 \frac{d}{dv}(v^2 f_1)\, dv \tag{5.100}$$

$$= neE\frac{4\pi}{3}\int_0^\infty v^3 f_1\, dv = neuE$$

(Here we used the general equation 5.21 for **u**.) The collision term that is determined by elastic electron–atom collisions (Eq. 5.32) with a Maxwellian electron velocity distribution (Eq. 5.99) is equal to

$$\begin{aligned} S_{0a} &= \frac{1}{2}\frac{\kappa_{ea}n}{v^2}\frac{d}{dv}\left[v^3\nu^t_{ea}f_0 + \frac{T_a}{m_e}v^2\frac{df_0}{dv_0}\right] \\ &= \frac{1}{2}\frac{\kappa_{ea}n}{v^2}\frac{d}{dv}\left[v^3\nu^t_{ea}f_0\left(1-\frac{T_a}{T_e}\right)\right] \end{aligned} \tag{5.101}$$

The integral in which it appears is calculated similarly to Eq. 5.100:

$$\begin{aligned} \int_0^\infty S_{0a}\frac{m_e v^2}{2}4\pi v^2\,dv &= -2\pi m_e n\kappa_{ae}\left(1-\frac{T_a}{T_e}\right)\int_0^\infty v^4\nu^t_{ea}f_0\,dv \\ &= -\frac{3}{2}n\kappa_{ea}\bar{\nu}^t_{ea}(T_e - T_a) \end{aligned} \tag{5.102}$$

where the following notation is introduced:

$$\bar{\nu}^t_{ea} = \frac{\int_0^\infty v^4\nu^t_{ea}f_0\,dv}{\int_0^\infty v^4 f_0\,dv} \tag{5.103}$$

and due allowance is made for the fact that

$$4\pi\int_0^\infty \frac{m_e v^2}{2}v^2 f_0\,dv = \frac{3}{2}T_e$$

The integral containing the collision term for inelastic collisions with small energy losses is calculated in the same way. Using Eq. 5.43, we find

$$\int_0^\infty S^l_{0a}\frac{m_e v^2}{2}4\pi v^2\,dv = -\frac{3}{2}n\overline{\kappa^l_{ea}\nu^l_{ea}}T_e \tag{5.104}$$

where the bar denotes, as before, averaging over v with a weight of v^2. The integral due to inelastic collisions with a large energy loss is found by using Eq. 5.48:

$$\begin{aligned} \int_0^\infty S^h_{0a}\frac{m_e v^2}{2}4\pi v^2\,dv &= -4\pi n\int_0^\infty \frac{m_e v^2}{2}\nu^h_{ea}f_0 v^2\,dv \\ &= -\frac{3}{2}n\bar{\nu}^h_{ea}T_e \end{aligned} \tag{5.105}$$

Finally, it is easy to show that the integral containing S_{0e} characterizes the summary change in electron energy due to electron–electron collisions (see p. 161). Therefore it vanishes irrespective of the type of the distribution function. Summing up the expressions 5.102–5.105, which are associated with electron–atom collisions, we find

$$\int_0^\infty \frac{m_e v^2}{2} S_{0a} 4\pi v^2\, dv = -\frac{3}{2} n\kappa_{ea}\bar{\nu}^t_{ea}(T_e - T_a) - \frac{3}{2} n\overline{\kappa^l_{ea}\nu^l_{ea}}T_e - \frac{3}{2} n\bar{\nu}^h_{ea}T_e$$
$$= -\frac{3}{2} n\kappa^s_{ea}\nu^s_{ea}(T_e - T_a) \tag{5.106}$$

where we introduce the following notation:

$$\kappa^s_{ea}\nu^s_{ea} = \kappa_{ea}\overline{\nu^t_{ea}} + \overline{\kappa^l_{ea}\nu^l_{ea}} + \overline{\nu^h_{ea}}; \qquad \nu^s_{ea} = \overline{\nu^t_{ea}} + \overline{\nu^l_{ea}} + \overline{\nu^h_{ea}} \tag{5.107}$$

and take into account that inelastic collisions are usually substantial only at $T_e \gg T_a$.

A similar expression is obtained for the integral containing the collision term S_{0i}:

$$\int_0^\infty S_{0i} \frac{m_e v^2}{2} 4\pi v^2\, dv = -\frac{3}{2} n\kappa^s_{ei}\nu^s_{ei}(T_e - T_i) \tag{5.108}$$

The energy balance equation is obtained by equating Eq. 5.100 to the sum of Eqs. 5.106 and 5.108:

$$-neuE = \frac{3}{2} n\kappa^s_{ea}\nu^s_{ea}(T_e - T_a) + \frac{3}{2} n\kappa^s_{ei}\nu^s_{ei}(T_e - T_i) \tag{5.109}$$

This represents the equality of the energy input per unit volume of the electron gas, and the energy lost by the electrons per unit volume. The input energy in the electric field is given by the conventional formula derived from the Joule–Lentz law $P_E = jE$, where j is the electron current density equal to $-neu$. The losses of energy in Eq. 5.109 are due to its transfer from electrons to atoms and ions in collisions.

For the balance equation 5.109 to be complete, one must obtain the expression for the directed velocity u. The relationship of the directed component of the distribution function f_1 with f_0 is defined by the second equation of Eqs. 5.49. For the conditions $\nu_{0e} \approx \nu_{ei} \ll \nu_{ea}$ one can use the previous expression for the collision term S_1, (Eq. 5.28), and, accordingly, the previous equation 5.55 for f_1. The directed velocity is given by Eq. 5.56, which takes the following form for the Maxwellian distribution (Eq. 5.99):

$$u = \frac{4\pi}{3}\frac{eE}{m_e}\int_0^\infty \frac{v^2}{\nu^t_{ea}}\frac{df_0}{dv}\, dv$$
$$= -\frac{4\pi}{3}\left(\frac{m_e}{2\pi T_e}\right)^{3/2}\frac{eE}{T_e}\int_0^\infty \frac{v^4}{\nu^t_{ea}}\exp\left(-\frac{m_e v^2}{2T_e}\right) dv \tag{5.110}$$

At $\nu_{ea}^t = \text{const}$, integration yields $u = eE/m_e\nu_{ea}^t$ (see Eq. 5.57).

The directed velocity can generally be expressed through averaged collision frequencies (Eq. 5.103) in a similar way:

$$u = -\frac{\gamma eE}{m_e \bar{\nu}_{ea}^t} \tag{5.111}$$

where γ is a numerical coefficient of the order of unity. In particular, when the collision frequency is proportional to the velocity $\nu_{ea} = v/\lambda_{ea}$ we find, with the aid of Eqs. 5.103 and 5.110,

$$\bar{\nu}_{ea} = \tfrac{8}{3}\sqrt{2/\pi}\,\frac{1}{\lambda_{ea}}\sqrt{T_e/m_e}, \qquad \gamma = \frac{32}{9\pi} \tag{5.112}$$

Substituting Eq. 5.111 into the balance equation 5.109, we transform it for conditions where $\nu_{ee} \approx \nu_{ei} \ll \nu_{ea}$ and the term S_{0i} can be neglected:

$$T_e - T_a = \frac{2\gamma}{3}\frac{e^2E^2}{m_e\kappa_{ea}^s(\nu_{ea}^s)^2} \tag{5.113}$$

At $\nu^s \approx \nu^t = \text{const}$ and $\kappa^s = \kappa^e = \text{const}$, the right-hand side of the equality is independent of the electron temperature and can be used directly for finding T_e (here $\gamma = 1$). Generally, when ν and κ depend on the electron velocity and, accordingly, the averaged values ν^s and κ^s (Eq. 5.107) depend on the electron temperature, the equality 5.113 is an equation in T_e. For instance, for $\nu^s = v/\lambda$ with $\lambda = \text{const}$ and $\kappa^s = \kappa^e = \text{const}$ we find, using Eqs. 5.113 and 5.112:

$$T_e - T_a = \frac{1}{6}\frac{e^2E^2\lambda_{ea}^2}{\kappa_{ea}T_e}$$

and for $T_e \gg T_a$

$$T_e \approx \frac{1}{\sqrt{6}}\frac{eE\lambda_{ea}}{\sqrt{\kappa_{ea}}} \tag{5.114}$$

When the inequality $\nu_{ea} \gg \nu_{ee}, \nu_{ei}$ fails one should consider the effect of collisions of electrons with electrons and ions on the function f_1. It is rather difficult to account for electron–electron collisions, because the collision term S_1 for them cannot be represented as a differential operator, and the equation for f_1 has the integro-differential form. To solve it the function f_1 is expanded in Laguerre polynomials. The calculation result can be represented in a form similar to Eq. 5.111:

$$u = -\frac{\gamma eE}{m_e(\bar{\nu}_{ea} + \bar{\nu}_{ei})} \tag{5.115}$$

where the coefficient γ is determined by the form of the dependence

$\nu_{ea}(v)$ and the ratio between ν_{ea} and ν_{ei}. For a highly ionized plasma, in which $\nu_{ei} \gg \nu_{ea}$,

$$u \approx -\frac{\gamma eE}{m_e \bar{\nu}_{ei}}$$

where $\gamma = 1.96$, ν_{ei} can be found from Eq. 5.103 with the aid of Eq. 2.69:

$$\bar{\nu}_{ei} = \frac{4\sqrt{2\pi}}{3} \frac{ne^4}{m_e^{1/2} T_e^{3/2}} L_e \tag{5.116}$$

Substituting Eq. 5.115 into the balance equation 5.109, we obtain the equation for the electron temperature, which is analyzed in Section 7.10. For a fully ionized plasma the field of application of the results obtained is generally very limited. Firstly, beginning with comparatively low electric field strengths the energy transferred to the electrons from the field does not have enough time to be retransferred to the ions, and the stationary state of balance described by Eq. 5.109 cannot be realized. Secondly, fast electrons switch to the continuous-acceleration regime, with the result that the distribution function becomes strongly anisotropic (see Section 7.11).

We have demonstrated that the electron velocity distribution with a predominant effect of electron–electron collisions is Maxwellian. With a sufficiently high degree of ionization at which the frequency of electron–electron collisions exceeds that of inelastic collisions ($\nu_{ee} \gg \nu_{ea}^h$), this distribution can be used for establishing the efficiency of inelastic processes, ionization in particular. By using the approximation equation (Eq. 2.86) for the ionization cross section we obtain

$$\begin{aligned} \nu^i &= \int_{v_i}^{\infty} n_a s_0^i v \left(\frac{v^2}{v_i^2} - 1\right) 4\pi v^2 f_0 \, dv \\ &= 2\sqrt{\frac{2}{\pi}}\, n_a s_0^i \sqrt{\frac{T_e}{m_e}} \left(1 + 2\frac{T_e}{\mathscr{E}_i}\right) \exp\left(-\frac{\mathscr{E}_i}{T_e}\right) \end{aligned} \tag{5.117}$$

where $v_i^2 = 2\mathscr{E}_i/m_e$. For $\mathscr{E}_i \gg T_e$ we find

$$\nu^i \approx 2\sqrt{\frac{2T_e}{\pi m_e}}\, n_a s_0^i \exp\left(-\frac{\mathscr{E}_i}{T_e}\right) \tag{5.118}$$

5.7 EFFECT OF MAGNETIC FIELD ON ELECTRON DISTRIBUTION FUNCTION

In the presence of constant electric and magnetic fields the electron distribution function should be sought in the form

$$f(\mathbf{v}) = f_0(v) + \frac{v_x}{v} f_{1x}(v) + \frac{v_y}{v} f_{1y}(v) + \frac{v_z}{v} f_{1z}(v)$$
$$= f_0(v) + \frac{\mathbf{v}\mathbf{f}_1(v)}{v} \qquad (5.119)$$

Assuming, as before, that the distribution function is independent of the time and coordinates, we can write Eqs. 5.22 for the functions $f_0, f_{1x}, f_{1y}, f_{1z}$ as follows:

$$-\frac{ne}{3m_e v^2} \frac{d}{dv} (v^2 \mathbf{E}\mathbf{f}_1) = S_0; \qquad (5.120)$$

$$\frac{ne\mathbf{E}}{m_e} \frac{df_0}{dv} + \frac{ne}{m_e c} [\mathbf{H} \times \mathbf{f}_1] = -\mathbf{S}_1 \qquad (5.121)$$

The second equation is the vector form of the three equations for the relationship between the directed distribution function components f_{1x}, f_{1y}, f_{1z} and the isotropic component f_0. The collision term $\mathbf{S}_1$ has the following components:

$$S_{1k} = \frac{3}{4\pi} \int_{(\omega)} \frac{\delta(nf)}{\delta t} \cos \Theta_k \, d\omega \qquad (5.122)$$

where Θ_k is the angle between the velocity vector and the k axis, and the integration is done over the solid angles enclosing all the directions of the velocity vector. Substituting the integral of elastic electron–atom collisions into Eq. 5.122 and assuming the pre- and postcollision electron velocities to be equal, we get

$$S_{1k} = \frac{3}{4\pi} nn_a \int_{(\omega)} \cos \Theta_k \, d\omega \int_{(\Omega)} \mathbf{f}_1(\mathbf{v}' - \mathbf{v}) \sigma \, d\Omega \qquad (5.123)$$

The velocity vector change on elastic collisions is related to the scattering angle by Eq. 2.16. Substituting it into the second integral of Eq. 5.123, we find

$$n_a \int_{(\Omega)} \mathbf{f}_1(v)(\mathbf{v}' - \mathbf{v}) \sigma \, d\Omega = -n_a(\mathbf{f}_1\mathbf{v}) \int_{(\Omega)} (1 - \cos \vartheta) \sigma \, d\Omega$$
$$= -n_a s^t(\mathbf{f}_1\mathbf{v}) = -\frac{\nu^t \mathbf{f}_1 \mathbf{v}}{v}$$

and, further,

$$S_{1k} = -\frac{3}{4\pi} \frac{n\nu^t}{v} \int_{(\omega)} \sum_l f_{1l} v_l \cos \Theta_k \, d\omega = -\frac{3}{4\pi} \frac{n\nu^t}{v} \sum_l f_{1l} \int_{(\omega)} v_l \cos \Theta_k \, d\omega$$

Obviously, the integral over the velocity directions is different from zero only for $l = k$:

$$\int_{(\omega)} v_l \cos \Theta_k \, d\omega = \delta_{lk} v \int \cos^2 \Theta_k \, d\omega = \frac{4\pi v}{3} \delta_{lk}$$

Substituting, we have $S_{1k} = -n\nu_{ea}^t f_{1k}$, or in vector form,

$$\mathbf{S}_1 = -n\nu_{ea}^t \mathbf{f}_1 \tag{5.124}$$

Similar expressions are obtained for the collision term associated with the inelastic electron–atom collisions (see Section 5.3). Adding them to Eq. 5.124, we find, for $\nu_{ea} \gg \nu_{ee}, \nu_{ei}$, that is, when electron–electron and electron–ion collisions are insignificant:

$$\mathbf{S}_1 = -n(\nu_{ea}^t + \nu_{ea}^l + \nu_{ea}^h)\mathbf{f}_1 = -n\nu_{ea}^s \mathbf{f}_1 \tag{5.125}$$

Substituting Eq. 5.125 into Eq. 5.121, we get

$$\frac{e\mathbf{E}}{m_e}\frac{df_0}{dv} + \frac{e}{m_e c}[\mathbf{H} \times \mathbf{f}_1] - \nu_{ea}^s \mathbf{f}_1 = 0 \tag{5.126}$$

Let us direct the 0z axis parallel to **H**, and the 0x axis so that the vector **E** lies in the xz plane. Then the projections of Eq. 5.126 on the coordinate axes take the form*

$$\frac{eE_x}{m_e}\frac{df_0}{dv} - \frac{eH}{m_e c} f_{1y} - \nu f_{1x} = 0;$$

$$\frac{eH}{m_e c} f_{1x} - \nu f_{1y} = 0;$$

$$\frac{eE_z}{m_e}\frac{df_0}{dv} - \nu f_{1z} = 0$$

Solving these equations for f_{1x}, f_{1y}, f_{1z}, we obtain

$$f_{1x} = \frac{eE_\perp \nu}{m_e(\omega_H^2 + \nu^2)}\frac{df_0}{dv};$$

$$f_{1y} = \frac{eE_\perp \omega_H}{m_e(\omega_H^2 + \nu^2)}\frac{df_0}{dv}; \tag{5.127}$$

$$f_{1z} = \frac{eE_\parallel}{m_e \nu}\frac{df_0}{dv}$$

where the following notation is introduced: $E_\parallel = E_z$ is the component of **E** parallel to the magnetic field, $E_\perp = E_x$ is the component perpendicular

*Hereafter, when the meaning of the quantity causes no doubt we omit the suffixes at ν.

to the magnetic field, and $\omega_H = eH/m_e c$ is the cyclotron frequency [it is known to determine the angular velocity of electron gyration in the magnetic field (see Section 8.1)].

From Eq. 5.127 it is possible, as before, to obtain general equations for the directed electron velocity. Substituting Eq. 5.127 into Eq. 5.21, we find

$$\begin{aligned} u_{1x} &= \frac{4\pi}{3}\frac{eE_\perp}{m_e}\int_0^\infty \frac{v^3\nu}{(\omega_H^2+\nu^2)}\frac{df_0}{dv}\,dv; \\ u_{1y} &= \frac{4\pi}{3}\frac{eE_\perp}{m_e}\int_0^\infty \frac{v^3\omega_H}{(\omega_H^2+\nu^2)}\frac{df_0}{dv}\,dv; \\ u_{1z} &= \frac{4\pi}{3}\frac{eE_\parallel}{m_e}\int_0^\infty \frac{v^3}{\nu}\frac{df_0}{dv}\,dv \end{aligned} \tag{5.128}$$

For $\nu = \text{const}$, integration can be done by parts. Then Eq. 5.128 is independent of the form of f_0:

$$\begin{aligned} u_{1x} &= -\frac{eE_\perp}{m_e}\frac{\nu}{(\omega_H^2+\nu^2)}; \\ u_{1y} &= -\frac{eE_\perp}{m_e}\frac{\omega_H}{(\omega_H^2+\nu^2)}; \\ u_{1z} &= -\frac{eE_\parallel}{m_e\nu} \end{aligned} \tag{5.129}$$

As can be seen, the magnetic field does not affect the velocity component u_z parallel to it. This is only natural, since the Lorentz force is proportional to the vector product of the field by the velocity and does not have a component parallel to **H**. The effect of the magnetic field on the perpendicular velocity components is substantial at $\omega_H > \nu$. Then the velocity in the direction of the component **E** perpendicular to **H** (u_x) decreases, and a velocity component perpendicular to **E** and **H** appears. This effect can be explained by gyration of electrons in a plane perpendicular to **H**. At $\omega_H \ll \nu$ the gyration does not have a chance to manifest itself, since during the intercollisional time the electrons move practically along a straight line. At $\omega_H \gg \nu$ the electrons make many revolutions between collisions. As a result the periods of electron acceleration under the effect of $E_\perp$ (during which $\mathbf{vE}_\perp < 0$) alternate with slowing-down periods (when $\mathbf{vE}_\perp > 0$), and the summary efficiency of electron acceleration in the direction of $\mathbf{E}_\perp$ drops off. A detailed analysis of the effect of the magnetic field on the direction of the electron motion is given in Chapter 9.

Using the obtained expressions for the components f_1, we consider the effect of the magnetic field on f_0. In the equation for f_0, (Eq. 5.120), the magnetic field affects the first term, which defines the energy acquisition by electrons under the effect of the electric field. This term, as noted in Section 5.4, represents the divergence of a spherically symmetric flux in the velocity space

$$\mathrm{div}_v\, \Gamma_{vE} = \frac{1}{v^2}\frac{d}{dv}(v^2\Gamma_{vE}) \tag{5.130}$$

where

$$\Gamma_{vE} = -\frac{ne}{3m_e}\mathbf{E}\mathbf{f}_1 = -\frac{ne}{3m_e}(E_\perp f_{1x} + E_\| f_{1z})$$

Substituting Eq. 5.127 into Eq. 5.130, we get

$$\Gamma_{vE} = -\frac{ne^2}{3m_e^2}\left(\frac{E_\|^2}{\nu} + \frac{E_\perp^2 \nu}{\omega_H^2 + \nu^2}\right)\frac{df_0}{dv} \tag{5.131}$$

Since the flux density is proportional to the derivative df_0/dv, the flux is of a diffusion nature and is caused by accidental "straying" of electrons on the velocity scale. As shown in Section 5.4 the diffusion coefficient is determined by the mean-square change in electron velocity within the intercollisional time (Eq. 5.60):

$$D_v = \tfrac{1}{2}\langle(\Delta v)^2\rangle\tau^{-1} = \tfrac{1}{6}(\Delta v_E)^2\nu^{-1}$$

In the absence of a magnetic field the velocity change is associated only with acceleration in the electric field ($\Delta v_E = eE/m_e\nu$); therefore the flux value has the form of Eq. 5.59. At $\mathbf{E} \| \mathbf{H}$ the value of Δv_E is the same and it remains practically the same at $E \perp H$ when $\omega_H < \nu$. But at $\omega_H > \nu$ the electron gyration leads, as noted above, to alternating periods of acceleration and slowing down of the electrons. Each acceleration period Δt_E is obviously equal to the time of half-revolution of an electron $\Delta t_E = \pi/\omega_H$. Accordingly, the maximum velocity change during the intercollisional time is $\Delta v_E \approx (eE_\perp/m_e)$, $\Delta t_E \approx \pi e E_\perp/m_e\omega_H$. It is this change that determines the diffusion coefficient for $\omega_H > \nu$ with an accuracy to a factor of the order of unity:

$$D_v = \frac{(\Delta v_E)^2}{6}\nu \approx \frac{e^2E_\perp^2}{m_e\omega_H^2}\nu$$

The equation obtained corresponds to the second term in Eq. 5.131. The expression 5.131 for $\mathbf{\Gamma}_E$ can be written in the form 5.59 by introducing an effective electric field:

$$E_{\mathrm{eff}}^2 = E_\|^2 + E_\perp^2\frac{\nu^2}{\omega_H^2 + \nu^2} \tag{5.132}$$

Then Eq. 5.120 for f_0 will be formally the same as in the absence of a magnetic field; the dependence of f_0 on **H** appears in it only in terms of E_{eff}. Therefore, to determine the effect of the magnetic field on f_0 we can use the overall results of Sections 5.4–5.6, replacing **E** by $\mathbf{E}_{\text{eff}}$. When establishing the specific form of f_0 we must remember that E_{eff} in Eq. 5.132 generally depends on v.

For $\nu = \text{const}$ there is no such dependence, and we can use the equations of Sections 5.4–5.6 directly. Thus when only elastic collisions are substantial the distribution d_0 is Maxwellian. The electron temperature is then obtained from Eq. 5.66, in which E must be replaced by E_{eff} as in Eq. 5.132:

$$T_e = T_a + \frac{2}{3}\frac{e^2}{m_e\kappa\nu^2}\left[E_\parallel^2 + \frac{\nu^2}{\omega_H^2+\nu^2}E_\perp^2\right] \tag{5.133}$$

Interestingly, in an electric field perpendicular to the magnetic field, at $\omega_H \gg \nu$ the function f_0 is altogether independent of ν. Indeed, in accordance with Eq. 5.131 the rate of energy accumulation by electrons, which is determined by Γ_{vE}, is proportional to ν, as is the collision term due to elastic collisions. Therefore in Eq. 5.120 the collision frequency cancels out, and it takes the form

$$\frac{1}{3}\frac{e^2E_\perp^2}{m_e^2\omega_H^2}\frac{df_0}{dv} = \frac{1}{2}\kappa v f_0 + \frac{T_a}{m_a}\frac{df_0}{dv} \tag{5.134}$$

The solution of this equation for $\kappa_{ea} = \text{const}$ acquires the form of a Maxwellian distribution with a temperature

$$T_e = T_a + \tfrac{3}{2}\frac{e^2E_\perp^2}{m_e\kappa\omega_H^2} \tag{5.135}$$

As indicated in Section 5.6, with a predominant effect of electron–electron collisions on the function f_0 (i.e., at $\nu_{ee} \gg \kappa_{ea}\nu_{ea}$) it is near-Maxwellian. The electron temperature is then found from the energy balance equation 5.109. The magnetic field affects the left-hand side of the balance equation. It includes the directed electron velocity, whose components can, at $\nu_{ei} \ll \nu_{ea}$, be obtained from Eq. 5.128. Then the balance equation can be reduced to a form similar to Eq. 5.113:

$$T_e - T_a = \frac{2\gamma}{3}\frac{e^2}{m_e\kappa\nu^2}E_{\text{eff}}^2 \tag{5.136}$$

where $E_{\text{eff}}^2 = E_\parallel^2 + \xi[\nu^2/(\omega_H^2+\nu^2)]E_\perp^2$. Here ν and κ are summary averaged values yielded by Eq. 5.107 and generally depending on T_e; and γ and ξ are numerical factors of the order of unity, which are determined by the dependence of ν_{ea} and κ_{ea} on v, with ξ additionally determined by the ratio ω_H/ν. The equality 5.136 also represents the equation for T_e. At $\nu_{ea} = \text{const}$, $\kappa_{ea} = \text{const}$ the right-hand side of Eq. 5.136 is independent of T_e, the factors γ and ξ are equal to unity, and the equality 5.136 changes to Eq. 5.133.

For a highly ionized plasma at $\nu_{ei} \gtrsim \nu_{ea}$ the results of numerical calculations of the directed velocity can also be represented in a form similar to Eq. 5.129 with factors of the order of unity. Accordingly, at $T_i = T_a$ the balance equation can be written in the form 5.136 if we use the summary collision frequency $\bar{\nu}_{ea} + \bar{\nu}_{ei}$ and the energy transfer coefficient averaged over all these collisions. Naturally, the values of the factors γ and ξ then differ from the preceding.

5.8 ELECTRON DISTRIBUTION FUNCTION IN ALTERNATING ELECTRIC FIELD

Consider now the electron velocity distribution in an alternating electric field, assuming it homogeneous in space. To determine the functions f_0, f_1, we use Eqs. 5.16 and 5.18:

$$\frac{\partial(nf_0)}{\partial t} - \frac{neE}{3m_e v^2}\frac{\partial}{\partial v}(v^2 f_1) = S_0;$$

$$\frac{\partial(nf_1)}{\partial t} - \frac{neE}{m_e}\frac{\partial f_0}{\partial v} = S_1 \qquad (5.137)$$

in which the collision terms S_0 and S_1 are obtained from equations of Section 5.3. When elastic electron–atom collisions play the most important part, they are of the order of

$$|S_0| \sim \kappa\nu(nf_0), \qquad |S_1| \sim \nu(nf_1)$$

Equations 5.137 differ from those considered in Section 5.4–5.6 by their "inertial" terms $\partial(nf_0)/\partial t$ and $\partial(nf_1)/\partial t$. The role of these terms can be evaluated by comparing the field alternation period with the time of relaxation (establishment) of the functions f_0 and f_1. Since the relaxation of the electron distribution function is caused by collisions, the relaxation times are of the following order: for the function $f_0 1/\tau_0 \approx |S_0|/nf_0 = \kappa\nu$,

$$\tau_0 = \frac{1}{\kappa\nu}$$

and for the function $f_1 1/\tau_1 = |S_1|/nf_1 \approx \nu$,

$$\tau_1 \approx \frac{1}{\nu}$$

The large difference between the relaxation times of the isotropic and directed components of the distribution function ($\tau_0 \gg \tau_1$) is obviously due to the difference in the effect of collisions on the value and direction of the electron velocity.

By comparing the field alternation period T with the relaxation times τ_0 and τ_1 we can isolate the low-frequency case, for which T greatly exceeds τ_0 and, all the more so, τ_1; that is, the following inequalities hold:

$$T \gg \frac{1}{\kappa\nu}; \qquad \omega \ll \kappa\nu \tag{5.138}$$

where $\omega = 2\pi/T$ is the angular frequency of field alternation. The inertial terms in both equations 5.137 are small compared with the collision terms, and they can be neglected, which means that the functions f_0 and f_1 are the same as in a constant magnetic field and are determined at each instant by the instantaneous field value.

For high frequencies, when the inequalities opposite to Eqs. 5.138 are fulfilled,

$$T \ll 1/\kappa\nu; \qquad \omega \gg \kappa\nu \tag{5.139}$$

that is, the field alternation period is much less than the relaxation time of f_0; this function cannot catch up with the field alternation. Here f_0 is nearly constant in time and is only slightly modulated by alternations with a frequency equal to the doubled frequency of the field. It is easy to ascertain with the aid of Eqs. 5.137 that the depth of this modulation is of the order of $\kappa\nu/\omega$. When considering the high-frequency case (Eq. 5.139) we neglect this modulation and assume f_0 to be time invariant, which simplifies the solution of the set of equations 5.137.

Consider first the equation for f_1. Assume that the electric field alternates according to the harmonic law

$$E = E_0 \cos \omega t \tag{5.140}$$

The collision term S_1 is represented in the form 5.28: $S_1 = -\nu_{ea}^s n f_1$, assuming that $\nu_{ea} \gg \nu_{ee}, \nu_{ei}$. Then

$$\frac{\partial f_1}{\partial t} + \nu f_1 = \frac{eE_0}{m_e}\frac{\partial f_0}{\partial v}\cos \omega t \tag{5.141}$$

The stationary solution of Eq. 5.141 can be sought in the form

$$f_1(v, t) = f_{11}(v) \cos \omega t + f_{12}(v) \sin \omega t \tag{5.142}$$

where f_{11} and f_{12} are time invariant. Substituting Eq. 5.142 into 5.141, we get

$$(\omega f_{12} + \nu f_{11}) \cos \omega t - (\omega f_{11} - \nu f_{12}) \sin \omega t - \frac{eE_0}{m_e}\frac{\partial f_0}{\partial v}\cos \omega t = 0$$

Equating the coefficients at cos ωt and sin ωt separately to zero, we find the expressions for f_{11} and f_{12}:

$$f_{11} = \frac{eE_0\nu}{m_e(\omega^2+\nu^2)}\frac{\partial f_0}{\partial v}; \qquad f_{12} = \frac{eE_0\omega}{m_e(\omega^2+\nu^2)}\frac{\partial f_0}{\partial v} \tag{5.143}$$

These expressions define the directed electron velocity in a high-frequency electric field. Substituting Eqs. 5.142 and 5.143 into Eq. 5.13, we obtain

$$u = u_1 \cos \omega t + u_2 \sin \omega t;$$

$$u_1 = \frac{4\pi}{3}\frac{eE_0}{m_e}\int_0^\infty \frac{\nu v^3}{\omega^2+\nu^2}\frac{\partial f_0}{\partial v}\,dv; \tag{5.144}$$

$$u_2 = \frac{4\pi}{3}\frac{eE_0}{m_e}\int_0^\infty \frac{\omega v^3}{\omega^2+\nu^2}\frac{\partial f_0}{\partial v}\,dv$$

At $\nu = \text{const}$ the integrals appearing in Eqs. 5.144 become, after integration by parts, normalization integrals, and

$$u_1 = -\frac{eE_0}{m_e}\frac{\nu}{(\omega^2+\nu^2)}; \qquad u_2 = -\frac{eE_0}{m_e}\frac{\omega}{(\omega^2+\nu^2)} \tag{5.145}$$

These equations yield two components of the directed velocity. The first one u_1 characterizes the electron motion in phase with the field (it changes with time in the same way as the field), and the second u_2, the motion shifted in phase by $\pi/2$ (the time dependence is determined by sin ωt instead of by cos ωt). At $\omega \ll \nu$ Eqs. 5.144 and 5.145 for the component u_1 change to the previously obtained expressions for the directed velocity in a constant field (Eqs. 5.56 and 5.57) and the component u_2 is much less than u_1. The condition $\omega \ll \nu$ corresponds to a large number of electron collisions during a field period, that is, to a small change in field during the intercollisional time. Naturally, the relationship between the directed velocities and the field is then the same as in a constant field. At $\omega \gg \nu$ the directed velocity component u_1, which is in phase with the field, is sharply reduced, whereas the component u_2, which is shifted in phase by $\pi/2$, becomes much larger than u_1. This is because within the intercollisional time many field periods take place, during which the electron accelerates and slows down alternately. Such a "reactive" energy exchange precisely corresponds to the velocity-field phase shift, which is close to $\pi/2$.

Note that Eqs. 5.144 and 5.145 for u_1 and u_2 coincide with the equations for the velocity components u_x and u_y in a constant electric field ($\mathbf{E} \parallel 0x$) perpendicular to the magnetic field if we replace ω by ω_H (see Eqs. 5.128 and 5.129 at $E_{11} = 0$). This coincidence reflects the identical, periodic nature of the effect of the electric field on the electrons. (In the presence of a constant magnetic field it is associated with electron gyration, and in a variable electric field, with the alternation of the field itself.)

Let us now find f_0. To this end we substitute the expression for f_1, determined by Eqs. 5.142 and 5.143, into the equation for f_0 (Eq. 5.137). Then it takes the form

$$\frac{\partial(nf_0)}{\partial t} = -\frac{1}{6}\frac{ne^2E_0^2}{m_e^2v^2}\left\{\left[1+\cos(2\omega t)\right]\frac{\partial}{\partial v}\left(\frac{\nu v^2}{\omega^2+\nu^2}\frac{\partial f_0}{\partial v}\right) + \sin(2\omega t)\frac{\partial}{\partial v}\left(\frac{\omega v^2}{\omega^2+\nu^2}\frac{\partial f_0}{\partial v}\right)\right\} = S_0 \tag{5.146}$$

As has already been noted for $\omega \gg \kappa\nu$ the function f_0 is modulated only slightly and can be assumed constant to a first approximation. We obtain the equation in this approximation by averaging Eq. 5.146 over the alternation period:

$$-\frac{1}{6}\frac{ne^2E_0^2}{m_e^2v^2}\frac{d}{dv}\left[\frac{\nu v^2}{\omega^2+\nu^2}\frac{df_0}{dv}\right] = S_0 \tag{5.147}$$

In the next approximation we can find the variable addition to f_0 with the aid of Eq. 5.146; it can be seen that this addition varies with a frequency of 2ω.

The left-hand side of Eq. 5.147 defines the energy acquired by the electrons due to their motion in the electric field. As before, it represents the divergence of a spherically symmetric flux

$$\Gamma_{vE} = -\frac{1}{6}\frac{ne^2E_0^2\nu}{m_e^2(\omega^2+\nu^2)}\frac{df_0}{dv} \tag{5.148}$$

which can be described as the result of the diffusion, and in velocity space

$$\Gamma_{vE} = -D_E n\frac{df_0}{dv}, \qquad D_E = \frac{1}{6}\frac{ne^2E_0^2\nu}{m_e^2(\omega^2+\nu^2)} \tag{5.149}$$

At $\omega \ll \nu$, when the field changes only slightly during the intercollisional time, the nature of electron acceleration is the same as in a constant field. Therefore the equation for the diffusion coefficient D_E at $\omega \ll \nu$ is the same (see Eq. 5.59); the quantity E^2 appearing in it is averaged over the field period ($\bar{E}^2 = E_0^2/2$). At $\omega \gg \nu$ the periods of electron acceleration and slowing down alternate because of the repeated changes in the direction of the field during the intercollisional time. The maximum velocity increment in a field of frequency ω is equal to $(\Delta v_e)_{max} \approx$

$eE_0/m_e\omega$. Then the coefficient of diffusion in velocity space (see Eq. 5.60),

$$D_E \approx \tfrac{1}{6}\langle(\Delta v_E)^2\rangle\nu \approx \tfrac{1}{6}\frac{e^2E^2\nu}{m_e^2\omega^2}$$

corresponds to Eq. 5.149.

The relation 5.149 for the flux density Γ_{vE} can be represented in a form similar to Eq. 5.59 if we introduce the effective field

$$E_{\text{eff}}^2 = \tfrac{1}{2}\frac{E_0^2\nu}{\nu^2+\omega^2} \tag{5.150}$$

Therefore the results of determination of f_0 for a constant electric field can be extended to the case of a high-frequency field. The situation here is precisely the same as for coexisting constant electric and magnetic fields (see Section 5.7). Moreover, because of the identical nature of the acceleration the equations for f_0 are themselves similar in the two cases (c.f. Eqs. 5.150 and 5.132) for $E_\parallel = 0$). Therefore we do not revert to the application of Eq. 5.150 in different conditions, but give the expression for f_0 when only elastic electron–atom collisions are substantial. In accordance with the above we can use Eq. 5.63, replacing E^2 by E_{eff}^2. Then we obtain

$$f_0 = A\exp\left(-\int_0^v \frac{m_e v\,dv}{T_a + \frac{1}{3}e^2E_0^2/m_e\kappa(\omega^2+\nu^2)}\right) \tag{5.151}$$

If we neglect T_α (the stong-field case), Eq. 5.151 acquires the form

$$f_0 = A\exp\left[-\frac{3m_e^2\kappa_{ea}}{e^2E_0^2}\left(\omega^2v^2 + 2\int_0^v v\nu_{ea}^2\,dv\right)\right] \tag{5.152}$$

At $\nu = \text{const}$ Eq. 5.151 leads to a Maxwellian velocity distribution with a temperature

$$T_e = T_a + \frac{e^2E_0^2}{3m_e\kappa_{ea}(\omega^2+\nu_{ea}^2)} \tag{5.153}$$

At sufficiently high frequencies ($\omega \gg \nu$) the distribution will be Maxwellian for any dependence $\nu(v)$. Then the electron temperature is altogether independent of ν:

$$T_e = T_a + \frac{e^2E_0^2}{3m_e\kappa_{ea}\omega^2}$$

5.9 ION DISTRIBUTION FUNCTION IN ELECTRIC FIELD

As shown in Section 5.1 the ion distribution function in a weakly ionized plasma, at electric fields much less than critical (Eq. 5.5), $E \ll E_{pi} =$

$T/e\lambda_{ia}$, is near-equilibrium (near-Maxwellian). The deviation of the distribution from equilibrium can be found by successive approximations, which makes it possible to obtain the directed velocity, the momentum and energy fluxes, the correction to the average energy, and other characteristics associated with the electric field (see Chapters 6 and 7). As the electric field approaches the critical value the ion distribution function increasingly deviates from equilibrium. The nature of this deviation depends on the ion–atom collisions. Since their masses are similar, and energy transfer as well as momentum transfer in collisions may be considerable, the directed ion velocity in a strong electric field is commensurate with the random velocity or exceeds it; that is, the distribution function is strongly anisotropic. Therefore, the method for solving the kinetic equation by expanding it in degrees of anisotropy, which was described in the earlier sections, is inapplicable for ions. For this reason the problem of determining the distribution function for ions in an electric field is much more complicated than for electrons, and it can be solved only in cases permitting simplification.

The distribution function can be found, for instance, for the "relay" model, which describes approximately the ion motion in a weakly ionized plasma containing atoms and ions of the same species at electric field strengths greatly exceeding the critical value. Here the principal role is played by ion–atom collisions of the type of resonance charge exchange, which are accompanied by insignificant energy transfer. Each such collision results in disappearance of the ion accelerated by the electric field in the intercollisional period (it turns into a neutral atom) and in the appearance of a slow ion with an energy practically equal to that of the neutral atom from which it was produced. Taking into account only such collisions, one can use, for approximate consideration, a model in which each collision reduces the ion energy to zero. Within the framework of this model the ion velocity depends exclusively on the acceleration due to the electric field in the period between two collisions. Accordingly, the directions of the motion and of the electric field coincide and the velocity distribution is determined by that of the intercollisional time.

To find the distribution function we can make use of the kinetic equation, which takes the following form for the model at hand:

$$\frac{eE}{m_i}\frac{df_i}{dv}=\frac{\delta f_i}{\delta t} \tag{5.155}$$

where f_i depends on just one velocity component parallel to the electric field. In accordance with the model used the collision term must include only charge exchange collisions, which lead to the removal of ions from the given velocity range. Therefore it can be written thus:

$$\frac{\delta f_i}{\delta t} = -v n_a s_{ia}(v) f_i(v) = -\nu_{ia}(v) f_i(v) \tag{5.156}$$

where s_{ia} is the total charge exchange cross section, and $\nu_{ia} = n_a s_{ia} v$ is the frequency of charge exchange collisions. We assume that the relative velocity on collision is practically equal to the ion velocity (that the velocity of the neutral atoms is much less than that of the ions). Substituting Eq. 5.156 into Eq. 5.155, we obtain $(eE/m_i)\, df_i/dv = -\nu_{ia}(v) f_i$. Integrating this equation yields the ion velocity distribution function along the field

$$f_i(v) = A \exp\left(-\frac{m_i}{eE}\int_0^v \nu_{ia}(v)\, dv\right) \tag{5.157}$$

where A is the normalization factor, which is found from the condition

$$\int_0^\infty f_i(v)\, dv = 1 \tag{5.158}$$

Note that within the framework of the model under discussion all the ions move in the direction of the field; that is, $f_i = 0$ at $v_i < 0$.

The resonance charge exchange cross section at low ion energies depends only slightly on velocity (see Section 2.6). Neglecting this dependence, that is, assuming $s_{ia} = \text{const}$, $\nu_{ia} = n_a s_{ia} v \sim v$, we obtain from Eq. 5.157 a distribution that coincides in shape with the Maxwellian at $v > 0$. It can be written as

$$f_i(v) = \left(\frac{2m_i}{\pi T_i^*}\right)^{1/2} \exp\left(-\frac{m_i v^2}{2T_i^*}\right) \tag{5.159}$$

where T_i^* is the effective temperature:

$$T_i^* = \frac{eE}{n_a s_{ia}} = eE\lambda_{ia} \tag{5.160}$$

The normalization factor was selected using the condition 5.158 (it is twice as large as the normalization factor for the total Maxwellian distribution since $f_i = 0$ at $v < 0$). From Eq. 5.159 it is easy to find the average velocity of the ions in the direction of the field and their average energy:

$$\begin{aligned} u_E &= \int_0^\infty v f_i(v)\, dv = \sqrt{\frac{2T_i^*}{\pi m_i}} = \sqrt{\frac{2eE}{\pi m_i n_a s_{ia}}}; \\ \langle K \rangle &= \int_0^\infty \frac{m_i v^2}{2} f_i(v)\, dv = \frac{T_i^*}{2} = \frac{eE}{2n_a s_{ia}} \end{aligned} \tag{5.161}$$

Let us now consider a highly ionized plasma in which

$$\nu_{ii} \gg \nu_{ia}; \qquad \eta \gg \frac{s_{ia}}{s_{ii}} \tag{5.162}$$

The role of the various collisions can be evaluated with the aid of the kinetic equation as in Section 5.6 for electrons. For ions in a homogeneous plasma it has the form

$$\frac{neE}{m_i}\,\mathrm{grad}_v f_i = S_{ia} + S_{ie} + S_{ii} \tag{5.163}$$

The balance of the average momentum and average energy depends on the first three terms of the equation: the first term characterizes the ion acceleration in the electric field, and the second and third, the momentum and energy exchange with electrons and neutral atoms. Therefore the term proportional to the field is of the same order as the sum of the first two collision terms. Ion–ion collisions obviously do not affect the average momentum or the average energy of the ions. They only lead to momentum and energy redistribution among the ions. At the same time, in the case under consideration the ion–ion collision term S_{ii} greatly exceeds the other terms in Eq. 5.163. Because of the small energy and momentum exchange on ion–electron collisions it is usually much greater than the ion–electron collision term:

$$\frac{S_{ii}}{S_{ie}} \approx \frac{m_i}{m_e}\frac{\nu_{ii}}{\nu_{ie}} \approx \sqrt{m_i/m_e}\left(\frac{T_e}{T_i}\right)^{3/2} \gg 1$$

If the inequality 5.162 holds, it is also much greater than the ion–atom collision term $S_{ii}/S_{ia} \approx \nu_{ii}/\nu_{ia} \gg 1$. Therefore the form of the distribution function is determined, to a first approximation, by the vanishing of the ion–ion collision term. As shown in Section 4.2, this condition leads to the Maxwellian distribution:

$$f_i(\mathbf{v}) = \left(\frac{m_i}{2\pi T_i}\right)^{3/2} \exp\left[-\frac{m_i(\mathbf{v}-u_i)^2}{2T_i}\right] \tag{5.164}$$

The directed velocity u_i and the ion temperature T_i can be defined by the average-momentum and average-energy balance equations. They are obtained by multiplying the kinetic equation 5.163 by the momentum $m_i v$ or by the energy $m_i v^2/2$, respectively, and term-by-term integration over the entire velocity space. This procedure is carried out for a more general case in Sections 6.3 and 6.4. We give here the corresponding equalities only for conditions where collisions of ions with electrons play an insignificant role compared with their collisions with neutral atoms, that is, at $S_{ie}/S_{ia} \approx (m_e/m_i)\nu_{ei}/\nu_{ia} \ll 1$. Under such conditions the momentum balance amounts to the equality of the electric force and the average momentum transferred on ion–atom collisions: $e\mathbf{E} = \mu_{ia}\nu_{ia}\mathbf{u}_i$. This equality gives the directed ion velocity

$$\mathbf{u}_i = \frac{e\mathbf{E}}{\mu_{ia}\nu_{ia}} = \frac{2e\mathbf{E}}{m_i\nu_{ia}} \tag{5.165}$$

at $m_i = m_a$. The ion energy balance equation amounts to the equality of the average energy acquired by ions in the electric field and the average energy lost by them in collisions with neutral atoms: $eEu_i = \frac{3}{2}\kappa_{ia}\nu_{ia}(T_i - T_a)$, where $\kappa_{ia} = 2m_i m_a/(m_i + m_a)^2 = \frac{1}{2}$ at $m_i = m_a$. It results in the following relation for the ion temperature:

$$T_i - T_a = \frac{8}{3}\frac{e^2E^2}{m_i\nu_{ia}^2} \tag{5.166}$$

At higher degrees of ionization, when the effect of electron–ion collisions is substantial, one should include in the energy balance of the ions not only their energy exchange with neutral atoms, but also the energy exchange between electrons and ions.

6

DISTRIBUTION FUNCTION MOMENTS EQUATIONS

6.1 DISTRIBUTION FUNCTION MOMENTS

Under given external conditions the kinetic equations make it possible, in principle, to find particle distribution functions and thus obtain the macroscopic characteristics of the plasma. In view of the complexity of kinetic equations it is not always possible to find their complete solution. Many problems, however, are solved using approximate equations for the distribution function moments, which can be derived from kinetic equations. The distribution function moments are combinations of particle velocity components averaged over the distribution; they may be linear (first-order moments), quadratic (second-order moments), and so on. Let us establish the physical quantities determined by these moments. Three first-order moments represent the average values of the velocity components.*

$$u_k = \langle v_k \rangle = \int_{(\mathbf{v})} v_k f(\mathbf{v})\, d^3v \tag{6.1}$$

where u_k and v_k are the projections of the vectors $\mathbf{u}$ and $\mathbf{v}$ on one of the coordinate axes. The equalities 6.1 can also be written in vector form:

$$\mathbf{u} = \int_{(\mathbf{v})} \mathbf{v} f(\mathbf{v})\, d^3v$$

The total velocity of each particle can be represented as the sum

$$\mathbf{v} = \mathbf{u} + \mathbf{w} \tag{6.2}$$

*Hereafter, in equations that can be applied to particles of different types we omit, for brevity, the suffix denoting the type of particle.

Obviously, the averaging of the second term over the velocities yields zero:

$$\langle \mathbf{w} \rangle = \int_{(\mathbf{v})} \mathbf{w} f(\mathbf{v})\, d^3 v = 0 \tag{6.3}$$

Therefore the vector **w** is called the *random velocity*, and the average velocity **u**, the *directed velocity*.

The second-order moments may be composed of the average values of the products of the velocity components $\langle v_k v_l \rangle$. Since subscript permutation does not affect the product, there are six independent moments of this type. The average particle energy is usually chosen as one of them:

$$\langle K \rangle = \left\langle \frac{mv^2}{2} \right\rangle = \int_{(\mathbf{v})} \frac{mv^2}{2} f(\mathbf{v})\, d^3 v \tag{6.4}$$

Representing **v** as the sum 6.2 and making allowance for Eq. 6.3, we get

$$\langle v^2 \rangle = u^2 + \mathbf{u}\langle \mathbf{w} \rangle + \langle w^2 \rangle = u^2 + \langle w^2 \rangle$$

Accordingly, the average energy is equal to the sum of the components associated with the directed and random motion:

$$\begin{aligned} \langle K \rangle &= K_u + K_w; \\ K_u &= \frac{mu^2}{2}; \\ K_w &= \frac{3}{2} T = \left\langle \frac{mw^2}{2} \right\rangle \end{aligned} \tag{6.5}$$

The temperature T is introduced here as the measure of the average energy of the random motion in conformity with the generally accepted definition.

With arbitrary k and l the moments $\langle v_k v_l \rangle$ can be related to the momentum flux. Let us, for instance, compute the flux density of the kth components of the momentum transferred across a unit area F in the l direction (Fig. 6.1). The number of particles with a velocity in the range of $\mathbf{v}\ d\mathbf{v}$ that cross the area F per unit time is equal to $nfvl\, d^3v$. Since the kth component of the momentum transferred by each particle is mv_k the momentum flux transferred by all the isolated particles is $mv_k n f v_l\, d^3\mathbf{v}$, and the desired average flux is equal to

$$P_{kl} = nm \int v_k v_l f\, d^3 v = nm \langle v_k v_l \rangle \tag{6.6}$$

The set of P_{kl} values is the *momentum flux density tensor* $\check{\mathbf{P}}$. In the reference system where the average particle velocity $\mathbf{u} = 0$ it is called the

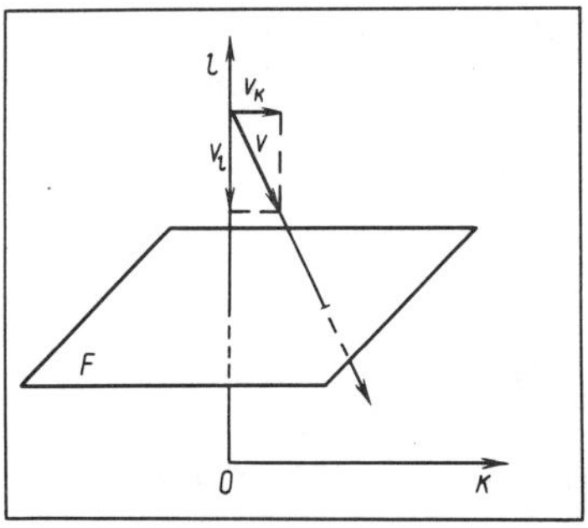

Fig. 6.1 Momentum transferred across a unit area.

pressure tensor $\check{\mathbf{p}}$. By the pressure is usually meant a force acting on an element of some surface and referred to its area. Since the change in momentum per unit time defines the force applied the two definitions are equivalent. It is easy to establish the relationship between P_{kl} and p_{kl}. By representing the total particle velocity in Eq. 6.6 as the sum of the directed and random velocities (see Eq. 6.2) we obtain

$$\begin{aligned} P_{kl} &= nm\langle(w_k + u_k)(w_l + u_l)\rangle \\ &= nmu_k u_l + nm\langle w_k w_l\rangle + nmu_k\langle w_l\rangle + nm\langle w_k\rangle u_l \end{aligned}$$

The last two terms are equal to zero on the strength of the equality 6.3. Therefore

$$P_{kl} = nmu_k u_l + p_{kl} \tag{6.7}$$

where

$$p_{kl} = nm\langle w_k w_l\rangle \tag{6.8}$$

The diagonal terms of the tensor $\check{\mathbf{p}}$ describe the pressure forces acting normally to areas perpendicular to the coordinate axes (for instance, p_{xx} is the normal pressure on an element of the yz plane, and so on). The nondiagonal components describe forces tangent to these areas. As seen from Eq. 6.8, p_{kl} are determined by the type of the particle velocity distribution function. If the distribution function over the random velocities is isotropic and the following relations hold:

$$\langle w_x^2\rangle = \langle w_y^2\rangle = \langle w_z^2\rangle = \frac{1}{3}\langle w^2\rangle$$

$$\langle w_x w_y\rangle = \langle w_x w_z\rangle = \langle w_y w_z\rangle = 0$$

the pressure tensor turns into a scalar $p_{kl} = p\delta_{kl}$, the scalar pressure being determined by the average energy of the random motion or by the temperature (see Eq. 6.5):

$$p = \tfrac{1}{3}nm\langle w^2\rangle = nT \tag{6.9}$$

The pressure tensor is generally divided into two terms: the normal scalar pressure (p), which is defined by Eq. 6.9, and the so-called *viscous stress tensor* $\check{\pi}$:

$$p_{kl} = p\delta_{kl} + \pi_{kl} \tag{6.10}$$

where

$$\pi_{kl} = nm\langle w_k w_l - \tfrac{1}{3}w^2\delta_{kl}\rangle \tag{6.11}$$

is the part of the tensor $\check{\mathbf{p}}$ due to the deviations of the distribution from spherical symmetry. As seen from Eq. 6.11, the diagonal elements of the tensor $\check{\pi}$ are related by the equation $\Sigma_k \pi_{kk} = 0$; that is, the tensor has five independent components.

Third-order moments consist of the averages of the products of the three velocity components $\langle v_k v_l v_m\rangle$. There are 10 independent moments of this type. The most important are the moments associated with the energy flux. Let us determine this value. The flux density of the energy transported in the kth direction by particles with a velocity in the range of $\mathbf{v}\,d\mathbf{v}$ is equal to $(mv^2/2)v_k nf\,d^3v$; the flux transported by all the particles is obtained by integrating this expression:

$$Q_k = n\int_{(\mathbf{v})} \frac{1}{2}mv^2 v_k f\,d^3v = \frac{1}{2}nm\langle v_k v^2\rangle \tag{6.12}$$

or, in vector form, $\mathbf{Q} = \frac{1}{2}nm\ \langle \mathbf{v}v^2\rangle$.

Representing the velocities as the sum of the random and directed components of Eq. 6.2 and taking into account that $\langle \mathbf{w}\rangle = 0$ (Eq. 6.3) we find for the kth component

$$Q_k = \frac{nm}{2}\langle (u_k + w_k)(\mathbf{u}+\mathbf{w})^2\rangle = \frac{nm}{2}\left(u_k u^2 + u_k\langle w^2\rangle + 2\sum_l u_l\langle w_l w_k\rangle + \langle w_k w^2\rangle\right)$$

Using Eqs. 6.5 and 6.8, this equality can be rewritten

$$Q_k = q_k + nK_u u_k + nK_w u_k + \sum_l p_{kl}u_l$$

or, in vector form,

$$\mathbf{Q} = \mathbf{q} + nK_u\mathbf{u} + nK_w\mathbf{u} + \check{\mathbf{p}}\mathbf{u} \tag{6.13}$$

The first term

$$\mathbf{q} = \frac{nm}{2}\langle \mathbf{w}w^2\rangle \tag{6.14}$$

characterizes the energy transfer in a system where the directed velocity is zero, that is, transfer of the energy associated exclusively with the random particle motion. Therefore the vector $\mathbf{q}$ is called the *heat flux density*. The other three terms in Eq. 6.13 describe the flux associated with the directed motion—the transfer of the energy of the directed motion itself (nK_u), the thermal energy (nK_w), and the energy determined by the work of the pressure forces.

Below we obtain approximate expressions for the moments of the charged particle distribution function. The application of such equations for describing the behavior of the electron and ion plasma components is somewhat similar to the approach used in hydrodynamics. Therefore these equations are sometimes called *equations of double-fluid hydrodynamics of the plasma.*

6.2 DERIVATION OF MOMENTS EQUATIONS

The equations of the distribution function moments can be obtained from the kinetic equation. To do this we must multiply the kinetic equation by the combination of the velocity projections corresponding to a certain time instant and integrate it term by term over the entire velocity space. Multiplying it by the projections of the velocity v_k, we obtain the first-moment equation; multiplying by the product $v_k v_l$, the second-moment equation; multiplying by $v_k v_l v_m$, the third-moment equation; and so on. The zero-moment equation is defined as the equation for the concentration obtained simply by term-by-term integration of the kinetic equation.

Let us find the general form of the moments equations. To this end we make use of the kinetic equation in the form 3.17:

$$\frac{\partial(nf)}{\partial t} - \mathbf{v}\,\mathrm{grad}(nf) + \frac{Ze}{m}\left\{\mathbf{E} + \frac{1}{c}[\mathbf{v}\times\mathbf{H}]\right\}\mathrm{grad}_v(nf) = \frac{\delta(nf)}{\delta t} \tag{6.15}$$

Denote by g some combination of velocity projections. The distribution function moment corresponding to this combination is $\langle g\rangle = \int gf(\mathbf{v})\,d^3v$. The equation for $\langle g\rangle$ is obtained by multiplying Eq. 6.15 by g and integrating over the velocities. The first term is transformed to

$$\int_{(\mathbf{v})} g\frac{\partial(nf)}{\partial t}\,d^3v = \frac{\partial}{\partial t}\left(n\int_{(v)} gf\,d^3v\right) = \frac{\partial}{\partial t}(n\langle g\rangle) \tag{6.16}$$

Here we change the order of differentiation and integration, taking advantage of the time t and the components v_k being independent variables in the kinetic equation. Since the coordinates in configuration space x_k and the velocity components v_k are also independent, we can similarly transform the second term:

$$\begin{aligned}\int_{(\mathbf{v})} g[\mathbf{v}\,\mathrm{grad}(nf)]\,d^3v &= \sum_k \int_{(\mathbf{v})} gv_k\frac{\partial(nf)}{\partial x_k}\,d^3v \\ &= \sum_k \frac{\partial}{\partial x_k}\left[n\int_{(\mathbf{v})} gv_k f\,d^3v\right] = \mathrm{div}(n\langle\mathbf{v}g\rangle)\end{aligned} \tag{6.17}$$

The term proportional to the electric field acquires the following form as a result of integration by parts:

$$\int_{(\mathbf{v})} g \frac{Ze}{m} \mathbf{E} \operatorname{grad}_v(nf)\, d^3v = \frac{Zen}{m} \int_{(\mathbf{v})} g \sum_k E_k \frac{\partial f}{\partial v_k}\, d^3v$$
$$= -\frac{Zen}{m} \sum_k E_k \int_{(\mathbf{v})} \frac{\partial g}{\partial v_k} f\, d^3v = -\frac{Ze}{m} n\mathbf{E} \langle \operatorname{grad}_v g \rangle \tag{6.18}$$

When integrating we took into account that the distribution function must vanish as $v_k \to \pm\infty$.

In a similar way, integrating by parts, we transform the term proportional to the magnetic field:

$$\frac{Ze}{mc} \oint_{(\mathbf{v})} \{[\mathbf{v} \times \mathbf{H}] \operatorname{grad}_v(nf)\} g\, d^3v = \frac{Zen}{mc} \sum_k \int_{(\mathbf{v})} [\mathbf{v} \times \mathbf{H}]_k \frac{\partial f}{\partial v_k} g\, d^3v$$
$$= -\frac{Zen}{mc} \sum_k \int_{(\mathbf{v})} \left\{ [\mathbf{v} \times \mathbf{H}]_k \frac{\partial g}{\partial v_k} + g \frac{\partial}{\partial v_k} [\mathbf{v} \times \mathbf{H}]_k \right\} f\, d^3v$$

The second term in the integrand is equal to zero since $[\mathbf{v} \times \mathbf{H}]_k$ depends only on the velocity components perpendicular to $\mathbf{v}_k$. Therefore we find

$$\frac{Ze}{mc} \int_{(\mathbf{v})} \{[\mathbf{v} \times \mathbf{H}] \operatorname{grad}_v(nf)\} g\, d^3v = -\frac{Zen}{mc} \sum_k \left\langle [\mathbf{v} \times \mathbf{H}]_k \frac{\partial g}{\partial v_k} \right\rangle$$
$$= -\frac{Zen}{mc} \langle [\mathbf{v} \times \mathbf{H}] \operatorname{grad}_v g \rangle \tag{6.19}$$

By taking into account Eqs. 6.16–6.19 the equation for $\langle g \rangle$ can be written thus:

$$\frac{\partial (n\langle g \rangle)}{\partial t} + \operatorname{div}(n\langle g\mathbf{v} \rangle) - \frac{Zen}{m} \mathbf{E} \langle \operatorname{grad}_v g \rangle$$
$$- \frac{Zen}{mc} \langle [\mathbf{v} \times \mathbf{H}] \operatorname{grad}_v g \rangle = \frac{\delta (n\langle g \rangle)}{\delta t} \tag{6.20}$$

The right-hand side of the equation is obtained by integrating the collision term

$$\int_{(\mathbf{v})} \frac{\delta (nf)}{\delta t} g\, d^3v = \frac{\delta (n\langle g \rangle)}{\delta t} \tag{6.21}$$

It can generally be represented as the sum

$$\frac{\delta (n_\alpha \langle g_\alpha \rangle)}{\delta t} = \sum_{\beta, j} \frac{\delta (n_\alpha \langle g_\alpha \rangle)}{\delta t} \bigg|_{\alpha\beta}^{(j)} \tag{6.22}$$

which extends to collisions of particles of this species with particles of all the other species, and to all the types of collision.

For elastic collisions, using the collision term in the form 3.23, we get

$$\left.\frac{\delta(n_\alpha\langle g_\alpha\rangle)}{\delta t}\right|_{\alpha\beta}^{(e)} = \int_{(\mathbf{v}_\alpha)} S_{\alpha\beta}^{(e)} g(\mathbf{v}_\alpha)\, d^3v_\alpha = n_\alpha n_\beta \int_{(\mathbf{v}_\alpha)}\int_{(\mathbf{v}_\beta)}\int_{(\Omega)} g(\mathbf{v}_\alpha)$$
$$\times [f_\alpha(v'_\alpha)f_\beta(v'_\beta) - f_\alpha(v_\alpha)f_\beta(v_\beta)]v_{\alpha\beta}\sigma_{\alpha\beta}\, d\Omega\, d^3v_\alpha\, d^3v_\beta \tag{6.23}$$

where $v_{\alpha\beta} = |v_\alpha - v_\beta|$, as before. The first of the integrals can be transformed by replacing the velocity components $\mathbf{v}'_\alpha$, $\mathbf{v}'_\beta$ in the integrand with $\mathbf{v}_\alpha$ and $\mathbf{v}_\beta$, and vice versa, since the collisions are reversible. Then, recalling that the value of the relative velocity remains unchanged on elastic collisions, $\mathbf{v}'_{\alpha\beta} = \mathbf{v}_{\alpha\beta}$, and $d^3v_\alpha\, d^3v_\beta = d^3v'_\alpha\, d^3v'_\beta$ (see p. 86), we obtain

$$\iiint g(\mathbf{v}_\alpha)f_\alpha(\mathbf{v}'_\alpha)f_\beta(\mathbf{v}'_\beta)v_{\alpha\beta}\sigma_{\alpha\beta}\, d\Omega\, d^3v_\alpha\, d^3v_\beta$$
$$= \iiint g(\mathbf{v}'_\alpha)f(\mathbf{v}_\alpha)f_\beta(\mathbf{v}_\beta)v_{\alpha\beta}\sigma_{\alpha\beta}\, d\Omega\, d^3v_\alpha\, d^3v_\beta$$

Substituting this equality into Eq. 6.23, we find

$$\left.\frac{\delta n_\alpha\langle g_\alpha\rangle}{\delta t}\right|_{\alpha\beta}^{(e)} = n_\alpha n_\beta \iiint (g'_\alpha - g_\alpha)f_\alpha f_\beta v_{\alpha\beta}\sigma_{\alpha\beta}\, d^3v_\alpha\, d^3v_\beta\, d\Omega \tag{6.24}$$

where we introduce the notation $g_\alpha = g(\mathbf{v}_\alpha)$, $g'_\alpha = g(\mathbf{v}'_\alpha)$.

The expression obtained determines the change in moment as a result of collisions summed over the collisions occurring per unit time and averaged over the velocities of the colliding particles.

The integral of Eq. 6.24 for collisions of identical particles vanishes if the sum of the values $g(\mathbf{v})$ for the colliding particles remains unchanged, that is, when it is the motion integral (a constant, momentum components, particles energy). Indeed, in this case the decrease of $g(\mathbf{v})$ in one of the colliding particles is offset by an equal increase of $g(\mathbf{v})$ in the other particle, and there is no change averaged over the particles of the type under review. We can ascertain this directly by simply transforming the integral 6.24 for collisions of identical particles, when f_α and f_β are identical. Since the velocity components $\mathbf{v}_1$ and $\mathbf{v}_2$ in the integrand can be permuted (because the integration is done with respect to them), we get

$$\iiint [g(\mathbf{v}'_1) - g(\mathbf{v}_1)]f(\mathbf{v}_1)f(\mathbf{v}_2)v\sigma\, d\Omega\, d^3v_1\, d^3v_2$$
$$= \iiint [g(\mathbf{v}'_2) - g(\mathbf{v}_2)]f(\mathbf{v}_1)f(\mathbf{v}_2)v\sigma\, d\Omega\, d^3v_1\, d^3v_2$$
$$= \frac{1}{2}\iiint [g(\mathbf{v}'_1) + g(\mathbf{v}'_2) - g(\mathbf{v}_1) - g(\mathbf{v}_2)]f(\mathbf{v}_1)f(\mathbf{v}_2)v\sigma\, d\Omega\, d^3v_1\, d^3v_2$$

where the subscript α is omitted for brevity. Since the sum $g(\mathbf{v}_1)+g(\mathbf{v}_2)$ remains unaffected on collisions, the expression in parentheses, and hence the integral of Eq. 6.24, vanishes.

Substituting different combinations of the velocity components $g = 1$, v_k, $v_k v_l$, $v_k v_l v_m$, and so on into the general equation 6.20, we obtain a set of equations for different moments. Each of these equations is related to the others. For instance, if g is a combination of nth-degree velocities, that is, if $\langle g\rangle$ is a moment of order n, the second term (Eq. 6.20) contains the quantities $\langle g v_k\rangle$, which are moments of order $(n+1)$, and the third term contains g/v_k, which are moments of order $(n-1)$. The expressions for the collision terms $\delta(n\langle g\rangle)/\delta t$ are relatively simple when the effective collision frequency is velocity independent. Here $\delta(n\langle g\rangle)/\delta t$ are related only to the moment $\langle g\rangle$ (see Sections 6.3 and 6.4). With an arbitrary dependence $\nu(v)$, a rigorous determination of the collision terms requires certain assumptions as to the form of the distribution function.

A regular method for obtaining moments equations with an arbitrary velocity dependence of the collision frequencies can be formulated for conditions where the anisotropy of the distribution function is small (see also Section 5.2). If the velocity distribution function in the absence of anisotropy $f_0(\mathbf{v})$ is assumed given, corrections to it can be represented as an expansion in the velocity components

$$f(\mathbf{v}) = f_0(v)\left(1 + \sum_k a_k v_k + \sum_{k,l} a_{kl} v_k v_l + \cdots\right) \tag{6.25}$$

It is easy to see that the expansion coefficients are associated with moments of different orders (a_k with first-order moments, a_{kl} with second-order moments, and so on). By using Eq. 6.25 in calculating the collision terms it is possible to obtain their relationship with different moments. Generally each collision term may include all the moments of the distribution function for the particles of the type discussed, as well as the moments of the distribution functions for other particles. Thus we obtain an infinite system of interrelated moments equations. It is fully equivalent to the kinetic equations for all the plasma components, and its solution is no less complicated than that of the kinetic equations. With a small anisotropy, however, we can truncate the equation chain, restricting ourselves to a moderate number of moments. The solution of the set of equations is thus substantially simplified.

This method for obtaining moments was developed in detail for the case where the particle velocity distribution functions can be assumed near-Maxwellian (*the Grad method*). Here it is natural to select the following local Maxwellian distribution as the initial function

$$f_0 = \left(\frac{m}{2\pi T}\right)^{3/2} \exp\left[-\frac{m(\mathbf{v}-\mathbf{u})^2}{2T}\right] \tag{6.26}$$

in which the temperature T and the directed velocity u are functions of the coordinates; they may be different for different types of particles. The corrections to f_0 for anisotropy can be conveniently found by using an expansion in three-dimensional Hermit-Chebyshev polynomials made up of the components of the dimensionless random velocity:

$$\zeta = (\mathbf{v}-\mathbf{u})\sqrt{m/T}; \qquad \zeta_k = (v_k - u_k)\sqrt{m/T} = w_k\sqrt{m/T} \tag{6.27}$$

These polynomials are defined by the relation

$$H^{(n)}_{k_1k_2\ldots k_n}(\xi) = (-1)^n \exp[\tfrac{1}{2}\zeta^2] \frac{\partial^n}{\partial\zeta_{k_1}\partial\zeta_{k_2}\cdots\partial\zeta_{k_n}} (\exp[(-\tfrac{1}{2})\zeta^2]) \tag{6.28}$$

where $k_1, k_2 \ldots, k_n$ can take values 1, 2, 3 corresponding to the three coordinate axes, in particular:

$$H^{(0)} = 1;\ H^{(1)}_k = \zeta_k;\ H^{(2)}_{kl} = \zeta_k\zeta_l - \delta_{kl};$$

$$H^{(3)}_{klm} = \zeta_k\zeta_l\zeta_m - \delta_{kl}\zeta_m - \delta_{km}\zeta_l - \delta_{lm}\zeta_k$$

The polynomials satisfy orthogonality conditions of the form

$$\frac{1}{(2\pi)^{3/2}} \int_{(\zeta)} \exp\left(-\frac{1}{2}\zeta^2\right) H^{(n)}_{k_1k_2\ldots k_n}(\zeta)\, H^{(p)}_{l_1l_2\ldots l_p}\, d^3\zeta = \delta_{np}\delta^{l_1l_2\ldots l_p}_{k_1k_2\ldots k_n} \tag{6.29}$$

where δ_{np} is the Kronecker delta function; and $\delta^{l_1l_2\ldots l_p}_{k_1k_2\ldots k_n}$ is different from zero and is equal to unity only when the upper sequence can be obtained from the lower one by simple permutation.

Using the polynomials (Eq. 6.28), we can represent the distribution function as

$$f = \left(\frac{m}{2\pi T}\right)^{3/2} \exp\left(-\frac{mw^2}{2}\right) \sum_{n=0}^{\infty} \sum_{k_1,k_2,\ldots k_n} C^{(n)}_{k_1k_2\ldots k_n} H^{(n)}_{k_1k_2\ldots k_n}\left(\sqrt{\frac{m}{T}}w\right) \tag{6.30}$$

Here the sum over the subscripts $k_1, k_2, \ldots, k_n$ only once includes combinations differing in subscript permutation. The normalization condition, with an allowance for Eq. 6.29, leads to the equality

$$\int_{(\mathbf{v})} f\, d^3v = C^{(0)} = 1 \tag{6.31}$$

The condition $\langle \mathbf{w} \rangle = 0$, which follows from the definition of the directed velocity (Eq. 6.3), reduces the components $C^{(1)}$ to zero:

$$\langle w_k \rangle = \sqrt{T/m}\, C^{(1)}_k = 0 \tag{6.32}$$

The other coefficients $C^{(n)}$ are related to nth-order moments. This relationship is readily established from Eq. 6.29. Using the definition of the viscous stress tensor 6.11 and the heat flux density 6.14, we obtain, in particular:

$$\pi_{ki} = nm\langle w_k w_l - \frac{1}{3} w^2 \delta_{kl}\rangle = nTC_{kl}^{(2)};$$

$$q_k = \frac{1}{2} nm\langle w_k w^2\rangle = nT\sqrt{T/m} \sum_l c_{kll}^{(3)}(1 + 2\delta_{kl}) \tag{6.33}$$

The possibility of solving the set of moments equations is based, as noted above, on the limitation of their number and, accordingly, of the number of terms in the expansion of the distribution function. Restricting the expansion to the first term, that is, adopting the isotropic distribution (Eq. 6.26) over the random velocities, we obtain the simplest approximation. In it, the distribution function is determined by five moments, including the concentration n (zero moment), the three components of the directed velocity u_k (first-order moments), and the temperature T (second-order moments). Taking into account the expansion terms defining the viscous stress tensor $\check{\pi}$ and the heat flux vector **q**, we can obtain the 13-moments approximation. The five indicated moments are supplemented by five independent components of the tensor and three components of the vector **q**. In this approximation, at $l = m$ we must assume $C^{(0)} = 1$, and $C_{kl}^{(2)}$ and $C_{klm}^{(3)}$ to be different from zero in Eq. 6.30. We also assume that the coefficients $C_{kll}^{(3)}$ are independent of l. With an allowance for the relations 6.33 the expansion 6.30 takes the form

$$f = \left(\frac{m}{2\pi T}\right\}^{3/2} \exp\left(-\frac{mw^2}{2T}\right)\left(1 + \sum_{k,l} C_{kl}^{(2)} H_{kl}^{(2)} + \sum_{k,l} C_{kll}^{(3)} h_{kll}^{(3)}\right)$$
$$= \left(\frac{m}{2\pi T}\right)^{3/2} \exp\left(-\frac{mw^2}{2T}\right)\left[1 + \sum_{k,l} \frac{\pi kl}{nT}\right.$$
$$\left. \times \left(\frac{mw_k w_l}{T} - \delta_{kl}\right) - \sum_k \frac{q_k m w_k}{nT^2}\left(1 - \frac{mw^2}{5T}\right)\right] \tag{6.34}$$

Hereafter we neglect the effects associated with viscosity, which are qualitatively discussed in Chapter 7. Then in place of Eq. 6.34 we can use the abridged expansion:

$$f = \left(\frac{m}{2\pi T}\right)^{3/2} \exp\left(-\frac{mw^2}{2T}\right)\left[1 - \sum_k q_k \frac{mw_k}{nT^2}\left(1 - \frac{mw^2}{5T}\right)\right] \tag{6.35}$$

6.3 EQUATIONS OF MOTION AND PARTICLE BALANCE OF PLASMA COMPONENTS

Let us now derive the equations of the zero and the first moments. To find the equation of the zero moment we must put $g = 1$ in Eq. 6.20. Then the averages of the quantities contained in Eq. 6.20 will be equal to

$$\langle g \rangle = 1; \qquad \langle g v_k \rangle = \langle v_k \rangle = u_k;$$

$$\langle \mathrm{grad}_v\, g \rangle = 0; \qquad \frac{\delta(n\langle g \rangle)}{\delta t} = \frac{\delta n}{\delta t} = \int_{(\mathbf{v})} \frac{\delta(nf)}{\delta t}\, d^3v$$

Substituting them, we obtain the zero moment equation:

$$\frac{\partial n}{\partial t} = -\,\mathrm{div}(n\mathbf{u}) + \frac{\delta n}{\delta t} \tag{6.36}$$

which defines the time variation of the concentration of particles of a given type. The first term on the right-hand side stands for the concentration change due to the particle motion, and the second, for the concentration change as a result of collisions leading to the appearance or disappearance of particles of a given type in the plasma volume. The appearance of charged particles is usually due to ionization on electron–atom collisions, and their disappearance, to recombination. The efficiency of these processes can be characterized by the frequencies

$$\nu^i = n_a v s^i(v); \qquad \nu^r = n_i s^r(v) v \tag{6.37}$$

depending on the relative velocity of the colliding particles, which is practically equal to the electron velocity. Accordingly, the collision term in Eq. 6.36 can be represented as

$$\frac{\delta n}{\delta t} = (\langle \nu^i \rangle - \langle \nu^r \rangle) n \tag{6.38}$$

where the quantities $\langle \nu^i \rangle$ and $\langle \nu^r \rangle$ are averaged over the electron velocities. In the case of the Maxwellian velocity distribution they naturally depend on the electron temperature.

We now pass over to the first-moment equations. To derive one of them, we put $g = v_k$ in Eq. 6.20. The corresponding moment is obviously equal to the kth component of the directed velocity $\langle g \rangle = \langle v_k \rangle = u_k$. The quantity $\langle g\mathbf{v} \rangle$, which appears in the second term of Eq. 6.20, can be expressed via the components of the momentum flux tensor 6.7 or via the pressure and the viscous stress tensor 6.11:

$$\langle g v_l \rangle = \langle v_k v_l \rangle = \frac{P_{kl}}{nm} = u_k u_l + \frac{p}{nm}\,\delta_{kl} + \frac{\pi_{kl}}{nm}$$

Using these relations we represent the second term of Eq. 6.20 thus:

$$\operatorname{div}(n\langle g\mathbf{v}\rangle) = \sum_l \frac{\partial}{\partial x_l}(n\langle gv_l\rangle)$$

$$= \sum_l \frac{\partial}{\partial x_i}(nu_k u_l) + \frac{1}{m}\sum_l \frac{\partial \pi_{kl}}{\partial x_l} + \frac{1}{m}\frac{\partial p}{\partial x_k}$$

At $g = v_k$, $\text{grad}_v\, g$, appearing in the third and fourth terms, is equal to a unit vector in the direction of the k axis. Therefore these terms take the form

$$\frac{Zen}{m}\mathbf{E}\langle \text{grad}_v\, g\rangle = \frac{Zen}{m}E_k;$$

$$\frac{Zen}{mc}\langle [\mathbf{v}\times\mathbf{H}]\,\text{grad}_v\, g\rangle = \frac{Zen}{mc}\langle [\mathbf{v}\times\mathbf{H}]_k\rangle = \frac{Zen}{mc}[\mathbf{u}\times\mathbf{H}]_k$$

Collecting the terms, we obtain

$$\frac{\partial(nu_k)}{\partial t} + \sum_l \frac{\partial}{\partial x_l}(nu_k u_l) + \frac{1}{m}\frac{\partial p}{\partial x_k} + \frac{1}{m}\sum_l \frac{\partial \pi_{kl}}{\partial x_l}$$
$$-\frac{Zen}{m}\left\{E_k + \frac{1}{c}[\mathbf{u}\times\mathbf{H}]_k\right\} = \frac{\delta(nu_k)}{\delta t} \qquad (6.39)$$

Substituting $\delta n/\partial t$ from Eq. 6.36 into the first term of Eq. 6.39,

$$\frac{\partial(nu_k)}{\partial t} = n\frac{\partial u_k}{\partial t} + u_k\frac{\partial n}{\partial t} = n\frac{\partial u_k}{\partial t} - u_k\sum_l \frac{\partial(nu_l)}{\partial x_l} + u_k\frac{\delta n}{\delta t}$$

we reduce the equation to

$$mn\frac{\partial u_k}{\partial t} + mn\sum_l u_l\frac{\partial u_k}{\partial x_l} = -\frac{\partial p}{\partial x_k} - \sum_l \frac{\partial \pi_{kl}}{\partial x_l}$$
$$+ nZeE_k + \frac{nZe}{c}[\mathbf{u}\times\mathbf{H}]_k + mn\frac{\delta u_k}{\delta t} \qquad (6.40)$$

where $n\delta u_k/\delta t = \delta(nu_k)/\delta t - u_k\delta n/\delta t$. Identical equations are obtained for the other two components of the vector $\mathbf{u}$. Together, they form the vector equation for the directed velocity

$$mn\left[\frac{\partial\mathbf{u}}{\partial t} + (\mathbf{u}\,\text{grad})\mathbf{u}\right] = Zen\mathbf{E} + \frac{Zen}{c}[\mathbf{u}\times\mathbf{H}] - \text{grad}\, p - \text{grad}\,\check{\pi} + nm\frac{\delta\mathbf{u}}{\delta t}$$
$$(6.41)$$

which is the equation describing the motion of a unit volume of gas consisting of a given type of particles. The quantity mn is the gas density. The bracketed expression in the left-hand side defines the "hydrodynamic acceleration." It is equal to the derivative of the directed velocity with respect to time $d\mathbf{u}/dt = \delta\mathbf{u}/\partial t + (\mathbf{u}\,\mathrm{grad})\mathbf{u}$, which is the average velocity change at a point and is composed of the change in average velocity at a fixed point $(\partial\mathbf{u}/\partial t)$ and the change due to the displacement of the gas:

$$\sum_l \frac{\partial\mathbf{u}}{\partial x_l}\frac{\partial \bar{x}_l}{\partial t} = \sum_l u_l \frac{\partial\mathbf{u}}{\partial x_l}$$

The right-hand side of Eq. 6.41 is equal to the sum of the forces acting on a given type of particles in a unit volume: the electric force $(Zen\mathbf{E})$, the Lorentz force $(Zen/c)[\mathbf{u}\times\mathbf{H}]$, the forces due to the gas pressure gradient $(-\mathrm{grad}\, p)$ and to viscous stresses $(-\mathrm{grad}\,\check{\pi})$, and the friction force related to particle collisions $(n\,\delta\mathbf{u}/\delta t)$. The forces associated with $\mathrm{grad}\,p$ and $\mathrm{grad}\,\pi$ are purely kinetic and are due to the momentum transfer in the thermal motion of the particles.

Consider now the collision term of the motion equation. For elastic collisions, we substitute $g_\alpha = v_{\alpha k}$ into Eq. 6.24 to obtain

$$\left(\frac{\delta n_\alpha u_{\alpha k}}{\delta t}\right)_{\alpha\beta} = n_\alpha\left(\frac{\delta u_{\alpha k}}{\delta t}\right)_{\alpha\beta}$$
$$= n_\alpha n_\beta \int_{(\Omega)}\int_{(\mathbf{v}_\alpha)}\int_{(\mathbf{v}_\beta)} (v'_{\alpha k} - v_{\alpha k}) f_\alpha(\mathbf{v}_\alpha) f_\beta(\mathbf{v}_\beta) v_{\alpha\beta}\sigma_{\alpha\beta}\, d\Omega\, d^3v_\alpha\, d^3v_\beta \tag{6.42}$$

Here we took into account that the number of particles remains unchanged on elastic collisions $(\delta n_\alpha/\delta t = 0)$. In accordance with Eq. 2.16, the change in velocity components on collisions, summed over the scattering angles, is equal to

$$\int_{(\Omega)} (v'_{k\alpha} - v_{k\alpha})\sigma_{\alpha\beta}\delta\Omega = -\frac{m_\beta}{m_\alpha + m_\beta} v_k \int \sigma_{\alpha\beta}(1-\cos\vartheta)\, d\Omega$$
$$= -\frac{m_\beta}{m_\alpha + m_\beta} v_k s^t_{\alpha\beta}$$

where $\omega_k = v_{k5} - v_{k\beta}$ is the kth component of the relative velocity. Substituting this relation into Eq. 6.42, we find

$$\left(\frac{\delta u_{\alpha k}}{\delta t}\right)_{\alpha\beta} = -\frac{m_\beta}{m_\alpha + m_\beta}\int_{(\mathbf{v}_\alpha)}\int_{(\mathbf{v}_\beta)} v_k \nu^t_{\alpha\beta}(v) f_\alpha f_\beta\, d^3v_a\, d^3v_\beta$$
$$= -\frac{m_\beta}{m_\alpha + m_\beta}\langle v_k \nu^t_{\alpha\beta}(v)\rangle \tag{6.43}$$

where the collision frequency $\nu^t_{\alpha\beta} = n_\beta v s^t(v)$ depends on the relative velocity $v = |v_\alpha - v_\beta|$, and the averaging is done over the velocities of both colliding particles.

Expression 6.43 yields the change in directed velocity as a result of elastic collisions. For the electrons, inelastic collisions are also of some importance. For processes in which the kinetic energy of the relative motion of the colliding particles greatly exceeds the inelastic losses of energy these losses can be neglected, and the expression for the collision term $\delta u_k/\delta t$ is the same as for elastic collisions. In the opposite case, when the inelastic losses are close to the energy of the relative motion, the change in velocity on each collision is due to the total loss of relative velocity. As a result (see Eq. 2.10),

$$\Delta v_{\alpha k} = \frac{m_\beta}{m_\alpha + m_\beta} \Delta v_k \approx -\frac{m_\beta}{m_\alpha + m_\beta} v_k$$

and the collision term retains the form 6.43, the only difference being that the frequency ν^t is replaced by the total frequency of inelastic collisions $\nu^h = n_\beta v \int \sigma^h \, d\Omega$. Thus the inclusion of inelastic collisions results in replacement of the transport frequency of elastic collisions in Eq. 6.43 by the summary collision frequency:

$$\nu^s(v) = \nu^t(v) + \nu^l(v) + \nu^h(v) \tag{6.44}$$

where ν^l is the summary transport frequency of inelastic collisions with a small energy loss and ν^h is the summary total frequency of inelastic collisions with a large energy loss. At an energy $K_e \gg \mathscr{E}_a, \mathscr{E}_i$ the corrections for inelastic collisions are small, since $\nu^l, \nu^h \ll \nu^t$. The corrections for ionization and recombination, which determine the change in n on collisions, are still smaller, of course. Their contribution to the collision term of the motion equation can be neglected. Substituting Eq. 6.44 into Eq. 6.43, we obtain

$$\frac{\delta(n_\alpha u_{\alpha k})}{\delta t} = n_\alpha \frac{\delta u_{\alpha k}}{\delta t} = -\frac{m_\beta}{m_\alpha + m_\beta} n_\alpha \langle \nu^s_{\alpha\beta}(v_{\alpha k} - v_{\beta k}) \rangle \tag{6.45}$$

The expressions 6.43 and 6.45 are especially simple when the collision frequency $\nu_{\alpha\beta}$ is velocity independent. Then it can be taken out of the averaging sign, and Eq. 6.45 becomes

$$\frac{\delta u_{\alpha k}}{\delta t} = -\frac{m_\beta}{m_\alpha + m_\beta} \nu_{\alpha\beta} \langle v_{\alpha k} - v_{\beta k} \rangle = -\frac{m_\beta}{m_\alpha + m_\beta} \nu_{\alpha\beta}(u_{\alpha k} - u_{\beta k}) \tag{6.46}$$

The collision term of Eq. 6.45 is generally defined by the dependence of $\nu_{\alpha\beta}$ on v and by the type of the distribution functions. We represent the relative velocity in it as the sum of the directed and random components:

$$\mathbf{v} = \mathbf{u} + \mathbf{w}; \qquad \mathbf{u} = \mathbf{u}_\alpha - \mathbf{u}_\beta; \qquad \mathbf{w} = \mathbf{w}_\alpha - \mathbf{w}_\beta \tag{6.47}$$

Assuming that $\mathbf{u} \ll \mathbf{w}$ and restricting ourselves to the expansion terms linear in **u**, we get

$$\frac{\delta u_{\alpha k}}{\delta t} = -\frac{m_\beta}{m_\alpha + m_\beta}\left(\langle w_k \nu_{\alpha\beta}(w)\rangle + \sum_l u_l \left\langle \frac{\partial(w_k \nu_{\alpha\beta})}{\partial w_l}\right\rangle\right) \tag{6.48}$$

If the distribution function over the random velocities can be considered isotropic, the first term and the terms of the sum appearing in the second sum vanish at $k \neq l$. In this case

$$\frac{\delta u_{\alpha k}}{\delta t} = -\frac{m_\beta}{m_\alpha + m_\beta} u_k \left\langle \frac{\partial(w_k \nu_{\alpha\beta})}{\partial w_k}\right\rangle = -\frac{m_\alpha}{m_\alpha + m_\beta}(u_{\alpha k} - u_{\beta k})\bar{\nu}_{\alpha\beta} \tag{6.49}$$

Equation 6.49 differs from Eq. 6.46 only in that it includes some averaged collision frequency. The averaging law is obtained from the relation

$$\bar{\nu}_{\alpha\beta} = \left\langle \frac{\partial(w_k \nu_{\alpha\beta})}{\partial w_k}\right\rangle = \int_{(w_\alpha)}\int_{(w_\beta)} \frac{\partial(w_k \nu_{\alpha\beta})}{\partial w_k} f_\alpha f_\beta \, d^3 w_\alpha \, d^3 w_\beta \tag{6.50}$$

Let us derive this law for the Maxwellian distribution of α- and β-type particles over the random velocities. Here

$$f_\alpha f_\beta = \left(\frac{m_\alpha}{2\pi T_\alpha}\right)^{3/2}\left(\frac{m_\beta}{2\pi T_\beta}\right)^{3/2} \exp\left[-\left(\frac{m_\alpha w_\alpha^2}{2T_\alpha} + \frac{m_\beta w_\beta^2}{2T_\beta}\right)\right] \tag{6.51}$$

To isolate the distribution over the relative velocities, we must switch, in Eqs 6.50 and 6.51, from the particle velocities $\mathbf{w}_\alpha$ and $\mathbf{w}_\beta$ to the relative velocity $\mathbf{w}$ and the velocity $\mathbf{w}_s$:

$$\mathbf{w} = \mathbf{w}_\alpha - \mathbf{w}_\beta; \qquad \mathbf{w}_s = \frac{(m_\alpha/T_\alpha)\mathbf{w}_\alpha + (m_\beta/T_\beta)\mathbf{w}_\beta}{m_\alpha/T_\alpha + m_\beta/T_\beta} \tag{6.52}$$

The velocities $\mathbf{w}_\alpha$ and $\mathbf{w}_\beta$ are related to $\mathbf{w}$ and $\mathbf{w}_s$ by the equations

$$\mathbf{w}_\alpha = \mathbf{w}_s + \frac{m_\beta T_\alpha \mathbf{w}}{m_\alpha T_\beta + m_\beta T_\alpha}; \qquad \mathbf{w}_\beta = \mathbf{w}_s - \frac{m_\alpha T_\beta \mathbf{w}}{m_\alpha T_\beta + m_\beta T_\alpha} \tag{6.53}$$

The coefficients in the equation for $\mathbf{w}_s$ are so selected that $f_\alpha f_\beta$ turns into the product of the distributions over the relative velocities and over the velocities $\mathbf{w}_s$. Indeed, substituting Eq 6.53 into Eq. 6.51, we obtain

$$\begin{aligned} &f_\alpha(\mathbf{w}_\alpha) f_\beta(\mathbf{w}_\beta) = f_r(\mathbf{w}) f_s(\mathbf{w}_s); \\ &f_r = \left(\frac{\mu_{\alpha\beta}}{2\pi T_{\alpha\beta}}\right)^{3/2} \exp\left(-\frac{\mu_{\alpha\beta} w^2}{2T_{\alpha\beta}}\right); \\ &f_s = \left(\frac{m_s}{2\pi T_s}\right)^{3/2} \exp\left(-\frac{m_s w_s^2}{2T_s}\right) \end{aligned} \tag{6.54}$$

where $\mu_{\alpha\beta} = m_\alpha m_\beta/(m_\alpha + m_\beta)$ is the reduced mass, $m_s = m_\alpha + m_\beta$, and

$$T_{\alpha\beta} = \frac{m_\alpha T_\beta + m_\beta T_\alpha}{m_\alpha + m_\beta}; \qquad T_s = \frac{T_\alpha T_\beta}{T_{\alpha\beta}} \tag{6.55}$$

When considering electron–heavy particle collisions, usually $m_a T_e \gg m_e T_a$ and $T_{\alpha\beta} \approx T_e$, $T_s \approx T_a$; for collisions of particles with a similar mass one can assume $T_\alpha \approx T_\beta$, $T_{\alpha\beta} \approx T_s \approx T_\alpha$. The first of these quantities characterizes the distribution over the relative velocities and the second, over the velocities $\mathbf{w}_s$. When substituting Eq. 6.54 into Eq. 6.50 we take into account that the Jacobian of the changeover from the velocities $\mathbf{w}_\alpha$ and $\mathbf{w}_\beta$ to $\mathbf{w}$ and $\mathbf{w}_s$ is equal to units, that is, $d^3w_\alpha\, d^3w_\beta = d^3w\, d^3w_s$. Then integrating with respect to $\mathbf{w}_s$ we get

$$\begin{aligned}\bar{\nu}_{\alpha\beta} &= \int_{(\mathbf{w})}\int_{(\mathbf{w})_s} \frac{\partial(w_k \nu_{\alpha\beta})}{\partial w_k} f_r(\mathbf{w}) f_s(\mathbf{w}_s)\, d^3w\, d^3w_s \\ &= \int_{(\mathbf{w})} \frac{\partial(w_k \nu_{\alpha\beta})}{\partial w_k} f_r(\mathbf{w})\, d^3w\end{aligned}$$

Integration with respect to the relative velocity component w_k is best performed by parts. As a result we find that

$$\begin{aligned}\bar{\nu}_{\alpha\beta} &= \left(\frac{\mu_{\alpha\beta}}{2\pi T_{\alpha\beta}}\right)^{3/2} \int_{(\mathbf{w})} \frac{\mu_{\alpha\beta} w_k^2}{T_{\alpha\beta}} \nu_{\alpha\beta}(w) \exp\left(-\frac{\mu_{\alpha\beta} w^2}{2T_{\alpha\beta}}\right) d^3w \\ &= \frac{1}{3}\sqrt{\frac{2}{\pi}} \left(\frac{\mu_{\alpha\beta}}{T_{\alpha\beta}}\right)^{5/2} \int_0^\infty w^4 \nu_{\alpha\beta} \exp\left(-\frac{\mu_{\alpha\beta} w^2}{2T_{\alpha\beta}}\right) dw\end{aligned} \tag{6.56}$$

or $\bar{\nu}_{\alpha\beta} = \langle \nu_{\alpha\beta} w^2 \rangle / \langle w^2 \rangle$, where the averaging is achieved with the aid of the relative velocity distribution (Eq. 6.54).

The result obtained, which is based on the assumption of isotropy of the random velocity distribution, corresponds to the five-moment approximation. If we take into account the corrections associated with moments of a higher order in the distribution functions, then in the expression for the collision term (Eq. 6.48) we have to make allowance for the other terms as well. Thus, assuming that the distribution function includes moments proportional to the heat flux components and hence is described by Eq. 6.35, we must take into account the first term in Eq. 6.48. For instance, for $w_\alpha^2 \gg w_\beta^2$ we obtain, by substituting Eq. 6.35 into this term,

$$\begin{aligned}\left(\frac{\delta u_{\alpha k}}{\delta t}\right)_q &= -\frac{m_\beta}{m_\alpha + m_\beta} \int \nu_{\alpha\beta} w_k f_{0\alpha}(w_\alpha)\left[1 - \sum_l \frac{q_{\alpha l}}{n_\alpha T_\alpha} \frac{n w_{\alpha l}}{T_\alpha}\left(1 - \frac{m_\alpha w_\alpha^2}{5T_\alpha}\right)\right] d^3w_\alpha \\ &= \frac{m_\beta}{m_\alpha + m_\beta} \frac{q_{\alpha k}}{3n_\alpha T_\alpha} \left\langle \nu_{\alpha\beta}\left(\frac{m_\alpha w_\alpha^2}{T_\alpha} - \frac{m_\alpha^2 w_\alpha^4}{5T_\alpha^2}\right)\right\rangle\end{aligned} \tag{6.57}$$

Here the averaging is done over the Maxwellian distribution of α-type particles. The components of the vector $q_{\alpha k}$ of the heat flux have terms proportional to the difference of the velocities $(u_{\alpha k} - u_{\beta k})$, as well as a term proportional to $\partial T_\alpha/\partial x_k$. Therefore the substitution of expressions of the type 6.57 into the components of the summary collision term $\delta u_{\alpha k}/\delta t = \Sigma_\beta\,(\delta u_{\alpha k}/\delta t)_{\alpha\beta}$ leads to two changes. Firstly, additional terms appear, which have the same structure as Eq. 6.49; they are proportional to the difference of the directed velocities $(u_{\alpha k} - u_{\beta k})$ and to some averaged collision frequency. The inclusion of these terms can be reduced to redetermination of the collision frequency—the effective collision frequency becomes different from Eq. 6.56. Secondly, terms appear which are proportional to the temperature gradients. As a result the summary collision term appearing in Eq. 6.41 can be written thus:

$$\frac{m_\alpha n_\alpha \delta u_\alpha}{\delta t} = n_\alpha \sum_\beta (\mathbf{R}_{\alpha\beta} + \mathbf{R}^T_{\alpha\beta}) \tag{6.58}$$

where

$$\mathbf{R}_{\alpha\beta} = -\,\mu_{\alpha\beta}\nu_{\alpha\beta}(\mathbf{u}_\alpha - \mathbf{u}_\beta); \qquad \mathbf{R}^T \sim \text{grad } T \tag{6.59}$$

The vector $\mathbf{R}_{\alpha\beta}$, which is proportional to the difference of the directed velocities $\mathbf{u}_\alpha - \mathbf{u}_\beta$ and to the effective collision frequency $\nu_{\alpha\beta}$, determines the friction force. The expression for the friction force is easily obtained from qualitative considerations as well, by considering the change in momentum on collisions. The maximum change in the velocity of the colliding particles occurs on head-on collision. In this case the relative velocity vector reverses its sign, and $\Delta\mathbf{v}_{\max} = \mathbf{v}' - \mathbf{v} = -2\mathbf{v}$. We can assume approximately that the collision-averaged change in relative velocity is equal to $\frac{1}{2}\Delta\mathbf{v}_{\max}$. Accordingly, the collision-averaged change in the momentum of one of the colliding particles is

$$\Delta(m_\alpha \mathbf{v}_\alpha) = -\,\mu_{\alpha\beta}\mathbf{v} = -\,\mu_{\alpha\beta}(\mathbf{v}_\alpha - \mathbf{v}_\beta)$$

Multiplying this by the collision frequency $\nu_{\alpha\beta}$ and averaging over the velocities, we obtain the average change in the momentum of the α-type particles because of their collisions with β-type particles per unit time. This change actually determines the friction force $R_{\alpha\beta} = -\,\mu_{\alpha\beta}\nu_{\alpha\beta}(\mathbf{u}_\alpha - \mathbf{u}_\beta)$.

If the collision frequency depends on the velocity and there is a temperature gradient, an additional term $\mathbf{R}^T$, for the temperature gradients, appears in the collision term expression. This term is called the thermal force. The presence of this term is essentially due to the collision frequency gradient. In the absence of a magnetic field its effect is that the number of collisions of the particles moving along the gradient (from the region with a lower collision frequency) is less than for the particles moving in the opposite direction. This gives rise to an average momentum change, which is directed along the gradient. Let us estimate this change. The different numbers of collisions of particles moving along and against the gradient are determined approximately by the change in collision frequency over the mean free path $\delta\nu \approx \lambda \operatorname{grad} \nu \approx (v/\nu) \operatorname{grad} \nu$. (It is assumed that each collision leads to randomization of the particle velocity direction). Accordingly, the collision-averaged change in momentum per unit time can be estimated as the product of the momentum change per one collision $\Delta(m_\alpha \mathbf{v}_\alpha) \approx \mu_{\alpha\beta}\mathbf{v}$ by $\delta\nu$. By averaging this product over the velocities we obtain the thermal force

$$\mathbf{R}^T_{\alpha\beta} = \mu_{\alpha\beta}\langle \mathbf{v}\delta\nu_{\alpha\beta}\rangle \approx \frac{T_{\alpha\beta}}{\nu_{\alpha\beta}} \operatorname{grad} \nu_{\alpha\beta} \approx \frac{T_{\alpha\beta}}{\nu_{\alpha\beta}} \frac{d\nu_{\alpha\beta}}{dT_{\alpha\beta}} \operatorname{grad} T_{\alpha\beta} \tag{6.60}$$

for $(T/\nu)\, d\omega/dT \approx 1$ $\mathbf{R}^T \sim \operatorname{grad} T$ [here $T_{\alpha\beta}$ is the temperature characterizing the relative-velocity distribution (Eq. 6.55)]. In a strong magnetic field, when the temperature gradient is perpendicular to the field, the thermal force is substantially smaller, since the transverse drift of charged particles between the collisions is much less than the mean free path. The estimate of the thermal force in this case is given in Chapter 9.

Substituting the expression for the collision term 6.58 into Eq. 6.41, we obtain the averaged equation of motion for α-type particles:

$$m_\alpha\left[\frac{\partial \mathbf{u}_\alpha}{\partial t} + (\mathbf{u}_\alpha \operatorname{grad})\mathbf{u}_\alpha\right] = Z_\alpha e\mathbf{E} + \frac{Z_\alpha e}{c}[\mathbf{u}_\alpha \times \mathbf{h}] - \frac{\operatorname{grad} p_\alpha}{n_\alpha}$$
$$- \frac{\operatorname{grad} \check{\pi}_\alpha}{n_\alpha} - \sum_\beta \mu_{\alpha\beta}\nu_{\alpha\beta}(\mathbf{u}_\alpha - \mathbf{u}_\beta) + \sum_\beta \mathbf{R}^T_{\alpha\beta} \tag{6.61}$$

Hereafter we neglect the viscosity effects.* In most cases one can also neglect the second term (quadratic with respect to **u**) on the left-hand side of the equation. Indeed, it is of the order of

$$|m_\alpha(\mathbf{u}_\alpha \operatorname{grad})\mathbf{u}_\alpha| \approx \frac{m_\alpha u_\alpha^2}{L}$$

and when the anisotropy of the velocity distribution is small, it is much less than the term proportional to the pressure gradient:

$$\operatorname{grad}\frac{p_\alpha}{n} \approx \frac{p_\alpha}{n_\alpha L} \approx \frac{T_\alpha}{L}$$

*The criterion for neglecting the viscosity as compared with the other terms in the transport equations in the absence of a magnetic field is explained in Section 7.3.

(L is the characteristic length over which the plasma parameters change substantially.) Here the equation of motion (Eq. 6.61) takes the form

$$m_\alpha \frac{\partial \mathbf{u}_\alpha}{\partial t} = Z_\alpha e\mathbf{E} + \frac{Z_\alpha e}{c}[\mathbf{u}_\alpha \times \mathbf{H}] - \frac{\text{grad}\, p_\alpha}{n_\alpha} - \sum_\beta \mu_{\alpha\beta}\nu_{\alpha\beta}(\mathbf{u}_\alpha - \mathbf{u}_\beta) + \sum_\beta \mathbf{R}^T_{\alpha\beta} \tag{6.62}$$

Note that it is linear in $\mathbf{u}$.

6.4 ENERGY BALANCE AND HEAT FLUX EQUATIONS

We now derive one of the equations of second-order moments, namely, the energy balance equation. To do this we assume $g = mv^2/2$ in Eq. 6.20. The average values of the quantities appearing in the equation are

$$\langle g\rangle = \left\langle\frac{mv^2}{2}\right\rangle = \langle K\rangle;$$

$$\langle g\mathbf{v}\rangle = \left\langle\frac{\mathbf{v}mv^2}{2}\right\rangle = \frac{\mathbf{Q}}{n};$$

$$\langle \text{grad}_v\, g\rangle = \langle m\mathbf{v}\rangle = m\mathbf{u};$$

$$\langle([\mathbf{v}\times\mathbf{H}]\,\text{grad}_v\, g)\rangle = \langle([\mathbf{v}\times\mathbf{H}]m\mathbf{v})\rangle = 0$$

where $\mathbf{Q}$ is the energy flux vector (Eq. 6.12). Substituting them into Eq. 6.20, we obtain

$$\frac{\partial(n\langle K\rangle)}{\partial t} = -\,\text{div}\,\mathbf{Q} = Zen\mathbf{uE} + \frac{\delta(n\langle K\rangle)}{\delta t} \tag{6.63}$$

Equation 6.63 shows that the change in the kinetic energy of particles of each type occurs for three reasons. The first reason is the energy transfer in particle motion. It is described by the flux divergence (div $\mathbf{Q}$). The second reason is the heating of the particles by the current in the plasma. The specific power of this heating is the product of the current density ($\mathbf{j} = Zen\mathbf{u}$) by the field strength ($\mathbf{E}$). The third reason is the energy change on collisions $\delta(n\langle K\rangle)/\delta t$.

We now transform the energy balance equations, representing the kinetic energy as the sum of the energies of the random and directed motion (Eq. 6.5): $K = mu^2/2 + \frac{3}{2}T$. Using also Eq. 6.13 for the energy flux density, we find

$$\frac{3}{2}n\frac{\partial T}{\partial t} + mn\mathbf{u}\frac{\partial \mathbf{u}}{\partial t} + \left(\frac{3}{2}T + \frac{mu^2}{2}\right)\frac{\partial n}{\partial t} + \text{div}\,\mathbf{q} + \text{div}\left(\frac{5}{2}\mathbf{u}nT\right)$$
$$+\,\text{div}(\check{\pi}\mathbf{u}) + \text{div}\left(n\mathbf{u}\frac{mu^2}{2}\right) + Zen\mathbf{uE} = \frac{\delta(n\langle K\rangle)}{\delta t} \tag{6.64}$$

With the aid of the expressions for $\partial n/\partial t$ and $\partial \mathbf{u}/\partial t$ from Eqs. 6.36 and 6.41 this equation is transformed to

$$n\frac{\partial T}{\partial t}+\frac{2}{3}\operatorname{div}\mathbf{q}+n\mathbf{u}\operatorname{grad}T+\frac{2}{3}nT\operatorname{div}\mathbf{u}+\frac{2}{3}\check{\pi}\operatorname{grad}\mathbf{u}=n\frac{\delta T}{\delta t} \tag{6.65}$$

where $n\delta T/\delta t=\frac{2}{3}\delta(n\langle K\rangle)/\delta t-T\delta n/\delta t-mn(\delta\mathbf{u}/\delta t)\mathbf{u}$.

Let us find the collision term for elastic collisions. Substituting $g_\alpha = m_\alpha v_\alpha^2/2$ into Eq. 6.24, we get

$$\frac{\delta(n\langle K_\alpha\rangle)}{\delta t}=n_\alpha n_\beta\int_{(\mathbf{v}_\alpha)}\int_{(\mathbf{v}_\beta)}\int_{(\Omega)}\left(\frac{m_\alpha v_\alpha'^2}{2}-\frac{m_\alpha v_\alpha^2}{2}\right)$$
$$+\,v_{\alpha\beta}\sigma\,d\Omega f_\alpha f_\beta\,d^3v_\alpha\,d^3v_\beta \tag{6.66}$$

In accordance with Eq. 2.18 the change in kinetic energy on collisions, summed over the scattering angles, can be found from the relation

$$\int_{(\Omega)}\left(\frac{m_\alpha v_\alpha'^2}{2}-\frac{m_\alpha v_\alpha^2}{2}\right)\delta\,d\Omega=-\frac{m_\alpha m_\beta}{m_\alpha+m_\beta}\int_{(\Omega)}\mathbf{v}_0\mathbf{v}_{\alpha\beta}(1-\cos\vartheta)\sigma\,d\Omega$$
$$=-\frac{m_\alpha m_\beta}{m_\alpha+m_\beta}\mathbf{v}_0\mathbf{v}_{\alpha\beta}s^t_{\alpha\beta}(v)$$
$$=-\frac{m_\alpha m_\beta}{(m_\alpha+m_\beta)^2}s^t_{\alpha\beta}[m_\alpha v_\alpha^2-m_\beta v_\beta^2+(m_\beta-m_\alpha)\mathbf{v}_\alpha\mathbf{v}_\beta]$$

Substituting this equation into 6.56, we obtain

$$\left(\frac{\delta n\langle T_\alpha\rangle}{\delta t}\right)_{\alpha\beta}=-\frac{m_\alpha m_\beta}{m_\alpha+m_\beta}n_\alpha\langle\mathbf{v}_0\mathbf{v}\nu^t_{\alpha\beta}(v)\rangle$$
$$=\frac{2m_\alpha m_\beta}{(m_\alpha+m_\beta)^2}n_\alpha\left\langle\nu_{\alpha\beta}\left[\frac{m_\perp v_\alpha^2}{2}-\frac{m_\beta v_\beta^2}{2}+\frac{(m_\beta-m_\alpha)\mathbf{v}_\alpha\mathbf{v}_\beta}{2}\right]\right\rangle \tag{6.67}$$

where averaging is done over the velocities of the α- and β-type particles. If the collision frequency $\nu_{\alpha\beta}$ is velocity independent, it can be taken out of the averaging sign.

Representing the velocity of each particle as the sum of the random and directed components, we find

$$\left(\frac{\delta n_\alpha\langle K_\alpha\rangle}{\delta t}\right)_{\alpha\beta}=-\frac{3}{2}\kappa_{\alpha\beta}n_\alpha\nu^t_{\alpha\beta}(T_\alpha-T_\beta)-\kappa_{\alpha\beta}\nu^t_{\alpha\beta}$$
$$\times\left[\frac{m_\alpha u_\alpha^2}{2}-\frac{m_\beta u_\beta^2}{2}+\frac{(m_\beta-m_\alpha)\mathbf{u}_\alpha\mathbf{u}_\beta}{2}\right] \tag{6.68}$$

The first term in Eq. 6.68 determines the random motion energy exchange on collisions, and the second, the change in directed energy. It is easy to see that it vanishes at $\mathbf{u}_\alpha=\mathbf{u}_\beta$, that is, when the two plasma components are, on the average, at rest with respect to one another, and there is no friction. To find the collision term appearing on the right-hand side of Eq. 6.65, we must add $-m\mathbf{u}_\alpha \delta(n_\alpha \mathbf{u}_\alpha)/\delta t$ to Eq. 6.68. Using Eq. 6.46 for $\nu_{\alpha\beta} = \text{const}$, we find

$$m_\alpha \mathbf{u}_\alpha \left(\frac{\delta n_\alpha \mathbf{u}_\alpha}{\delta t}\right)_{\alpha\beta} = -\frac{m_\alpha m_\beta}{m_\alpha + m_\beta} n_\alpha \overset{+}{\nu}_{\alpha\beta} \mathbf{u}_\alpha (\mathbf{u}_\alpha - \mathbf{u}_\beta)$$

$$\frac{3}{2}\frac{\delta(n_\alpha T_\alpha)}{\delta t} = \frac{\delta(n_\alpha \langle K_\alpha \rangle)}{\delta t} - m_\alpha \mathbf{u}_\alpha \frac{\delta(n_\alpha \mathbf{u}_\alpha)}{\delta t} \tag{6.69}$$

$$= -\frac{3}{2}\kappa_{\alpha\beta} n_\alpha \overset{t}{\nu}_{\alpha\beta}(T_\alpha - T_\beta) + \frac{m_\alpha m_\beta^2}{(m_\alpha + m_\beta)^2} n_\alpha \overset{t}{\nu}_{\alpha\beta}(\mathbf{u}_\alpha - \mathbf{u}_\beta)^2$$

When the collision frequency is velocity dependent, averaging according to Eq. 6.67 will be more complicated. Let us do it for the Maxwellian particle velocity distribution. As before, we represent in Eq. 6.67 the particle velocities as the sum of the directed and random components. Then

$$\langle \mathbf{v}_0 \mathbf{v} \nu(v) \rangle = \sum_k \langle v_{0k} v_k \nu(v) \rangle$$

$$\approx \sum_k \langle (u_{0k} + w_{0k}) \rangle \left[w_k \nu(w) + \sum_l \frac{\partial (w_k \nu)}{\partial w_l} u_l \right]$$

$$= \sum_k \langle w_{0k} w_k \nu(w) \rangle + \sum_k u_{0k} \left\langle \frac{\delta(w_k \nu)}{\partial w_k} \right\rangle u_k \tag{6.70}$$

Here $\mathbf{u}_0$ and $\mathbf{w}_0$ are the center-of-mass velocity components:

$$\mathbf{u}_0 = \frac{m_\alpha \mathbf{u}_\alpha + m_\beta \mathbf{u}_\beta}{m_\alpha + m_\beta}; \qquad \mathbf{w}_0 = \frac{m_\alpha \mathbf{w}_\alpha + m_\beta \mathbf{w}_\beta}{m_\alpha + m_\beta}$$

In Eq. 6.70 the random velocity distribution is treated as isotropic. To obtain the averages for the quantities appearing in Eq. 6.70 we must multiply them by the distribution functions of the α- and β-type particles and integrate over the velocities. As shown in Section 6.3 (see 6.54), the product of the Maxwellian distribution functions of the α- and β-type particles $f_\alpha(\mathbf{w}_\alpha) f_\beta(\mathbf{w}_\beta)$ can be transformed into the product of the Maxwellian over the relative velocities $\mathbf{w}$ and over the velocities $\mathbf{w}_s$ defined by the equality 6.52. Therefore, when averaging (Eq. 6.70) we must switch from $\mathbf{w}_0$ to $\mathbf{w}_s$. The relationship between them is found from Eq. 6.53:

$$\mathbf{w}_0 = \frac{m_\alpha \mathbf{w}_\alpha + m_\beta \mathbf{w}_\beta}{m_\alpha + m_\beta} = \mathbf{w}_s + \frac{m_\alpha m_\beta}{m_\alpha + m_\beta} \frac{(T_\alpha - T_\beta)}{T_{\alpha\beta}} \mathbf{w}$$

where $T_{\alpha\beta} = (m_\alpha T_\beta + m_\beta T_\alpha)/(m_\alpha + m_\beta)$ is the temperature characterizing the relative velocity distribution. Using this equation and bearing in mind the independence of the distributions over $\mathbf{w}$ and $\mathbf{w}_s$, we find

$$\langle w_{0k} w_k \nu(w) \rangle = \langle w_{sk} \rangle \langle w_k \nu(w) \rangle + \frac{m_\alpha m_\beta}{(m_\alpha + m_\beta)^2} \frac{(T_\alpha - T_\beta)}{T_{\alpha\beta}} \langle w_k^2 \nu(w) \rangle$$
$$= \frac{1}{3} \frac{m_\alpha m_\beta (T_\alpha - T_\beta)}{(m_\alpha + m_\beta)^2 T_{\alpha\beta}} \langle w^2 \nu(w) \rangle$$

Substituting the result into Eq. 6.70 and using the averaged collision frequency as defined by Eqs. 6.50 and 6.56, we get

$$\langle \mathbf{v}_0 \mathbf{v} \nu_{\alpha\beta} \rangle = \frac{T_\alpha - T_\beta}{m_\alpha + m_\beta} \bar{\nu}_{\alpha\beta} + \frac{m_\alpha u_\alpha^2 - m_\beta u_\beta^2 + (m_\beta - m_\alpha) u_\alpha u_\beta}{m_\alpha + m_\beta} \bar{\nu}_{\alpha\beta} \tag{6.71}$$

Equation 6.71 gives the value of $\delta(n\langle K \rangle)/\delta t$ (see Eq. 6.67). Allowing also for Eq. 6.49, we find the expression for the collision term of Eq. 6.65:

$$\frac{3}{2} n_\alpha \frac{\delta T_\alpha}{\delta t} = \frac{\delta(n_\alpha \langle K_\alpha \rangle)}{\delta t} - m_\alpha n_\alpha \mathbf{u}_\alpha \frac{\delta \mathbf{u}_\alpha}{\delta t}$$
$$= -\frac{m_\alpha m_\beta}{m_\alpha + m_\beta} n_\alpha [\langle \mathbf{v}_0 \mathbf{v} \nu_{\alpha\beta} \rangle - \mathbf{u}_\alpha \langle \mathbf{v} \nu_{\alpha\beta} \rangle]$$
$$= -\frac{3}{2} \kappa_{\alpha\beta} n_\alpha (T_\alpha - T_\beta) \bar{\nu}_{\alpha\beta} + \frac{m_\alpha m_\beta^2}{m_\alpha^2 + m_\beta^2} n_\alpha (\mathbf{u}_\alpha - \mathbf{u}_\beta)^2 \bar{\nu}_{\alpha\beta} \tag{6.72}$$

This differs from Eq. 6.69 only in that the collision frequency is replaced by an averaged value determined by Eq. 6.56. Note that if we include the anisotropy of the random velocity distribution function (see Eq. 6.35), the collision term acquires components proportional to those of the heat flux (cf. Eq. 6.57).

For the electrons, the energy balance equation must cover inelastic collisions. They can be divided into two groups, as before. Collisions with small energy losses can be treated, with the aid of Eq. 6.66, similarly to elastic collisions. The change of the kinetic energy in this case is practically equal to the excitation energy $m_e v'^2/2 - m_e v^2/2 = -\mathscr{E}_j$. Substituting this change into Eq. 6.66 and summing over all the processes with a small energy loss, we find

$$\left(\frac{\delta n_e \langle K_e \rangle}{\delta t} \right)_{ea}^{l} = -n_e \sum_j \int_{(\mathbf{v}_a)} \int_{(\mathbf{v}_e)} \mathscr{E}_j \nu_{ea}^{(j)} f_e f_a \, d^3 v_a \, d^3 v_e$$
$$= -n_e \int_{(\mathbf{v}_a)} \int_{(\mathbf{v}_e)} \kappa_{ea}^{l} \nu_{ea}^{l} \frac{m_e v^2}{2} f_e f_a \, d^e v^e \, d^3 v_a \tag{6.73}$$

where we introduced, as in Section 5.3, the summary frequency of inelastic collisions with a small energy loss ν_{ea}^{l} and the average energy transfer coefficient κ_{ea}^{l}. Assuming that the kinetic energy of the electrons greatly exceeds that of the atoms, we can neglect the dependence of $\kappa^l \nu^l$ on the atom velocity and integrate Eq. 6.73 with respect to $\mathbf{v}_a$. Then the collision term will take the form

$$\left(\frac{\delta n_e \langle K_e \rangle}{\delta t}\right)^l_{ea} = -n_e \int_{(\mathbf{v}_e)} \kappa^l_{ea} \nu^l_{ea} \frac{m_e v_e^2}{2} f_e(\mathbf{v}_e)\, d^3 v_e$$

$$= -n_e \langle \kappa^l_{ea} \nu^l_{ea} K_e \rangle \approx -\frac{3}{2} \kappa^l_{ea} \nu^l_{ea} n_e T_e \qquad (6.74)$$

(In the last equality, assuming that $m_e u^2/2 \ll T_e$, we omitted the term associated with the directed motion.)

It will be easy to include collisions with large energy losses if we assume the postcollision electron energy to be near zero. Then we can use the collision integral in the form 5.45, and the collision term becomes

$$\left(\frac{\delta n_e \langle K_e \rangle}{\delta t}\right)^h_{ea} = -n_e \sum_j \int_{(\mathbf{v}_e)} \frac{m_e v_e^2}{2} \nu^{(j)}_{ea} f_e\, d^3 v_e$$

$$= -n_e \langle \nu^h_{ea} K_e \rangle \approx -\frac{3}{2} \bar{\nu}^h_{ea} n_e T_e \qquad (6.75)$$

Here we introduce the summary frequency of inelastic collisions with a large energy loss $\nu^h_{ea} = \Sigma_j\, \nu^{(j)}_{ea}$.

The equalities 6.74 and 6.75 show that inclusion of inelastic collisions results in replacement of $\kappa \nu_{\alpha\beta}$ in the expression for the collision term (Eq. 6.72) by a summary value, which takes into account the energy losses on both elastic and inelastic collisions:

$$\overline{\kappa^s_{ea} \nu^s_{ea}} = \overline{\kappa_{ea} \nu_{ea} + \kappa^l_{ea} \nu^l_{ea} + \nu^h_{ea}} \qquad (6.76)$$

It is easy to see that the correction in the second term of Eq. 6.72 for inelastic collisions amounts to replacement of the collision frequency ν_{ea} by the summary frequency ν^s_{ea} (Eq. 6.44). This correction is small when the average frequency of inelastic collisions is much less than that of elastic collisions (see Sections 2.5 and 2.7).

On substituting the expression 6.72 for the collision term into the energy balance equation (Eq. 6.65) we get

$$n_\alpha \frac{\partial T_\alpha}{\partial t} + \frac{2}{3} \operatorname{div} \mathbf{q}_\alpha + n_\alpha \mathbf{u}_\alpha \operatorname{grad} T_\alpha + \frac{2}{3} n_\alpha T_\alpha \operatorname{div} \mathbf{u}_\alpha$$

$$+ \frac{2}{3} \check{\pi}_\alpha \operatorname{grad} \mathbf{u}_\alpha = -n_\alpha \sum_\beta \overline{\kappa_{\alpha\beta} \nu_{\alpha\beta}} (T_\alpha - T_\beta)$$

$$+ \frac{2}{3} n_\alpha \sum_\beta \langle \nu_{\alpha\beta} \rangle \frac{m_\alpha m_\beta^2}{m_\alpha^2 + m_\beta^2} (\mathbf{u}_\alpha - \mathbf{u}_\beta)^2 \qquad (6.77)$$

We subsequently neglect the viscosity effect, omitting the corresponding term of the equation. For heavy particles (ions and atoms) we can also often omit the collision term components related to the directed motion, because at a small anisotropy (for $mu^2/2 \ll T$) they are much smaller than the terms related to the random motion. Then the energy balance equation for heavy particles takes the form

$$n_\alpha \frac{\partial T_\alpha}{\partial t} + \frac{2}{3} \operatorname{div} \mathbf{q}_\alpha + n_\alpha \mathbf{u}_\alpha \operatorname{grad} T_\alpha + \frac{2}{3} n_\alpha T_\alpha \operatorname{div} \mathbf{u}_\alpha = - n_\alpha \sum_\beta \overline{\kappa_{\alpha\beta} \nu_{\alpha\beta}} (T_\alpha - T_\beta) \tag{6.78}$$

where $\kappa_{\alpha\beta} = 2m_\alpha m_\beta/(m_\alpha + m_\beta)^2$. The energy balance equation for the electrons can be written thus:

$$n_e \frac{\partial T_e}{\partial t} + \frac{2}{3} \operatorname{div} \mathbf{q}_e + n_e \mathbf{u}_e \operatorname{grad} T_e + \frac{2}{3} n_e T_e \operatorname{div} \mathbf{u}_e$$

$$= -n_e \sum_\beta \overline{\overset{s}{\kappa}_{e\beta} \overset{s}{\nu}_{e\beta}} (T_e - T_\beta) + \frac{2}{3} n_e \sum_\beta \nu_{e\beta} m_e (\mathbf{u}_e - \mathbf{u}_b)^2 \tag{6.79}$$

where, in conformity with Eq. 6.76, the product $\kappa\nu$ is replaced by the summary value, which takes into account inelastic collisions.

Equations 6.77–6.79 include third-order moments—the components of the heat flux vector $\mathbf{q}$. Equations for them can be derived similarly to those for lower moments. When the random velocity distribution functions are near-Maxwellian and can be represented in the form 6.35, and the energy of the directed motion is much less than the random energy ($mu^2/2 \ll T$), the heat flux vector equations take the form

$$\frac{\partial \mathbf{q}}{\partial t} + \frac{5}{2} \frac{nT}{m} \operatorname{grad} \mathrm{T} - \frac{Ze}{mc} [\mathbf{q} \times \mathbf{H}] = \frac{\delta \mathbf{q}}{\delta t} \tag{6.80}$$

Generally, the collision terms of these equations are rather complicated. The expressions for the collision terms associated with electron–atom and ion–atom elastic collisions at a velocity-independent collision frequency can be represented as

$$\left(\frac{\delta \mathbf{q}_\alpha}{\delta t}\right)_{ab} = \frac{m_\beta}{m_\alpha + m_\beta} \nu_{\alpha\beta} \left[- \mathbf{q}_\alpha \left(1 + \frac{m_\alpha}{m_\beta} \kappa_{\alpha\beta}\right) \right.$$

$$\left. + \mathbf{q}_\beta \kappa_{\alpha\beta} \frac{n_\alpha}{n_\beta} + \frac{5}{2} n_\alpha \kappa_{\alpha\beta} (T_\alpha - T_\beta)(\mathbf{u}_\alpha - \mathbf{u}_\beta) \right] \tag{6.81}$$

where the first term is due to the "scattering" of the heat flux of charged particles on collisions, the second describes the transfer of the heat flux from the atoms to the charged particles, and the third is related to energy exchange and the directed motion on collisions. For electron–atom collisions, expression 6.81 is simplified, since $\kappa_{ea} = 2m_e/m_a \ll 1$.

Neglecting the quantities proportional to κ, we find

$$\left(\frac{\delta \mathbf{q}_e}{\delta t}\right)_{ea} = -\nu_{ea}\mathbf{q}_e \tag{6.82}$$

If the frequencies ν_{ea} and ν_{ia} are velocity dependent, the expressions defining $\delta\mathbf{q}/\delta t$ differ from those given above by factors of the order of unity and also by additional terms proportional to the directed velocity. For electron–ion collisions, whose frequency is inversely proportional to the cube of the velocity, calculations lead to the following equation:

$$\left(\frac{\delta \mathbf{q}_e}{\delta t}\right)_{ei} = -\frac{13}{10}\bar{\nu}_{ei}\mathbf{q}_e + \frac{3}{3} n_e T_e \bar{\nu}_{ei}(\mathbf{u}_e - \mathbf{u}_i) \tag{6.83}$$

where $\bar{\nu}_{ei}$ is the averaged collision frequency given by Eq. 5.116. The collision term of Eq. 6.80 must, in distinction to the terms considered earlier, take into account collisions of identical particles, since the heat flux is not a quantity retained on collisions. When the collision frequency $\nu_{\alpha\alpha}$ is velocity independent, calculations result in the equality

$$\left(\frac{\delta \mathbf{q}_\alpha}{\delta t}\right)_{\alpha\alpha} = -\tfrac{1}{4}\nu_{\alpha\alpha}\mathbf{q}_\alpha \tag{6.84}$$

Inclusion of the velocity dependence of $\nu_{\alpha\alpha}$ changes the numerical factor in the equation. In particular, the collision term determined by electron–electron and ion–ion collisions is equal to

$$\left(\frac{\delta \mathbf{q}_\alpha}{\delta t}\right)_{\alpha\alpha} = -\tfrac{4}{5}\bar{\nu}_{\alpha\alpha}\mathbf{q}_\alpha \tag{6.85}$$

where

$$\nu_{\alpha\alpha} = \frac{4\sqrt{\pi}}{3\sqrt{m\alpha}} \frac{e^4 n_\alpha L_\alpha}{T_\alpha^{3/2}}$$

is the frequency of Coulomb collisions of α-type particles averaged over the Maxwellian distribution in conformity with Eq. 6.56.

The above expressions enable one to obtain the general form of the collision terms of Eq. 6.80 for different cases. In the equation for the heat flux of the electrons of a three-component plasma consisting of electrons, single-charged ions, and atoms, the collision term can be written as follows, using Eqs. 6.82–6.84:

$$\frac{\delta \mathbf{q}_e}{\delta t} = -(\nu_{ea} + 1.87\bar{\nu}_{ei})\mathbf{q}_e + \frac{3}{2} n_e T_e \bar{\nu}_{ei}(\mathbf{u}_e - \mathbf{u}_i) \tag{6.86}$$

(It was taken into consideration that $\nu_{ee} = \sqrt{2}\nu_{ei}$.) In the equation for the ion heat flux we find the term $\delta\mathbf{q}_i/\delta t$ from Eqs. 6.81 and 6.85, assuming that $m_i = m_a$:

$$\frac{\delta \mathbf{q}_i}{\delta t} = -\left(\frac{3}{4}\nu_{ia} + \frac{4}{5}\nu_{ii}\right)\mathbf{q}_i + \frac{1}{4}\frac{n_i}{n_a}\nu_{ia}\mathbf{q}_a + \frac{5}{8}\nu_{ia}(T_i - T_a)(\mathbf{u}_i - \mathbf{u}_a)n_i \tag{6.87}$$

This does not cover the contribution of ion–electron collisions $(\delta\mathbf{q}_i/\delta t)_{ie} \sim (m_e/m_i)\nu_{ei}\mathbf{q}_i$, since it is always much less than the contribution of ion-ion collisions. The collision term determining the atom heat flux has a similar form. With the aid of Eqs. 6.81 and 6.84 we obtain

$$\frac{\delta \mathbf{q}_a}{\delta t} = -\left(\frac{3}{4}\nu_{ai} + \frac{1}{4}\nu_{aa}\right)\mathbf{q}_a + \frac{1}{4}\frac{n_a}{n_i}\bar{\nu}_{ai}\mathbf{q}_i + \frac{5}{8}\nu_{ai}(T_a - T_i)(\mathbf{u}_a - \mathbf{u}_i)n_a \tag{6.88}$$

7

TRANSPORT PROCESSES IN PLASMA WITHOUT MAGNETIC FIELD

7.1 CHARGED PARTICLE DIRECTED MOTION AND ENERGY TRANSPORT IN WEAKLY IONIZED PLASMA

This chapter discusses the processes governing the balance of charged particles and their energies in a plasma without a magnetic field. The main processes here are the directed motion of the charged particles (particle transport), the directed transport of their energies, and the particle energy exchange in collision. In considering transport processes we use the moments equations obtained and analyzed in the preceding chapter. In most cases we assume stationary conditions and neglect terms proportional to the time derivatives. As before, we consider a three-component plasma consisting of electrons, single-charged positive ions, and neutral atoms. The transport processes in such a plasma depend on the degree of ionization. Two extremes can therefore be distinguished, namely, a *weakly ionized plasma*, in which the frequency of collisions of electrons and ions with atoms greatly exceeds that of collisions of these particles with one another,

$$\nu_{ea} \gg \nu_{ei}, \nu_{ee}; \qquad \nu_{ia} \gg \nu_{ii}, \frac{m_e}{m_i} \nu_{ei} \tag{7.1}$$

and a *highly ionized plasma*, which is described by the reverse inequalities. The degree of ionization characteristic of transition between the two cases is illustrated for a hydrogen plasma in Fig. 7.1.

We first consider the transport processes in a weakly ionized plasma. The directed motion of the charged particles is defined by the first-moment equations. When the viscosity effects are negligible (see Section 7.2), these equations reduce to the vector equality 6.62. In the absence of

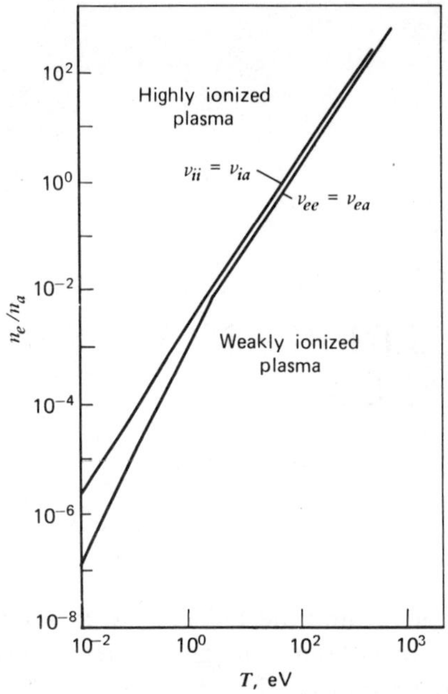

Fig. 7.1 Degree of ionization characteristics of transitions between weakly and highly ionized plasmas.

a magnetic field it takes the form

$$\frac{m_\alpha \partial \mathbf{u}_\alpha}{\partial t} = Z_\alpha e\mathbf{E} - \frac{1}{n}\,\text{grad}(nT_\alpha) + \frac{m_\alpha \delta \mathbf{u}_\alpha}{\delta t} \tag{7.2}$$

where for the electrons $\alpha = e$, $Z_e = -1$, and for the ions $\alpha = i$, $Z_i = 1$, $n_e = n_i = n$. The collision term for a weakly ionized plasma, that is, when the inequalities 7.1 are valid, includes only the terms due to the collisions of the charged particles with neutrals. When the collision frequency is velocity independent, the collision term is equal to the friction force satisfying Eq. 6.59:

$$\frac{m_\alpha \delta \mathbf{u}_\alpha}{\delta t} = \mathbf{R}_{\alpha a} = -\mu_{\alpha a}\nu_{\alpha a}(\mathbf{u}_\alpha - \mathbf{u}_a) \tag{7.3}$$

where $\nu_{\alpha a} = \nu_{\alpha a}^t$ is the transport frequency of collisions, and $\mu_{\alpha a} = m_\alpha m_a/(m_\alpha + m_a)$ is the reduced mass (for electrons $\mu_{ea} = m_e$ and for ions $\mu_{ia} = \frac{1}{2}m_i$ at $m_i = m_a$). In a weakly ionized plasma the directed velocity of the neutral atoms is usually much lower than that of charged particles. Here the friction force is equal to*

$$\mathbf{R}_{\alpha a} = -\mu_{\alpha a}\nu_{\alpha a}\mathbf{u}_\alpha \tag{7.4}$$

*Note that when the directed velocity of the atoms cannot be neglected Eq. 7.4 defines the friction force in a moving frame of reference in which this velocity is zero.

Hereafter we neglect the terms of the moments equations proportional to the time derivatives. The condition of such neglect in Eq. 7.2 can easily be obtained by comparing the first term with the friction force (Eq. 7.4). This condition has the form

$$\left|\frac{m_\alpha \partial u_{\alpha k}}{\partial t}\right| \approx \frac{m_\alpha u_{\alpha k}}{\tau} \ll |R_k| \approx m_\alpha u_{\alpha k} \nu_{\alpha a}$$

or

$$\nu_{\alpha a}\tau \gg 1 \tag{7.5}$$

This means that the use of the stationary solution of Eq. 7.2 is legitimate when the characteristic time of variation in plasma parameters τ greatly exceeds the time between collisions i/ν. Dropping the first term in Eq. 7.2 and substituting Eq. 7.4 for the collision term, we reduce Eq. 7.2 to

$$Z_{\alpha e}\mathbf{E} - \frac{1}{n}\operatorname{grad}(nT_\alpha) - \mu_{\alpha a}\nu_{\alpha a}\mathbf{u}_\alpha = 0 \tag{7.6}$$

The equality obtained enables us to find the directed velocity,

$$\mathbf{u}_\alpha = \frac{Z_\alpha e}{\mu_\alpha \nu_{\alpha a}}\mathbf{E} - \frac{T_\alpha}{\mu_\alpha \nu_{\alpha a}}\frac{\operatorname{grad} n}{n} - \frac{1}{\mu_\alpha \nu_{\alpha a}}\operatorname{grad} T_\alpha \tag{7.7}$$

which is a sum of three terms:

$$\mathbf{u}_\alpha = \mathbf{u}_{\alpha E} + \mathbf{u}_{\alpha n} + \mathbf{u}_{\alpha T} \tag{7.8}$$

The first of these, $\mathbf{u}_{\alpha E}$, determines the directed velocity associated with the acceleration of the charged particles in an electric field. The proportionality factor between velocity and field b_α is called mobility,

$$\mathbf{u}_{\alpha E} = Z_\alpha b_a \mathbf{E}, \qquad b_a = \frac{e}{\mu_{\alpha a}\nu_{\alpha a}} \tag{7.9}$$

The second term describes diffusion due to the inhomogeneity of the charged particle density. The directed diffusive velocity is proportional to the relative density gradient

$$\mathbf{u}_{\alpha n} = -D_\alpha \frac{\operatorname{grad} n}{n}, \qquad D_\alpha = \frac{T_\alpha}{\mu_{\alpha a}\nu_{\alpha a}} \tag{7.10}$$

The proportionality factor D_α is referred to as the *diffusion coefficient.*

The last term in Eq. 7.8 describes the diffusion due to the temperature gradient—the so-called *thermal diffusion.* The directed thermal diffusion velocity $\mathbf{u}_{\alpha T}$ can be represented in a form similar to Eq. 7.10:

$$\mathbf{u}_{\alpha T} = -D_\alpha^T \frac{\operatorname{grad} T_\alpha}{T_\alpha}, \qquad D_\alpha^T = \frac{T_\alpha}{\mu_{\alpha a}\nu_{\alpha a}} \tag{7.11}$$

where D_α^T is the *thermal diffusion coefficient.*

There exists a definite relationship between the coefficients b, D, and D^T, which determine the directed motion. In particular, using Eqs. 7.9 and 7.10 we obtain the relation between the diffusion and mobility coefficients

$$\frac{D_\alpha}{b_\alpha} = \frac{T_\alpha}{e} \tag{7.12}$$

which is termed the *Einstein relation.* For particle velocity distributions close to the equilibrium one it is valid for any velocity dependence of particle collision frequency.

The total directed velocity of charged particles can be expressed via the above transport coefficients. Substituting Eqs. 7.9–7.11 into Eq. 7.8, we find for the electrons

$$\mathbf{u}_e = -b_e\mathbf{E} - D_e\frac{\text{grad } n}{n} - D_e^T\frac{\text{grad } T_e}{T_e} \tag{7.13}$$

where $b_e = e/m_e\nu_{ea}$, $D_e = T_e/m_e\nu_{ea}$, and for the ions

$$\mathbf{u}_i = b_i\mathbf{E} - D_i\frac{\text{grad } n}{n} - D_i^T\frac{\text{grad } T_i}{T_i} \tag{7.14}$$

where $b_i = 2e/m_i\nu_{ia}$, $D_i = 2T_i/m_i\nu_{ia}$. Using these expressions, we can also find the current density in the plasma, which is the sum of the electron and ion components:

$$\mathbf{j} = ne\mathbf{u}_i - ne\mathbf{u}_e = en(b_e + b_i)\mathbf{E} + e(D_e - D_i)\,\text{grad } n + en\left(D_e^T\frac{\text{grad } T_e}{T_e} - D_i^T\frac{\text{grad } T_i}{T_i}\right) \tag{7.15}$$

The ion components in this sum can usually be neglected, since the transport coefficients contained in it are inversely proportional to the masses. The first term in Eq. 7.15 determines the plasma conductivity σ in a constant electric field:

$$\mathbf{j}_E = \sigma\mathbf{E}, \qquad \sigma \approx enb_e = \frac{ne^2}{m_e\nu_{ea}} \tag{7.16}$$

One can see that it is proportional to the electron density.

We have obtained expressions describing the directed velocity of the charged particles when the frequency of their collisions with the neutral particles is velocity independent. Generally, with an arbitrary velocity dependence the expressions for the directed velocity components determined by the electric field and density and temperature gradients are similar to Eqs. 7.9–7.11. Here the transport coefficients contain certain collision frequencies averaged over velocity distributions; the frequencies differ by a factor of the order of unity in the various coefficients.

For a near-Maxwellian charged particle velocity distribution the directed velocity can be obtained by substituting the expression for the collision term derived in Section 6.3 into the first-moment equation. The expression consists of two terms, the friction force and the thermal force (see Eq. 6.58):

$$m_\alpha \frac{\delta \mathbf{u}_\alpha}{\delta t} = \mathbf{R}_{\alpha a} + \mathbf{R}^T_{\alpha a} \tag{7.17}$$

As before, the friction force can be represented in the form 7.4 if we use the averaged collision frequency. In the first approximation the averaging law is given by the following equality (see Eq. 6.56):

$$\begin{aligned}\nu_{\alpha a} &= \frac{1}{3}\sqrt{\frac{2}{\pi}}\left(\frac{\mu_{\alpha a}}{T_{\alpha a}}\right)^{5/2} \int_0^\infty w^4 \nu^t_{\alpha a}(w) \exp\left(\frac{\mu_{\alpha a} w^2}{2T_{\alpha a}}\right) dw \\ &= \frac{1}{3}\sqrt{\frac{2}{\pi}} \int_0^\infty x^4 \nu^t_{\alpha a}\left(x\sqrt{\frac{T_{\alpha a}}{\mu_{\alpha a}}}\right) \exp\left(-\frac{x^2}{2}\right) dx\end{aligned} \tag{7.18}$$

where the temperature, $T_{\alpha a} = (m_a T_\alpha + m_\alpha T_a)/(m_a + m_\alpha)$, corresponds to the relative-velocity distribution. In particular, if the collision frequency is proportional to the velocity $\nu^t_{\alpha a} \sim w$, the averaged value of Eq. 7.18 is equal to

$$\nu_\alpha = \frac{8}{3}\sqrt{\frac{2}{3\pi}}\, \nu^t_\alpha(v_T)$$

and if the collision frequency is inversely proportional to the velocity it is equal to

$$\nu_\alpha = 2\sqrt{\frac{2}{3\pi}}\, \nu^t_\alpha(v_T)$$

where $v_T = \sqrt{3T_{\alpha a}/\mu_{\alpha a}}$.

A thermal force is induced by a temperature gradient and is equal, without a magnetic field, to

$$\mathbf{R}^T_{\alpha a} \approx g_{Ta} \operatorname{grad} T_{\alpha a} \tag{7.19}$$

where the coefficient $g_{T\alpha}$ is determined by the velocity dependence of $\nu_{\alpha a}$. The value of g_{T_α} coincides, with an accuracy to a factor of the order of unity, with Eq. 6.60:

$$g_{T_\alpha} \approx \frac{T_{\alpha a}}{\nu_{\alpha a}} \frac{d\nu_{\alpha a}}{dT_{\alpha a}}$$

To find a more accurate value of the effective collision frequency, which determines the friction force, and to calculate the value of the coefficient g_{T_α}, which defines the thermal force, we have to solve the motion and heat flux equations simultaneously (see Section 6.3).

Substitution of the collision term 7.17 into Eq. 7.2 leads back to the former expressions for the velocity components (Eqs. 7.9–7.11). The coefficients of mobility (b) and diffusion (D) also satisfy the former expressions 7.9 and 7.10 if the collision frequency is replaced by its averaged value. Accordingly the relation between D and b is given by Eq. 7.12 as before. In determining the thermal diffusion coefficient we must consider the effect of the thermal force in Eq. 7.18. Adding it to the force related to the pressure gradient, we obtain

$$D_\alpha^T = \left(\frac{T_\alpha}{\mu_{\alpha a}\nu_{\alpha a}}\right)(1 - g_{T_\alpha}) \tag{7.20}$$

This equation shows that the thermal diffusion coefficient may be either greater (at $g \sim d\nu/dT < 0$) or less (at $g \sim d\nu/dT > 0$) than the diffusion coefficient.

Let us now obtain heat fluxes that characterize the charged particle energy transport. The equations for the heat flux vector $\mathbf{q}$ (third-moment equations) were given in Section 6.4. For our stationary case without a magnetic field they have the following form (see Eq. 6.80):

$$\frac{5}{2}\frac{n_\alpha T_\alpha}{m_\alpha}\,\text{grad}\ T_\alpha = \frac{\delta \mathbf{q}_\alpha}{\delta t} \tag{7.21}$$

In a weakly ionized plasma only collisions between charged and neutral particles are substantial. For electrons the corresponding collision term at a velocity-independent collision frequency is defined by Eq. 6.82: $\delta \mathbf{q}_e/\delta t = -\nu_{ea}\mathbf{q}_e$. Substituting it into Eq. 7.21, we obtain the expression for the electron heat flux

$$\mathbf{q}_e = -\frac{5}{2}\frac{nT_e}{m_e\nu_{ea}}\,\text{grad}\ T_e \tag{7.22}$$

The proportionality factor between heat flux and temperature gradient is referred to as the *thermal conductivity coefficient*, and its ratio to the density as the *temperature conductivity coefficient*. For the case at hand these coefficients are equal to, respectively,

$$\begin{gathered}\mathcal{K}_e = \frac{5}{2}\frac{nT_e}{m_e\nu_{ea}};\\ \chi_e = \frac{\mathcal{K}_e}{n} = \frac{5}{2}\frac{T_e}{m_e\nu_{ea}} = \frac{5}{2}D_e\end{gathered} \tag{7.23}$$

Note that the temperature conductivity coefficient differs from the electron diffusion coefficient (Eq. 7.13) only by a factor of the order of unity. If the electron collision frequency is velocity dependent, the heat flux expression becomes more complicated. In addition to the term similar to Eq. 7.22 it contains a term proportional to the directed electron velocity. As a result $\mathbf{q}_e$ takes the form

$$\mathbf{q}_e = -\frac{5}{2} g_{qe} \frac{nT_e}{m_e \nu_{ea}} \text{grad } T_e + g_{T_e} n T_e \mathbf{u}_e \tag{7.24}$$

The numerical values of the coefficients g_q and g_T are obtained from the dependence of ν_{ea} on v.

The heat flux of ions is often associated with that of the neutral atoms, since they effectively exchange energy on collision. For this reason the ion and atom heat flux equations often must be considered simultaneously. Substituting the collision terms 6.87 and 6.88 into Eqs. 7.21 for the ions and the neutral atoms (these terms are valid for velocity-independent collision frequencies ν_{ia}, ν_{aa}), neglecting the ion–ion collisions, and taking into account that $\nu_{ai} = (n_i/n_a)\nu_{ia}$ and $m_i = m_a$, we get

$$\frac{5}{2}\frac{n_i T_i}{m_i} \text{grad } T_i = \nu_{ia}\left[-\frac{3}{4}\mathbf{q}_i + \frac{1}{4}\frac{n_i}{n_a}\mathbf{q}_a + \frac{5}{8} n_i (T_i - T_a)(\mathbf{u}_i - \mathbf{u}_a)\right];$$

$$\frac{5}{2}\frac{n_a T_a}{m_a} \text{grad } T_a = \frac{n_i}{n_a}\nu_{ia}\left[-\frac{3}{4}\mathbf{q}_a + \frac{1}{4}\frac{n_a}{n_i}\mathbf{q}_i + \frac{5}{8} n_a (T_i - T_a)(\mathbf{u}_i - \mathbf{u}_a)\right] - \frac{1}{4}\nu_{aa}\mathbf{q}_a$$

Solving these equations for $\mathbf{q}_i$ and $\mathbf{q}_a$ and taking into consideration that for a weakly ionized plasma $n_i \ll n_a$, we find

$$\begin{aligned} \mathbf{q}_i &= -\frac{10}{3}\frac{n_i T_i}{m_i \nu_{ia}} \text{grad } T_i + \frac{5}{6} n_i (T_i - T_a)(\mathbf{u}_i - \mathbf{u}_a) \\ &\quad -\frac{10}{3}\frac{n_i T_a}{m_a \nu_{\alpha a}} \text{grad } T_a; \\ \mathbf{q}_a &= -10\frac{n_a T_a}{m_a \nu_{\alpha a}} \text{grad } T_a - \frac{10}{3}\frac{n_i T_i}{m_i \nu_{ia}} \text{grad } T_i \\ &\quad -\frac{10}{3} n_i \frac{\nu_{ia}}{\nu_{\alpha a}} (T_i - T_a)(\mathbf{u}_i - \mathbf{u}_a) \end{aligned} \tag{7.25}$$

In the expression for the neutral atom heat flux the ion terms can be neglected if $n_i T_i^2 \ll n_a T_a^2$. This flux is controlled only by atom–atom collisions. In the expression for the ion heat flux at $T_a \ll T_i$ the last term can be neglected, and then it takes a form similar to Eq. 7.24:

$$\mathbf{q}_i = -\frac{10}{3}\frac{n_i T_i}{m_i \nu_{ia}} \operatorname{grad} T_i + \frac{5}{6} n_i T_i \mathbf{u}_i \tag{7.26}$$

The structure of the expression remains unchanged for a velocity-dependent collision frequency as well, but the numerical coefficients will change, of course.

7.2 ELECTRON MOBILITY, DIFFUSION, AND THERMAL CONDUCTIVITY COEFFICIENTS

The coefficients of mobility, diffusion, and thermal conductivity of electrons in a weakly ionized plasma can be obtained by a method using the expansion of the distribution function in the parameters characterizing its anisotropy. Such a method was described in Chapter 5. As demonstrated in Sections 5.1 and 5.2, this expansion is of a rapidly converging type. Restricting ourselves to the first two terms, we can represent the distribution function as in Eq. 5.20:

$$f(\mathbf{v}) = f_0(v) + \frac{\mathbf{v}}{v}\mathbf{f}_1(v) \tag{7.27}$$

The first term in the sum is the isotropic distribution function component depending solely on v. The directed component appearing in the second term is responsible for the directed electron velocity. In accordance with Eq. 5.13, we find

$$\mathbf{u} = \int_{(\mathbf{v})} \mathbf{v} f(\mathbf{v})\, d^3v = \frac{1}{3}\int_{(\mathbf{v})} v\mathbf{f}_1(v)\, d^3v = \frac{4\pi}{3}\int_0^\infty v^3 \mathbf{f}_1(v)\, dv \tag{7.28}$$

Substituting Eq. 7.27 into the kinetic equation and making allowance for the smallness of the second term, we can easily obtain the equation for both components f_0 and $\mathbf{f}_1$ (see Section 5.2). The equation defining $\mathbf{f}_1$ (see Eq. 5.22) has the following form for the stationary case ($\partial(n\mathbf{f}_1)/\partial t = 0$) without a magnetic field:

$$-\frac{ne\mathbf{E}}{m_e}\frac{\partial f_0}{\partial v} + v \operatorname{grad}(nf_0) = \mathbf{S}_1 \tag{7.29}$$

The collision term $\mathbf{S}_1$ was obtained in Section 5.3. Taking into account elastic and inelastic collisions, we can write it as

$$\mathbf{S}_1 = -n\mathbf{f}_1 \nu_{ea}^s(v) \tag{7.30}$$

where $\nu_{ca}^s = \nu^t + \nu^l + \nu^h$ is the summary frequency of electron–atom collisions. Equation 7.29 makes it possible to find the function $\mathbf{f}_1$, due account being taken of Eq. 7.30:

$$\mathbf{f}_1 = \frac{e\mathbf{E}}{m_e \nu_{ea}^s} \frac{df_0}{dv} - \frac{v}{\nu_{ea}^s} \frac{\operatorname{grad}(nf_0)}{n} \tag{7.31}$$

Substituting Eq. 7.31 into Eq. 7.28, we obtain the directed electron velocity:

$$\mathbf{u} = \frac{4\pi}{3} \frac{e\mathbf{E}}{m_e} \int_0^\infty \frac{v^3}{\nu_{ea}^s} \frac{df_0}{dv} dv - \frac{4\pi}{3n} \operatorname{grad}\left(n \int_0^\infty \frac{v^4}{\nu_{ea}^s} f_0 \, dv\right) \tag{7.32}$$

(here the operator grad is taken out of the integral). In conformity with the definition given in Section 7.1 (see Eqs. 7.8–7.11) it can be expressed via the mobility and diffusion coefficients:

$$\mathbf{u} = -b_e \mathbf{E} - \frac{1}{n} \operatorname{grad}(D_e n) = -b_e \mathbf{E} - D_e \frac{\operatorname{grad} n}{n} - D_e^T \frac{\operatorname{grad} T_e}{T_e} \tag{7.33}$$

where

$$b_e = -\frac{4\pi}{3} \frac{e}{m_e} \int_0^\infty \frac{v^3}{\nu_{ea}^s} \frac{df_0}{dv} dv; \tag{7.34}$$

$$D_e = \frac{4\pi}{3} \int_0^\infty \frac{v^4}{\nu_{ea}^s} f_0 \, dv; \tag{7.35}$$

$$D_e^T = T_e \frac{\partial D_e}{\partial T_e} \tag{7.36}$$

Equations 7.34 and 7.35 for the coefficients b_e and D_e can be written in a form similar to Eqs. 7.9 and 7.10:

$$b_e = \frac{e}{m_e \nu_{ea}^b}; \qquad D_e = \frac{T_e}{m_e \nu_{ea}^D} \tag{7.37}$$

if we appropriately determine the effective collision frequencies ν_{ea}^b and ν_{ea}^D. Generally, with an arbitrary distribution function f_0 these frequencies are different. Comparing Eqs. 7.34 and 7.35 with Eq. 7.37, we find for the collision frequency determining the mobility:

$$\nu_{ea}^b = \left[-\frac{4\pi}{3} \int_0^\infty \frac{v^3}{\nu_{ea}^s} \frac{df_0}{dv} dv\right]^{-1} = \left[\frac{4\pi}{3} \int_0^\infty \frac{d}{dv}\left(\frac{v^3}{\nu_{ea}^s}\right) f_0 \, dv\right]^{-1} \tag{7.38}$$

and for the collision frequency appearing in the diffusion coefficient:

$$\nu_{ea}^D = \frac{T_e}{m_e} \left[\frac{4\pi}{3} \int_0^\infty \frac{v^4}{\nu_{ea}^s} f_0 \, dv\right]^{-1} \tag{7.39}$$

It is easy to ascertain that when ν_{ea}^s is independent of v, both these values are equal to ν_{ea}^{s}. With a Maxwellian electron velocity distribution

$$f_0(v) = \left(\frac{m_e}{2\pi T_e}\right)^{3/2} \exp\left(-\frac{m_e v^2}{2T_e}\right)$$

the collision frequencies ν_{ea}^{b} and ν_{ea}^{D} in the mobility and diffusion coefficients are also equal. This leads to the relationship 7.12 between b_e and D_e. The expression for the effective collision frequency, which is derived by substituting the Maxwellian distribution into Eqs. 7.38 or 7.39, is of the form

$$\begin{aligned}\nu_{ea}(T_e) &= \left[\frac{1}{3}\sqrt{\frac{2}{\pi}}\left(\frac{m_e}{T_e}\right)^{5/2}\int_0^\infty \frac{v^4}{\nu_{ea}^{s}(v)}\exp\left(-\frac{m_e v^2}{2T_e}\right)dv\right]^{-1} \\ &= \left[\frac{1}{3}\sqrt{\frac{2}{\pi}}\int_0^\infty \frac{x^4\exp(-x^2/2)}{\nu_{ea}^{s}(x\sqrt{T_e/m_e})}\,dx\right]^{-1}\end{aligned} \tag{7.40}$$

For instance, when the collision frequency is proportional to the velocity $\nu_{ea}^{s} \sim v$ the effective frequency is equal to

$$\nu_{ea}(T_e) = \sqrt{3\pi/8}\,\nu_{ea}^{s}(v_T)$$

and when the collision frequency is inversely proportional to the velocity ($\nu_{ea}^{s} \sim 1/v$)

$$\nu_{ea}(T_e) = \frac{3}{8}\sqrt{\frac{3\pi}{2}}\,\nu_{ea}^{s}(v_T)$$

where $v_T = \sqrt{3T_e/m_e}$. Note that with a weak dependence $\nu(v)$ the difference of the effective collision frequency (Eq. 7.40) from the approximate expression 7.18 given in Section 7.1 is small. For $\nu \sim v$ and $\nu \sim 1/v$ this difference is about 13%.

The relationship between the thermal-diffusion and diffusion coefficients is given by Eq. 7.36. In accordance with it and using the determination of D_e via the effective collision frequency ν_{ea}^{D} (Eq. 7.37), we obtain

$$D_e^T = T_e\frac{\partial}{\partial T_e}\left(\frac{T_e}{m_e\nu_{ea}^{D}}\right) = \frac{T_e}{m_e\nu_{ea}^{D}}(1 - g_{T_e}) \tag{7.41}$$

where $g_{T_e} = (T_e/\nu_{ea}^{D})\partial\nu_{ea}^{D}/\partial T_e$. This equation corresponds to Eq. 7.20 given in Section 7.1. It enables us to find the relationship between the coefficients D_a and D_e^T. As mentioned previously, when the collision frequency is velocity independent $D_e = D_e^T$ (see Eq. 7.11). With a collision frequency proportional to the velocity $\nu_{ea}^{s} \sim v$, Eq. 7.41 results in the relation $D^T = \frac{1}{2}D$ and with $\nu_{ea}^{s} \sim 1/v$, in the relation $D^T = \frac{3}{2}D$.

Let us now obtain the electron heat flux. We first find the electron energy flux $\mathbf{Q} = \frac{1}{2}nm_e\langle\mathbf{v}v^2\rangle$ (Eq. 6.12) by expanding Eq. 7.27. It is clearly determined by the directed component of the distribution function

$$\mathbf{Q}_e = \frac{1}{2} nm_e \int_{(\mathbf{v})} \mathbf{v}v^2 f(\mathbf{v})\, d^3v = \frac{1}{6} nm_e \int_{(\mathbf{v})} v^3 \mathbf{f}_1(v)\, d^3v$$
$$= \frac{4\pi}{6} nm_e \int_0^\infty v^5 \mathbf{f}_1(v)\, dv \tag{7.42}$$

With low directed velocities the heat flux is related to the energy flux by the equation

$$\mathbf{q}_e = \frac{1}{2} nm_e \langle (\mathbf{v}-\mathbf{u})(\mathbf{v}-\mathbf{u})^2 \rangle \approx \tfrac{1}{2} nm_e [\langle \mathbf{v}v^2 \rangle - \mathbf{u}\langle v^2 \rangle$$
$$-2\langle \mathbf{v}(\mathbf{vu}) \rangle] = \mathbf{Q}_e - \frac{5}{2} nT_e \mathbf{u}_e \tag{7.43}$$

Substituting Eq. 7.42 for Q and Eq. 7.28 for u into Eq. 7.43, we get

$$\mathbf{q}_e = \frac{2\pi}{3} nm_e \int_0^\infty \left(v^5 - 5\frac{T_e}{m_e} v^3 \right) \mathbf{f}_1\, dv \tag{7.44}$$

Using Eq. 7.31, we find

$$\mathbf{q}_e = -\frac{4\pi}{6} ne\mathbf{E} \int_0^\infty \frac{v^5 - 5(T_e/m_e)v^3}{\nu_{ea}^s} \frac{df_0}{dv}\, dv - \frac{4\pi}{6} m_e$$
$$\times \text{grad}\left[n \int_0^\infty \frac{v^6 - 5(T_e/m_e)v^4}{\nu_{ea}^s} f_0\, dv \right] - \frac{10\pi}{3} (n \text{ grad } T_e) \int_0^\infty \frac{v^4}{\nu_{ea}^s} f_0\, dv$$

This expression can be represented as

$$\mathbf{q}_e = -\frac{5}{2} \frac{nT_e}{m_e \nu_{ea}^D} \text{grad } T_e - g_{u1} \frac{e\mathbf{E}}{m_e} \frac{nT_e}{\nu_{ea}^D} + g_{u2} \text{ grad} \left(\frac{nT_e^2}{m_e \nu_{ea}^D} \right) \tag{7.45}$$

where ν_{ea}^D is the collision frequency determining the diffusion coefficient (Eq. 7.38):

$$\frac{g_{u1}}{\nu_{ea}^D} = \frac{4\pi}{6} \frac{m_e}{T_e} \int_0^\infty \left(v^5 - 5\frac{T_e}{m_e} v^3 \right) \frac{1}{\nu_{ea}^s} \frac{df_0}{dv}\, dv;$$
$$\frac{g_{u2}}{\nu_{ea}^D} = \frac{-4\pi}{6} \left(\frac{m_e}{T_e} \right)^2 \int_0^\infty \left(v^6 - 5\frac{T_e}{m_e} v^4 \right) \frac{f_0}{\nu_{ea}^s}\, dv \tag{7.46}$$

With the Maxwellian distribution function f_0 we find

$$\frac{g_{u1}}{\nu_{ea}^D} = \frac{g_{u2}}{\nu_{ea}^D} = -\frac{1}{3\sqrt{2\pi}} \int_0^\infty \frac{(x^6 - 5x^4)}{\nu_{ea}^s \left(x\sqrt{\frac{T_e}{m_e}} \right)} \exp\left(-\frac{x^2}{2} \right) dx$$
$$= \frac{1}{3\sqrt{2\pi}} \int_0^\infty \frac{\partial}{\partial x} \left(\frac{1}{\nu_{ea}^s} \right) x^5 \exp\left(-\frac{x^2}{2} \right) dx$$

Comparison of this equation with Eq. 7.40, makes it easy to see that $g_u/\nu_{ea} = -T_e[\partial(1/\nu_{ea})/\partial T_e]$ and hence

$$g_u = \frac{T_e}{\nu_{ea}} \frac{\partial \nu_{ea}}{\partial T_e} \tag{7.47}$$

Bearing in mind that with the Maxwellian distribution $g_{u1} = g_{u2} = g_u$, we transform Eq. 7.46 to

$$\mathbf{q} = -\left(\frac{5}{2} - g_u\right) \frac{nT_e}{m_e \nu_{ea}} \text{grad } T_e - g_u n T_e \mathbf{u} \tag{7.48}$$

The first term in Eq. 7.48 defines the electron thermal conductivity coefficient with an arbitrary dependence $\nu_{ea}^s(v)$, which is equal to

$$\mathcal{K}_e = (\tfrac{5}{2} - g_u) \frac{nT_e}{m_e \nu_{ea}} \tag{7.49}$$

The second term describes the heat transfer due to the directed motion. Note that the coefficient g_u, which determines this transfer (Eq. 7.47), is equal to g_T appearing in the expression for the thermal-diffusion flux (see Eqs. 7.33 and 7.41). It can be shown that the equality of these coefficients at a near-Maxwellian velocity distribution results from the principle of symmetry of the kinetic coefficient (*Onsager principle*), which is well known in thermodynamics.

7.3 MECHANISM OF TRANSFER PROCESSES

Let us now consider the physical picture of the transfer of charged particles and of their energy under the effect of the electric field, the density gradients, and the temperature.

The charged particle motion in the electric field is a superposition of the random, Brownian motion and the directed motion due to acceleration by the field. When the directed velocity is, on the average, much less than the random [it is shown in Section 5.1 that such conditions are fulfilled for ions in a weak field (Eq. 5.5), and for electrons in any field], the motion pattern differs little from Brownian. The total velocity of each particle consists of the velocity of the random (thermal) motion ($\mathbf{w}$) and the velocity under the effect of the electric field ($\mathbf{v}_E$):

$$\mathbf{v} = \mathbf{w} + \mathbf{v}_E \tag{7.50}$$

For simplicity we assume that collisions of charged particles with atoms lead to complete isotropy of velocities; their average velocity immediately upon collision is zero. Then for each particle that experienced a collision at an instant t_0 we can assume $\mathbf{v}_E(t_0) = 0$, $\mathbf{v}(t_0) =$

w. In the interval between collisions a charged particle in the electric field undergoes acceleration. Its velocity at instant t is equal to $\mathbf{v} = \mathbf{w} + (Ze\mathbf{E}/m)(t - t_0)$ on the assumption that the field strength $\mathbf{E}$ changes but slightly between collisions. Averaging the velocity over the particle group and taking into account that the average random velocity is zero, while the average time from the last collision is of the order of the reciprocal collision frequency, we obtain the equation for the directed velocity

$$\mathbf{u}_E = \langle \mathbf{v}_E \rangle \approx \frac{Z_e \mathbf{E}}{m\nu} \tag{7.51}$$

which corresponds to Eq. 7.7. Here ν is the collision frequency averaged over collisions and velocities (naturally, the averaging law remains uncertain in a qualitative consideration). It can be seen that the obtained dependence of the directed velocity on the collision frequency ($\mathbf{u}_E \sim 1/\nu$) is due to the fact that the collisions limit the time of acceleration in the electric field.

Let us now take a look at the diffusion mechanism. The diffusion results from the random motion of the particles. In a homogeneous plasma, in the absence of an electric field the heat flux of the particles across any area is offset by the reverse flux. With a density gradient, there appears an uncompensated particle flux in the direction opposite to the gradient. This flux is caused by the different particle density on the two sides of the area perpendicular to the gradient. Assuming that the density gradient is directed along the x axis, we compute the flux across an area perpendicular to this axis. We assume, as before, that the collisions result in complete randomization of the velocities. After a collision the velocity of a particle in the direction of the isolated area is equal to $w_x = w \cos \theta$, where w is the random velocity and θ is the angle between the velocity vector and the x axis. The density of the flux of particles of a given velocity and direction across the area is equal to

$$d\Gamma_x = nw \cos \Theta \, f(\mathbf{w}) \, d^3w \tag{7.52}$$

where n is the density of the particles in the area where they experienced the last collision, and $f(\mathbf{w}) \, d^3w$ is the fraction of particles with the given velocity value and direction. The flux $d\Gamma_x$ is controlled by collisions at a point with a coordinate $x' = x - l \cos \theta$, where l is the mean free path from the point of collision to the area. Assuming that the particle density changes only slightly over the length l, we find

$$d\Gamma_x = n(x - l \cos \Theta) w \cos \Theta f(\mathbf{w}) \, d^3w$$

$$\approx [nw \cos \Theta - \left(\frac{\partial n}{\partial x}\right) wl \cos^2 \Theta] f(\mathbf{w}) \, d^3w$$

Averaging this expression over the directions, summing over the values of the random velocity, and taking into account that $\langle \cos \theta \rangle = 0$ and $\langle \cos^2 \theta \rangle = \frac{1}{3}$, we get

$$\Gamma_x = nu_n = -\tfrac{1}{3}\langle wl \rangle \frac{\partial n}{\partial x} \tag{7.53}$$

Equation 7.53 enables us to find the diffusion coefficient. Assuming that the mean path to the last collision is determined by the collision frequency $\langle l \rangle \approx \lambda \approx w/\nu$, we have

$$D = \frac{1}{3}\langle wl \rangle \approx \frac{1}{3}\left\langle \frac{w^2}{\nu} \right\rangle \approx \frac{T}{m\nu}$$

which agrees with Eq. 7.10. The dependence of the diffusion flux on the temperature and collision frequency is due to the nature of the diffusion transfer caused by the thermal motion of the particles; it is proportional to the thermal velocity and to the difference of densities over the mean free path.

The diffusion due to the gradient of the temperature of the charged particles can be considered in a similar way. In this case the difference of the flux along and against the gradient is due to the difference in the average values of the random particle velocity and the mean free path on the different sides of the isolated area perpendicular to the temperature gradient. Making use of Eq. 7.52 for the flux density of particles with a given value and direction of random velocity and assuming that the random velocity distribution of the particles is determined by the site of the last collision, we get

$$\begin{aligned} d\Gamma_x &= nw \cos \Theta f(\mathbf{w}, x - l \cos \Theta)\, d^3w \\ &= \left(nwf \cos \Theta - nw \frac{\partial f}{\partial x} l \cos^2 \Theta\right) d^3w \end{aligned}$$

Averaging this equation over the angles θ and summing over the random velocities, we find the density of the flux associated with the temperature gradient:

$$\begin{aligned} \Gamma_x = nu_T &= -\frac{n}{3}\int_{(\mathbf{w})} wl \frac{\partial f}{\partial x}\, d^3w \\ &= -\frac{1}{3} n \frac{\partial}{\partial x}\int_{(\mathbf{w})} wlf\, d^3w = -n \frac{\partial}{\partial x}\left(\frac{1}{3}\langle wl \rangle\right) \end{aligned} \tag{7.54}$$

and further, assuming $l = w/\nu$, we obtain

$$u_T = -\frac{\partial}{\partial x}\left(\frac{T}{m\nu}\right) = -\frac{T}{m\nu}\left(1 - \frac{T}{\nu}\frac{d\nu}{dT}\right)\frac{\text{grad } T}{T} \tag{7.55}$$

This equation for the directed velocity is similar to Eq. 7.11 (for $\nu = \text{const}$) and Eq. 7.20 (for the general case). As seen from Eq. 7.54, at a constant (velocity-independent) mean free path the thermal diffusion flux is determined by the difference of the thermal velocities over this path, which leads to the decomposition of the particle fluxes along and against the temperature gradient. The additional effect is due to the change of the mean free path itself along the gradient.

The energy transfer under the effect of the temperature gradient is also determined by the thermal motion of the particles. It also exists when no particle flux is observed (for instance, in a reference system where the directed velocity is zero). Indeed, particles with a higher average energy must come from a region with a higher temperature to a region with a lower temperature. Therefore, even if the forward and reverse particle fluxes are equal, there must exist an uncompensated flux of their energy. Let us find this flux. We assume, as before, that the temperature gradient is directed along the x axis. The density of the energy flux carried over by particles with a given velocity is obtained by multiplying the flux density of the particles (Eq. 7.52) by their energy:

$$dQ_x = \frac{mw^2}{2} nw \cos \Theta f(\mathbf{w})\, d^3w$$

Assuming, as before, that the velocity distribution of the particles is determined by the site of their last collision and that it changes only slightly over the distance between collisions, we find

$$dQ_x = \frac{nmw^3}{2} \cos \Theta f(\mathbf{w}, x - l \cos \Theta)\, d^3w$$

$$= \frac{nmw^3}{2} \cos \Theta \left(f - l \cos \Theta \frac{\partial f}{\partial x} \right) d^3w$$

where $l \approx w/\nu$. Averaging the flux over the angles θ and summing over the velocities, we obtain

$$Q_x = -\frac{1}{6} nm \int_{(\mathbf{w})} \frac{w^4}{\nu} \frac{\partial f}{\partial x}\, d^3w = -\frac{1}{6} nm \frac{\partial}{\partial x} \left(\int_{(\mathbf{w})} \frac{w^4}{\nu} f d^3w \right) \tag{7.56}$$

In particular, with a Maxwellian velocity distribution for the case where the collision frequency is velocity independent

$$Q_x = -\frac{1}{6} \frac{nm}{\nu} \frac{\partial}{\partial x} \left[4\pi \left(\frac{m}{2\pi T} \right)^{3/2} \right]$$

$$\times \int_0^\infty w^6 \exp\left(-\frac{mw^2}{2T} \right) dw = -5 \frac{uT}{m\nu} \frac{\partial T}{\partial x}$$

To find the heat flux density q_x we must deduct from Q_x the energy flux associated with the directed motion:

$$q_x = \frac{1}{2} nm\langle(v_x - u_x)(\mathbf{v} - \mathbf{u})^2\rangle = Q_x - \frac{5}{2} nu_x T$$

Using Eq. 7.54, we find, for $\nu = \text{const}$,

$$q_x = -\frac{5}{2} \frac{nT}{m\nu} \frac{\partial T}{\partial x} \tag{7.57}$$

The expression 7.57 yields the thermal conductivity coefficient, whose value coincides with the product of the diffusion coefficient by the average particle energy, with an accuracy to a numerical factor. This reflects the analogy in the mechanisms of diffusion transfer of particles and energy transfer, both related to the random motion of the particles.

Note that the electric field and the concentration gradient may lead to an additional energy transfer. The expressions for the corresponding fluxes, which are similar to Eqs. 7.24 and 7.45, are readily obtained from consideration of the difference in the energy fluxes along and against the gradients. The difference in such fluxes in a reference system where the directed velocity is zero is associated with the dependence of the collision frequency on the particle energy. Owing to this dependence the time of acceleration in the electric field and the concentration gradient over the mean free path are different for particles of different energy. As a result, in a reference system where the particle fluxes along and against the gradients are compensated the energy fluxes are also different.

In conclusion we estimate the components of the viscous stress tensor, which determines the anisotropic part of the momentum flux (see Section 6.1). Let us estimate, for instance, the transfer of the y component of the momentum along the x axis associated with the gradient of the directed velocity $\partial u_y/\partial x$ and described by the component π_{yx} of the viscosity tensor. The momentum flux arises because in the presence of the directed velocity gradient $\partial u_y \partial x$ the momenta mu_y, transferred as a result of the thermal motion along the x axis in both directions, are uncompensated. This flux is determined by the product of the average momentum at the site of the last collision and the particle flux. For a group of particles with a given thermal velocity we find

$$d\pi_{yx} = mu_y(x - l\cos\Theta)\, d\Gamma_x \approx nm\left(u_y - \frac{\partial u_y}{\partial x} l \cos\Theta\right) w\cos\Theta f(\mathbf{w})\, d^3w$$

Summing over the velocities, we get

$$\pi_{yx} \approx -\frac{1}{3} nm\langle lw\rangle \frac{\partial u_y}{\partial x} \approx -n\frac{T}{\nu}\frac{\partial u_y}{\partial x} \tag{7.58}$$

Similar relations are derived for the other tensor components. One can show that the tensor is generally defined by the relation

$$\pi_{kl} = -\eta \left(\frac{\partial u_k}{\partial x_l} - \frac{\partial u_l}{\partial x_k} - \frac{2}{3}\, \delta_{lk} \operatorname{div} \mathbf{u} \right) \tag{7.59}$$

where the coefficient $\eta \approx nT/\nu$ is called the *viscosity coefficient.*

To establish the conditions for neglecting the viscosity effects in the transfer equations we compare the components of the viscosity tensor, which characterizes the anisotropic part of the total pressure, with the scalar pressure. Then the corresponding criterion takes the form $|\pi_{\alpha\beta}| = (nT/\nu)u/L \ll p = nT$, or $u \ll \nu L = \langle w \rangle L/\lambda$, where L is the characteristic scale of directed-velocity change. Since the conditions $\lambda \ll L, u \ll \langle w \rangle$ are necessary for the entire consideration this inequality is fulfilled *a priori*; hence the viscosity effect on the transfer processes must be small.

7.4 AMBIPOLAR DIFFUSION

In the preceding sections we obtained expressions for the directed velocity of charged particles under the effect of the electric fields, the density gradients, and the temperature. The mobility and diffusion coefficients appearing in these expressions are inversely proportional to the mass; they are much larger for the electrons than for the ions. But in view of the quasi-neutrality condition, independent motion of the electrons and ions in a plasma is impossible. A rapid removal of electrons from some element of the plasma volume will inevitably give rise to an electric field, which will prevent their further removal from this volume and will speed up the removal of ions.

Let us see, for instance, how charged particles diffuse in a long cylindrical tube, assuming that the principal mechanism of their removal is recombination on the tube walls. A typical radial distribution of the density of charged particles in a volume is given in Fig. 7.2. With this distribution the diffusion occurs from the axis to the walls (against the density gradient). Suppose that at some initial moment the quasi-neutrality condition is rigorously fulfilled throughout the volume. Then, in the subsequent period the electron diffusion flux greatly exceeds the ion flux (because $D_e \gg D_i$). As a result the walls become negatively charged and an excess positive charge accumulates in the volume. The charge separation leads to the formation of a radial electric field, which speeds the ions toward the walls and retards the electrons. The field must increase until the electron and ion fluxes become equal. The space charge then no longer changes; that is, a quasi-stationary state sets in. This diffusion regime is called *ambipolar.* For the ambipolar motion to

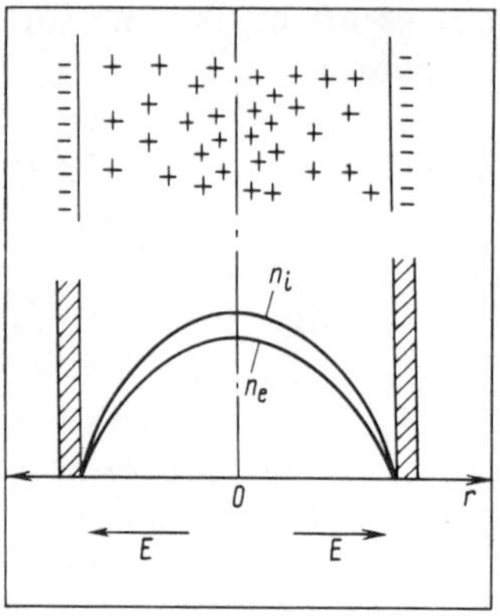

Fig. 7.2 Radial distribution of the density of charged particles.

be maintained, the negative potential of the plasma boundaries (walls) must become high enough to considerably decrease the electron flux. The corresponding potential energy obviously must exceed the average thermal energy of the electrons; that is, $c(\varphi_g - \varphi_0) > T_e$. The space charge that causes such a potential difference is ensured by the stationary difference in the electron and ion densities ($\Delta n = n_i - n_e$). In the plasma, the value of $\Delta n/n$ is low ($\Delta n \ll n$) throughout the volume, with the exception of the wall layers. Naturally, the presence of the walls is by no means necessary for the ambipolar diffusion. The charge separation required for the ambipolar regime is the result of plasma inhomogeneity due to any reasons.

Let us now describe quantitatively the ambipolar diffusion in a plasma containing electrons and single-charged ions. The quasi-neutrality condition for such a plasma amounts to the equality of the electron and ion densities throughout the volume, with the exception of the wall layers, whose thickness is of the order of the Debye length,

$$|n_e - n_i| \ll n_e, \qquad n_e \approx n_i \approx n \tag{7.60}$$

To maintain quasi-neutrality, the changes in electron and ion concentrations in each volume element must be equal, that is, $\partial n_e/\partial t = \partial n_i/\partial t$ or, in conformity with the particle balance equations (Eq. 6.36),

$$-\text{div}(n_e \mathbf{u}_e) + \frac{\delta n_e}{\delta t} = -\text{div}(n_i \mathbf{u}_i) + \frac{\delta n_i}{\delta t}$$

The volume processes of appearance and removal of particles (ionization and recombination) in a three-component plasma cannot lead to a disturbance of quasi-neutrality—an electron and an ion appear or disappear in each of these processes simultaneously. Accordingly, $\delta n_e/\delta t =$

$\delta n_i/\delta t$ and the condition for maintaining quasineutrality amounts to the equality of the flux divergences

$$\operatorname{div}(n_e \mathbf{u}_e) = \operatorname{div}(n_i \mathbf{u}_i) \tag{7.61}$$

This equation is equivalent to the expression $n\mathbf{u}_i = n\mathbf{u}_e + \mathbf{\Gamma}_j$, where $\operatorname{div} \mathbf{\Gamma}_j = 0$, that is, the flux $\mathbf{\Gamma}_j$ does not affect the density. It evidently determines the current in the plasma $\mathbf{j} = ne(\mathbf{u}_i - \mathbf{u}_e) = e\mathbf{\Gamma}_j$. One can assume that this current is associated with the electric field induced by external sources. In the unidimensional case (when the flux nu depends on a single coordinate) this is obvious, since the current can close only through external electrodes. Thus, in the absence of an external field, when there is no current in the plasma either and $\mathbf{\Gamma}_j = 0$, the condition 7.61 leads to the equality of the directed velocities of the electrons and ions

$$\mathbf{u}_e = \boldsymbol{u}_i \tag{7.62}$$

The directed motion defined by Eq. 7.62 is actually the ambipolar diffusion.

In the presence of an external field the directed electron and ion velocity can be represented as a sum of the ambipolar velocity and the current velocity related to the field of the external sources $\mathbf{E}_0$ via the mobility coefficients:

$$\mathbf{u}_e = \mathbf{u}_A - b_e \mathbf{E}_0; \qquad \mathbf{u}_i = \mathbf{u}_A + b_i \mathbf{E}_0 \tag{7.63}$$

The condition 7.61 results in the equality

$$\operatorname{div}[n(b_e + b_i)\mathbf{E}_0] = 0$$

which yields the distribution of the field E_0 in the plasma.

Let us find the characteristics of the ambipolar motion. To this end we use Eqs. 7.13 and 7.14 for the directed velocity. When $|\operatorname{grad} T/T| \ll |\operatorname{grad} n/n|$ and the thermal diffusion is insignificant the directed electron and ion velocity is equal to

$$\begin{aligned} \mathbf{u}_e &= -D_e \frac{\operatorname{grad} n}{n} - b_e \mathbf{E}; \\ \mathbf{u}_i &= -D_i \frac{\operatorname{grad} n}{n} + b_i \mathbf{E} \end{aligned} \tag{7.64}$$

By equating these velocities in conformity with Eq. 7.62 it is easy to find the *ambipolar electric field*, which forms automatically in the plasma to equalize the fluxes of oppositely charged particles:

$$\mathbf{E}_A = \frac{D_i - D_e}{b_e + b_i} \frac{\operatorname{grad} n}{n} \tag{7.65}$$

Considering that $D_e \gg D_i$ and $b_e \gg b_i$ and taking advantage of the relationship between D_e and b_e (Eq. 7.12), we find the approximate expression for $\mathbf{E}_A$:

$$\mathbf{E}_A \approx -\frac{D_e}{b_e}\frac{\text{grad}\, n}{n} = -\frac{T_e}{e}\frac{\text{grad}\, n}{n} \tag{7.66}$$

The electric field is directed oppositely to the density gradient. Therefore, as would be expected, it hinders the diffusion of the electrons and increases the ion flux. Since $\mathbf{E} = -\text{grad}\, \varphi$, we obtain from Eq. 7.66 the potential distribution

$$\varphi - \varphi_0 = \frac{T_e}{e} \ln \frac{n}{n_0} \tag{7.67}$$

This equation determines, in particular, the potential difference between the central region of the plasma and its boundaries. It is seen that at a low density at the boundary $n_g \ll n_0$ this difference greatly exceeds T_e/e. The distribution obtained $n = n_0 \exp[e(\varphi - \varphi_0)/T_e]$ corresponds to the Boltzmann equation (Eq. 4.18). The equilibrium Boltzmann distribution in an ambipolar electric field arises because the ambipolar field leads to a near-total reflection of the electron flux from the walls (as demonstrated in Section 4.1 the absence of a directed motion is the condition for the existence of the equilibrium distribution).

Knowing the electric field strength, we can find the directed particle velocity. Substituting Eq. 7.65 into Eq. 7.64, we obtain the velocity of the joint (ambipolar) directed motion of the charged particles under the effect of the density gradient:

$$\mathbf{u}_e = \mathbf{u}_i = -D_A \frac{\text{grad}\, n}{n}; \tag{7.68}$$

$$D_A = \frac{D_e b_i + D_i b_e}{b_e + b_i} \tag{7.69}$$

The expression 7.68 formally coincides with the expression for the diffusion flux velocity. The coefficient D_A is called the *ambipolar diffusion coefficient*. Since $D_e \gg D_i$, $b_e \gg b_i$, we obtain from Eq. 7.69 $D_a \approx D_i + b_i D_e / b_e$ and further, from Eqs. 7.10 and 7.12,

$$D_A \approx D_i \left(1 + \frac{T_e}{T_i}\right) = \frac{T_e + T_i}{\mu_{ia} \nu_{ia}} \tag{7.70}$$

It follows that the ambipolar diffusion coefficient is much less than the coefficient of free (unipolar) electron diffusion and greater than the ion diffusion coefficient $D_i < D_A \ll D_e$. Thus an ambipolar electric field greatly reduces the directed electron velocity.

In deriving equations of the ambipolar directed velocity we assumed the absence of temperature gradients. It is easy to include the thermal diffusion in the same way as was done above. Equating the complete expressions for the electron and ion fluxes, that is, Eqs. 7.13 and 7.14, one can find the ambipolar flux with an allowance for the temperature gradients. Then the electric field strength will be equal to

$$\mathbf{E}_A = -\frac{D_e - D_i}{b_e + b_i}\frac{\text{grad}\, n}{n} - \frac{D_e^T}{b_e + b_i}\frac{\text{grad}\, T_e}{T_e} + \frac{D_i^T}{b_e + b_i}\frac{\text{grad}\, T_i}{T_i} \tag{7.71}$$

or, since $b_i \ll b_e$, $D_i \ll D_e$:

$$\mathbf{E}_A \approx -\frac{D_e}{b_e}\frac{\text{grad}\, n}{n} - \frac{D_e^T}{b_e}\frac{\text{grad}\, T_e}{e}$$

$$\approx -\frac{T_e}{e}\left[\frac{\text{grad}\, n}{n} + (1 - g_{T_e})\frac{\text{grad}\, T_e}{T_e}\right]$$

where the relation 7.9, 7.10, and 7.20 for b_e, D_e, and D_e^T are used.

The expression for the ambipolar directed velocity takes the form

$$\mathbf{u}_A = -D_A\frac{\text{grad}\, n}{n} - D_A^{T_e}\frac{\text{grad}\, T_e}{T_e} - D_A^{T_i}\frac{\text{grad}\, T_i}{T_i} \tag{7.72}$$

where the ambipolar diffusion and thermal-diffusion coefficients are equal to

$$D_A = \frac{b_i D_e + b_e D_i}{b_i + b_e} \approx D_i\left(1 + \frac{T_e}{T_i}\right);$$

$$D_A^{T_e} = D_e^T\frac{b_i}{b_e + b_i} \approx D_i\frac{T_e}{T_i}(1 - g_{T_e}); \tag{7.73}$$

$$D_A^{T_i} = D_i^T\frac{b_e}{b_e + b_i} \approx D_i(1 - g_{T_i})$$

Here, in conformity with 7.11 and 7.20, $D_i = T_i/\mu_{ia}\nu_{ia}$; $g_{T_\alpha} \approx (T_{\alpha a}/\nu_{\alpha a})d\nu_{\alpha a}/dT_{\alpha a}$.

When the collision frequencies are velocity independent, $g_{T_e} = g_{T_i} = 0$. In this case the expressions for the ambipolar velocity and the ambipolar field strength can be represented as

$$\mathbf{u}_A = -\frac{\text{grad}(D_A n)}{n} = -\frac{\text{grad}[n(T_e + T_i)]}{n\mu_{ia}\nu_{ia}}; \qquad \mathbf{E}_A = -\frac{\text{grad}(nT_e)}{en} \tag{7.74}$$

7.5 CHARGED PARTICLE AND ENERGY BALANCE EQUATIONS FOR WEAKLY IONIZED PLASMAS

The expressions obtained for the directed velocity and the heat flux of charged particles make it possible to finally establish the particle and energy balance equations considered in Chapter 6. The stationary values of the transfer coefficients can be used if the characteristic times of change in the concentration and temperature of the charged particles greatly exceed the collision frequencies (see Eq. 7.5). We begin with the particle balance equation (see Chapter 6):

$$\frac{\partial n}{\partial t} + \operatorname{div}(n\mathbf{u}) = \frac{\delta n}{\delta t} \tag{7.75}$$

As demonstrated in the preceding sections, the change in concentration is determined by the ambipolar component of the directed velocity. In the general case it is given by Eq. 7.72. When the frequencies of collisions of electrons and ions with atoms are velocity independent, we obtain, by substituting Eq. 7.74 into Eq. 7.75,

$$\frac{\partial n}{\partial t} - \nabla(D_A n) = \frac{\delta n}{\delta t} \tag{7.76}$$

where $D_A = (T_e + T_i)/\mu_{ia}\nu_{ia}$. This equation includes the density and temperature as unknowns. Therefore it must be solved simultaneously with the energy balance equations. The relative temperature gradients, however, are usually much lower than the relative gradients of charged particle density. In this case the directed velocity is defined by Eq. 7.68, and the charged particle balance equation 7.75 takes the form

$$\frac{\partial n}{\partial t} - D_A \nabla n = \frac{\delta n}{\delta t} \tag{7.77}$$

It is an equation in partial derivatives for the density and can be solved independently of the energy balance equations. To solve Eq. 7.77 one must know the initial density distribution and the boundary conditions.

Let us dwell on the latter. More often than not, these conditions result from the removal of charged particles to the dielectric or metal walls enclosing the plasma. In the wall regions one can isolate layers in which the quasi-neutrality conditions are not fulfilled (Fig. 7.3). Their thickness (Δr) is of the order of the Debye radius; it is usually much less than the mean free path. Since on ambipolar removal the walls acquire a negative charge, the electric field in the layer is directed toward the wall. The flux of charged particles toward the wall is evidently determined by their density near the boundary layer n_g, the average velocity in the direction

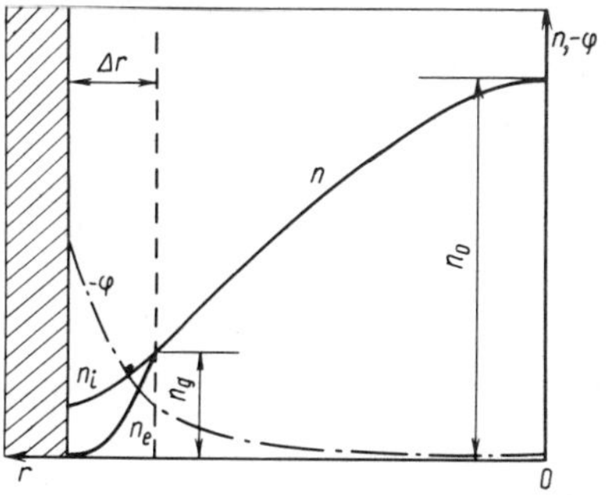

Fig. 7.3 Wall regions in ambipolar diffusion.

of the layer $v_{\alpha g}$, and the coefficient of particle reflection from the layer η_α:

$$\Gamma_{\alpha g} = (1 - \eta_\alpha) n_g v_{\alpha g} \tag{7.78}$$

The average velocity of the wall-bound ions at $T_e > T_i$ may exceed their random velocity because of their acceleration in the ambipolar electric field (Eq. 7.66). At distances of several mean free paths from the wall this field sets up a potential difference of the order of T_e/e; accordingly, near the layer boundary the ions accelerate to an energy of the order of T_e, and their velocity $v_{ig} \sim (T_e/m_i)^{1/2}$. In the electric field of the layer, the ions move toward the wall with acceleration, and since the collisions in the layer are insignificant, practically all of them reach the wall. Besides, most of the ions reflected from the wall are returned by the field of the layer. Therefore the coefficient of reflection of the ions from the layer is low, $\eta_i \ll 1$. Bearing this in mind, we estimate the ion flux toward the wall at $T_e \gg T_i$:

$$\Gamma_{ig} \approx (1 - \eta_i) n_g v_{ig} \approx n_g \sqrt{T_e/m_i} \tag{7.79}$$

When estimating the flux of electrons toward the wall one must remember that their average random velocity always greatly exceeds the directed velocity. Accordingly, their average velocity in the direction of the wall is determined by the random motion $v_{eg} \approx \frac{1}{4}\langle w_e \rangle \approx \sqrt{T_e/m_e}$, and the electron flux toward the wall is equal to $\Gamma_{eg} \approx (1 - \eta_e) n_g \sqrt{T_e/m_e}$. In the absence of a current toward the wall the electron flux must be equal to that of the ions. This equality is possible only when the reflection coefficient of the electrons η_e is close to unity: $1 - \eta_e \approx \sqrt{m_e/m_i}$. It is easy to estimate the potential difference in the layer $\Delta\varphi_l$ ensuring the necessary reflection coefficient. Obviously, only those electrons whose energy of motion toward the wall exceeds $e\Delta\varphi_l$ can reach the wall through the layer. The flux of such electrons toward the layer boundary is equal to

$$\Gamma_{eg} = n_g \int_{(2e\Delta\varphi_l/m_e)^{1/2}}^{\infty} w_x f_{ex}(w_x)\, dw_x$$

where the x axis is directed along the normal to the layer. With a Maxwellian distribution $f_{ex} = (m_e/2\pi T_e)^{3/2} \exp(-m_e w_x^2/2T_e)$ we obtain

$$\Gamma_{eg} = \frac{1}{\sqrt{2\pi}} n_g \sqrt{\frac{T_E}{m_e}} \exp\left(-\frac{e\Delta\varphi_l}{T_e}\right) \tag{7.80}$$

Equating this flux to that of the ions (Eq. 7.79), we find the potential difference ensuring ambipolar removal of particles from the plasma:

$$\Delta\varphi_l \approx \frac{T_e}{e} \ln \sqrt{m_i/m_e} \approx (4\text{–}7)\frac{T_e}{e} \tag{7.81}$$

To estimate the density of charged particles at the plasma/layer boundary (in the quasi-neutrality region) it is necessary to equate the charged particle flux at the layer boundary to the diffusion flux of the particles from the plasma. The flux from the plasma can be determined with the aid of equations for the ambipolar diffusion rate (Eq. 7.68)

$$\Gamma_{ag} = D_A|\text{grad}\, n| \approx \frac{T_e}{m_i \nu_{ia}} \frac{n_0}{L}$$

where n_0 is the charged particle density in the central region of the plasma and L is the characteristic dimensions of the plasma. Equating this flux to the ion flux toward the layer (Eq. 7.79), we obtain the ratio between the densities of the charged particles at the plasma boundary and in the center

$$\frac{n_g}{n_0} \approx \frac{1}{L\nu_{ia}} \sqrt{\frac{T_e}{m_i}} \approx \frac{\lambda_{ia}}{L} \sqrt{\frac{T_e}{T_i}} \tag{7.82}$$

In the region of conditions where the transport equations are applicable the change in plasma parameters over the mean free path must be small. Therefore it is usually possible to put $\lambda_{ia} \ll L$, $n_g \ll n_0$ and approximately assume the boundary density to be equal to zero, $n_g = 0$. The zero boundary conditions are commonly employed in solving diffusion problems.

Now we pass over to the charged particle energy balance equations obtained in Section 6.4 from the second-moment equations. For a weakly ionized plasma, in which only collisions of charged particles with neutrals are substantial, the electron and ion energy balance equations

6.77 can be written thus:

$$\frac{\partial T_\alpha}{\partial t} + \mathbf{u}_\alpha \operatorname{grad} T_\alpha + \frac{2}{3} T_\alpha \operatorname{div} \mathbf{u}_\alpha + \frac{2}{3} \operatorname{div} \mathbf{q}_\alpha$$
$$= -\kappa_{\alpha a}\nu_{\alpha a}(T_\alpha - T_a) + \frac{2}{3}\nu_{\alpha a}\frac{m_\alpha m_a^2}{(m_\alpha + m_a)^2} u_\alpha^2 \qquad (7.83)$$

The expressions obtained above for the directed velocity **u** and the heat flux **q** must be sustituted into these equations. As a result, we get nonlinear partial differential equations. Simultaneous solution of the electron and ion energy balance equations and the particle balance equation makes it possible, in principle, to find the distribution of the density and the electron and ion temperatures in the plasma volume. In the general case this problem is very complicated, of course.

Consider now the electron energy balance equation, assuming that the frequencies of collisions of electrons and ions with atoms are velocity independent. As shown in Section 6.4, the directed electron velocity can be represented as a sum of the ambipolar velocity (Eq. 7.74) and the current velocity due to the external field (Eq. 7.63):

$$\mathbf{u}_e = -\frac{1}{n}\operatorname{grad}(D_A n) - b_e \mathbf{E}_0$$

where $D_A = (T_e + T_i)/\mu_{ia}\nu_{ia}$, $b_e = e/m_e\nu_{ea}$, and the field distribution $\mathbf{E}_0$ must meet the condition $\operatorname{div}(n\mathbf{u}_E) = b_e \operatorname{div}(n\mathbf{E}_0) = 0$. The electron heat flux is determined by the equality 7.22:

$$\mathbf{q}_e = -\mathcal{K}_e \operatorname{grad} T_e = -\chi_e n \operatorname{grad} T_e$$

where $\chi_e = \frac{5}{2}T_e/m_e\nu_{ea} = \frac{5}{2}D_e$. When substituting these expressions into Eq. 7.83 we take into account that $D_A \ll D_e$ and the ambipolar term in the expression for $\mathbf{u}_a$ is usually much less than the current velocity $D_A/L \ll b_e E_0$. Neglecting the corresponding small terms, we obtain

$$\frac{\partial T_e}{\partial t} - \frac{5}{3n}\operatorname{div}(D_e n \operatorname{grad} T_e) - \frac{2}{3} D_A \operatorname{div}\left(\frac{\operatorname{grad} n}{n}\right)$$
$$-\frac{2}{3} e D_e \mathbf{E}_0 \frac{\operatorname{grad} n}{n} - e D_e \mathbf{E}_0 \frac{\operatorname{grad} T_e}{T_e} = -\kappa_{ea}\nu_{ea}(T_e - T_a) + \frac{2}{3}\frac{e^2 E_0^2}{m_e \nu_{ea}} \qquad (7.84)$$

Here the average coefficient of energy transfer on collisions κ_{ea} and the collision frequency ν_{ea} satisfy Eqs. 6.44 and 6.76; that is, they include both elastic and inelastic collisions of electrons with atoms. The nonlinear equation 7.84 is rather complicated but can often be simplified, because some of the terms on the left-hand side are negligibly small. Thus in a gas-discharge plasma in long cylindrical tubes the external field is directed along the axis, and the gradients along the radius; therefore,

$\mathbf{E}_0 \perp \text{grad}\, n$, grad T_e, and the two last terms on the left-hand side of Eq. 7.84 vanish. If grad T/T and grad n/n are comparable, one can neglect the third term compared with the second, since $D_A \ll D_e$. Finally, with sufficiently large ratios of the characteristic dimensions to the mean free paths one can neglect all the terms on the left-hand side that define the energy transfer. Comparing them with the first term on the right-hand side, which describes the electron energy transfer on collisions, we get the conditions for such neglect:

$$L_T^2; \qquad L_T L_n \gg \frac{\lambda_{ea}^2}{\kappa_{ea}}; \qquad L_n^2 \gg \left(\frac{m_e T_E}{m_i T_i}\right)^{1/2} \frac{\lambda_{ea}\lambda_{ia}}{\kappa_{ea}} \tag{7.85}$$

where $L_T = (\text{grad}\, T_e/T_e)^{-1}$, $L_n = (\text{grad}\, n/n)^{-1}$ are the characteristic scales of change in density and temperature. If these conditions hold, Eq. 7.84 takes the simplest form:

$$\frac{\partial T_e}{\partial t} = -\kappa_{ea}\nu_{ea}(T_e - T_a) + \frac{2}{3}\frac{e^2 E_0^2}{m_e \nu_{ea}} \tag{7.86}$$

It defines the local relationship of the electron temperature with the strength of the field induced by external sources.

To solve the electron energy balance equation, with due regard for energy transfer, one must, along with the boundary conditions for the concentration, determine the boundary conditions for the electron temperature by estimating the energy transferred by the electrons leaving the plasma. As has been shown above (see Fig. 7.3), the wall layer has a potential barrier for electrons $e\Delta\varphi_l$ (Eq. 7.81), which greatly reduces their flux toward the wall and thus ensures the ambipolarity of their removal. The electrons with an energy below $e\Delta\varphi_l$ are reflected from the layer and return to the plasma without a change in energy. The electrons with an energy above $e\Delta\varphi_l$ pass through the layer and move away to the walls. They carry part of their energy (remaining after slowing down in the layer) to the walls. The remainder is spent on maintaining the potential difference in the layer; it is carried to the walls by ions accelerating in the layer. Thus the electron energy flux toward the layer boundary is carried by fast particles with an energy exceeding $e\Delta\varphi_l$. Let us determine it for the Maxwellian electron velocity distribution.

Assuming that the x axis is parallel to the normal to the layer, we find

$$\begin{aligned} q_{eg} &= n_g \int w_x \frac{m_e w^2}{2} f_{eg}(\mathbf{w})\, d^3w \\ &= n_g \left(\frac{m}{2\pi T_{eg}}\right)^{3/2} \int_{(2e\Delta\varphi_l/m_e)^{1/2}}^{\infty} w_x\, dw_x \int_{-\infty}^{\infty}\int_{-\infty}^{\infty} dw_y\, dw_z \\ &\quad \times \frac{m_e w^2}{2} \exp\left(-\frac{m_e w^2}{2T_{eg}}\right) \end{aligned}$$

where the quantities n_g, T_{eg}, and q_{eg} are found near the layer boundary, and integration is done over the velocities w_x, at which the barrier can be penetrated ($m_e w_x^2/2 > e\Delta\varphi_l$), and over all the velocities w_y and w_z. Integration yields

$$q_{eg} = \frac{1}{\sqrt{2\pi}} \frac{n_g T_{eg}^{3/2}}{m_e^{1/2}} \left(2 + \frac{e\Delta\varphi_l}{T_{eg}}\right) \exp\left(-\frac{e\Delta\varphi_l}{T_{eg}}\right) \tag{7.87}$$

The ratio of the heat flux to the particle flux at the layer boundary (Eq. 7.80) is equal to the average energy carried away from the plasma by a single electron:

$$\mathscr{E}_0 = \frac{q_{eg}}{\Gamma_{eg}} = T_{eg}\left(2 + \frac{e\Delta\varphi_l}{T_{eg}}\right) \approx T_{eg}(2 + \ln\sqrt{m_i/m_e}) \tag{7.88}$$

To obtain the boundary conditions the heat flux toward the layer boundary must be equated to the heat flux from the plasma. Taking into account Eqs. 7.22 and 7.68, we obtain

$$\frac{5}{2} n_g D_e |\text{grad } T_e| = T_{eg}\left(2 + \ln\sqrt{\frac{m_i}{m_e}}\right)\Gamma_{eg}$$

$$\approx T_{eg}\left(2 + \ln\sqrt{\frac{m_i}{m_e}}\right) D_A |\text{grad } n|$$

From this equality we find the ratio of the temperature and density gradients at the boundary

$$\left(\frac{\text{grad } T_e}{T_e}\right)_g \approx \left(2 + \ln\sqrt{\frac{m_i}{m_e}}\right)\frac{D_A}{D_e}\left(\frac{\text{grad } n}{n}\right)_g$$

$$\approx \left(2 + \ln\sqrt{\frac{m_i}{m_e}}\right)\frac{m_e \nu_{ea}}{m_i \nu_{ia}}\left(\frac{\text{grad } n}{n}\right)_g \tag{7.89}$$

It is seen that near the boundary $(\text{grad } T_e/T_e) \ll (\text{grad } n/n)$, and we can approximately assign the boundary conditions for T_e, assuming $(\text{grad } T_e)_g = 0$. This condition stems from the substantial difference between the coefficient of ambipolar diffusion D_A and the coefficient of thermal conductivity $\chi_e = \frac{5}{2}D_e$. Since $D_A \ll \chi_e$ the transport of the thermal energy of the electrons from the central part to the periphery of the plasma is much faster than the ambipolar transport of the electrons themselves. At the same time the energy transport to the walls is associated with the removal of electrons to the walls, and therefore it occurs at a rate only slightly exceeding that of the ambipolar particle transport. It is the faster exchange of electron energy inside the plasma volume that equalizes the electron temperature in the volume.

The ion energy balance equation can be obtained by substituting Eqs. 7.63 and 7.74 for the directed velocity and Eq. 7.25 for the heat flux into

Eq. 7.83. As a result we derive a still more intricate equation than the one for the electrons, since the ion heat flux is, generally speaking, related to that of the atoms. We do not consider the equation here, but indicate the conditions under which heat transfer is insignificant. They result from a comparison of the equation terms proportional to the density and temperature gradients with the term defining the transfer of the ion energy to the neutral atoms on collisions. It is easy to see that these conditions (for $\mathbf{E}_0 \perp \text{grad}\, n, \text{grad}\, T$) amount to the inequality

$$L \gg \sqrt{T_e/T_i}\, \lambda_{in} \tag{7.90}$$

If this holds, the ion energy balance equation takes the form

$$\frac{\partial T_i}{\partial t} = -\tfrac{1}{2}\nu_{ia}(T_i - T_a) + \tfrac{1}{6}\nu_{ia} m_i u_i^2 \tag{7.91}$$

where, in accordance with Eq. 7.63,

$$\mathbf{u}_i = b_i \mathbf{E}_0 - \frac{\text{grad}(D_A n)}{n} = \frac{2}{m_i \nu_{ia}}\left[e\mathbf{E}_0 - \frac{1}{n}\,\text{grad}\, n(T_e + T_i)\right]$$

(here we took into account that $m_i = m_a$, $\mu_{ia} = \tfrac{1}{2}m_i$).

7.6 CHARGED PARTICLE AND ENERGY BALANCE IN STATIONARY GAS DISCHARGE PLASMAS

To illustrate the application of the equations obtained we consider the charged particle and energy balance in a plasma of a stationary gas discharge maintained by a longitudinal electric field in a long cylindrical chamber. If the length of the chamber greatly exceeds its diameter, we can assume that the plasma parameters are independent of the longitudinal coordinate; that is, the gradients are perpendicular to the axis. The external electric field in such a plasma must be practically homogeneous. According to the analysis carried out in Section 7.5, as a result of the ambipolar regime of charged particle removal to the walls the relative electron temperature gradient near the walls is much lower than the relative density gradient. Therefore when the electron heating is uniform throughout the cross section the electron temperature can also be assumed constant, as a first approximation. Owing to the intensive energy exchange between the ions and the atoms, the ion temperature is usually much lower than the electron temperature and also changes insignificantly over the cross section. The distribution of charged particle densities is then described accurately enough by Eq. 7.77. For the stationary case, assuming $\partial n/\partial t = 0$, we have

$$D_A \nabla n + \frac{\delta n}{\delta t} = 0 \tag{7.92}$$

Recall that in a cylindrically symmetric plasma the density depends exclusively on the radius. Then

$$D_A \frac{1}{r}\frac{d}{dr}\left(r\frac{dn}{dr}\right) + \frac{\delta n}{\delta t} = 0$$

The collision term of the equation $\delta n/\delta t$ determines the efficiency of the ionization and recombination processes in the volume. It may generally include direct ionization on electron–atom collisions, stepwise ionization, electron–ion recombination, and electron capture with subsequent ion–ion recombination. Let us first discuss the simplest situation, when the only volume process appreciably affecting the particle balance is direct ionization. Here $\delta n/\delta t = \nu^i n$, and the equation takes the form

$$D_A \frac{1}{r}\frac{d}{dr}\left(r\frac{dn}{dr}\right) + \nu^i n = 0 \tag{7.93}$$

where $\nu^i = \langle n_a s^i_{ea} v \rangle$ is the average ionization frequency defined by the electron energy distribution function. If the distribution function of the electrons is independent of their density Eq. 7.93 is found to be linear. Its particular solution, which is finite at $r = 0$, is known to be a zero-order Bessel function. It can be written as

$$n = n_0 J_0\left(\frac{r}{\Lambda}\right); \qquad \Lambda = \sqrt{D_A/\nu^i} \tag{7.94}$$

For the density to vanish at the boundary (at $r = a$) the argument of the Bessel function at this point must be its root. The number of roots of the Bessel function is infinite, but it is only the solution corresponding to the first root ($\zeta = 2.405$) that has a physical meaning, since only this solution is positive throughout the region $r < a$. Bearing this in mind, we find

$$\Lambda = \frac{a}{2.405} \tag{7.95}$$

The relation 7.95 determines the value of Λ and hence the density distribution (Eq. 7.94). This distribution, which is called the *diffusion distribution*, is depicted in Fig. 7.4 (curve 1). The characteristic length Λ is called the *diffusion length*. It defines the ratio between ν^i and D_A.

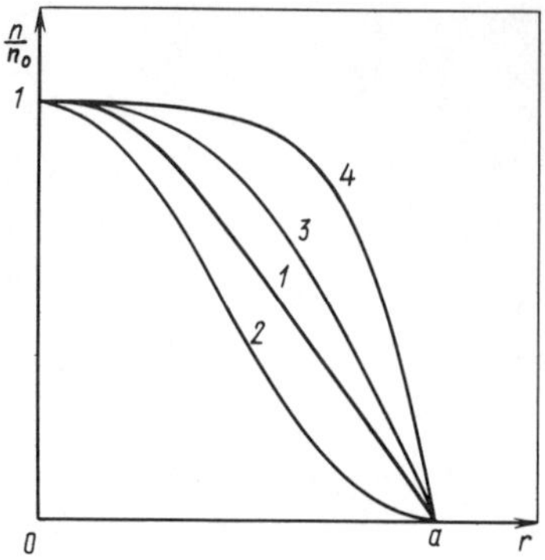

Fig. 7.4 Density distributions as a function of radius.

In accordance with Eqs. 7.94 and 7.95 we find

$$\nu^i = \frac{D_A}{\Lambda^2} \approx \frac{5.8 D_A}{a^2} \tag{7.96}$$

This ratio is the condition for the equilibrium of density, that is, for the equality of the rates of production of charged particles and their diffusion removal from the plasma volume. The right-hand side of Eq. 7.96 is sometimes called the diffusion frequency of removal ν_D, and its reciprocal, the diffusion time τ_D:

$$\tau_D = \frac{1}{\nu_D} = \frac{\Lambda^2}{D_A} \tag{7.97}$$

When solving the charged particle balance equation 7.93, we assumed that the ionization frequency ν^i is density independent. In many cases, however, this is not so. At low densities the dependence of ν^i on n may be due to the effect of electron–electron collisions. As shown in Section 5.6 these collisions tend to "maxwellize" the distribution, increasing the number of fast electrons, that is, the ionization frequency. At high concentrations stepwise ionization also plays an important part; its efficiency depends quadratically on the density. If the dependence of ν^i on n is allowed for, the balance equation 7.93 becomes nonlinear. The density distribution described by this equation depends on the form of the dependence of ν^i on n. Figure 7.4 (curve 2) presents, by way of example, the distribution $n(r)$ for $\nu^i \sim n$. The particle balance condition obtained by solving the equation can be written in a form similar to Eq. 7.96: $\nu^i(n)_0 = D_A/\Lambda^2$, where $\nu^i(n_0)$ is the ionization frequency at the maximum electron density. The ratio between the diffusion length Λ and the plasma radius naturally differs from Eq. 7.95. At $\nu^i n$, for instance, $\Lambda \approx a/3.5$.

The equality 7.96 makes it possible to find the electron temperature. It can be found if the dependence of ν^i on T_e is known. For the Maxwellian distribution this dependence is given by Eq. 5.117. Substituting it into Eq. 7.96, we obtain the transcendental equation for T_e. At $T_e \ll \mathscr{E}_i$ it has the form

$$2\sqrt{\frac{2}{\pi}}\sqrt{\frac{T_e}{m_e}}\, n_0 s_0^i \exp\left(-\frac{\mathscr{E}_i}{T_e}\right) = \frac{T_e}{m_i \nu_{ia} \Lambda^2} \tag{7.98}$$

where s_0^i is determined in conformity with Eq. 2.86. Its approximate solution yields

$$\frac{T_e}{\mathscr{E}_i} = \left[\ln\left(\frac{\Lambda^2}{\lambda_{ea}\lambda_0^i}\frac{m_i \nu_{ia}}{m_e \nu_{ea}}\right)\right]^{-1}$$

(here $\lambda_0^i = 1/n_0 s_0^i$).

The solution of the particle balance equation 7.93 made it possible to find the radial density distribution, but did not yield its absolute value, that is, n_0. In the case at hand the concentration depends on the longitudinal discharge current. The current density is related to the external field via the electron mobility (see Eq. 7.16) $\mathbf{j} = enb_e\mathbf{E}_0$. Accordingly, the total current can be calculated by integrating over the plasma cross section $\mathbf{I} = \int \mathbf{j}\, dS = 2\pi e b_e \mathbf{E}_0 \int_0^a n(r) r\, dr$. Substituting Eq. 7.94, we find

$$n_0 \approx \frac{2.3 I}{\pi a^2 e b_e E_0} = \frac{2.3 m_e \nu_{ea}}{\pi a^2 e^2}\frac{I}{E_0} \tag{7.99}$$

So far we have neglected the effect of volume recombination. When it is necessary to allow for the quadratic recombination the particle balance equation 7.92 takes the form

$$D_A \frac{1}{r}\frac{d}{dr}\left(r\frac{dn}{dr}\right) + \nu^i n - \alpha n^2 = 0 \tag{7.100}$$

where $\alpha = \langle s^r v \rangle$ is the recombination coefficient. If the recombination effect is predominant, that is, $\alpha n_0 \gg \nu_D = D_A/\Lambda^2$, we can neglect the effect of the diffusion term of the equation in the central region of the plasma. The resulting particle balance condition will be

$$\nu^i = \alpha n_0 \tag{7.101}$$

It means that the charged particle production rate as a result of ionization is equal to the rate of their recombination removal. The right-hand side of (7.101) can be called the recombination removal frequency, and the reciprocal quantity, the recombination time

$$\tau_r = \frac{1}{\nu_r} = \frac{1}{\alpha n_0} \tag{7.102}$$

Since ν^i and α depend on T_e, the equality 7.101 defines the electron temperature relative to the density at the center. It is disturbed only near the plasma boundaries, where the diffusion removal of particles is substantial. It is easy to estimate the width of the region (b), where the diffusion is considerable, by comparing the first and last terms of (7.100). Assuming $\partial n/\partial r \sim n/b$ in this region, we obtain

$$\frac{D_A}{b^2} \approx \alpha n_0, \qquad b \approx \sqrt{\frac{D_A}{\alpha n_0}} \ll \Lambda$$

In this case the radial density distribution is plane in the central region and has steep gradients near the boundaries (see Fig. 7.4, curves 3, $\alpha n_0 = 2D_A/\Lambda^2$; and 4, $\alpha n_0 = 3D_A/\Lambda^2$). As the ratio of the diffusion coefficient to the recombination coefficient increases and so does ν_D/ν_r, the density gradients at the plasma boundaries flatten out, and at $\nu_D > \nu_r$ the density distribution approaches the diffusion distribution (Eq. 7.94). Quantitatively, this transition can be traced using Eq. 7.100.

We now pass to the electron energy balance. The balance equation 7.84 for the case in question ($\mathbf{E} \perp \operatorname{grad} n$; $\operatorname{grad} T = 0$, $\partial T/\partial t = 0$) takes the form

$$-\operatorname{div}(\mathcal{K}_e \operatorname{grad} T_e) - D_A n T_e \operatorname{div}\left(\frac{\operatorname{grad} n}{n}\right) = -\frac{3}{2} n\kappa_{ea}\nu_{ea}(T_e - T_a) + \frac{ne^2E_0^2}{m_e\nu_{ea}} \tag{7.103}$$

With a given concentration distribution this equation is a nonlinear differential equation for the electron temperature. As noted above, the relative gradient of the electron temperature can usually be assumed much less than the relative density gradient. Assuming also that the conditions 7.85 are met, we neglect the terms on the left-hand side, as a first approximation. The energy balance equation then amounts to the equality of the average energy acquired by the electrons in the longitudinal electric field and the energy lost by them on elastic and inelastic collisions (Sections 5.4 and 5.6):

$$\frac{ne^2E_0^2}{m_e\nu_{ea}} = \tfrac{3}{2} n\kappa_{ea}\nu_{ea}(T_e - T_a)$$

whence we find the relationship between the strength of the external electric field and the electron temperature:

$$E_0^2 = \frac{3}{2}\frac{m_e}{e^2}\kappa_{ea}\nu_{ea}^2(T_e - T_a) \tag{7.104}$$

Note that this relationship is determined not only by the factor $T_e - T_a$;

the average energy transfer coefficient κ_{ea} and the average collision frequency ν_{ea} may also depend on the electron temperature.

A more accurate condition of the balance of average energies can be obtained from Eq. 7.103. To do this, we must integrate the equation of the plasma cross section. In all the terms, except the first one, we can, as before, assume the electron temperature to be radius independent. We integrate, assuming the radial density distribution to be a diffusion distribution (Eq. 7.94). The integral of the first term is found from the boundary condition (Eq. 7.88):

$$J_1 = \int \operatorname{div} \mathbf{q}_e \, dS = 2\pi a q_{eg} \approx 2\pi a T_e \left(2 + \ln \sqrt{\frac{m_i}{m_e}}\right) \Gamma_g$$

and, further, since the flux of the particles at the boundary Γ_g is determined by their diffusion $2\pi a \Gamma_g = (D_A/\Lambda^2) \int n \, dS$, we obtain

$$J_1 = \frac{D_A}{\Lambda^2} \bar{n} T_e \pi a^2 \left(2 + \ln \sqrt{\frac{m_i}{m_e}}\right)$$

($\bar{n} = 1/\pi a^2 \int n \, dS$ is the density averaged over the cross section).

The integral of the second term transforms to

$$J_2 = D_A T_e \int \operatorname{div}\left(\frac{\operatorname{grad} n}{n}\right) dS = D_A T_e \int \left[\Delta n - \frac{(\operatorname{grad} n)^2}{n}\right] dS$$

$$= -\frac{D_A T_e}{\Lambda^2} \bar{n} \pi a^2 - D_A T_e 2\pi \int_0^a \frac{1}{n} \left(\frac{dn}{dr}\right)^2 r \, dr$$

When calculating the integral appearing in the second term we take into account that it increases logarithmically at $r \to a$, when $n \to 0$. Therefore we can write approximately

$$2\pi \int_0^a \frac{1}{n} \left(\frac{dn}{dr}\right)^2 r \, dr \approx 2\pi a \left(\frac{dn}{dr}\right)_{r=a} \ln \frac{n_0}{n_g} = \frac{\pi a^2 \langle n \rangle}{\Lambda^2} \ln \frac{n_0}{n_g}$$

The integrals of the third and fourth terms include the quantity $\int n \, dS = \pi a^2 \langle n \rangle$.

Collecting the four terms, we obtain the average–energy balance condition:

$$\frac{e^2 E_0^2}{m_e \nu_{ea}} \approx \kappa_{ea} \nu_{ea} (T_e - T_a) + \frac{D_A}{\Lambda^2} T_e \left(2 + \ln \sqrt{\frac{m_i}{m_e}} + \ln \frac{n_0}{n_g}\right) \qquad (7.105)$$

This equality differs from Eq. 7.104 by the second term, which takes into account the energy losses due to the removal of particles to the walls. Since the factor $D_A/\Lambda^2 = \nu_D$ determines the reciprocal diffusion time, the average energy losses per one removed electron are equal to

$$\mathscr{E} = 3T_e + T_e \ln\left(\frac{n_0}{n_g}\right) + T_e \ln \sqrt{m_i/m_e} \tag{7.106}$$

This sum includes, firstly, the energy carried away by the electrons directly to the walls. As shown in Section 7.5, the average value of this energy with a Maxwellian distribution is $2T_e$. Secondly, it includes the energy spent on maintaining the ambipolar field in the plasma. This energy is approximately equal to $e(\varphi_g - \varphi_0) = T_e \ln(n_0/n_g)$, where $\varphi_g - \varphi_0$ is the ambipolar potential difference (Eq. 7.66); it accelerates the ions in their motion toward the boundaries. Finally, the sum 7.106 includes the energy expended on maintaining the wall-layer potential difference $\Delta\varphi_g = (T_e/e) \ln \sqrt{m_i/m_e}$ (Eq. 7.81). As noted above (see p. 207), this energy goes to accelerate the ions, which carry it to the wall.

Let us now obtain the ion temperature, assuming that the condition 7.90 is met and the energy transfer by the ions is insignificant. Making use of the energy balance equation 7.91 for a stationary plasma, we get

$$T_i - T_a = \frac{m_i}{3} b_i^2 E_0^2 + \frac{1}{3} m_i \left[\frac{1}{n} \frac{\partial(D_A n)}{\partial r}\right]^2$$

and with a velocity-independent ion collision frequency,

$$\begin{aligned} T_i - T_a &= \frac{4}{3} \frac{e^2 E_0^2}{m_i \nu_{ia}^2} + \frac{4}{3} \frac{T_e^2}{m_i \nu_{ia}^2} \left(\frac{1}{n} \frac{dn}{dr}\right)^2 \\ &= \frac{4}{3} \frac{e^2}{m_i \nu_{ia}^2} (E_0^2 + E_A^2) \end{aligned} \tag{7.107}$$

where $E_A = (T_e/en)\, dn/dr$ is the ambipolar field (Eq. 7.66).

The ion temperature is determined by the balance between the energy received by the ions in the electric field and the energy transferred by them to the neutral atoms on elastic collisions. Expression 7.107 shows that the ion heating is accomplished both by the external field E_0 and the ambipolar field E_A. It is easy to estimate, with the aid of equality 7.107, the relationship between these fields:

$$\frac{E_A}{E_0} \approx \frac{T_e}{eE_0} \left|\frac{1}{n} \frac{dn}{dr}\right| \approx \frac{\lambda_{ia}}{\sqrt{\kappa_{ia}}} \left|\frac{1}{n} \frac{dn}{dr}\right| \approx \frac{1}{\sqrt{\kappa_{ia}}} \frac{\lambda_{ia}}{\Lambda} \frac{n_0}{n}$$

This estimate shows that at a sufficiently high value of λ_{ia} the ambipolar field can be compared with the longitudinal one. Then the ion temperature must rise from the central region to the plasma boundaries, since the ambipolar field $E_A \sim 1/n$ increases in that direction.

It is easy to ascertain that ion heating, which is determined by the expression 7.107, is much weaker than that of the electrons. Comparing Eqs. 7.107 and 7.104, we obtain

$$\frac{T_i - T_a}{T_e - T_a} \approx \frac{\kappa_{ea} m_e \nu_{ea}^2}{m_i \nu_{ia}^2} \approx \kappa_{ea} \frac{\lambda_{ia}^2}{\lambda_{ea}^2}$$

The difference in heating is due primarily to the difference in energy transfer in collisions: the fraction of energy transferred by the electron ($\kappa_{ea} \ll 1$) is always much less than the ion energy losses ($\kappa_{ia} \approx \frac{1}{2}$).

Thus the particle and energy balance equations make it possible to establish the main characteristics of a gas-discharge plasma. When direct ionization of the atoms by the electrons is the basic ionization process, charged particles are mainly removed through ambipolar diffusion, and the transfer only slightly affects the particle energy balance, we obtain expressions for these characteristics. Thus the density distribution $n(r)$ (Eq. 7.94) is found by solving the particle balance equation, the concentration n_0 is determined by the discharge current (Eq. 7.99), and the electron temperature T_e by the particle balance condition (Eq. 7.96); the strength of the field in the discharge E_0 is obtained from the electron energy balance (Eq. 7.104), and the ion temperature T_i from the ion energy balance (Eq. 7.107).

7.7 IONIZATION INSTABILITY

The balance of the charged particles in a plasma of a stationary gas discharge considered in Section 7.6 may turn out to be unstable. The instability of the ionization balance is caused by the dependence of the ionization frequency on the electron density. As noted above, this dependence may be due to the effect of electron–electron collisions on the distribution function and to stepwise ionization. If the effect of these processes is substantial, an accidental increase in electron density in some section of the plasma column may lead to a local increase of ionization frequency and to a further growth of density. As a result it will increase exponentially with time. This increase of a small perturbation points to instability of the ionization equilibrium with respect to the perturbation.

Let us consider the conditions for *ionization instability*. To this end we use the particle balance equation (see Eq. 7.77):

$$\frac{\partial n}{\partial t} - D_A \Delta n = \nu^i n \tag{7.108}$$

Assume that a plasma exhibits a small density perturbation depending harmonically on the longitudinal coordinate. With this perturbation the density distribution can be represented as

$$\begin{aligned} n(r, z, t) &= n^{(0)}(r) + n^{(1)}(r, z, t) \\ &= n_0(r) + \mathrm{Re}[n_k^{(1)}(r, t)\exp(-\mathrm{i}\,kz)] \end{aligned} \tag{7.109}$$

where $n^{(0)}(r)$ is the unperturbed density, $\mathrm{Re}[n_k^{(1)}\exp(-ikz)]$ is the density perturbation expressed in a complex form to simplify the analysis, and k determines the longitudinal scale or the wavelength of the perturbation $\lambda_k = 2\pi/k$. The density perturbation changes the electric field strength distribution and causes nonuniformity in electron heating. Accordingly, the electron temperature is also perturbed. Since, however, the electron thermal conductivity of the plasma is high, this perturbation is small in a definite range of conditions. We neglect it for the time being.

Substituting the density of Eq. 7.109 into Eq. 7.108 and assuming the radial distributions of the stationary density and of the perturbation to be equal, we obtain the following equation for the density perturbation:

$$\frac{\partial n^{(1)}}{dt} + k^2 D_A n^{(1)} = n^{(0)}(\nu_*^i - \nu^i) \tag{7.110}$$

where ν_*^i is the average ionization frequency in the presence of a perturbation, and ν^i, in its absence. Assuming the perturbation to be small and restricting ourselves to the first term of the expansion ν_*^i in its powers, we get

$$\nu_*^i - \nu^i \approx \frac{\partial \nu^i}{\partial n} n^{(1)}$$

The perturbation equation then takes the form

$$\frac{\partial n^{(1)}}{\partial t} = \left(\frac{n\partial \nu^i}{\partial n} - k^2 D_A\right) n^{(1)} \tag{7.111}$$

Its solution, with a positive coefficient on the right-hand side, yields exponentially increasing perturbation

$$n^{(1)} = n_0^{(1)}\exp(\gamma t); \qquad \gamma = n\frac{\partial \nu^i}{\partial n} - k^2 D_A \tag{7.112}$$

The quantity γ is called the *instability increment.* Thus the criterion of instability in the model under consideration amounts to the inequality

$$\frac{n\partial \nu^i}{\partial n} > k^2 D_A \tag{7.113}$$

With the opposite inequality the power exponent becomes negative. This means that an accidental perturbation decreases with time and hence the plasma is stable with respect to the perturbation.

It is easy to understand the physical meaning of Eq. 7.113. Its left-hand side describes the increase in ionization frequency which perturbs the ionization equilibrium in the presence of a perturbation, and the right-hand side determines the rate of diffusion spread of the perturbation ($\nu_D \sim D_A/\lambda_k^2 \sim D_A k^2$). Therefore the inequality 7.113 means that the rate of ionization increase in perturbation exceeds that of its diffusion spread. The criterion 7.113 limits the perturbation from below. Taking into account the stationary relationship between the ambipolar diffusion coefficient and the ionization frequency (Eq. 7.96), we reduce this criterion to

$$k^2\Lambda^2 < \frac{n}{\nu^i}\frac{\partial \nu^1}{\partial n} \tag{7.114}$$

Since the right-hand side of Eq. 7.114 is of the order of unity, it indicates that instability can arise only at perturbation wavelengths ($\lambda_z = 2\pi/k$) exceeding the plasma column radius.

Let us now consider the effect of a perturbation in electron temperature on the conditions of development of the instability. This effect may be considerable even for relatively small perturbations $T^{(1)}/T \ll n^{(1)}/n$, because the dependence of the ionization frequency on the electron temperature is usually much stronger than its dependence on the density. The perturbation of the electron temperature is due to the longitudinal electric field, which determines the electron heating. In accordance with Eq. 7.63 the longitudinal field consists of two components: the current component E_j and the ambipolar one E_A. The current field is given by the condition of constancy of the longitudinal current along the length

$$j_z = enb_eE_j \tag{7.115}$$

Assuming for simplicity that the collision frequency of the electrons and, hence, their mobility $b_e = e/m_e\nu_{ea}$ are independent of temperature, we obtain

$$nE_j = (n^{(0)} + n^{(1)})(E_0 + E_j^{(1)})$$

where the field E_j is represented as a sum of the unpertured component E_0 and the perturbation $E_j^{(1)}$. Assuming the perturbations to be small, $n^1 \ll n^{(0)}$ and $E_j^{(1)} \ll E_0$, we find the relationship of the perturbations of the field E_j and the density n:

$$E_j^{(1)} = -\left(\frac{n^{(1)}}{n^{(0)}}\right)E_0 \tag{7.116}$$

which characterizes the distribution of $E_j^{(1)}$ along the length.

As can be seen, the perturbation of $E_j^{(1)}$ is in counterphase with the density perturbation: an increase in density reduces the field (Fig. 7.5). The ambipolar component of the longitudinal electric field is given by Eq. 7.66; $E_A = -(T_e/en)\partial n/\partial z$. Substituting Eq. 7.109, we have

$$E_A^{(1)} = \mathrm{i}k \frac{T_e}{en} n^{(1)} \tag{7.117}$$

The longitudinal distribution $E_A^{(1)}(z)$ is shifted in phase by a quarter wavelength with respect to the density distribution $n^{(1)}(z)$—the field is at a maximum in the region where the density gradient is at a maximum (see Fig. 7.5). The relations obtained determine the power of heating of the electron gas with an allowance for the field perturbations. Accordingly, the power released in a unit volume,

$$\begin{aligned} P_E = j_z E_z &\approx enb_e E_0 (E_0 + E_j^{(1)} + E_A^{(1)}) \\ &= enb_e E_0^2 - en^{(1)} b_e E_0^2 + \mathrm{i}kb_e T_e E_0 n^{(1)} \end{aligned} \tag{7.118}$$

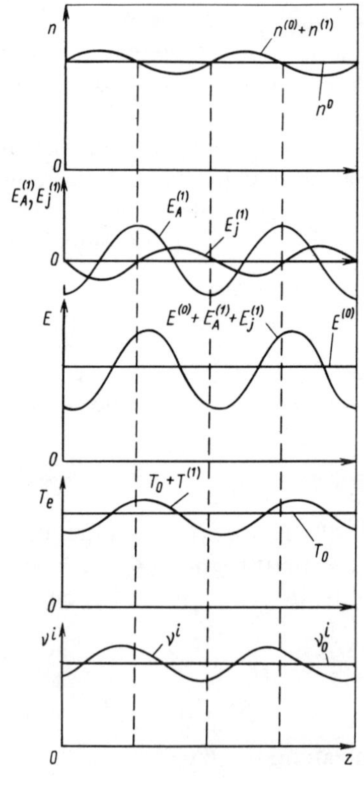

Fig. 7.5 Plasma parameters under ionization instability conditions.

where the constancy of the longitudinal current (Eq. 7.115) is taken into account. The first term of this equation describes longitudinally uniform heating in the absence of perturbations and the second and third, the change in heating due to the density perturbation. The resulting perturbations in electron temperature can be found from the energy balance equation 7.84, which we write approximately in the form

$$\frac{3}{2} n \frac{\partial T_e}{\partial t} - \operatorname{div}(\mathcal{K}_e \operatorname{grad} T_e) = P_E - \frac{3}{2} \kappa_{ea} \nu_{ea} n T_e \tag{7.119}$$

where at $\nu = \text{const}$, $\mathcal{K}_e = \frac{5}{2}(T_e n / m_e \nu_{ea})$. Without perturbations the left-hand side of the equation vanishes, and the energy balance condition amounts to the equality

$$P_{E_0} = e n_0 b_e E_0^2 = \tfrac{3}{2} \kappa_{ea} \nu_{ea} n T_e$$

It yields the relationship of the strength of the longitudinal field with the temperature (Eq. 7.104):

$$e E_0 = \left(\frac{3}{2} \kappa_{ea} m_e \nu_{ea}^2 T_e\right)^{1/2} = \sqrt{\frac{3}{2}} \frac{T_e}{\lambda T_e} \tag{7.120}$$

where

$$\lambda_{T_e} = \left(\frac{T_e}{\kappa_{ea} m_e \nu_{ea}^2}\right)^{1/2} \approx \frac{\lambda_e}{\sqrt{\kappa_{ea}}}$$

is the thermal relaxation length, that is, the length over which energy is exchanged between the electrons and atoms.

The temperature perturbation must be sought in a form similar to Eq. 7.109:

$$T^{(1)}(z, t) = \operatorname{Re}[(T^{(1)}(t) \exp(-ikz)]$$

When substituting it into Eq. 7.119, we take into account that owing to the high thermal conductivity of the electron gas the first term describing the change in perturbation $T^{(1)}$ with time is usually much less than the second term $n|\partial T^{(1)}/\partial t| \ll \mathcal{K}_e k^2 |T^{(1)}|$ and the relative temperature perturbation is much less than the relative density perturbation $|T^{(1)}/T| \ll |n^{(1)}/n|$.* Neglecting the corresponding small terms, we get

$$k^2 \mathcal{K}_e T_e^{(1)} = P_E^{(1)} - \tfrac{3}{2} \kappa_{e0} \nu_{e0} T_{e0} n^{(1)} \tag{7.121}$$

Using the relation 7.118 for P_E and taking into account Eq. 7.120, we find

*As follows from Eq. 7.122, the smallness conditions $|T^{(1)}/T|$ amount to the inequality $k \gg eE_0/T_e = 1/\lambda_{T_e}$.

the relationship of the temperature and density perturbation:

$$\frac{T_e^{(1)}}{T_{e0}} = -\left(\frac{4}{5}\frac{e^2E_0^2}{k^2T_e^2} - \mathrm{i}\frac{eE_0}{kT_E}\right)\frac{n^{(1)}}{n}$$
$$= -\left[\frac{6}{5}\frac{1}{(k^2\lambda_{T_e}^2)} - \frac{\sqrt{6}}{5}\frac{\mathrm{i}}{k\lambda_{T_e}}\right]\frac{n^{(1)}}{n} \tag{7.122}$$

which allows for the fact that at $\nu_{ea} = \text{const}$, $\mathcal{K}_e/b_e = \frac{5}{2}(nT_e/e)$.

The perturbation of the electron temperature, as well as the density perturbation, changes the ionization frequency. Taking both perturbations into consideration, we obtain

$$\nu_*^i - \nu^i = \frac{\partial\nu}{\partial n}n^{(1)} + \frac{\partial\nu}{\partial T_e}T_e^{(1)}$$

where $T_e^{(1)}$ satisfies Eq. 7.122. Substituting this difference into the particle balance equation 7.110, we reduce it to the form

$$\frac{\partial n^{(1)}}{\partial t} = \mathrm{i}\frac{\sqrt{6}}{5}\frac{T_e}{k\lambda_{T_e}}\frac{\partial\nu^i}{\partial T_e}n^{(1)} + \left(n\frac{\partial\nu^i}{\partial n} - D_A k^2 - \frac{6}{5}\frac{T_e}{k^2\lambda_{T_e}}\frac{\partial\nu^i}{\partial T_e}\right)n^{(1)} \tag{7.123}$$

The solution of the equation yields the time dependence of $n^{(1)}$

$$n^{(1)} = n_0^{(1)}\exp(\mathrm{i}\omega t)\exp(\gamma t) \tag{7.124}$$

where

$$\omega = \frac{\sqrt{6}}{5}\frac{T_e}{k\lambda_{T_e}}\frac{\partial\nu_{\mathrm{i}}}{\partial T_e}$$

$$\gamma = n\frac{\partial\nu^i}{\partial n} - D_A k^2 - \frac{6}{5}\frac{T_e}{k^2\lambda_{T_e}^2}\frac{\partial\nu^i}{\partial T_e}$$

Accordingly we obtain

$$n^{(1)} = \mathrm{Re}[n^{(1)}\exp(-\mathrm{i}kz)] = n_{10}\exp(\gamma t)\cos(\omega t - kz) \tag{7.125}$$

As seen from this equation, the density perturbation (as well as the associated field temperature perturbations) is a wave of frequency ω propagating in the direction of the z axis. It is easy to understand its propagation mechanism. As has already been noted, the ambipolar electric field proportional to grad n is shifted in phase by a quarter wavelength relative to the density perturbation (see Fig. 7.5). The electron heating and the ionization frequency increase with the electric force eE. Accordingly, the density increases in the region where $eE > eE_0$ and the ionization frequency exceeds that of diffusion $\nu^i > \nu_D$ and decreases in the region where $eE < eE_0$ and $\nu' < \nu_D$. Since the regions of the maximum and minimum fields are shifted at each given instant relative to the density maxima and minima, the change of density in them results in a displacement of the maxima and minima, that is, in ionization wave propagation (see Fig. 7.5).

The equation for the instability increment γ (Eq. 7.124), with due regard for the temperature perturbation, differs from Eq. 7.112 by an additional term, which reduces the increment. It is associated with the perturbation of the current electric field E_j (Eq. 7.116). Since this perturbation is in counterphase with the density perturbation, it reduces the electron temperature and the ionization frequency in the region of the density maxima and increases them in the region of the density minima. Such changes reduce the initial perturbation. This effect is the greater, the larger the perturbation wavelength, because with decreasing wavelength the electron thermal conductivity decreases the temperature perturbation (as can be seen from Eq. 7.124 the last term is proportional to $1/k^2$).

An analysis of the expression for γ enables us to determine the conditions for ionization instability. Since the diffusion term in γ is proportional to k^2 and the temperature term is inversely proportional to it, the maximum of γ is obtained at a k such that these terms are equal:

$$k_0^2 \approx \left(\frac{6}{5}\frac{T_e}{D_A\lambda_{T_e}^2}\frac{\partial\nu^i}{\partial T_e}\right)^{1/2} = \frac{1}{\Lambda\lambda_{Te}}\left(\frac{6}{5}\frac{T_e}{\nu_i^i}\frac{\partial\nu^i}{\partial T_e}\right)^{1/2} \qquad (7.126)$$

Here γ is positive if $n\partial\nu^i/\partial n > 2D_A k_0^2 = 2\nu^i\Lambda^2 k_0^2$, or

$$\Lambda < \frac{1}{2}\lambda_{T_e}\left(\frac{n}{\nu^i}\frac{\partial\nu^i}{\partial n}\right)\left(\frac{6}{5}\frac{T_e}{\nu^i}\frac{\partial\nu^i}{\partial T_e}\right)^{-1/2} \qquad (7.127)$$

The inequality 7.127 shows that instability is possible only with a sufficiently small plasma radius:

$$a \lesssim \lambda_{T_e} = \left(\frac{T_e}{\kappa_{ea}m_e\nu_{ea}^2}\right)^{1/2}$$

If Eq. 7.127 is fulfilled with a large safety margin, there is a wide range of k values within which ionization instability can arise. On the side of large k this region is bounded by Eq. 7.114, which takes into account the effect of diffusion spread of the perturbation. On the side of small k it is bounded by the effect of temperature perturbation. In accordance with Eq. 7.124

$$n\frac{\partial\nu^i}{\partial n} > \frac{T_e}{k^2\lambda_{T_e}^2}\frac{\partial\nu^i}{\partial T_e} \qquad (7.128)$$

or

$$k^2\lambda_{T_e}^2 > \left(\frac{T_e}{\nu^i}\frac{\partial\nu^i}{\partial T_e}\right)\left(\frac{n}{\nu^i}\frac{\partial\nu^i}{\partial n}\right)^{-1}$$

The maximum increment is determined by the first term in the expression for γ (see Eq. 7.124):

$$\gamma_{\max} \approx \frac{n\partial \nu^i}{\partial n} \tag{7.129}$$

The foregoing analysis makes it possible to determine the conditions for the stability of the ionization balance of a gas-discharge plasma with respect to small perturbations. It is based on the linearization of the balance equations, which is valid while the perturbations of the plasma parameters are much less than their stationary values. In the parameter region where the ionization equilibrium is unstable an increase in perturbation may lead to a new stationary state characterized by a substantial change in plasma parameters. When considering such changes no simplifications associated with the linearization of the equations are allowed, and a much more complicated nonlinear analysis is necessary. Without dwelling on it, we just note that ionization instability may result in a strong longitudinal modulation of the main parameters of a gas-discharge plasma. As a consequence, standing or traveling nonlinear waves are formed in the plasma column, which are called *striations.*

7.8 PLASMA DECAY

We now consider, with the aid of the balance equations, the decay (deionization) of a plasma set up in a long cylindrical chamber under the effect of an external electric field or some other ionization sources. After these are cut off, the temperature of the charged particles decreases, since the energy losses are not offset by external sources. With decreasing temperature the ionization efficiency drops sharply, and diffusion and recombination reduce the concentration of the charged particles (Fig. 7.6). The plasma at this stage is described as *decaying.*

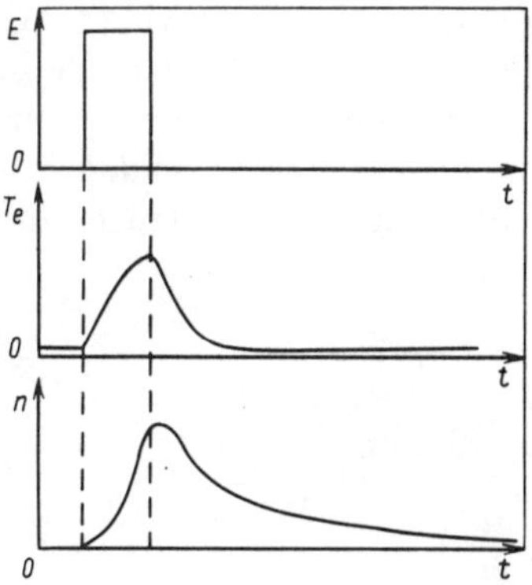

Fig. 7.6 Decay of a plasma in a long cylinder.

To obtain the decay characteristics we take advantage of the particle and energy balance equations. As in the preceding section we assume that the main mechanism of charged particle removal is the ambipolar diffusion and that energy transfer by electrons and ions is insignificant (i.e., the inequalities 7.85 and 7.90 are valid). The balance equations 7.77, 7.86, and 7.91 then take the form

$$\frac{\partial n}{\partial t} - D_A \nabla n = 0;$$

$$\frac{\partial T_e}{\partial t} = -\kappa_{ea}\nu_{ea}(T_e - T_a); \qquad (7.130)$$

$$\frac{\partial T_i}{\partial t} = -\frac{1}{2}\nu_{ia}(T_i - T_a)$$

In the first one we omitted the term $\nu^i n$, assuming that $\nu^i \ll D_A/\Lambda^2$, which means that during decay ionization practically does not affect the particle balance. At velocity-independent collision frequencies ν_{ea} and ν_{ia} with a coefficient κ_{ea} the solutions of the energy balance equations can be written thus:

$$T_e = T_a + (T_{e0} - T_a)\exp(-\kappa_{ea}\nu_{ea}t);$$
$$T_i = T_a + (T_{i0} - T_a)\exp\left(-\frac{1}{2}\nu_{ia}t\right) \qquad (7.131)$$

where T_{e0} and T_{i0} are the electron and ion temperatures at $t = 0$.

It is seen from Eq. 7.131 that after the external energy sources are cut off the electron and ion temperatures vary exponentially. As a result of collisions with neutral atoms their temperatures in the course of decay approach (relax to) the temperature of the neutral gas. The time constants of the temperature drop are equal to

$$\tau_{T_e} = \frac{1}{\kappa_{ea}\nu_{ea}}; \qquad \tau_{T_i} = \frac{2}{\nu_{ia}} \qquad (7.132)$$

The relaxation time of the electron temperature greatly exceeds that of the ion temperature ($\tau_{T_e} \gg \tau_{T_i}$). This difference is due to the low efficiency of electron energy transfer on collisions with atoms ($\kappa_{ea} \ll 1$, whereas $\kappa_{ia} = \frac{1}{2}$).

When ν_{ea}, ν_{ia}, and κ_{ea} are temperature dependent the relaxation law naturally differs from the exponential one. For electrons, say, at κ_{ea}, $\nu_{ea} \sim T_e^s$, and $T_e \gg T_a$, the energy balance equation takes the form

$$\frac{\partial T_e}{\partial t} = -(\kappa_{ea}\nu_{ea})_0\left(\frac{T_e}{T_{e0}}\right)^s T_e$$

Its solution leads to the following $T_e(t)$ dependence:

$$T_e = T_{eo}[1 + s(\kappa_{ea}\nu_{ea})_0 t]^{1/s} \tag{7.133}$$

where $(\kappa_{ea}\nu_{ea})_0$ is the value of $\kappa_{ea}\nu_{ea}$ at $t = 0$.

We now consider the solution of the particle balance equation. For a cylindrically symmetric plasma it can be written as

$$\frac{\partial n(r,t)}{dt} - D_A(t)\frac{1}{r}\frac{\partial}{\partial r}\left[r\frac{\partial n(r,t)}{\partial r}\right] = 0 \tag{7.134}$$

where the diffusion coefficient $D_A = 2(T_e + T_i)/m_i\nu_{ia}$ is, generally speaking, a function of time.

This linear partial differential equation is sometimes called the diffusion equation. Its solution is well known. A partial solution of the equation has the form

$$g(r,t) = \exp\left[-\frac{\zeta^2}{a^2}\int_0^t D_A(t)\,dt\right] J_0\left(\zeta\frac{r}{a}\right)$$

This solution meets the boundary conditions, that is, it vanishes at $r = a$ if ζ is a root of the Bessel function $J_0(\zeta) = 0$. Since the number of such roots is infinite we obtain an infinite number of independent solutions. The general solution can be represented as their linear combination:

$$n(r,t) = \sum_k C_k \exp\left[-\frac{\zeta_k^2}{a^2}\int_0^t D_A\,dt\right] J_0\left(\zeta_k\frac{r}{a}\right) \tag{7.135}$$

The constant coefficients C_k are determined by the initial density distribution:

$$C_k = \int_0^a n(r,0) J_0\left(\zeta_k\frac{r}{a}\right) r\,dr$$

In particular, if the initial distribution is a diffusion distribution (see Eq. 7.94), $n = n_0 J_0(r/\Lambda)$, where $\Lambda = a/\zeta_1$ and $\zeta_1 = 2.405$ is the first root of the Bessel function, the sum 7.135 retains only one term.

Thus the diffusion distribution established in the stationary discharge remains unchanged on plasma decay. The arbitrary initial distribution is deformed in the course of decay. Indeed, the power exponents in Eq. 7.135 depend on the number k, since ζ_k increases with k. Therefore, after a sufficiently long time $t > \tau_D = \Lambda^2/\bar{D}_A$ the first term with the lowest index greatly exceeds the others, and the solution can be written approximately as follows:

$$n(r,t) = C_1 \exp\left[-\frac{1}{\Lambda^2}\int_0^t D_A\,dt\right] J_0\left(\frac{r}{\Lambda}\right) \tag{7.136}$$

This equation shows that irrespective of the initial density distribution, after a time of the order of τ_D a diffusion concentration distribution described by the zero-order Bessel function is established.

In accordance with the equations obtained the decrease in charged particle density in the course of plasma decay is determined by the ambipolar diffusion coefficient $D_A = 2(T_e + T_i)/m_i\nu_{ia}$. Its time dependence is associated with the time variation in electron and ion temperatures. For example, when ν_{ea}, ν_{ia}, and κ_{ea} are velocity independent, one can use Eq. 7.131 to find the dependence $D_A(t)$. The charged particle diffusion time τ_D usually greatly exceeds the relaxation time of the electron temperature τ_{T_e}:

$$\frac{\tau_D}{\tau_{T_e}} = \frac{\Lambda^2}{D_A}\kappa_{ea}\nu_{ea} \approx \sqrt{\frac{m_i}{m_e}}\frac{\kappa_{ea}\Lambda^2}{\lambda_{ea}\lambda_{ia}} \gg 1$$

The time τ_{T_e}, in turn, exceeds the relaxation time of the ion temperature (Eq. 7.132). Therefore in the first stage of decay the electron and ion temperatures decrease, while the density remains practically the same. In the next stage (at $t \gg \tau_{T_e}$) the temperature of the charged particles is virtually equal to that of the neutral gas, and the density decreases relatively slower. In this stage the ambipolar diffusion coefficient is practically constant $D_A \approx 4T_a/m_i\nu_{ia}$. Substituting it into Eq. 7.135 and 7.136, we obtain the law of density variation. In a sufficiently late stage of decay $t > \tau_D$, when the diffusion distribution of densities is established, this law becomes exponential. From Eq. 7.136 we obtain

$$n_0(t) = n_0(0)\exp\left(-\frac{D_A t}{\Lambda^2}\right) \tag{7.137}$$

The time constant of plasma decay in this stage is equal to the diffusion time (Eq. 7.97):

$$\tau_n = \frac{D_A}{\Lambda^2} = \frac{23.2}{a^2}\frac{T_a}{m_i\nu_{ia}} \tag{7.138}$$

Under conditions where, along with the diffusion, the quadratic recombination substantially affects the plasma decay, the particle balance equation takes the form

$$\frac{\partial n}{\partial t} - D_A\frac{1}{r}\frac{\partial}{\partial r}\left(r\frac{\partial n}{\partial r}\right) = -\alpha n^2 \tag{7.139}$$

and in the late stage of decay we can assume that the coefficients of diffusion D_A and recombination α are constant and correspond to $T_e = T_i = T_a$.

It can be shown that, with constant coefficients D_A and α, Eq. 7.139 has a partial solution in which the radial density distribution is, at each given instant, close to the radial distribution in a stationary plasma, which is determined by the solution of Eq. 7.100. This distribution depends on the ratio of the diffusion and recombination removal frequencies $\nu_D = D_A/\Lambda^2$ (Eq. 7.97) and $\nu_r = \alpha n_0$ (Eq. 7.102) and changes in the course of decay from a "flattened" distribution at $\nu_r > \nu_D$ to a diffusion distribution at $\nu_D \gg \nu_r$ (see Fig. 7.4). The change in maximum density n_0 with time, corresponding to this solution, is approximately described by the equation

$$\frac{\partial n_0}{\partial t} = -\rho_D \frac{D_A}{\Lambda^2} n_0 - \rho_\alpha \alpha n_0^2$$

where the coefficients ρ_D and ρ_α (of the order of unity) depend on ν_D/ν_r. The solution of this equation has the form

$$n_0(t) = \frac{\rho_D \dfrac{D_A}{\Lambda^2} n_0(0) \exp\left(-\rho_D \dfrac{D_D}{\Lambda^2} t\right)}{\rho_D \dfrac{D_A}{\Lambda^2} + \rho_D \alpha n_0(0) \left[1 - \exp\left(-\rho_D \dfrac{D_A}{\Lambda^2} t\right)\right]} \tag{7.140}$$

At $D_A/\Lambda^2 \gg \alpha n_0$ the solution 7.140 changes into an exponential dependence characteristic of diffusion decay. At $\alpha n_0 \gg D_A/\Lambda^2$ the particle balance equation can be written as $\partial n/\partial t = -\alpha n^2$. It gives the change in density on plasma recombination decay

$$n(t) = \frac{n(0)}{1 + \alpha n(0) t} \tag{7.141}$$

or $1/n(t) = 1/n(0) + \alpha t$.

Thus the change in charged particle density in the late stage of plasma decay is characterized by ambipolar diffusion and recombination coefficients at a temperature of charged particles close to that of the neutral gas. Therefore the plasma decay is often used for measuring these coefficients.

7.9 DIRECTED MOTION IN HIGHLY IONIZED PLASMAS

In a highly ionized plasma one must take into account not only collisions of charged particles with neutrals, but also their collisions with each other. Then the quasi-stationary equations of electron and ion motion can be written as follows (see Eq. 6.62):

$$\begin{aligned} -e\mathbf{E} - \frac{1}{n}\,\mathrm{grad}(nT_e) + \mathbf{R}_{ea} + \mathbf{R}_{ea}^T + \mathbf{R}_{ei} + \mathbf{R}_{ei}^T = 0; \\ +e\mathbf{E} - \frac{1}{n}\,\mathrm{grad}(nT_i) + \mathbf{R}_{ia} + \mathbf{R}_{ia}^T + \mathbf{R}_{ie} + \mathbf{R}_{ie}^T = 0 \end{aligned} \tag{7.142}$$

They hold for the conditions 7.5 when the changes in directed velocities during the time between collisions can be neglected. In Eq. 7.142 $\mathbf{R}_{\alpha\beta}$ is the friction force acting on α-type particles as a result of their collisions with β-type particles; in reference systems where the directed velocity of the neutral atoms is zero,

$$\mathbf{R}_{ea} = -m_e \nu_{ea} \mathbf{u}_e; \qquad \mathbf{R}_{ia} = -\mu_{ia} \nu_{ia} \mathbf{u}_i;$$
$$\mathbf{R}_{ei} = -\mathbf{R}_{ie} = -m_e \nu_{ei} (\mathbf{u}_e - \mathbf{u}_i) \tag{7.143}$$

The quantities $\mathbf{R}^T_{\alpha\beta} = g^T_{\alpha\beta}$ grad T, where $g^T_{\alpha\beta} \approx (T_{\alpha\beta}/\nu_{\alpha\beta})\, d\nu_{\alpha\beta}/dT_{\alpha\beta}$ are the components of the thermal force acting on the α-type particles due to their collisions with β-type particles. As follows from the momentum conservation law on collisions, $\mathbf{R}^T_{ei} = -\mathbf{R}^T_{ie}$.

Let us find the numerical values of the effective collision frequencies and the coefficients g^T for the case where the frequencies of collisions of electrons and ions with neutral atoms are velocity independent and these collisions make no contribution to the thermal force ($\nu_{ea} = \text{const}$; $\nu_{ia} = \text{const}$; $g^T_{en} = g^T_{in} = 0$). The frequency of electron–ion collisons (see Eq. 2.69),

$$\nu^t_{ei} = \frac{4\pi n e^4}{m_e^2 v^3} L_e$$

can be assumed inversely proportional to the cube of the electron velocity, neglecting the contribution of the ions to the relative velocity and the weak velocity dependence of the coulomb logarithm. For this case the collision term representing the effect of electron–ion collisions on the directed velocity of the electrons when their distribution is near-Maxwellian can be found using Eqs. 6.48, 6.49, and 6.57 (in the eight-moments approximation):

$$\left(\frac{\delta \mathbf{u}_e}{\delta t}\right)_{ei} = -\frac{m_e}{3T_e} \langle \nu^t_{ei} v^2 \rangle (\mathbf{u}_e - \mathbf{u}_i) + \frac{1}{3} \frac{\mathbf{q}_e}{nT_e} \left\langle \nu^t_{ei} \left(\frac{m_e v^2}{T_e} - \frac{m_e^2 v^4}{5T_e^2} \right) \right\rangle \tag{7.144}$$

where $\langle\ \rangle$ means averaging over the Maxwellian electron distribution. Carrying out this averaging, we obtain (see Eq. 5.116)

$$\begin{aligned}\bar{\nu}_{ei} &= \frac{m_e}{3T_e} \langle \nu_{ei} v^2 \rangle \\ &= 4\pi \frac{n e^4}{m_e^2} L_e \frac{1}{3} \sqrt{\frac{2}{\pi}} \left(\frac{m_e}{T_e}\right)^{5/2} \int_0^\infty v \exp\left(-\frac{m_e v^2}{2T_e}\right) dv \\ &= \frac{4\sqrt{2\pi}}{3} \frac{n e^4}{m_e^{1/2} T_e^{3/2}} L_e \end{aligned} \tag{7.145}$$

$$\left\langle \nu_{ei}\left(\frac{m_e v^2}{T_e}-\frac{m_e^2 v^4}{5T_e^2}\right)\right\rangle = \frac{4\pi n e^4 L_e}{m_e^2}\sqrt{\frac{2}{\pi}}\left(\frac{m_e}{T_e}\right)^{5/2}$$

$$\times\int_0^\infty dv\left(v-\frac{m_e}{5T_e}v^3\right)\exp\left(-\frac{m_e v^2}{2T_e}\right)$$

$$=\frac{9}{5}\bar{\nu}_{ei}$$

Equation 7.144 includes the electron heat flux. Substituting its value obtainable from Eq. 7.160, which takes into account the effect of electron–ion, electron–electron, and electron–atom collisions, we get

$$m_e\left(\frac{\delta \mathbf{u}_e}{\delta t}\right)_{ei} \approx -m_e\bar{\nu}_{ei}\frac{\nu_{ea}+0.97\bar{\nu}_{ei}}{\nu_{ea}+1.87\bar{\nu}_{ei}}(\mathbf{u}_e-\mathbf{u}_i)-\frac{3}{2}\frac{\bar{\nu}_{ei}}{(\nu_{ea}+1.87\bar{\nu}_{ei})}\operatorname{grad} T_e$$

The first term of this equation yields the friction force,

$$\mathbf{R}_{ei} = -m_e\nu_{ei}(\mathbf{u}_e-\mathbf{u}_i); \qquad \nu_{ei} = \bar{\nu}_{ei}\frac{\nu_{ea}+0.97\bar{\nu}_{ei}}{\nu_{ea}+1.87\bar{\nu}_{ei}} \tag{7.146}$$

and the second term the thermal force,

$$\mathbf{R}_{ei}^T = -g_T \operatorname{grad} T_e; \qquad g_T = \frac{3}{2}\frac{\bar{\nu}_{ei}}{\nu_{ea}+1.87\bar{\nu}_{ei}} \tag{7.147}$$

It can be seen that the effective frequency of electron–ion collisions, which determines the friction force, and the coefficient g^T, which determines the thermal force, depend not only on $\bar{\nu}_{ei}$, but also on ν_{ea}. In particular, if $\nu_{ea} \gg \nu_{ei}$, then $\nu_{ei} = \bar{\nu}_{ei}$ and $g_T = \frac{3}{2}\bar{\nu}_{ei}/\nu_{ea} \ll 1$; at $\nu_{ei} \gg \nu_{ea}$ and for a fully ionized plasma, $\nu_{ei} \approx 0.52\bar{\nu}_{ei}$; $g_T \approx 0.80$. More accurate calculations (corresponding to inclusion of the higher moments of the distribution function) give somewhat different numerical coefficients. For $\nu_{ei} \gg \nu_{ea}$ they yield

$$\nu_{ei} = 0.51\bar{\nu}_{ei}; \qquad g_T = 0.71 \tag{7.148}$$

It should be noted that the above values of ν_{ei} and g^T cover the effect of electron–electron and electron–ion collisions. It is easy to understand the mechanism of this effect. Because of the strong velocity dependence of electron–ion collision frequency the main contribution to the directed velocity is made by the fast electrons, whose collision frequencies are substantially lower than those of the slow electrons. Therefore the effective collision frequency is much lower than the average (considering only electron–ion collisions, it is equal to $0.3\bar{\nu}_{ei}$). Electron–electron collisions lead to energy exchange between the fast and slow particles and therefore reduce this effect; the reduction is comparable with the effect, since $\nu_{ee} = \sqrt{2}\nu_{ei}$. Its additional decrease at $\nu_{ea} \gtrsim \nu_{ei}$ is due to electron–atom collisions, which at $\nu_{ea} = \text{const}$ equalize the contribution to the directed velocity from the fast and slow particles. As shown in Section 6.3 the thermal force is due to the velocity dependence of the collision frequency, that is, arises from the more frequent collision with ions of electrons moving from the lower-temperature region than of the particles moving in the opposite direction. With electron–ion collisions only, this effect leads to a coefficient $|g_T| \approx |(T/\nu)d\nu/dT| = \frac{3}{2}$ (see Eq. 7.41). Electron–electron and electron–atom collisions reduce the effect, since the associated change in velocity direction causes mixing of the particles moving along and against the gradient.

Thus at velocity-independent frequencies of collisions of electrons and ions with atoms the equations of motion 7.142 take the form

$$-e\mathbf{E} - \frac{1}{n}\,\text{grad}(nT_e) - g_T\,\text{grad}\,T_e - m_e\nu_{ea}\mathbf{u}_e - m_e\nu_{ei}(\mathbf{u}_e - \mathbf{u}_i) = 0$$

$$+e\mathbf{E} - \frac{1}{n}\,\text{grad}(nT_i) + g_T\,\text{grad}\,T_e - \mu_{ia}\nu_{ia}\mathbf{u}_l - m_e\nu_{ei}(\mathbf{u}_i - \mathbf{u}_e) = 0 \qquad (7.149)$$

where the collision frequency ν_{ei} and the coefficient g_T are determined by Eqs. 7.146–7.148. It is taken into consideration here that in the electron and ion equations the collision terms due to electron–ion collisions are equal in magnitude and opposite in sign. Equations 7.149 enable us to find the directed velocity of the charged particles. As in a weakly ionized plasma, it can usually be represented as the sum of the ambipolar velocity and the velocity determined by the external electric field. Let us first find the directed velocity due to the field. To do this, we put grad n = grad T_e = grad $T_i = 0$ in our equations. Then, adding them up, we obtain the ratio between $\mathbf{u}_{eE}$ and $\mathbf{u}_{iE}$:

$$\mathbf{u}_{iE} = -\frac{m_e\nu_{ea}}{\mu_{ia}\nu_{ia}}\,\mathbf{u}_{eE} \qquad (7.150)$$

which indicates that $\mathbf{u}_{iE} \ll \mathbf{u}_{eE}$. Bearing this in mind, we find the directed electron velocity

$$\mathbf{u}_{eE} = -\frac{e\mathbf{E}}{m_e(\nu_{ea} + \nu_{ei})} \qquad (7.151)$$

with the aid of Eq. 7.151 we write the current density and the plasma conductivity

$$\mathbf{j} = -ne\mathbf{u}_{eE} = \sigma E; \qquad \sigma = \frac{ne^2}{m_e(\nu_{ea} + \nu_{ei})} \qquad (7.152)$$

The expression 7.152 is similar to Eq. 7.16 for the conductivity of a weakly ionized plasma, the only difference being that it contains the summary collision frequency $\nu_{ea} + \nu_{ei}$. For a highly ionized plasma, in which $\nu_{ei} \gg \nu_{ea}$, we get, taking into account Eqs. 7.145 and 7.148,

$$\sigma \approx \frac{2ne^2}{m_e \bar{\nu}_{ei}} = \frac{3}{2\sqrt{2\pi}} \frac{T_e^{3/2}}{e^2 m_e^{1/2} L_e} \tag{7.153}$$

The conductivity of such a plasma is practically independent of the charged particle density. It is easy to understand the cause. On the one hand, the conductivity is proportional to the number of current carriers, that is, to the electron density. On the other, it is inversely proportional to the density of ions, collisions which hinder electron acceleration. Since the electron and ion densities are equal, these dependences offset each other. The conductivity of a strongly ionized plasma depends only on the electron temperature, permiting the use of conductivity measurements for temperature determination.

Let us now find the ambipolar component of the directed velocity, assuming $\mathbf{u}_e = \mathbf{u}_i = \mathbf{u}_A$ in Eq. 7.149. The friction force due to the electron–ion collisions reduces to zero, and the equations take a form similar to the equations of motion in a weakly ionized plasma (see Eq. 7.6). Adding up the equations of motion of the electrons and ions, we get the expression for the ambipolar velocity, which is similar to Eq. 7.74:

$$\mathbf{u}_A = -\frac{\text{grad}[n(T_e + T_i)]}{\mu_{ia}\nu_{ia} n} = -\frac{1}{n}\,\text{grad}(D_A n) \tag{7.154}$$

(this takes into account that $m_e \nu_{ea} \ll \mu_{ia}\nu_{ia}$). The value of the ambipolar electric field will be found from the difference of the equations of motion (Eqs. 7.149):

$$\begin{aligned} \mathbf{E}_A &= -\frac{\text{grad}(nT_e)}{en} - \frac{g_T}{e}\,\text{grad}\,T_e \\ &= -\frac{T_e}{e}\frac{\text{grad}\,n}{n} - \frac{(1+g_T)}{e}\,\text{grad}\,T_e \end{aligned} \tag{7.155}$$

Since Eqs. 7.151 and 7.154, which define the directed velocity of charged particles in a highly ionized plasma, are similar to the corresponding equations for a weakly ionized plasma, the particle balance equations are also similar, and we do not discuss them here.

The obtained expressions for the directed velocity fail when the mean free paths of the charged particles relative to their collisions with neutral atoms are comparable with or greater than the characteristic size of the plasma. Under such conditions the motion of the charged particles under the effect of the density and temperature gradients no longer bears a diffusion character; their directed velocity in the presence of gradients can be compared with the thermal velocity. Switching to the limits

$\nu_{ia} \to 0$, $\nu_{ea} \to 0$ in Eq. 7.149 and adding them up, we find that the total-pressure gradient must be equal to zero:

$$\text{grad}[n(T_e + T_i)] = 0; \qquad p = n(T_e + T_i) = \text{const}$$

The constancy of the total plasma pressure is the condition for the applicability of Eqs. 7.149 in describing the directed motion in a fully ionized plasma. In the absence of gradients of the ion and electron pressure $[\text{grad}(nT_e) = \text{grad}(nT_i) = 0]$ the two equations amount to the equality:

$$e\mathbf{E} + 0.71 \text{ grad } T_e = -0.51\, m_e \bar{\nu}_{ei}(\mathbf{u}_e - \mathbf{u}_i) \tag{7.156}$$

where we have used the relations 7.148 for ν_{ei} and g_T. It can be seen that Eq. 7.156 determines exclusively the relative electron and ion velocity. For the ambipolar regime ($\mathbf{u}_e = \mathbf{u}_i$) Eq. 7.156 yields the strength of the electric field ensuring ambipolarity:

$$\mathbf{E}_A = -\frac{g_T}{e} \text{ grad } T_e = -\frac{0.71}{e} \text{ grad } T_e \tag{7.157}$$

Representing in the general case the field as the sum of the ambipolar field and the current-inducing field $\mathbf{E}_j$, we find from Eq. 7.156 the conductivity of a fully ionized plasma in the field $\mathbf{E}_j$:

$$\mathbf{j} = -ne(\mathbf{u}_e - \mathbf{u}_i) \approx 2\frac{ne^2}{m_e \bar{\nu}_{ei}} \mathbf{E}_j; \qquad \sigma \approx \frac{2ne^2}{m_e \bar{\nu}_{ei}} \tag{7.158}$$

The conductivity equation coincides with Eq. 7.153.

7.10 ENERGY TRANSFER IN HIGHLY IONIZED PLASMAS

Let us now find the expression for the electron heat flux in a highly ionized plasma. The stationary heat flux equation for the case where the frequency of electron–atom collisions is velocity independent can be obtained by substituting the collision term 6.86 into 6.80. It has the form

$$\frac{5}{2}\frac{nT_e}{m_e} \text{ grad } T_e = -(\nu_{ea} + 1.87\bar{\nu}_{ei})\mathbf{q}_e + \frac{3}{2}\bar{\nu}_{ei} n T_e(\mathbf{u}_e - \mathbf{u}_i) \tag{7.159}$$

where the first term on the right-hand side covers the electron–atom, electron–ion, and electron–electron collisions. Equation 7.159 defines the electron heat flux

$$\mathbf{q}_e = -\mathcal{K}_e \text{ grad } T_e + g_T n T_e(\mathbf{u}_e - \mathbf{u}_i) \tag{7.160}$$

where

$$\mathcal{K}_e = \frac{5}{2}\frac{nT_e}{m_e(\nu_{ea} + 1.87\bar{\nu}_{ei})}; \qquad g_T = \frac{3}{2}\frac{\bar{\nu}_{ei}}{\nu_{ea} + 1.87\bar{\nu}_{ei}}$$

(because the kinetic coefficients are symmetrical, the coefficient g_T coincides with the coefficient appearing in the thermal force expression 7.147). The expression obtained differs from the corresponding equation for a weakly ionized plasma (see Eq. 7.22) in that the collision frequency ν_{ea} is replaced by the summary frequency $\nu_{ea} + 1.87\nu_{ei}$ in the coefficient of thermal conductivity and that a term appears which is proportional to the directed velocity (it is associated with the velocity dependence of the collision frequency ν_{ei}^t). At a sufficiently high degree of ionization, when $\nu_{ei} \gg \nu_{ea}$, the thermal conductivity coefficient takes the form

$$\mathcal{K}_e = \frac{5}{2} g_q \frac{nT_e}{m_e\bar{\nu}_{ei}} = \frac{15}{8\sqrt{2\pi}} \frac{g_q T_e^{5/2}}{e^4 m_e^{1/2} L_e} \tag{7.161}$$

where $g_q = 0.54$. The thermal conductivity coefficient strongly depends on the electron temperature ($\sim T_e^{5/2}$) and is practically independent of the density (as in the conductivity equation 7.153, there is no such dependence because the frequency of electron–ion collisions ν_{ei} is proportional to the density).

A more accurate calculation with the inclusion of the higher moments of the distribution function leads to a considerable change of the numerical coefficients appearing in Eqs. 7.160 and 7.161. For a fully ionized plasma it yields a coefficient $g_q = 1.22$. Note that this value is obtained by considering both the electron–ion and electron–electron collisions. When only electron–ion collisions are taken into account, the coefficient g_q is much larger ($g_q = 5$). The value of electron thermal conductivity is high because the energy is principally transported by the fast electrons, which collide with ions more seldom than the slow. Electron–electron collisions greatly reduce the difference between the contribution of the fast and slow electrons to the transport, since they result in effective energy exchange between them.

The thermal flux of ions and atoms can be obtained from Eq. 6.80 with the collision terms 6.87 and 6.88. We give here the expression for the thermal ion flux in a plasma with a high degree of ionization, in which the frequency of ion–ion collisions is much higher than for ion–atom collisions, $\nu_{ii} \gg \nu_{ia}$. The collision term of the equation of the thermal flux of ions is practically determined by the ion–ion collisions alone (ion–electron interaction plays a much lesser role because of the low electron mass). Accordingly, the thermal flux is equal to

$$\mathbf{q}_i = -\mathcal{K}_i \operatorname{grad} T_i,$$
$$\mathcal{K}_i = \frac{5}{2} g_{qi} \frac{nT_i}{m_i\nu_{ii}} = \frac{15}{8\sqrt{\pi}} g_{qi} \frac{T_i^{5/2}}{\sqrt{m_i} e^4 L_i} \tag{7.162}$$

The use of the collision term 6.85 in the thermal flux equation 6.80 leads to $g_{qi} = 1.25$. A more accurate calculation yields $g_{qi} = 1.56$. Note that at similar temperatures T_e and T_i the coefficient of ion thermal conductivity (Eq. 7.162) is much less than for the electrons (Eq. 7.161):

$$\frac{\mathcal{K}_i}{\mathcal{K}_e} \approx \left(\frac{T_i}{T_e}\right)^{5/2} \left(\frac{m_e}{m_i}\right)^{1/2} \tag{7.163}$$

The expressions obtained for the heat fluxes make it possible to establish the equations of balance of the average energies of the charged particles in a highly ionized plasma. At a sufficiently high degree of ionization, when $\nu_{ei} \gg \nu_{ea}$ and $\nu_{ii} \gg \nu_{ia}$, these equations result from substitution of Eqs. 7.161 and 7.162 into Eq. 6.79. Without an external electric field (at $\mathbf{u}_e = \mathbf{u}_i = 0$) they take the following form:

$$\begin{aligned} n\frac{\partial T_e}{\partial t} &= \frac{5}{3} g_{qe} \operatorname{div}\left(\frac{nT_e}{m_e\bar{\nu}_{ei}} \operatorname{grad} T_e\right) - \kappa_{ei}\bar{\nu}_{ei}n(T_e - T_i); \\ n\frac{\partial T_i}{\partial t} &= \frac{5}{3} g_{qi} \operatorname{div}\left(\frac{nT_i}{m_i\bar{\nu}_{ii}} \operatorname{grad} T_i\right) - \kappa_{ie}\bar{\nu}_{ei}n(T_i - T_e) - \frac{1}{2}\nu_{ia}n(T_i - T_a) \end{aligned} \tag{7.164}$$

where it is taken into consideration that the coefficient of energy transfer on collisions of ions and atoms $\kappa_{ia} = \frac{1}{2}$, since $m_i = m_a$; the energy transfer coefficient for elastic electron–ion collisions $\kappa_{ei} = 2m_e/m_i$ and, in principle, κ_{ei} can also cover the effect of inelastic losses of electron energy. The first term on the right-hand side of the equations describes the heat transfer. The second term determines the energy exchange between the electrons and ions. The third term on the right-hand side of the ion equation determines the energy losses on collisions of ions with neutral particles; these collisions can be substantial even in a highly ionized plasma, since $\kappa_{ia} = \frac{1}{2}$, while $\kappa_{ie} = 2m_e/m_i \ll 1$.

To evaluate the relative role of the different terms we can introduce characteristic times determining the efficiency of the corresponding processes $\tau_{\alpha q} = T_\alpha(\partial T_\alpha/\partial t)_q^{-1}$: the heat transfer times for electrons and ions,

$$\tau_{eq} \approx \frac{nL_T^2}{\mathcal{K}_e} \approx \frac{m_e\bar{\nu}_{ei}}{T_e} L_T^2; \qquad \tau_{iq} \approx \frac{nL_T^2}{\mathcal{K}_i} \approx \frac{m_i\bar{\nu}_{ii}}{T_i} L_T^2 \tag{7.165}$$

(here $L_T \approx [(1/T) \operatorname{grad} T]^{-1}$), and the energy exchange times,

$$\tau_{ei}^T \approx \frac{1}{\kappa_{ei}\bar{\nu}_{ei}}; \qquad \tau_{ia}^T \approx \frac{2}{\nu_{ia}} \tag{7.166}$$

For a fully ionized plasma we can derive the summary heat transfer equation by adding up Eqs. 7.164. Neglecting the thermal conductivity, since $\mathcal{K}_i \ll \mathcal{K}_e$ (Eq. 7.163), and assuming that the inelastic electron energy losses are insignificant ($\kappa_{ei} = \kappa_{ie} = 2m_e/m_i$), we have

$$n\,\frac{\partial(T_e + T_i)}{\partial t} - \frac{5}{3}\,g_{qe}\,\mathrm{div}\left(\frac{nT_e}{m\bar{\nu}_{ei}}\,\mathrm{grad}\,T_e\right) = 0 \qquad (7.167)$$

The ratio between the electron and ion temperatures depends on the efficiency of energy exchange between the electrons and ions. With a relatively high efficiency, when $\tau^T_{ei} \ll \tau_{eq}$, it can be assumed that $T_e = T_i$. In the opposite case the ratio depends on the initial and boundary conditions and can be obtained from the electron equation of energy balance.

7.11 ELECTRON "RUNAWAYS"

In a highly ionized plasma, where the frequency of collisions of electrons with ions greatly exceeds that of their collisions with neutral atoms $\nu_{ei} \gg \nu_{ea}$, the mean free path of the electrons rapidly increases with energy. In a sufficiently strong electric field, where the electrons accumulate, between collisions, an energy comparable with the random, they can go into the continuous-acceleration regime. This transition is called the *effect of runaway* of electrons (they "run away" from collisions).

The conditions for transition of electrons to the runaway regime can be evaluated with the aid of the averaged equation of motion of electrons in an electric field. In accordance with this equation the stationary value of the directed velocity can be found by equating the electric force and the friction force due to the electron–ion collisions:

$$e\mathbf{E} = \mathbf{R}_{ei} = -m_e\langle \nu^t_{ei}(v)\mathbf{v}\rangle \qquad (7.168)$$

where

$$\nu_{ei} = \frac{4\pi n e^4}{m_e^2 v^3}\,L_e \approx \beta\,\frac{n}{v^3} \qquad (7.169)$$

Here $\mathbf{v}$ is the relative velocity, which is practically equal to the electron velocity, and $\langle\ \ \rangle$ means averaging over the velocities. For a weak electric field, in which the directed electron velocity is much less than the random one $\mathbf{u} \ll \mathbf{w}$, the averaging results in an equation corresponding to Eq. 7.146:

$$\mathbf{R}_{ei} = -0.51 m_e \bar{\nu}_{ei}\mathbf{u}_e \approx -0.14\beta m_e n\left(\frac{m_e}{T_e}\right)^{3/2}\mathbf{u}_e \qquad (7.170)$$

Here the friction force $\mathbf{R}_{ei}$ increases linearly with $\mathbf{u}_e$.

For a strong field in which the energy accumulated by the electrons between collisions is much higher than the thermal energy ($u_e \gg w$), $v \approx u_e$ and the collision frequency is practically independent of the random velocity. In this case the friction force

$$\mathbf{R}_{ei} = -m_e \mathbf{u}_e \nu_{ei}(u_e) \approx -\frac{\beta m_e n \mathbf{u}}{u^3} \tag{7.171}$$

is inversely proportional to the square of the directed velocity. Thus the dependence of R_{ei} on u_e represents a curve with a maximum at $u \approx w$. To find its approximate estimates, we can use the following equation, which is correct to within a numerical factor of the order of unity:

$$\mathbf{R}_{ei} = -\frac{\beta m_e n \mathbf{u}}{(u^2 + v_T^2)^{3/2}}; \qquad v_T^2 = \frac{3T_e}{m_e} \tag{7.172}$$

A more accurate calculation of R_{ei} at $u \approx w$ can be accomplished by averaging Eq. 7.168 over the distribution function. The result of the averaging carried out, with the assumption that the random velocity distribution is Maxwellian, is presented in Fig. 7.7. From Eq. 7.172 it is seen that the maximum value of the friction force at $u \approx v_T$ is equal to

$$\begin{aligned} R_{\max} &\approx 0.2\beta \frac{m_e^2 n}{T_e} \approx 0.2 \frac{4\pi n e^4}{T_e} L_e \\ &\approx 0.75 m_e^{1/2} T_e^{1/2} \bar{\nu}_{ei} \end{aligned} \tag{7.173}$$

At $|eE| > R_{\max}$ the friction force cannot offset the electric force at any directed velocity. Under the effect of the electric field the electrons will go into the continuous-acceleration regime. The critical field strength, which determines the transition boundary, is*

$$E_c \approx 0.2 \frac{4\pi n e^3}{T_e} L_e \approx 0.2 \frac{e L_e}{r_D^2} \tag{7.174}$$

It is easy to ascertain that the average energy accumulated by the electrons in such a field along the mean free path is of the order of the

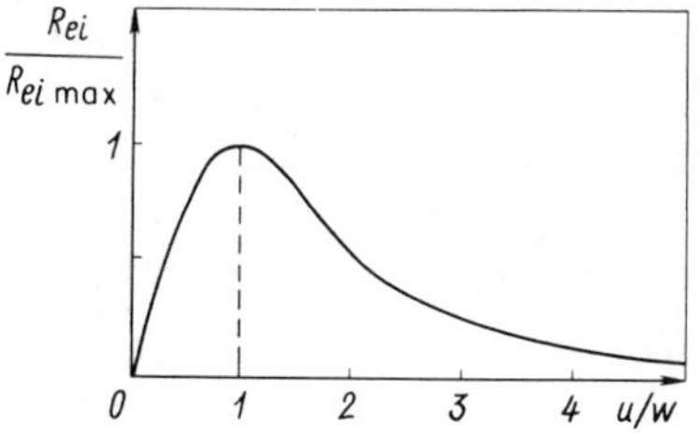

Fig. 7.7 Average friction force on runaway electrons as a function of energy.

*The condition 7.174 is sometimes called the *Dreicer regime.*

thermal energy

$$W_\lambda = \overline{eE_c\lambda_{ei}} \approx \left(\frac{eE_c}{\bar{\nu}_{ei}}\right)\sqrt{\frac{T_e}{m_e}} \approx T_e$$

Although at typical plasma parameters the critical field (Eq. 7.174) is not high, it is usually difficult to set up such a field in a fully ionized plasma possessing a high conductivity. But at lower field strengths "partial runaway" of electrons is also possible; that is, transition to the acceleration regime of fast electrons, for which the collision frequency, determining the friction, is lower than for the bulk of the electrons.

Let us consider, for instance, a group of electrons moving in the direction of the external electric force with similar velocities substantially exceeding the thermal velocities ($v_p \gg v_{T_e}$). The friction force acting on these electrons is due to their collisions with ions and with the bulk of the electrons. We can readily estimate this force by assuming that the relative velocity on collisions is equal to the velocity of the particles of the isolated group:

$$R(v_p) = R_{ei}(v_p) + R_{ee}(v_p) \approx \mu_{ei}\nu^t_{ei}(v_p)v_p + \mu_{ee}\nu^t_{ee}(v_p)v_p \qquad (7.175)$$

This estimate neglects collisions with electrons whose velocity exceeds v_p, but at $v_p \gg v_{T_e}$ the number of such electrons is small. The electrons of the isolated group will be accelerated if the electric force acting on them is larger than the friction force. Using Eq. 2.69 for electron collision frequencies, we write this condition as

$$eE > m_e v_p \left[\nu^t_{ei}(v_p) + \frac{1}{2}\nu^t_{ee}(v_p)\right] \approx \frac{12\pi e^4}{m_e v_p^2} L_e n$$

which takes into account that $\mu_{ei} \approx m_e$; and $\mu_{ee} = m_e/2$. With a given electric field it determines the velocity from which electrons moving in the direction of the field are accelerated

$$\frac{m_e v_p^2}{T_e} > \frac{12\pi n e^3 L_e}{E T_e} \approx \frac{15E_c}{E} \qquad (7.176)$$

Here E_c is the critical field strength (see Eq. 7.174). Thus at $E < E_c$ the bulk of the electrons in the electric field move with a quasi-stationary directed velocity, whereas the electrons from the "tail end" of the distribution function find themselves in the runaway regime. In this region of the velocity space the distribution function is no longer at equilibrium. With time, the electrons acquire an ever-increasing energy, and the distribution function is "stretched" along the axis in the direction of the electric force. This process reduces the distribution function below the equilibrium near the runaway boundary. As a result there arises an excess diffusion flux of electrons from the low (thermal)—velocity region to the runaway boundary, which is due mainly to electron–electron collisions. If there were no obstacles to unlimited acceleration, all the plasma electrons would gradually find themselves in the runaway regime. Actually, however, a number of limitations hinder acceleration. In the first place the acceleration time is limited because of fast electron runaway from the plasma volume. Further, in a nonfully ionized plasma, acceleration is limited by electron–neutral particle collisions. In the presence of complex ions a limitation is imposed because at a small distance the ion potential differs considerably from the coulomb, and at high velocities the collision frequency ceases to reduce. Another factor that may hinder acceleration is electron emission during the motion in the plasma ("radiation friction"). Finally, mention should be made of the possible kinetic instabilities during the passage of fast electrons through the plasma.

We discuss only the limitation of acceleration associated with collisions of electrons with neutral particles. Allowing for these, the friction force acting on the electron gas as a whole is equal to

$$\begin{aligned} R &= R_{ei} + R_{ea} = m_e\langle \nu^t_{ei}(v)v\rangle + m_e\nu^t_{ea}u \\ &\approx \frac{4\pi n e^4 L_e}{m_e(u^2+v_T^2)^{3/2}}\,u + m_e\nu^t_{ea}u \end{aligned} \tag{7.177}$$

where we assumed that ν^t_{ea} is velocity independent. For the group of fast electrons moving in the direction of **E** with the isolated velocity $\mathbf{v}_p$ we obtain the friction force by adding R_{ea} to Eq. 7.175:

$$R(v_p) = R_{ei} + R_{ee} + R_{ea} = \frac{12\pi n e^4 L_e}{m_e v_p^2} + m_e\nu_{ea}v_p \tag{7.178}$$

The dependence of the values of R on u and v_p is illustrated in Fig. 7.8. It can be seen that the curves $R(u)$ have a minimum at $u > v_T$ and $R(v_p)$. The minimum of $R(v_p)$ corresponds to $\nu_{ea} = 2(\nu_{ei} + \nu_{ee})$ and is equal to

$$R_{\min} = m_e\nu_{ea}v_{pm} = (9\sqrt{2\pi})^{1/3}\sqrt{m_e T_e}\,\nu_{ea}^{2/3}\bar{\nu}_{ei}^{1/3}$$

In electric fields where the electron-accelerating force is less than the minimum friction force $eE < R_{\min}(v_p)$ electron runaway is clearly impossible. The value of the threshold field characterizing the runaway boundary at $\nu_{ea} \ll \nu_{ei}$ can be found from the relation

$$\frac{E_t}{E_c} = \frac{R_{\min}(v_p)}{R_{\max}} \approx 4\left(\frac{\nu_{ea}}{\bar{\nu}_{ei}}\right)^{2/3} \tag{7.179}$$

In fields less than E_t the stationary electron velocity distribution may be

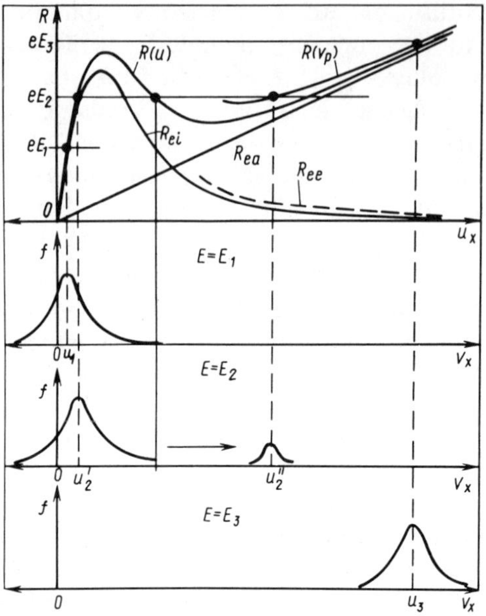

Fig. 7.8 Dependence of friction force on energy and velocity.

near-Maxwellian. In high fields $E_t < E < E_c$ there is a velocity region within which the electrons must be entrained in the acceleration regime (see Fig. 7.8). This region, however, is bounded on the high-velocity side by the condition of equality of the electric force to the force of friction against the neutral particles

$$m_e \nu_{ea} u = eE$$

Near the boundary of the acceleration region, at $\mathbf{v} \approx \mathbf{u}$, electron accumulation must take place; a distribution function maximum must then form, whose width depends on the ratio between the electron acceleration in the electric field and the diffusion in the velocity space due to electron–electron collisions. At $E > E_c$, when the bulk of the electrons go into acceleration, the entire distribution must shift into the velocity field in which the electric force is offset by the friction of the electrons against the neutral gas.

We now consider another peculiarity in the behavior of a highly ionized plasma in an electric field, which is sometimes called the electron "*energy runaway.*" This may occur when the principal mechanism of energy losses of the electrons is their cooling on collisions with ions. The stationary electron energy balance equation (Eq. 6.79) in this case amounts to the equilibrium between the energy acquired by the electrons in the electric field P_E and the energy lost by them on collisions with ions P_{ei}:

$$P_E = \sigma E^2 \approx 2\frac{ne^2E^2}{m_e\bar{\nu}_{ei}}; \qquad P_{ei} = \frac{3}{2}\kappa_{ei}n\bar{\nu}_{ei}(T_e - T_i) \tag{7.180}$$

Recalling that $\nu_{ei}(T_e) \sim T_e^{-3/2}$, we reduce the energy balance equation to

$$\frac{4}{3}\frac{e^2E^2}{m_e} = \kappa_{ei}[\bar{\nu}_{ei}(T_i)]^2T_i\left(\frac{T_i}{T_e}\right)^3\left(\frac{T_e}{T_i} - 1\right) \tag{7.181}$$

It is easy to see that at κ_{ei} = const the dependence of the right-hand side of the equality on T_e has a maximum at $T_e = 1.5T_i$. Obviously, the stationary electric field in the case at hand cannot exceed the value corresponding to this maximum:

$$E_k = 0.4\frac{\sqrt{\kappa_{ei}m_eT_i}}{e}\,\bar{\nu}_{ei}(T_i) \approx 0.53\sqrt{\kappa_{ei}}\,E_c(T_i) \tag{7.182}$$

where $E_c(T_i)$ is the critical field at $T_e = T_i$ (see Eq. 7.174). The dependence $T_e(E)$, determined by Eq. 7.181, is illustrated in Fig. 7.9. It is seen that a gradual increase in field strength E from zero to E_k raises the stationary electron temperature from T_i to $1.5T_i$. With a further increase in field the stationary solution of the energy balance equation vanishes. This means that the energy accumulated by the electrons in the electric field exceeds the energy losses on collisions, and the electron temperature rises with time. The temperature growth leads to a further drop in collision frequency T_e and to a greater "unbalance." Thus the field E_k yields the boundary of transition to the regime of increasing heating (energy runaway).

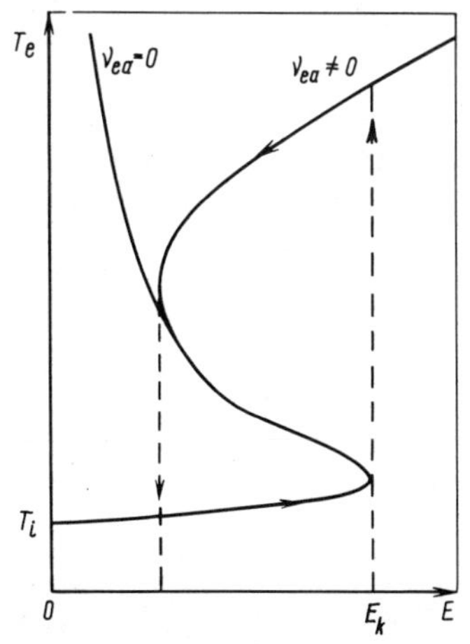

Fig. 7.9 Dependence of electron temperature on electric field.

Under actual conditions the energy runaway effect is limited by other loss mechanisms. In a nonfully ionized plasma the losses due to electron–neutral particle collisions may be considerable. Their inclusion amounts to adding the quantity $P_{ea} = (\frac{3}{2})\kappa_{ea} n \nu_{ea}(T_e - T_a)$ on the right-hand side of the balance equation (Eq. 7.180). At $\kappa_{ea}\nu_{ea}(T_i) \ll \kappa_{ei}\nu_{ei}(T_i)$ these losses are relatively small in the temperature range of $T_e \approx T_i$. But with increasing temperature at $T_e \gg T_i$ they become predominant, because $\nu_{ei} \sim 1/T_e^{3/2}$, while $\nu_{ea} \approx \text{const}$. Accordingly, on the curve $E(T_e)$ there appears, after the maximum, a growing branch determined by electron–atom collisions (see Fig. 7.9). Its position can be found from the balance condition

$$P_E \approx P_{ea}; \qquad T_e \approx \frac{\frac{4}{3}e^2E^2}{m_e\kappa_{ea}\nu_{ea}^2} \tag{7.183}$$

In this case, as can be seen from Fig. 7.9, the increase in electric field from zero to E_k, which results in transition to the nonstationary heating regime, terminates with a jump to the region of $T_e \gg T_i$. Note that the reverse jump with a decrease in field occurs at a lower value of the field than the forward jump, that is, a kind of a heating hysteresis is observed.

Another essential limitation of the energy runaway effect can be imposed by the thermal conductivity of the plasma, since the thermal conductivity coefficient increases rapidly with electron temperature, $\mathcal{K}_e \sim T_e^{5/2}$. A quantitative consideration of this limitation requires the solution of the nonlinear balance equation, and here we restrict ourselves to a crude estimate. It can be obtained by equating the value of the energy accumulated by the electrons in the field P_e (Eq. 7.180) to the average energy losses due to thermal conductivity:

$$P_q = \operatorname{div}(\mathcal{K}_e \operatorname{grad} T_e) \sim \frac{\mathcal{K}_e T_e}{L_T^2} \approx \frac{nT_e^2}{m_e\bar{\nu}_{ei}L_T^2}$$

where $L_T = [(1/T_e)\operatorname{grad} T_e]^{-1}$. As a result of the estimate we find the temperature in the electric field where the thermal conductivity is the principal source of losses:

$$T_e \approx eEL_T \tag{7.184}$$

8

MOTION OF CHARGED PLASMA PARTICLES IN MAGNETIC FIELD

8.1 SOME DATA ON STATIC MAGNETIC FIELDS

The effect of a magnetic field on the charged particle motion depends on its spatial distribution. Therefore we first discuss some quantities characterizing the spatial properties of a magnetic field.

It is common knowledge that the strength distribution of a static magnetic field **H**(**r**) is described by the equations

$$\text{curl}\,\mathbf{H} = \frac{4\pi}{c}\,\mathbf{j} \tag{8.1}$$

$$\text{div}\,\mathbf{H} = 0 \tag{8.2}$$

With given current distribution and boundary conditions these equations determine the field **H**(**r**) unambiguously. Field variation in space can be characterized by a vector derivative **H**, which is a tensor. In the Cartesian coordinates it has the form

$$\nabla\mathbf{H} = \begin{Vmatrix} \partial H_x/\partial x, & \partial H_x/\partial y, & \partial H_x/\partial z \\ \partial H_y/\partial x, & \partial H_y/\partial y, & \partial H_y/\partial z \\ \partial H_z/\partial x, & \partial H_z/\partial y, & \partial H_z/\partial z \end{Vmatrix} \tag{8.3}$$

The tensor components are related by Eqs. 8.1 and 8.2. In particular, the sum of the diagonal terms, which is equal to div **H**, is obviously zero. In current-free areas curl **H** = 0 and

$$\frac{\partial H_y}{\partial x} = \frac{\partial H_x}{\partial y}; \qquad \frac{\partial H_x}{\partial z} = \frac{\partial H_z}{\partial x}; \qquad \frac{\partial H_z}{\partial y} = \frac{\partial H_y}{\partial z}$$

In describing the spatial behavior of **H** the *line-of force* concept is often used. By a magnetic line of force is meant some imaginary spatial curve at each point of which the direction of the tangent coincides with that of **H**. The line-of-force equation can be written on the basis of the fact that each element of the line $d\mathbf{l}$ is parallel to the field **H**:

$$\frac{\partial l_x}{H_x} = \frac{dl_y}{H_y} = \frac{dl_z}{H_z} \tag{8.4}$$

The number of lines of force intersecting a unit area normal to the vector **H** is chosen to be proportional to the field strength. Since by definition lines of force can be passed through any point in space, their "density" is a purely conventional notion.

From Eq. 8.2 it follows that the flux of the vector **H**

$$\Phi = \int_{(S)} H_n \, dS \tag{8.5}$$

through any closed surface is equal to zero. Indeed, using the Gauss theorem, we find

$$\int_{(V)} \operatorname{div} \mathbf{H} \, dV = \int_{(S)} H_n \, dS = 0 \tag{8.6}$$

The condition 8.6 also means that there are no points in space where lines of force begin or terminate; that is, the number of lines entering and leaving any volume is the same. Thus a line of force is either closed or infinite. In analyzing configurations of a magnetic field in a finite volume one can distinguish two types of line of force. First, the lines intersecting the surface of the given volume form configurations often called *open*. The length of a line in a volume may be different (in particular, it may tend to infinity). Second, there are lines of force that do not leave the volume. As in the former case, their length may greatly exceed the characteristic dimensions of the volume and tend to infinity in the limit. To put it differently, a line of force may be an intricate, but not self-intersecting, closed spatial curve. If the lines of force inside a volume do not intersect its surface, one speaks of *closed magnetic configurations*.

Consider a small area intersected by a set of lines of force. Let us move it along the lines of force, deforming them so that the same lines of force will always remain inside the contour of this area. As a result we obtain a spatial figure, which is termed a *flux tube*. The surface of the tube is clearly formed by the set of boundary lines of force, whereas the total flux inside the tube remains unchanged $\delta\Phi = \int_{(\delta S)} H_n \, dS = \text{const}$. Like lines of force, magnetic tubes may be closed or open. It follows from the above definitions that a line of force is actually the limit to which the flux tube tends as its cross-section area reduces to zero. An important characteristic to be considered in analyzing the stability of a plasma in a magnetic field is *the specific volume of the flux tube* V_Φ, by which is meant the ratio of the geometrical volume of the tube δV to the magnetic flux $\delta\Phi$. The value of δV can be obtained from the curvilinear integral $\delta V = \int \delta S\, dl$ taken along the tube axis. Since $\delta\Phi = H\delta S = \text{const}$ we have

$$\delta V = \int \frac{H\delta S}{H}\, dl = \delta\Phi \frac{dl}{H}$$

and

$$V_\Phi = \int_{(l)} \frac{dl}{H} \tag{8.7}$$

If the lines of force are closed, the integral is taken along the entire length of the tube. If not, the limits of integration must be chosen from the particular conditions of the problem.

If the line of force does not go to infinity and does not close on itself, then under certain conditions the space may contain some surface areas whose entire multitude of points are formed by elements of the given line of force. In other words, within these areas any two points lie on the same line of force. The areas may close between themselves and then the line of force fills continuously the surface of some closed volume. In such cases one speaks of *magnetic surfaces*. Let us consider the formation of magnetic surfaces, say, in the case of fields with toroidal geometry. Assume that the magnetic field has two components, $\mathbf{H}_\Theta$ and $\mathbf{H}_\varphi$, at each point (see Fig. 8.1). The component $\mathbf{H}_\Theta$ can be induced either by a toroidal coil or by a linear current I_z. The component $\mathbf{H}_\varphi$ can easily be induced by a current I_Θ along a circular contour. The line of force of the field $\mathbf{H} = \mathbf{H}_\Theta + \mathbf{H}_\varphi$ is a helical line winding around a toroidal surface. To describe such lines of force one introduces the notion of *the*

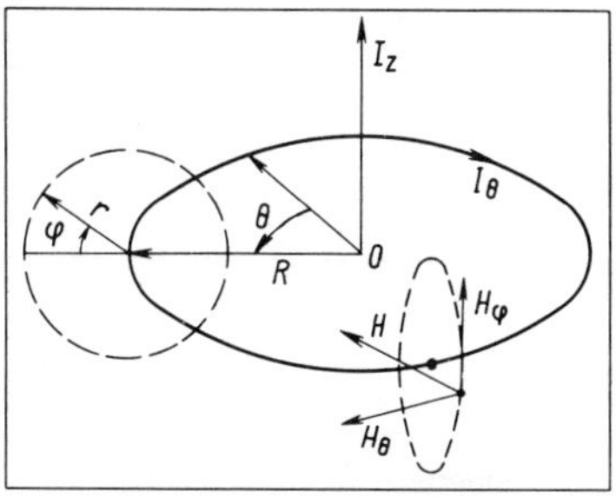

Fig. 8.1 Magnetic surfaces in fields with toroidal geometry.

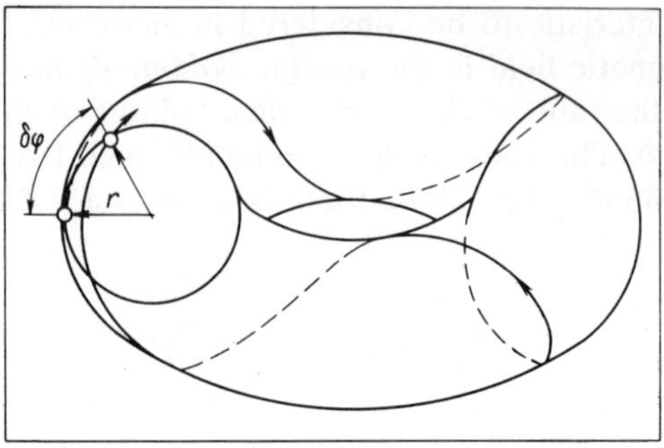

Fig. 8.2 Trace of a line of force.

rotational transform angle ψ which is equal to $d\varphi/dl_\Theta$. If $d\varphi/dl_\Theta$ is independent of Θ this value actually determines the pitch of the helical line winding around the toroidal surface. From the line-of-force equation it follows that $dl_\Theta/H_\Theta = dl_\varphi/H_\varphi = rd\varphi/H_\varphi$. Therefore

$$\psi = \frac{d\varphi}{dl_\Theta} = \frac{H_\varphi}{rH_\Theta} \tag{8.8}$$

On a complete sweep of Θ the trace of the line of force is displaced in the cross section of the torus by some $\delta\varphi$ (Fig. 8.2). Obviously, if $\delta\varphi/2\pi$ is irrational, the line of force does not return to its origin after any finite number of turns and upon an infinite number of turns fills the entire surface of the torus. Each line of force forms a magnetic surface of its own. The complete magnetic configuration here represents a system of magnetic surfaces in the form of toruses inserted in each other (Fig. 8.3).

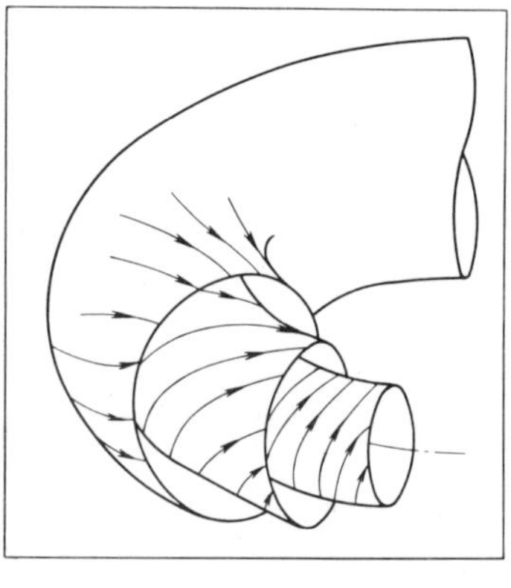

Fig. 8.3 Complete magnetic configuration.

If $\delta\varphi/2\pi$ is equal to a rational number of m/n, then after n full turns the line of force closes without any magnetic surface being formed.

Note that such characteristics of a magnetic field as a line of force, a flux tube, and a magnetic surface are nonlocal, as distinct from the field strength **H**. Their nature may change substantially even on small perturbations of the magnetic field. In a toroidal field, for instance, at $\mathbf{H}_\varphi = 0$, the lines of force have the shape of closed concentric circles. If even a very low field $\mathbf{H}_\varphi$ appears the line of force becomes helical and may extend to infinity: a magnetic surface is formed. Magnetic surfaces are naturally still less stable to perturbations.

Consider now the relationship between the behavior of lines of force in the vicinity of some point and the components of the tensor for the field derivative 8.3. We choose the coordinate system so that at the given point the field is directed along the axis z ($\mathbf{H} = H\mathbf{z}_0$).

Assume that the derivatives $\delta H_z/\delta y$ are the only nonzero components of the tensor 8.3. From Eq. 8.1 it follows that this is possible only in the presence of currents within the volume. Then the lines of force are parallel straight lines, their density varying in the x and y directions (Fig. 8.4). The tensor component $\delta H_z/\delta z$ can differ from zero only simultaneously with the other diagonal terms (see Eq. 8.2). Then the lines of force in the vicinity of the given point are nonparallel, since at least one of the derivatives $\partial H_x/\partial x$, $\partial H_y/\partial y$ must be nonzero. Figure 8.5 depicts the lines of force for $\partial H_z/\partial z < 0$ and $\partial H_x/\partial x > 0$. If $\partial H_y/\partial y = 0$ and $\partial H_x/\partial x = -\partial H_z/\partial z$ the configuration of the magnetic field is independent (the plane case). In the case of cylindrical symmetry Fig. 8.5 must be regarded as a section of a spatial configuration symmetrical about the z axis (a divergent beam of lines of force). For $\partial H_z/\partial_z = 0$ and $\partial H_y/\partial y = -\partial H_x/\partial x \neq 0$ the lines of force in the vicinity of the given point are not only nonparallel, but lie in different planes. The sections of the flux tube for different values of z and the projection of the lines of force on the xy plane for this case are illustrated in Fig. 8.6.

The terms $\partial H_x/\partial z$ and $\delta H_y/\partial z$ describe the bending of the lines of force. To ascertain this, we first recall the well-known characteristics of

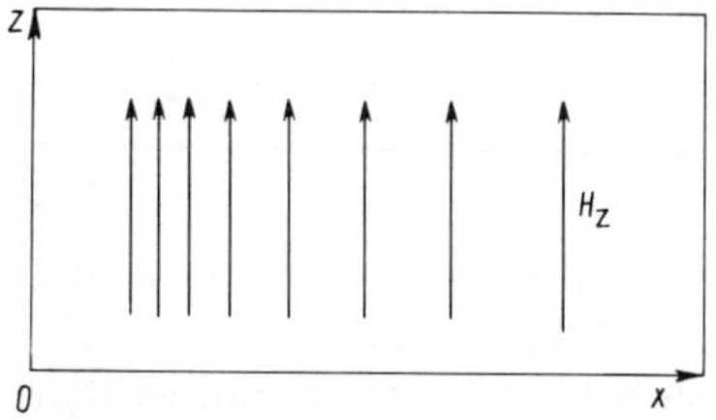

Fig. 8.4 Density of lines of force.

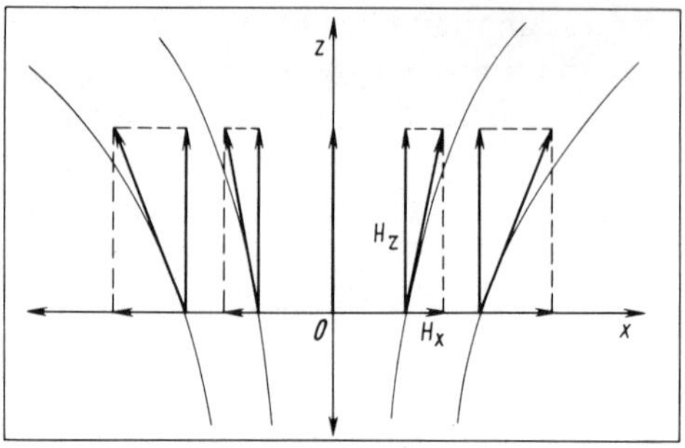

Fig. 8.5 Configuration of magnetic field, plane case.

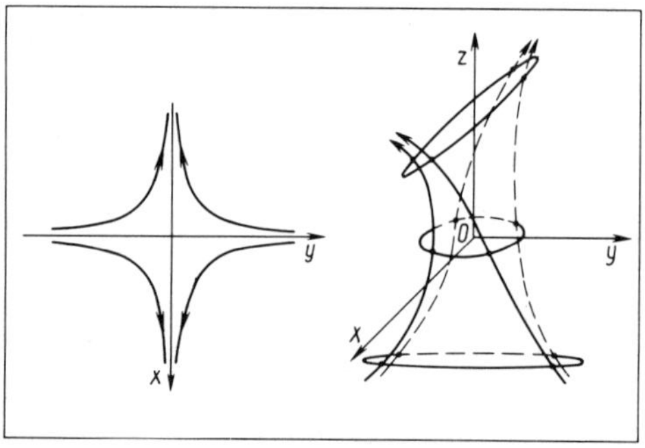

Fig. 8.6 Configuration of magnetic field, cylindrical case.

the spatial configuration of a curve. The direction of a spatial curve at a preassigned point is given by a unit vector of the tangent. For the lines of force it is clearly equal to $\mathbf{H}/H = \mathbf{h}$. Accordingly, the change in the direction of the curve in space during the transition from point 1 to 2 along the curve is determined by $\Delta\mathbf{h}$ (Fig. 8.7). The vector $\Delta\mathbf{h}$ can be written as $\Delta\mathbf{h} = \Delta l(d\mathbf{h}/\mathrm{dl})$. Hence the bending of the line is given by $d\mathbf{h}/\mathrm{dl}$, which is called the curvature and is the direction derivative of the vector $\mathbf{h}$. This derivative can be written in the form $d\mathbf{h}/dl = (\mathbf{h}\,\mathrm{grad})\mathbf{h}$, since the operator $(\mathbf{h}\,\mathrm{grad})$ determines increments along the tangent to the curve (l).

To characterize the bending of a spatial curve, one also introduces the notion of the *curvature radius* $\mathbf{R}$. The direction of the vector $\mathbf{R}$ coincides with that of the main normal, which is, in turn, determined by the direction of $d\mathbf{h}$. The length of $\mathbf{R}$ is given by the point of intersection (the curvature center) of the two main normals drawn at points 1, 2, with Δl_{12} tending to zero. The geometrical meaning of the curvature radius is that

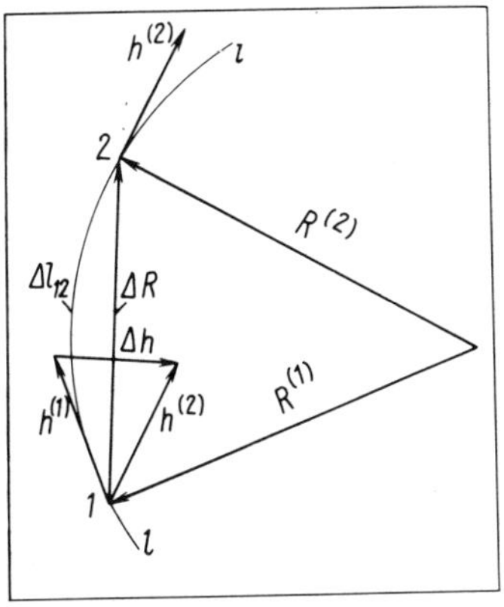

Fig. 8.7 Geometrical meaning of the radius of curvature.

for $\Delta l \to 0$ a segment of any curve can be replaced by a segment of a circle of radius R with its center lying at the curvature center. From Fig. 8.7 it is easy to establish the relation between the curvature and the curvature radius:

$$\frac{|\Delta \mathbf{R}|}{R} = \frac{|\Delta \mathbf{h}|}{h}$$

Dividing both sides by Δl, we obtain

$$\lim_{\Delta l \to 0} \frac{|\Delta \mathbf{R}|}{\Delta l} \frac{1}{R} = \frac{1}{R} = \left| \frac{d\mathbf{h}}{dl} \right|$$

or, in vector form,

$$\frac{\mathbf{R}}{R^2} = -\left(\frac{\mathbf{H}}{H} \operatorname{grad}\right) \frac{\mathbf{H}}{H} \tag{8.9}$$

Using Eq. 8.9, we revert to the analysis of the lines of force in the local coordinate system, where $\mathbf{H} = H z_0$. In this system

$$\frac{\mathbf{R}}{R^2} = -\frac{\partial}{\partial z}\left(\frac{\mathbf{H}}{H}\right) = -\frac{1}{H}\frac{\partial \mathbf{H}}{\partial z} + \frac{\mathbf{H}}{H^2}\frac{\partial H}{\partial z} = -\frac{1}{H} \times \left(\frac{\partial H_x}{\partial z}\mathbf{x}_0 + \frac{\partial H_y}{\partial z}\mathbf{y}_0\right) \tag{8.10}$$

Thus the derivatives $\partial H_x/\partial z$ and $\partial H_y/\partial z$ indeed determine unambiguously the curvature radius of the lines at the given point.

The given line of force is often a plane curve, and no currents are observed at the point under discussion. Then the curvature radius is inversely proportional to the derivative of the magnetic field strength

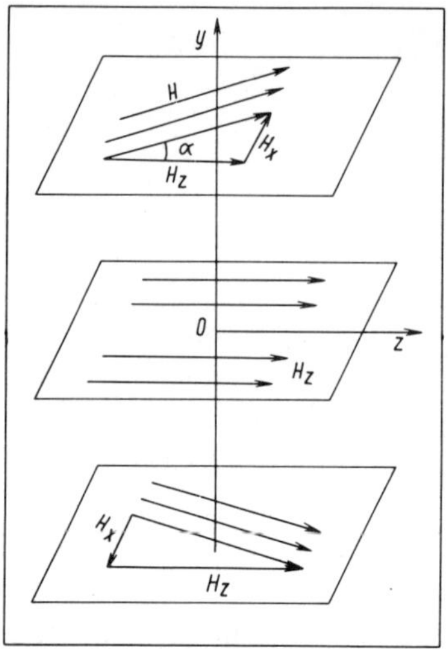

Fig. 8.8 Shear in magnetic lines of force.

along the direction of the curvature radius. Indeed, let us select the coordinate system so that the x axis is directed along **R**. Then Eq. 8.9 takes the form $1/R = (1/H)\partial H_x/\partial z$. From the condition curl $\mathbf{H} = 0$ we have $\partial H_z/\partial x - \partial H_x/\partial z = 0$, and hence

$$\frac{1}{R} = \frac{1}{H}\frac{\partial H_z}{\partial x} \tag{8.11}$$

In conclusion we consider the specific features of magnetic configurations associated with the components $\partial H_x/\partial y$ and $\partial H_y/\partial z$. Assume that the derivative $\partial H_x/\partial y$ is different from zero. Then in each plane $y = \text{const.}$ the lines of force are straight lines, whose direction depends on y (Fig. 8.8). This crossing and twisting of the lines of force relative to closely spaced magnetic surfaces is termed *shear*. The quantitative characteristic of shear in the example just discussed may be the quantity determining the rate of variation of the field **H** in the direction along the y axis

$$\frac{d\alpha}{dy} = \frac{1}{H_z}\frac{dH_x}{dy}$$

The shear concept is particularly often applied to toroidal magnetic configurations, in which the direction of the lines of force is characterized by the rotational transform angle ψ (see Fig. 8.2). The existence of a shear means that this angle, and hence the pitch of the lines of force, is different for different r (different magnetic surfaces). Therefore the shear is given here by

$$\frac{d\psi}{dr} = \frac{d}{dr}\left(\frac{H_\varphi}{rH_\Theta}\right) \tag{8.12}$$

8.2 CHARGED PARTICLE MOTION IN HOMOGENEOUS MAGNETIC FIELD

Before proceeding to the behavior of a plasma in a magnetic field we consider the motion of individual charged particles in the fields of given forces. We begin with the simplest case of particle motion in a constant homogeneous magnetic field. Here the particles are subjected only to the Lorentz force

$$\mathbf{F}_L = \frac{Ze}{c}[\mathbf{v} \times \mathbf{H}] \tag{8.13}$$

(as before, we assume that the charge number Z is $+1$ for the ions and -1 for the electrons). The equation of particle motion under the effect of the Lorentz force has the form*

$$m\frac{d\mathbf{v}}{dt} = \frac{Ze}{c}[\mathbf{v} \times \mathbf{H}] \tag{8.14}$$

It defines two independent integrals of motion. The first one is obtained by projecting the equation onto the magnetic field. Since the projection of $\mathbf{F}_L$ on $\mathbf{H}$ is zero, the velocity component parallel to the magnetic field ($v_\parallel$) remains constant

$$\frac{d\mathbf{v}_\parallel}{dt} = 0; \qquad \mathbf{v}_\parallel = \text{const} \tag{8.15}$$

Further, since the Lorentz force is perpendicular to the velocity ($\mathbf{F}_L \perp \mathbf{v}$), it does not affect the particle energy. Performing scalar multiplication of Eq. 8.14 by $\mathbf{v}$, we get

$$\frac{d}{dt}\left(\frac{mv^2}{2}\right) = 0; \qquad K = \frac{mv^2}{2} = \text{const} \tag{8.16}$$

Thus, while in motion, the total velocity of a particle v, its projections on the direction of the magnetic field $v_\parallel$, and its projection on a plane perpendicular to the field $v_\perp = (v^2 - v_\parallel^2)^{1/2}$ remain constant.

The particle trajectory (Eq. 8.14) in a plane perpendicular to the field is clearly a circle. Its radius ρ_H can be found from the equality of the

*We ignore the relativistic effects, assuming that the particle velocity is much less than that of light ($v \ll c$).

Lorentz and centrifugal forces:

$$\frac{e}{c}v_{\perp}H = \frac{mv_{\perp}^2}{\rho_H}$$

Hence

$$\rho_H = \frac{v_{\perp}}{\omega_H} \tag{8.17}$$

$$\omega_H = \frac{eH}{mc} \tag{8.18}$$

The quantity ω_H is obviously the angular velocity of circular revolution of a particle. The direction of revolution is determined by the condition that the Lorentz force must be centripetal. It depends on the sign of the particle charge: a positively charged particle revolves clockwise, and a negatively charged one, counterclockwise (looking in the direction of the field). The circle representing the projection of the particle trajectory on a plane perpendicular to the field is called the *Larmor circle*; its radius, ρ_H, the *Larmor radius*; and the angular velocity, ω_H, the *cyclotron* (or *Larmor*) *frequency*. The numerical expressions of ω_H and ρ_H for the electrons are as follows:

$$\omega_{He} \approx 1.7 \times 10^7\, H; \qquad \rho_{He} \approx \frac{3.5\sqrt{K_{\perp}}}{H}$$

and for the hydrogen ions (protons):

$$\omega_{Hp} \approx 10^4 H; \qquad \rho_{Hp} \approx \frac{150\sqrt{K_{\perp}}}{H}$$

where ω is given in $\sec^{-1}$, ρ_H in cm, H in Oe, and $K_{\perp} = mv_{\perp}^2/2$ in eV.

When the interaction of the charged particles with the magnetic field is analyzed, it is convenient to regard the Larmor circle as a current-carrying loop (Fig. 8.9). Here one assumes averaging over times greatly exceeding the period of revolution of the particle $T_H = 2\pi/\omega_H$. The value of the average current along the Larmor circle is obviously equal to the charge transferred per unit time across an area Q perpendicular to the particle trajectory, that is, to the ratio of the particle charge to the period of revolution:

$$I = \frac{e}{T_H} = \frac{e\omega_H}{2\pi} \tag{8.19}$$

The current-carrying turn can be characterized by the magnetic moment. As is well known, the magnetic moment of a current is equal to

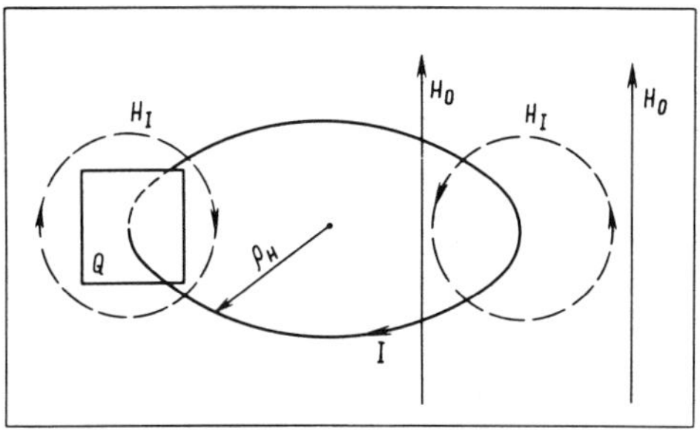

Fig. 8.9 Larmor circle as a current-carrying turn.

$$\mu = \frac{1}{c} IS = \frac{1}{2c} e\omega_H \rho_H^2$$

(Here $S = \pi\rho_H^2$ is the area of the Larmor circle). Substituting the expressions for ρ_H and ω_H (see Eqs. 8.17 and 8.18), we obtain

$$\mu = \frac{mv_\perp^2}{2H} = \frac{K_\perp}{H} \tag{8.20}$$

or, in vector form, $\boldsymbol{\mu} = -(K_H/H)\mathbf{h}$, because for a particle of either sign the direction of the magnetic moment is opposite to that of an external magnetic field. Thus inside the Larmor circle the magnetic field of the current is subtracted from the external field H_0, whereas outside the circle the two fields are added together. This means that the Larmor loop is diamagnetic.

Thus the trajectory of a charged particle in a constant homogeneous magnetic field is a helical line composed of uniform motion of the particle along the field and uniform revolution along the Larmor circle in a plane perpendicular to the field (Fig. 8.10). The radius of this helical line is clearly the Larmor radius 8.17, while the pitch of the helical line is determined by the longitudinal velocity of the particle and is equal to

$$\xi_H = \frac{2\pi v_\parallel}{\omega_H} \tag{8.21}$$

Under more complex conditions, when the magnetic field is inhomogeneous in space and time dependent, an electric field is present, or other forces operate, exact integration of the equation of motion is possible only in special cases. There exists, however, a simple approximate method for solving the problem if the Larmor radius and the pitch of the helical trajectory are much less than the characteristic

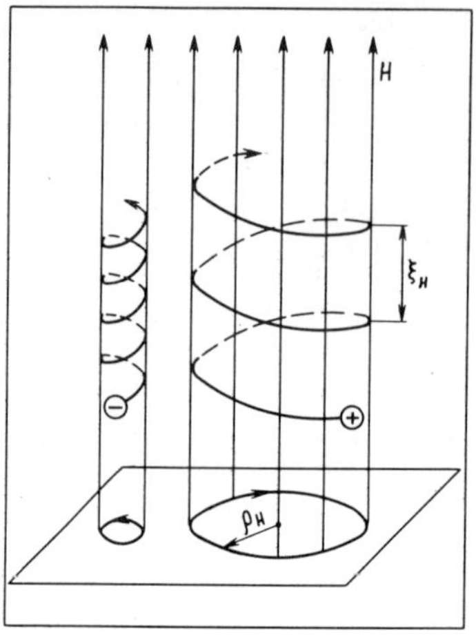

Fig. 8.10 Trajectory of a charged particle in a constant homogeneous magnetic field.

dimensions of inhomogeneity,

$$\rho_H \ll L; \qquad \xi_H \ll L \tag{8.22}$$

and if the period of revolution of charged particles along the Larmor circle is much greater than the characteristic time of field variation,

$$T_H \ll \tau; \qquad \omega_H \tau \gg 1 \tag{8.23}$$

Under such conditions the particle trajectory in a plane perpendicular to the field can be represented as a superposition of the Larmor revolution and the drift of the Larmor circle center (it is called the *guiding, or Larmor, center*). Then one can obtain comparatively simple approximate expressions for both velocity components, that is, for the drift velocity and the velocity of Larmor revolution. This is called the *drift approximation.* It is widely used for describing the behavior of a plasma in a strong magnetic field.

8.3 CHARGED PARTICLE DRIFT IN HOMOGENEOUS MAGNETIC FIELD

Let us first consider the motion of charged particles in the presence of a force **F**, constant and homogeneous in space, in addition to the magnetic field.

Then the equation of motion takes the form

$$m\frac{d\mathbf{v}}{dt} = \mathbf{F} + \frac{Ze}{c}[\mathbf{v}\times\mathbf{H}] \tag{8.24}$$

The motion of a particle along the magnetic field is described by the projection of Eq. 8.24 on **H**:

$$\frac{m d\mathbf{v}_{\parallel}}{dt} = \mathbf{F}_{\parallel}$$

Since the Lorentz force has no longitudinal component, the magnetic field does not affect the longitudinal motion, and it is uniformly accelerated. The motion in a plane perpendicular to the field can be represented as the sum of Larmor rotation and a constant-velocity drift. Indeed, let us represent the transverse velocity as the sum

$$\mathbf{v}_{\perp} = \mathbf{v}_{\omega} + \mathbf{v}_{d} \tag{8.25}$$

assuming that $\mathbf{v}_{\omega}$ is the velocity of Larmor revolution (i.e., it is described by Eq. 8.14), and $\mathbf{v}_d$ is the constant drift velocity. Substituting Eq. 8.25 into Eq. 8.24, we obtain the vector equation for $\mathbf{v}_d$:

$$\mathbf{F}_{\perp} + \frac{Ze}{c}[\mathbf{v}_d \times \mathbf{H}] = 0$$

Multiplying it vectorially by **H** and allowing for the well-known equation for the double vector product $[[\mathbf{v} \times \mathbf{H}] \times \mathbf{H}] = -H^2\mathbf{v}_{\perp}$, we find

$$\mathbf{v}_d = \frac{c}{Zeh^2}[\mathbf{F} \times \mathbf{H}] = \frac{c}{ZeH}[\mathbf{F} \times \mathbf{h}] \tag{8.26}$$

where $\mathbf{h} = \mathbf{H}/H$ is a unit vector in the direction of the magnetic field.

Thus the trajectory of the motion in a plane perpendicular to the magnetic field is formed by summing the rotation along the Larmor circle and the motion (drift) of the center of this circle with the velocity (Eq. 8.26) in a direction perpendicular to the acting force.

It is easy to find out the origin of the drift by qualitative consideration of the motion of a charged particle in a plane perpendicular to the magnetic field. In the absence of the force **F** the particle moves along a circle of radius ρ_H (Fig. 8.11*a*). The force $f_{\perp}$ accelerates the particle on the

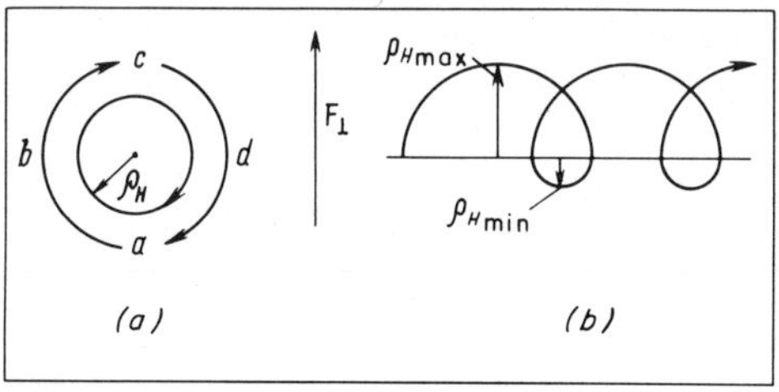

Fig. 8.11 Motion of a charged particle in a plane perpendicular to the magnetic field, *a* in the absence of the force **F**, *b* in the presence of the force $F_{\perp}$.

trajectory section *abc* and slows it down on the section *cda*. The velocity is maximum at the point *c*. Therefore the particle traverses the section *bcd* with an average velocity exceeding the one in the absence of the force. The average curvature radius of the trajectory is larger as well ($\rho_H \sim v_\perp$). On the lower section (*dab*) the situation is reversed. In the long run the trajectory acquires the shape shown in Fig. 8.11*b*. This curve is called a trochoid.

If the force acting on the particle is due to the electric field, the drift direction and velocity are independent of the charge and generally of the particle properties. Indeed, substituting $\mathbf{F} = Ze\mathbf{E}$ into Eq. 8.26, we obtain the equation for the velocity of the electric drift in the form*

$$\mathbf{v}_{dE} = \frac{c}{H}[\mathbf{E} \times \mathbf{h}] \tag{8.27}$$

The drift velocity is charge independent because both forces (electric and Lorentz), whose balance determines the drift, are proportional to the charge.

The equality of the drift velocity for particles of different nature can also be explained on the basis of Lorentz transformations for the electric field. At $v \ll c$ they yield the relationship between the fields in a moving ($\mathbf{E}'$) and a fixed ($\mathbf{E}$) reference system:

$$\mathbf{E}' = \mathbf{E} + \frac{[\mathbf{v} \times \mathbf{H}]}{c}$$

It can be seen that in a reference system moving with a velocity $\mathbf{v}_d$ we have $\mathbf{E}' = 0$. Therefore in this system the charged particles do not experience any drift. Then in the laboratory reference system the drift velocity of charged particles of any type is the same.

Consider now the charged particle motion in a field of slowly varying forces, assuming that Eqs. 8.22 and 8.23 are fulfilled, which means that the characteristic time of change in force greatly exceeds the Larmor period and the characteristic scale is much greater than the Larmor radius. Thus it is possible to find the solution of the motion equation (Eq. 8.24) for the velocity component perpendicular to the magnetic field,

$$m\frac{d\mathbf{v}_\perp}{dt} = \mathbf{F}_\perp(\mathbf{r}, t) + \frac{Ze}{c}[\mathbf{v}_\perp \times \mathbf{H}] \tag{8.28}$$

*It should be remembered that Eqs. 8.26 and 8.27 were obtained in the nonrelativistic approximation. Therefore they are valid only at $v_d \ll c$, that is, at $E \ll H$.

by the method of successive approximations, neglecting the force variation in the zero approximation.

In conformity with this approach we represent the particle velocity $\mathbf{v}_\perp$ as the sum

$$\mathbf{v}_\perp = \mathbf{v}_\omega + \mathbf{v}_d^{(0)} + \mathbf{v}_d^{(1)} + \mathbf{v}_d^{(2)} + \cdots \tag{8.29}$$

where $\mathbf{v}_\omega$ is, as before, the velocity of Larmor revolution described by Eq. 8.14; $\mathbf{v}_d^{(0)}$ is the drift velocity in the zero approximation, determined by Eq. 8.26; and $\mathbf{v}_d^{(1)}$ and $\mathbf{v}_d^{(2)}$ are corrections to the drift velocity corresponding to the first, second, etc., approximation. Substituting Eqs. 8.29 into 8.28 we obtain, as a first approximation,

$$m\frac{d\mathbf{v}_d^{(0)}}{dt} + \cdots = \frac{Ze}{c}[(\mathbf{v}_d^{(1)} + \cdots) \times \mathbf{H}]$$

where only the first nonvanishing terms are left on the right- and left-hand sides, because $\mathbf{v}_d^{(0)} \gg \mathbf{v}_d^{(1)} \gg \mathbf{v}_d^{(2)}$. Multiplying this equation vectorially by $\mathbf{H}$, we find the ratio between $\mathbf{v}_d^{(1)}$ and $\mathbf{v}_d^{(0)}$

$$\mathbf{v}_d^{(1)} = -\frac{mc}{ZeH}\left[\frac{d\mathbf{v}_d^{(0)}}{dt} \times \mathbf{h}\right] \tag{8.30}$$

Similarly, we can find higher-order corrections. In this case the ratio of $\mathbf{v}_d^{(1)}$ to $\mathbf{v}_d^{(0)}$ (as well as the ratio of higher-order corrections to $\mathbf{v}_d^{(1)}$) is naturally small. It is equal in its order of magnitude to $\mathbf{v}_d^{(1)}/\mathbf{v}_d^{(0)} \approx 1/\omega_H\tau_F \ll 1$, where τ_F is the characteristic time of variation in velocity $\mathbf{v}_d^{(0)}$ or in force $\mathbf{F}$ along the trajectory. It should, however, be borne in mind that the direction of $\mathbf{v}_d^{(1)}$ differs from that of $\mathbf{v}_d^{(0)}$, as seen from Eq. 8.30. Therefore the correction $\mathbf{v}_d^{(1)}$ must be included in many cases. At the same time corrections of higher approximations are less significant, since they change the velocity components but slightly.

Substituting the expression 8.26 into Eq. 8.30, we obtain the following equation:

$$\mathbf{v}_d^{(1)} = \frac{mc^2}{Z^2e^2H^2}\frac{d\mathbf{F}_\perp}{dt} = \frac{1}{m\omega_H^2}\frac{d\mathbf{F}_\perp}{dt} \tag{8.31}$$

The derivative $d\mathbf{F}_\perp/dt$ yields the change in force $\mathbf{F}_\perp$ along the trajectory corresponding to the zero approximation. It can be represented as

$$\frac{d\mathbf{F}_\perp}{dt} = \frac{\partial \mathbf{F}_\perp}{\partial t} + (\mathbf{v}\ \mathrm{grad})\mathbf{F}_\perp \approx \frac{\partial \mathbf{F}_\perp}{\partial t} + (\mathbf{v}_\parallel\ \mathrm{grad})\mathbf{F}_\perp + (\mathbf{v}_d^{(0)}\ \mathrm{grad})\mathbf{F}_\perp \tag{8.32}$$

We omitted here the rapidly oscillating term $(\mathbf{v}_\omega\ \mathrm{grad})\mathbf{F}_\perp$. The corresponding oscillation of the force can affect only the Larmor rotation velocity. Its ratio to the Lorentz force is of the order of

$$\frac{\frac{\rho_H F_\perp}{L}}{\frac{e}{c} H v_\omega} \approx \frac{v_d^{(0)}}{v_\omega} \frac{\rho_H}{L}$$

and this force can usually be neglected.

Equation 8.31 shows that the velocity $\mathbf{v}_d^{(1)}$ is proportional to the total derivative of the force $\mathbf{F}_\perp$ with respect to time, which determines the force variation along the particle trajectory. With a constant direction of $\mathbf{F}_\perp$ the velocity $\mathbf{v}_d^{(1)}$ is parallel to the force, as distinct from the velocity of the main drift $\mathbf{v}_d^{(0)}$, which is perpendicular to the force $\mathbf{F}_\perp$. In conformity with Eq. 8.30 the motion component determined by the velocity $\mathbf{v}_d^{(1)}$ can be regarded as a drift under the action of the inertia force $m d\mathbf{v}_d^{(0)}/dt$ (cf. Eqs. 8.26 and 8.30). Therefore this component is termed the *inertial drift.*

When an electric force $\mathbf{F}_\perp = Ze\mathbf{E}_\perp$ acts on a charged particle Eq. 8.31 takes the form

$$\mathbf{v}_{dE}^{(1)} = \frac{mc^2}{ZeH^2} \frac{d\mathbf{E}}{dt} \tag{8.33}$$

It describes the inertial drift of electrons and ions in a slowly varying electric field. It is worth noting that, in contrast to the main velocity component of the electric drift $\mathbf{v}_{dE}^{(0)}$ the velocities of the inertial drift of ions and electrons are opposite in sign and differ widely in value; the velocity of the inertial drift of electrons is negligibly small compared with the ion drift, because $\mathbf{v}_{dE}^{(1)} \sim m$.

It is easy to establish the physical picture of the inertial drift. A change in force $\mathbf{F}_\perp$ results in distortion of the motion trajectory. Let, for instance, the force increase either with time, or in space in the direction of the drift $\mathbf{v}_d^{(0)}$ (Fig. 8.12). Then the particle that has accumulated energy in region I will lose it in region II on a smaller section of trajectory,

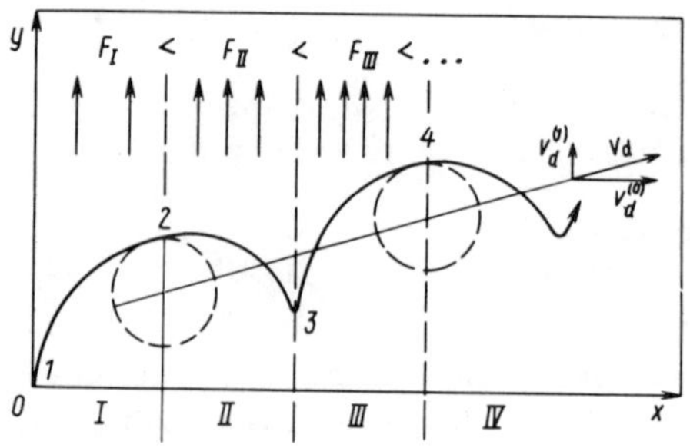

Fig. 8.12 Inertial drift of charge particles in a slowly varying magnetic field.

since $F_{\mathrm{I}} < F_{\mathrm{II}}$. The turning point 3 will shift in the direction of the force with respect to point 1. Point 4 is displaced in the same direction with respect to point 2, because $F_{\mathrm{III}} > F_{\mathrm{II}}$. Since $F_{\mathrm{IV}} > F_{\mathrm{III}}$, and so on, the process repeats itself and the trajectory is inclined with respect to the equipotential. It is easy to see that with an increase in mass and hence in the period of revolution of the particle along the Larmor circle the trajectory becomes steeper.

Thus in the course of drift the guiding center shifts toward the increase in force. The particle gradually increases its transverse energy, since some work is done during the motion along the force. A peculiarity of this process is that with a constant grad F or a constant $\partial F/\partial t$ the energy of the drift motion of the particle varies linearly with time. The energy increment within the time δt is determined by the displacement along the force $\delta \mathbf{r}_F$. It is equal to

$$\delta \mathbf{r}_F = \mathbf{v}_d^{(1)} \delta t = \frac{1}{m\omega_H^2} \delta \mathbf{F}_\perp$$

where $\delta \mathbf{F}_\perp$ is the total change in force $\mathbf{F}_\perp$ on displacement of the guiding center along the trajectory within the time δt. Accordingly,

$$\delta U = \mathbf{F}_\perp \cdot \delta \mathbf{r}_F = \frac{1}{m\omega_H^2} \mathbf{F}_\perp \cdot \delta \mathbf{F}_\perp = \frac{1}{m\omega_H^2} \delta\left(\frac{F_\perp^2}{2}\right) = \delta\left[\frac{1}{2} m (v_d^{(0)})^2\right] \tag{8.34}$$

Note that the kinetic energy associated with the rotational motion depends, in the approximation discussed, only on the initial conditions and therefore remains unchanged in a field of varying forces. Correspondingly, the magnetic moment $\mu = K_\perp / H$ also remains unchanged.

8.4 CHARGED PARTICLE MOTION IN SLOWLY VARYING MAGNETIC FIELD

Let us consider the motion of charged particles in a slowly varying magnetic field, assuming that the characteristic time of its variation greatly exceeds the period of revolution in the Larmor orbit and the characteristic scale is much larger than the Larmor radius (Eqs. 8.22 and 8.23). First of all we find out how the rate of revolution of a particle along the Larmor circle varies in this motion. Assume, to begin with, that the magnetic field is homogeneous in space but varies with time. Variation in magnetic field is known to induce an eddy electric field, which is defined by one of Maxwell's equations:

$$\operatorname{curl} \mathbf{E} = -\frac{1}{c} \frac{\partial \mathbf{H}}{\partial t}$$

Under the effect of this field the particle changes its transverse energy. The change in energy per revolution along the Larmor circle is obviously equal to

$$\delta K_\perp = Ze \oint_{(l)} E_l \, dl = -e \oint_{(S)} (\mathbf{h} \cdot \text{curl } \mathbf{E}) \, dS = \frac{e}{c} \int_{(S)} \frac{\partial H}{\partial t} \, dS \qquad (8.35)$$

where the curvilinear integral $\oint$ is taken along the Larmor circle and E_l is the electric field component parallel to the revolution velocity (here we use the Stokes equation for transition to the integral over the area and take into account that the direction of the normal is opposite to that of the magnetic field). Since the time dependence of the magnetic field is weak we can approximate it by a linear function on an interval equal to one period of revolution of the particle along the Larmor circle, that is, we can take it that $\partial H/\partial t = \text{const}$. Assuming also that ρ_H changes only slightly during one period, we find

$$\delta K_\perp = \frac{e}{c} \int_{(S)} \frac{\partial H}{\partial t} \, dS \approx \frac{e}{c} \frac{\partial H}{\partial t} \pi \rho_H^2 = \frac{2\pi K_\perp}{\omega_H} \frac{1}{H} \frac{\partial H}{\partial t} \qquad (8.36)$$

The change in kinetic energy per unit time is obtained by multiplying $\delta K_\perp$ by the number of revolutions per unit time $\omega_H/2\pi$. Then

$$\frac{\partial K_\perp}{\partial t} = \frac{K_\perp}{H} \frac{\partial H}{\partial t} \qquad (8.37)$$

Equation 8.37 shows that the revolution energy of the particle $K_\perp = mv_\perp^2/2$ varies proportionally to the magnetic field. Since, in accordance with the definition of the magnetic moment μ (Eq. 8.20), $K_\perp = \mu H$, Eq. 8.37 indicates the constancy of the magnetic moment in the course of motion:

$$\frac{d\mu}{dt} = 0 \qquad \mu = \text{const} \qquad (8.38)$$

The result obtained also extends to the case where the magnetic field is inhomogeneous in space. Indeed, in a coordinate system moving together with the guiding center the inhomogeneity of the magnetic field in space transforms to the time dependence of **H**. Hence in weakly inhomogeneous fields the magnetic moment of a particle is also motion invariant. It is called the *adiabatic invariant*. Naturally, the conclusion about the constancy of the magnetic moment μ is only approximate. But the accuracy of conservation of μ is very high, provided the conditions 8.22 and 8.23 are fulfiilled. A more detailed analysis shows that its variation is an exponentially small function of $1/\omega_H\tau$ or ρ_H/L.

The constancy of the magnetic moment enables one to find the variation in transverse energy during the motion of a charged particle in a slowly varying magnetic field $K_\perp = \mu H$. As already noted, when H

varies in time the corresponding variation in $K_\perp$ is due to the supply of energy from the external electric field. If the magnetic field is constant in time but inhomogeneous in space, the total energy of the particle in the laboratory system of coordinates remains unchanged, since there is no eddy electric field, and the only force acting on the particle (the Lorentz force) is perpendicular to the velocity. Variation in transverse energy is due to the variation in longitudinal energy provided $K = K_\perp + K_\parallel =$ const.

In a magnetic field a particle always moves so that the plane of the Larmor circle is normal to the line of force through the center of the circle. If the magnetic field is longitudinally inhomogeneous (see, for instance, Fig. 8.5), then the lines of force on a Larmor circle of finite radius are not perpendicular to the plane under consideration. Accordingly, the Lorentz force has a component parallel to the z axis in addition to the radial component (Fig. 8.13):

$$|\mathbf{F}_{Lz}| = \left|\frac{e}{c}\, v_\varphi H_r\right|$$

This force is directed toward the weaker magnetic field. It changes the longitudinal component of the velocity v_z by rotating the total velocity $\mathbf{v}$. If the particle moves toward the increasing field, the longitudinal velocity component decreases, and v_z vanishes at some point. Since $\mathbf{F}_{Lz}$ does not reduce to zero, the particle begins to move in the opposite direction, being reflected from the increasing magnetic field. Therefore such a configuration is called a *magnetic mirror*. If the magnetic field has only a transverse gradient (see Fig. 8.4), the velocity component v_z remains constant in the absence of the external forces. Since $K =$ const the transverse energy of the particle must also be constant. Then it follows

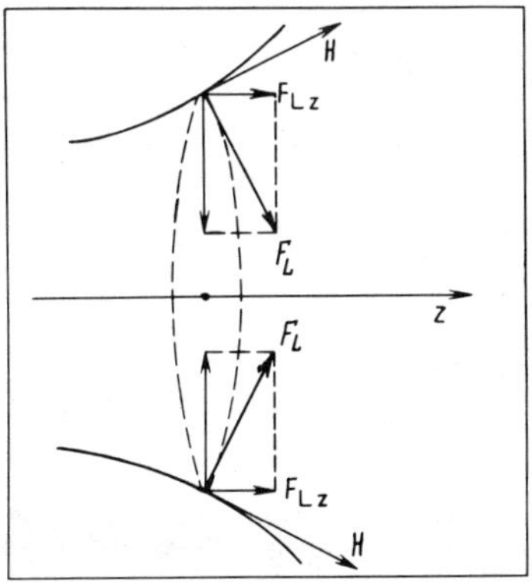

Fig. 8.13 Components of the Lorentz force in a slowly varying magnetic field.

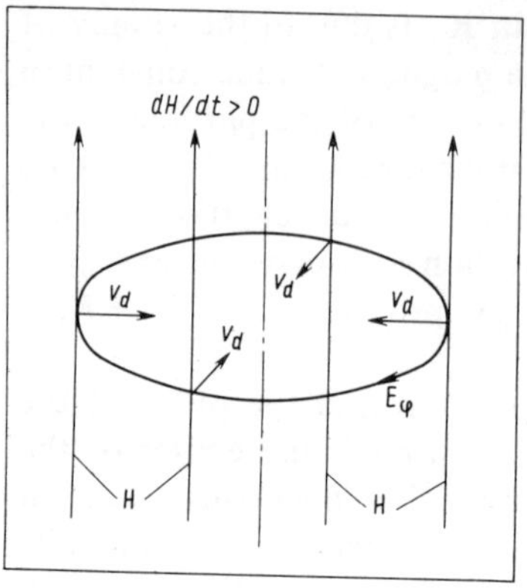

Fig. 8.14 Drift of charged particles in an inhomogeneous magnetic field varying with time.

from the condition $|\mu| = \text{const}$ that the trajectory of the guiding center must lie on the surface $\mathbf{H} = \text{const}$. It can move only normally to the magnetic field gradient.

Let us consider the drift of charged particles in a slowly varying magnetic field. In a homogeneous magnetic field varying with time the drift is due to the eddy electric field induced by this variation. Thus in a cylindrically symmetric region, where the lines of force of the eddy field are circles, the drift is directed along the radius (Fig. 8.14). The electric field strength is then expressed in terms of the induction law

$$2\pi r E_\varphi = -\frac{1}{c}\frac{d\Phi}{dt} = -\frac{1}{c}\pi r^2 \frac{dH}{dt}$$

(r is the distance from the symmetry axis, and $\Phi = \pi r^2 H$ is the magnetic flux). Hence

$$E_\varphi = -\frac{1}{2c} r \frac{dH}{dt} \tag{8.39}$$

This field determines the drift velocities (see Eq. 8.27)

$$\mathbf{v}_d = \frac{c}{H}[\mathbf{E} \times \mathbf{h}]; \qquad v_{dr} = -\frac{r}{2H}\frac{dH}{dt}$$

The equation for the drift trajectory takes the form

$$\frac{dr}{dt} = -\frac{r}{2H}\frac{dH}{dt} \tag{8.40}$$

Integration yields

$$r^2 H = \text{const}; \qquad \frac{r}{r_0} = \left(\frac{H_0}{H}\right)^{1/2} \tag{8.41}$$

These relations show that with an increase in magnetic field a charged particle drifts toward the symmetry axis along the radius, and with a decrease, in the opposite direction. In either case the drift occurs so that the magnetic flux $\Phi = \pi r^2 H$ remains constant.

Let us now discuss the drift for an inhomogeneous but time-constant field. This drift can be attributed to two average forces acting upon a particle revolving along the Larmor circle. One of them is associated with the transverse, and the other with the longitudinal velocity component. The force associated with the transverse components can be obtained from the magnetic moment. It is equivalent to the force acting on the magnetic dipole, which is equal to

$$\mathbf{F}_\mu = \text{grad}\,(\mu \mathbf{H})$$

Bearing in mind that μ is antiparallel to $\mathbf{H}$, we get

$$\begin{aligned}\mathbf{F}_\mu &= -\,\text{grad}\,(\mu H)\\ &= -\,\mu\,\text{grad}\,H \end{aligned}\tag{8.42}$$

since $\mu = \text{const}$. The transverse component of this force leads to a drift, whose velocity is found by using Eq. 8.26:

$$\mathbf{v}_d^\mu = \frac{c\,[\mathbf{F}_\mu \times \mathbf{h}]}{ZeH} = \frac{\mu c\,[\mathbf{h} \times \text{grad}\,H]}{ZeH}$$

From Eq. 8.20 for μ we have

$$\mathbf{v}_d^\mu = \frac{cK_\perp}{ZeH^2}\,[\mathbf{h} \times \text{grad}\,H] \tag{8.43}$$

The value of the velocity v_d^μ is

$$v_d^\mu = \frac{cmv_\perp^2}{2eH}\,\frac{\text{grad}_\perp H}{H} \approx \frac{1}{2}\,v_\perp \rho_\perp \frac{\text{grad}_\perp H}{H} \tag{8.44}$$

It can be seen that when Eq. 8.22 holds, the drift velocity is much lower than that of Larmor revolution $v_d^\mu / v_\perp \approx \rho_H / L \ll 1$. Note that the drift associated with Larmor revolution in an inhomogeneous magnetic field can also be considered without introducing the diamagnetic force; it is determined by a slow continuous change in the curvature radius of the trajectory in a plane perpendicular to the field. Since an increase in magnetic field reduces the curvature radius $\rho_H \sim v_\perp/\omega_H$, in the course of Larmor revolution the particles alternately traverse sections with a larger and a smaller radius (Fig. 8.15). This change leads to a drift in a direction perpendicular to the field gradient, just as it does in the presence of the force (see Fig. 8.11).

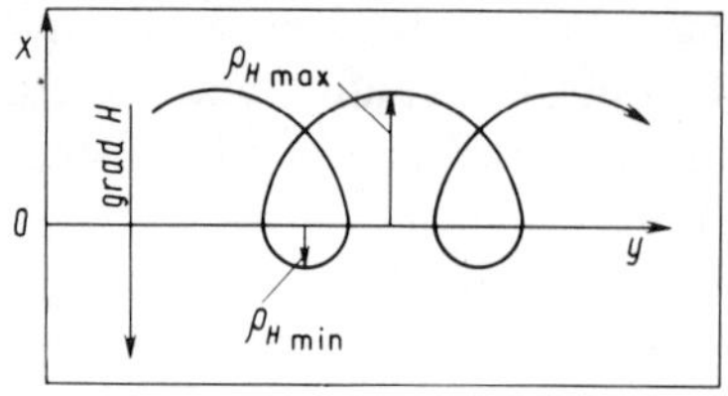

Fig. 8.15 Drift of charged particles in an inhomogeneous time-constant magnetic field.

The drift associated with the motion along the magnetic field arises when the lines of force are curved. A particle moving along a line of force is affected by a centrifugal force $\mathbf{F}_R = mv_\parallel^2\mathbf{R}/R^2$, which induces a drift with a velocity

$$\mathbf{v}_d^R = mv_\parallel^2 \frac{c[\mathbf{R}\times\mathbf{H}]}{ZeR^2H^2} \tag{8.45}$$

The curvature radius is related to the magnetic field gradient by Eq. 8.9. Hence $\mathbf{v}_d^R$ can be written as

$$\mathbf{v}_d^R = \frac{cmv_\parallel^2}{ZeH^2}\left[\mathbf{H}\times\left(\frac{\mathbf{H}}{H}\,\text{grad}\right)\frac{\mathbf{H}}{H}\right] \tag{8.46}$$

This drift velocity component is equal to

$$v_d^R = \frac{cmv_\parallel^2}{eHL_\parallel} \approx \frac{v_\parallel^2}{v_\perp}\frac{\rho_H}{L_\parallel}$$

where $L_\parallel = |(\mathbf{h}\,\text{grad}_\parallel)\mathbf{h}|^{-1}$. As in the preceding case, when Eq. 8.22 holds, the velocity v_d is much lower than that of Larmor revolution.

If the lines of force are plane curves and there are no currents in the volume, Eq. 8.46 can be simplified by using Eq. 8.11:

$$\mathbf{v}_d^R = \frac{cmv_\parallel^2}{Ze}\left[\frac{\mathbf{H}}{H^3}\times\text{grad}\,H\right] \tag{8.47}$$

With such a magnetic field configuration the total particle drift velocity associated with the magnetic field gradient is given by the sum of Eqs. 8.43 and 8.47:

$$\mathbf{v}_d^H = \frac{c}{ZeH^2}(K_\perp + 2K_\parallel)[\mathbf{h}\times\text{grad}\,H] \tag{8.48}$$

8.5 CONFINEMENT OF CHARGED PARTICLES BY SOME MAGNETIC CONFIGURATIONS

In the analysis of the possibility of utilizing various magnetic field configurations as magnetic traps for confining a plasma, the primary concern is the motion of individual charged particles. The trajectories of the charged particles in a magnetic field obviously depend not only on the local, but also on the integral characteristics of the field: the structure of the lines of force, their closed or open state, the presence of magnetic surfaces, and so on. We consider briefly the conditions of charged particle retention for two configurations used as magnetic traps (open and closed).

The simplest open magnetic trap has an axially symmetric configuration with a longitudinal field, which strengthens toward both ends. This configuration can be achieved with the aid of two turns or two solenoids. The distribution of the lines of force is illustrated in Fig. 8.16. The magnetic field regions near the solenoids are *magnetic mirrors*. The trajectory of a charged particle in an open magnetic trap is a superposition of the motion along the line of force, Larmor revolution, and the drift of the Larmor center. Since the drift is perpendicular to **H** and grad H, it is directed along the azimuth. During the motion of particles from the central region toward the increasing magnetic field the longitudinal component of the Lorentz force reduces the longitudinal component of the velocity and accordingly increases its transverse component. As a result, particles with a sufficiently high transverse velocity cannot reach the plane of the maximum magnetic field and are reflected from magnetic mirrors. At the same time particles with a low $v_\perp/v_\parallel$ ratio pass through the magnetic mirrors and escape from the trap. The typical trajectories of the Larmor centers of a trapped (solid line) and transit (dashed line) particle are shown in Fig. 8.16.

Quantitatively, the conditions for the reflection of a charged particle from a magnetic mirror can be obtained in terms of the total kinetic energy conservation law and of the magnetic moment conservation law during the motion of particles in a magnetic field:

$$K = K_\perp + K_\parallel = \text{const}; \qquad \mu = \frac{K_\perp}{H} = \text{const}$$

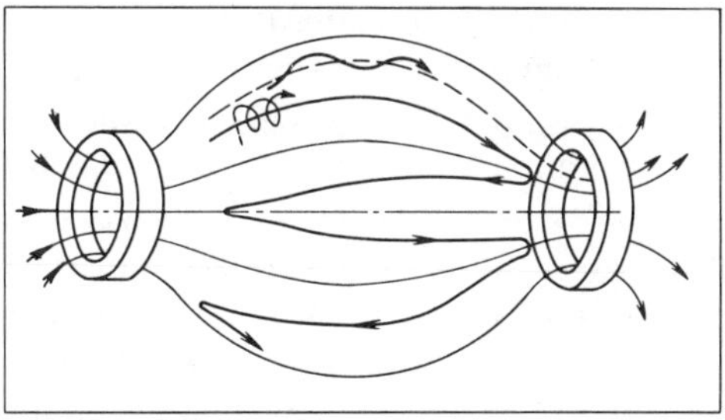

Fig. 8.16 Distribution of lines of force in a magnetic mirror.

Suppose in the center of a system, where the magnetic field strength is minimum (H_0), a particle has components of the total velocity $v_{\|0}$ and $v_{\perp 0}$, that is, the velocity vector $\mathbf{v}$ makes a certain angle $\vartheta_0(\sin^2\vartheta_0 = v_{\perp 0}^2/v^2 = K_{\perp 0}/K)$ with the direction of the line of force. When the particle moves toward the mirror, because $|\mu| = K_\perp/H = \text{const}$ the value of $k_\perp$ increases, since the angle between $\mathbf{v}$ and the line of force gradually changes. At the turning point the entire kinetic energy of the particle changes into the energy of transverse motion, and the angle ϑ becomes equal to $\pi/2$. The field strength at the turning point can be obtained from the equality

$$H = H_0 \frac{K}{K_{\perp 0}} = \frac{H_0}{\sin^2\vartheta_0} \tag{8.49}$$

This particle will not leave the bounds of the mirrors if the field strength in each of the mirrors H_1 exceeds $H_0/\sin^2\vartheta_0$.

Thus with a given value of H_0/H_1 only those particles are trapped for which the angle ϑ_0 satisfies the inequality

$$\sin^2\vartheta_0 > \frac{H_0}{H_1} \tag{8.50}$$

The quantity H_0/H_1 is called the *mirror ratio.*

From Eqs. 8.49 and 8.50 it is seen that the particle retention condition is independent of the absolute velocity and is determined exclusively by the ratio between $v_\|$ and $v_\perp$ at the central cross section of the trap. All the particles for which Eq. 8.50 is not satisfied pass freely through the mirrors of the system. In other words, a finite motion in the trap is accomplished only by those particles whose velocity vectors lie in the range of angles ϑ_0 from ϑ_{0m} to $\pi - \vartheta_{0m}$, where ϑ_{0m} is equal to $\vartheta_{0m} = \arcsin(H_0/H_1)^{1/2}$. The cone in velocity space corresponding to the angles ϑ_{0m} is called the *loss cone.*

Let us now discuss the charged particle motion in a closed magnetic trap. The simplest closed configuration is formed by concentric circular lines of force in the absence of currents in the plasma volume. Such a configuration can be obtained, for instance, with the aid of a solenoid whose axis is bent into a ring [a toroidal solenoid, (Fig. 8.17)]. This annular axis is usually called the minor axis of the torus (Θ axis) as distinct from major Z axis, which is the axis of symmetry of the whole system. The quantities R_0 and r_0 in Fig. 8.17 are called, respectively, the major and the minor radii of the toroidal surface. It is readily seen that such a configuration of the magnetic field is not a trap. Indeed, both the curvature of the lines of force and the associated inhomogeneity of the magnetic field (see Eq. 8.11) lead to a drift perpendicular to the plane in which the line of force lies. This drift is not restricted in any way.

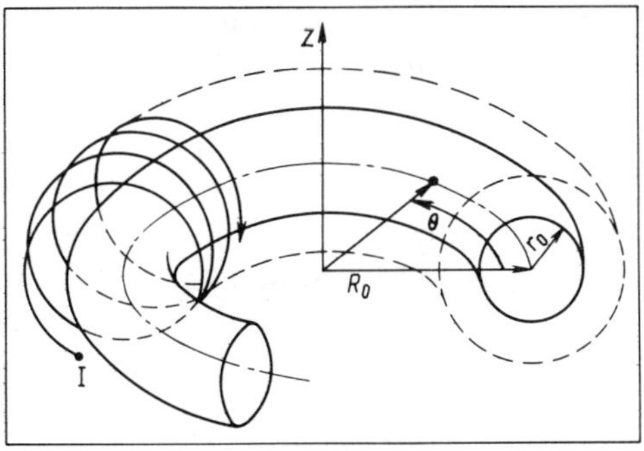

Fig. 8.17 Toroidal solenoid configuration.

The situation reverses if the system contains rotational transformation (see Section 8.1). Consider, for instance, the motion of particles in some toroidal system in which the magnetic field has a configuration shown in Fig. 8.2.* This field is formed due to the superposition of two components H_Θ and H_φ. As follows from Eq. 8.11, the field component H_Θ reduces with increasing R. The component H_φ is, generally speaking, also nonuniform in space. But we neglect the dependence of H_φ on the coordinates, assuming the rotational transformation angle to be small, that is, we put $H_\varphi \ll H_\Theta$. Accordingly, the main contribution to the particle drift will be made by the inhomogeneity of H_Θ. Obviously, the helical lines of force characteristic of the configuration under study become denser on the internal sections of the toroidal surface (Fig. 8.18, points 1, 2). Therefore the particle motion along the lines of force is similar to that in traps with magnetic mirrors. During the motion of the

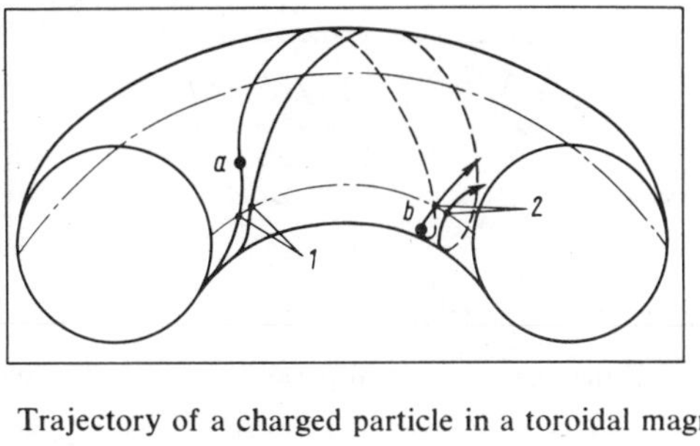

Fig. 8.18 Trajectory of a charged particle in a toroidal magnetic field.

*Such a magnetic field configuration is characteristic, in particular, of tokamak magnetic traps widely used in investigations on controlled fusion.

particles from the outside to the inside surface of the torus (i.e., into the region of the large field) their longitudinal velocity decreases, and particles with a sufficiently large $v_{\perp 0}/V_{\parallel 0}$ ratio are reflected; that is, they find themselves trapped on a certain section of the line of force [for instance, between points a and b (Fig. 8.18)]. Therefore the particles can be divided into two groups, depending on the $v_{\perp 0}/v_{\parallel 0}$ ratio. Those in the first group can make an unlimited number of revolutions about the Z axis inside the torus (so-called *transit particles*). The second group consists of *trapped particles.*

Let us consider the trajectory of a transit particle. Since the toroidal configuration is symmetrical about the major torus axis, no motions along the minor axis affect the retention of particles within the closed volume. Therefore we are interested in the projection of the particle trajectory onto a plane perpendicular to the minor torus axis; that is, we consider the change in the r and φ coordinates of the guiding center* (Fig. 8.19). The projections of the helical lines of force onto this plane are determined by the magnitude and direction of the field component H_φ and, with the above assumptions of the coordinate independence of H_φ, are concentric circles, depicted in Fig. 8.19 as dashed lines. If the drift associated with grad H were zero the projections of the trajectories would also coincide with those of the lines of force. To explain the effect of the drift let us choose the sign of the particle charge and the direction of the magnetic field so that the velocity $\mathbf{v}_d^H$ is directed upward (see Fig. 8.19). Let us also assume that the particle begins its motion

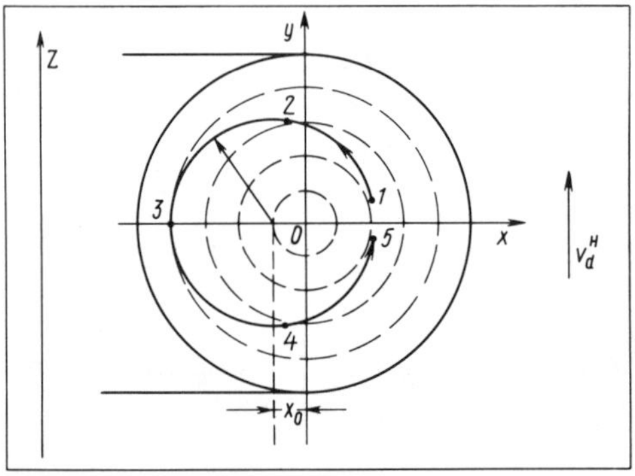

Fig. 8.19 Trajectory of a transit particle in toroidal configuration.

*The term projection here implies a curve formed by the r and φ coordinates irrespective of the toroidal Θ coordinate.

from point 1, and the direction $\mathbf{v}_\parallel$ determines the counterclockwise motion of the guiding center in the projection under review. The drift results in a continuous upward movement of the trajectory. As can be seen the upward movement on the trajectory section 1, 2, 3 means transition to lines of force external to the initial line. In the lower half (the trajectory section 3, 4, 5) the particle continues drifting upward. But now this corresponds to a return to the internal lines of force. As follows from the symmetry about the plane in which the minor torus axis lies, the drift displacements in the upper and lower parts balance each other; that is, the projection of the trajectory must be a closed curve.

Let us now find the equation of the trajectory projection. To do this we introduce a local rectangular system of coordinates (see Fig. 8.19). The motion of the guiding center of the particle consists in displacement along the line of force with a velocity v and a drift along the y axis. The projection of the first velocity component on the plane under study is equal to $v_\parallel H_\varphi/H = v_\varphi$. It is directed along the tangent to the projection of the line of force. Accordingly, the velocities of the guiding center along the x and y axes are defined by the expressions

$$v_x = \frac{dx}{dt} = -v_\varphi \sin\varphi = -v_\parallel \frac{H_\varphi}{H}\frac{y}{r};$$

$$v_y = \frac{dy}{dt} = v_\varphi \cos\varphi + v_d^H = v_\parallel \frac{H_\varphi}{H}\frac{x}{r} + v_d^H \tag{8.51}$$

Here $r = (x^2 + y^2)^{1/2}$.

Eliminating time from Eq. 8.51, we obtain

$$x\,dx + y\,dy = \frac{1}{2}\,d(r^2) = -\frac{Hv_d^H}{H_\varphi v_\parallel}\,r\,dx$$

and further,

$$\frac{dr}{dx} = -\frac{v_d^H H}{H_\varphi v_\parallel} \tag{8.52}$$

If H, H_φ, and $v_\parallel$ change relatively little in the region of motion the right-hand side of the equation can be considered approximately constant. Then the solution of the equation takes the form $r = -\alpha x + r_0$, where $\alpha = (H/H_\varphi)v_d^H/v_\parallel = \text{const}$. It is obvious that at $\alpha < 1$ the trajectory projection described by this equality is a closed curve. It is easy to find, in particular at $\alpha \ll 1$. From the relation $r^2 = x^2 + y^2$, neglecting the quantities proportional to α^2, we then obtain

$$(x + \alpha r_0)^2 + y^2 = r_0^2 \tag{8.53}$$

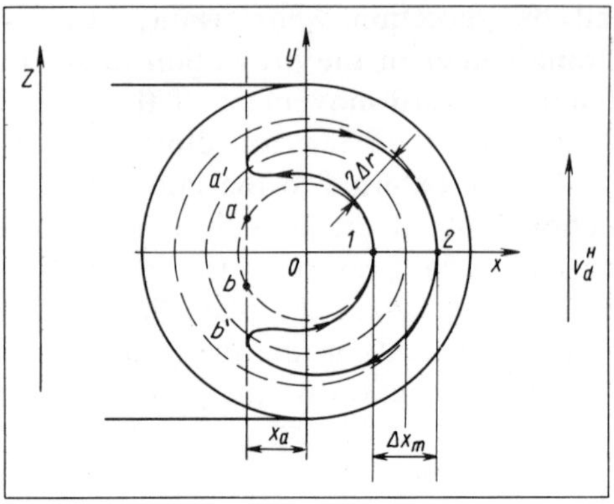

Fig. 8.20 Projection of the trajectory of trapped particles onto the x-y plane.

This is the equation of a circle displaced from the center by a distance of $x_0 = \alpha r_0$. Since x_0 is much less than r_0, one can speak of a small distortion of the trajectory by the toroidal drift.

We now analyze the trajectory of trapped particles. As in the case of transit particles we consider the projection of the trajectory onto the xy plane (Fig. 8.20). Let the particle move from point 1, as shown in the figure. Assume that the ratio of the longitudinal and transverse velocities is such that in the absence of a drift the particle would be reflected at points a and b. In the presence of a drift, as in the case of a transit particle, along with longitudinal displacement, the particle continuously moves upward relative to the initial line of force. Since the drift does not affect the $v_\parallel/v_\perp$ ratio, the particle is now reflected at point a', where the field strength is equal to that at point a. After the turning point the direction of the longitudinal velocity becomes opposite, whereas that of the drift remains unchanged. Therefore, returning to the initial plane xy, the particle continues to shift toward the external lines of force. On the lower section of the trajectory the drift "presses" the particle against the x axis, that is, causes its displacement toward the inner lines of force. As a result, a closed curve is formed, which is referred to as a "banana trajectory."

The width of the banana $\Delta r(x)$ can be found by using Eq. 8.52, which must cover the change in $v_\parallel$ along the trajectory. Let the field strength at the reflection point be H_a. Then from the condition μ = const we have

$$\frac{K_\perp}{H} = \frac{K}{H_a} = \frac{K - K_\parallel}{H}; \qquad v_\parallel = \pm v\sqrt{1 - H/H_a}$$

The dependence of the magnetic field on the coordinates at $H_\varphi \ll H_\Theta$ is determined by the component H_Θ. In a field induced by the external toroidal solenoid:

$$H_\Theta = H_0 \frac{R_0}{R_0 + x} \approx H_0\left(1 - \frac{x}{R_0}\right)$$

at $x \ll R_0$. Using this dependence, we get

$$v_\parallel \approx \pm v\sqrt{1 - \frac{1 - x/R_0}{1 - x_a/R_0}} \approx \pm v\frac{\sqrt{x - x_a}}{\sqrt{R_0}}$$

where x_a is the coordinate of the turning point a. Since $v_\parallel \ll v_\perp$ for trapped particles, $v_\perp \sim v$ and v_d^H can be considered constant. Substituting $v_\parallel$ into Eq. 8.52 and integrating, we find

$$r - r_1 = \Delta r(x) = \pm\frac{2v_d^H H}{vH_\varphi}\sqrt{R_0(x - x_a)} \tag{8.54}$$

The total width of the banana is equal to $2\Delta r_m$. Denoting by r_1 the radius of the projection of the line of force on which the turning points a' and b' lie, we can express the coordinates x and x_a in terms of the polar angle φ assuming $x = r_1 \cos\varphi$. Then in the widest part the value $2\Delta r_m = \Delta x_m$ will satisfy the relation

$$\Delta x_m = \frac{4v_d^H H}{vH_\varphi}\sqrt{r_1 R_0(1 - \cos\varphi_a)} \tag{8.55}$$

Let us compare x_m and x_0—the displacement of the transit particle trajectory center. At equal kinetic energies, assuming that for a transit particle $v_\parallel \sim v$, we obtain

$$\frac{\Delta x_m}{x_0} \approx 4\sqrt{R_0/r_1}$$

that is, the width of the banana appreciably exceeds x_0.

The following condition must be met for the trajectory of a charged particle not to cross the surface of radius a, that is, for the particle to be retained by the trap:

$$\frac{v_d^H H}{vH_\varphi}\sqrt{r_1 R_0} \ll a \tag{8.56}$$

8.6 DIAMAGNETIC EFFECT IN PLASMAS

The motion of individual charged particles in a magnetic field yields information on the behavior of the plasma as a whole. This generalization is most easily accomplished for the so-called collisionless regime, that is, when the effect of collisions of charged particles on their motion can be neglected. Under such conditions analysis of the plasma behavior requires averaging the motion of individual particles. We consider the results of such averaging with respect to the charged particle motion in a plane perpendicular to the magnetic field when the lines of force are parallel to each other, that is, grad $H \perp \mathbf{H}$. If the magnetic and electric fields are constant in time the motion of the charged particles in a plane perpendicular to $\mathbf{H}$ is (as demonstrated above) the sum of Larmor gyration and the drift of the Larmor center. The velocity of the drift associated with the electric field is obtained from Eq. 8.27. It is the same for all particles and does not produce an electric current. The velocity of the drift due to the inhomogeneity of the magnetic field is given by Eq. 8.48. Its averaging over the velocities for α-type particles (electrons or ions) leads to the relation

$$\mathbf{u}_{d\alpha} = \langle \mathbf{v}_d^H \rangle = \frac{2cT_\alpha}{Z_\alpha e H^2} [\mathbf{h} \times \text{grad } H] \tag{8.57}$$

where it is assumed that $\langle K_\perp \rangle = 2\langle K_\| \rangle = T_\alpha$; that is, the velocity distribution is treated as isotropic.

The averaging of the Larmor gyration of particles is slightly more complicated. As noted above, each Larmor circle is equivalent to the current producing a diamagnetic moment. Within a homogeneous plasma the diamagnetic currents of the individual particles naturally compensate each other. Figure 8.21 schematizes the Larmor gyration of particles of like sign. At each point inside the plasma the number of particles moving in different directions is seen to be equal, on the average. There is no such compensation at the boundary. On the plasma surface all the particles move in the same direction. This induces a *diamagnetic current*

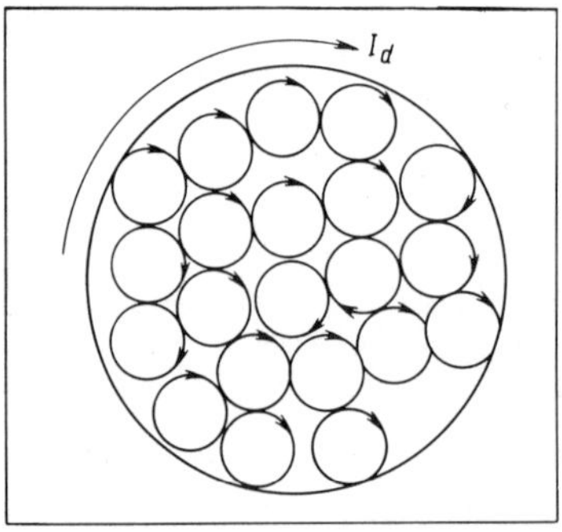

Fig. 8.21 Larmor gyration of particles of like sign.

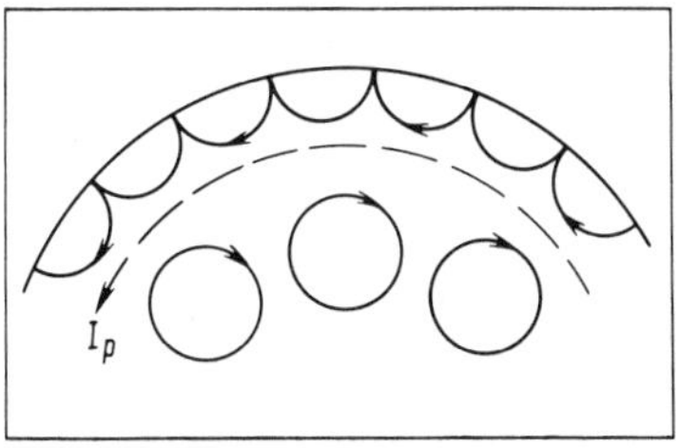

Fig. 8.22 Generation of paramagnetic currents.

I_d, which flows over the surface (it is easy to see that the diamagnetic currents of the electrons and ions add up). Note that when the plasma touches the walls and particles are reflected, a current I_p also appears on the surface, which is opposite to the diamagnetic current (it is called a *paramagnetic current.* Its origin is readily understood from Fig. 8.22. Upon successive reflections from the wall the Larmor center is seen to shift in a direction opposite to that of particle gyration. If elastic reflection is experienced by all charged particles, then in a homogeneous plasma the diamagnetic and paramagnetic currents completely offset each other.

In an inhomogeneous plasma diamagnetic currents exist in the plasma volume as well. Consider, for instance, the formation of a diamagnetic flux in the presence of a concentration gradient (see Fig. 8.23). To find the current density one must consider the motion of the particles on both sides of a plane perpendicular to the density gradient (plane 1 in Fig. 8.23). The flux arises because the numbers of particles on the left and on the right of this plane are not the same. A contribution to the flux is clearly made only by the particles whose Larmor centers are less than one Larmor radius distant from plane 1. The flux density of particles with a given Larmor radius and a given distance from the Larmor center to plane 1 in a direction perpendicular to the density gradient is obtained from the equality

$$\Gamma = v \sin \varphi n(x - \rho_H \sin \varphi) \approx nv \sin \varphi - v\rho_H \sin^2 \varphi \frac{dn}{dx}$$

where φ is the angle between the velocity vector at the moment the particle crosses plane 1 and the density gradient (see Fig. 8.23), $v \sin \varphi$ is the projection of the velocity onto a direction perpendicular to the density gradient, and $\rho_H \sin \varphi$ is the distance from plane 1 to the Larmor center (the density is given at the point $x - \rho_H \sin \varphi$). Averaging this flux

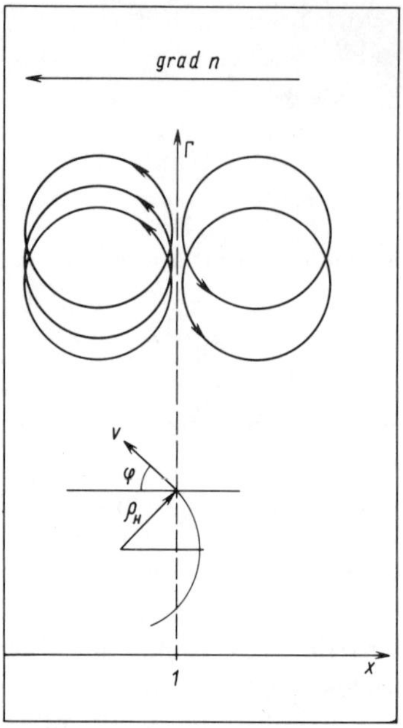

Fig. 8.23 Formation of a diamagnetic flux in the presence of a concentration gradient.

over the angles φ and over the velocities v, we get

$$\bar{\Gamma} = \frac{1}{2}\langle v_{\perp}\rho_H\rangle \frac{\partial n}{\partial x} = \frac{1}{2}\left\langle \frac{cmv_{\perp}^2}{eH} \right\rangle \frac{\partial n}{\partial x} = \frac{cT}{eH}\frac{\partial n}{\partial x} \tag{8.58}$$

Since the flux is perpendicular to **H** and grad n, we have for the α-type particles in vector form

$$\mathbf{\Gamma}_\alpha = n\mathbf{u}_\alpha = \frac{cT_\alpha}{Z_\alpha eH}[\mathbf{h} \times \text{grad } n] \tag{8.59}$$

We can similarly obtain the expression for the fluxes associated with the temperature and magnetic field gradients.

It is easier, however, to find the general expression for the average velocity of the charged particles by using the averaged equations of motion given in Section 6.3. When the collisions can be neglected they can be written as

$$m_\alpha \frac{\partial \mathbf{u}_\alpha}{\partial t} = Z_\alpha e\mathbf{E} + \frac{Z_\alpha e}{c}[\mathbf{u}_\alpha \times \mathbf{H}] - \frac{1}{n}\text{grad } p_\alpha \tag{8.60}$$

If the elactric and magnetic fields are constant, the left-hand side of the equation reduces to zero, and it becomes the equation of equilibrium of averaged forces:

$$Z_\alpha e\mathbf{E}+\frac{Z_\alpha e}{c}[\mathbf{u}_\alpha\times\mathbf{H}]-\frac{1}{n}\operatorname{grad} p_\alpha=0 \tag{8.61}$$

Multiplying Eq. 8.61 vectorially by $\mathbf{h}=\mathbf{H}/H$, as in Section 8.3, we obtain the following expression for the directed velocity:

$$\mathbf{u}_\alpha=\frac{c}{H}[\mathbf{E}\times\mathbf{h}]+\frac{c}{Z_\alpha eHn}[\mathbf{h}\times\operatorname{grad} p_\alpha] \tag{8.62}$$

The first term is due to the drift of the charged particles in the electric field. The second term can be regarded formally as the drift velocity in the field of a force equal to $-(\operatorname{grad} p_\alpha)/n$. This velocity yields the averaged flux associated with the Larmor gyration of the particles and also with their drift in an inhomogeneous magnetic field (it is seen below that $\operatorname{grad} H$ is uniquely related to $\operatorname{grad} p$). In particular, if the density gradient is the only source of the flux, the directed velocity determined by Eq. 8.62 corresponds to Eq. 8.59.

It is easy to find from Eq. 8.62 the density of the diamagnetic flux in a plasma consisting of electrons and single-charged ions:

$$\mathbf{j}=ne(\mathbf{u}_i-\mathbf{u}_e)=\frac{c}{H}[\mathbf{h}\times\operatorname{grad} p] \tag{8.63}$$

where $p=p_e+p_i$ is the summary pressure of the charged particles. This current is seen to be directed perpendicularly to the pressure gradient. For a cylindrically symmetrical plasma in a longitudinal magnetic field, for example, it is directed along the azimuth. The direction of the current is such that the magnetic field induced by it is opposed to the external field. The value of the diamagnetic field due to the current (see Eq. 8.63) can be obtained from the magnetostatic equation $\operatorname{curl}\mathbf{H}=(4\pi/c)\mathbf{j}$. Substituting Eq. 8.63 into it and multiplying vectorially by H, we get

$$-\operatorname{grad}_\perp p=\frac{1}{4\pi}[\mathbf{H}\times\operatorname{curl}\mathbf{H}] \tag{8.64}$$

The vector product on the right-hand side is transformed to

$$[\mathbf{H}\times\operatorname{curl}\mathbf{H}]=\frac{1}{2}\operatorname{grad} H^2-(\mathbf{H}\operatorname{grad})\,\mathbf{H}$$

Since in this case $\mathbf{H}_\perp \operatorname{grad} H$, the relation describing $\operatorname{grad} p$ can be written thus:

$$\operatorname{grad}_\perp p=-\frac{\operatorname{grad}_\perp H^2}{8\pi} \tag{8.65}$$

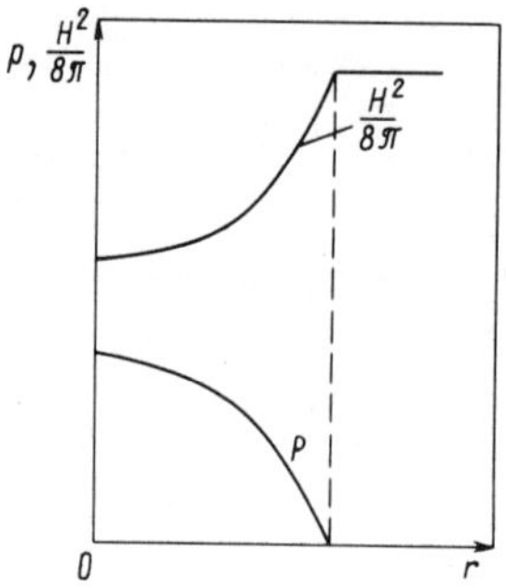

Fig. 8.24 Magnetic pressure as a function of radius.

Whence we find the relationship between the pressure and the magnetic field in a plasma:

$$p + \frac{H^2}{8\pi} = \text{const} \tag{8.66}$$

The quantity $H^2/8\pi$ is called the *magnetic pressure.* The equality 8.66 shows that as the pressure of the charged particles increases from the perimeter to the central regions of the plasma the magnetic pressure, and hence the magnetic field, reduces (Fig. 8.24). The maximum field reduction depends on the maximum pressure. This reduction is usually characterized by the coefficient β, which is equal to the ratio of the kinetic to the magnetic pressure:

$$\beta = \frac{8\pi p}{H^2} \tag{8.67}$$

We recall once again that the results given in this section have been obtained for the case where the lines of force of the magnetic field are parallel and grad $H \perp \mathbf{H}$. If the lines of force are curved, relation 8.66 fails, and the magnetic field distribution in the plasma is determined not only by the pressure gradient but also by the magnetic field tension (see Section 10.2).

8.7 PLASMA POLARIZATION IN ELECTRIC FIELD PERPENDICULAR TO MAGNETIC FIELD

As we have already demonstrated, with a constant electric field the charged particles drift in a direction perpendicular to the electric and magnetic fields with a velocity of $\mathbf{v}_d = c[\mathbf{E} \times \mathbf{h}]/H$.

This drift does not induce any current. Thus in the absence of collisions in a homogeneous plasma a constant electric field does not produce a current; that is, the plasma behaves as a dielectric. If the electric field is not constant, an additional, inertial drift arises, whose velocity is given by Eq. 8.33, which holds true when the characteristic time of field variation greatly exceeds the Larmor period. The direction of this drift is opposite for the ions and the electrons, and its velocity is much higher for the ions than for the electrons. Therefore in a plasma it leads to the appearance of a current practically equal to the ion current:

$$\mathbf{j} = ne(\mathbf{v}_{di}^{(1)} - \mathbf{v}_{de}^{(1)}) \approx \frac{nm_i c^2}{H^2} \frac{\partial \mathbf{E}}{\partial t} \tag{8.68}$$

This current obviously adds up to the vacuum displacement current $\mathbf{j}_v = (1/4\pi)\partial \mathbf{E}/\partial t$. Their sum yields the total current in the plasma under the effect of the variable electric field:

$$\mathbf{j}_s = \mathbf{j} + \mathbf{j}_v = \frac{1}{4\pi}\left(1 + \frac{4\pi m_i c^2 n}{H^2}\right)\frac{\partial \mathbf{E}}{\partial t} \tag{8.69}$$

It can obviously be characterized by the effective dielectric constant of the plasma:

$$\epsilon = 1 + \frac{4\pi n m_i c^2}{H^2} \tag{8.70}$$

Thus a plasma responds to a slowly varying electric field perpendicular to the magnetic field as a dielectric constant given by Eq. 8.70. In particular, "switching on" of an electric field results in polarization of the plasma layer (Fig. 8.25*a*).

Suppose that in some volume, which has a plasma inside it, the external electric field gradually increases from zero to E_0. Then the field in the plasma increases to E_p. Using Eq. 8.33 for the drift velocity, we

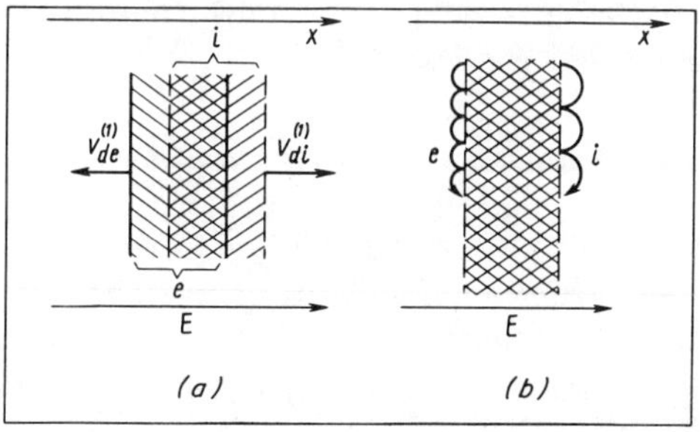

Fig. 8.25a,b The results of switching on an external electric field.

find the displacement of the ions and electrons with increasing field

$$\Delta x_{e,i} = \int_0^t \frac{m_{e,i}c^2}{Z_{e,i}eH^2} \frac{\partial E}{\partial t}\, dt = \frac{m_{e,i}c^2 E_p}{Z_{e,i}eH^2} \tag{8.71}$$

Since the electrons and ions drift in opposite directions (see Fig. 8.25*a*), their relative displacement is equal to $\Delta x = \Delta x_i + \Delta x_e \approx \Delta x_i$ ($\Delta x_i \gg \Delta x_e$, since $m_i \gg m_e$).

In a homogeneous plasma layer displacement leads to the release of the charge at its boundaries. The surface density of the charge released is equal to

$$P \approx ne\Delta x_i = \frac{nm_ic^2}{H^2} E_p \tag{8.72}$$

The field induced by this charge is opposite to the external field and is obviously equal to $\delta E = -4\pi P$.

Adding this field to the external one, we obtain

$$E_p = E_0 - 4\pi P = E_0 - \frac{4\pi nm_ic^2}{H^2} E_p \tag{8.73}$$

Thus the relationship of the field in the plasma with the external field is determined by the expression

$$E_p = \frac{E_0}{1 + 4\pi nm_ic^2/H^2} = \frac{E_0}{\epsilon} \tag{8.74}$$

where ϵ is the dielectric constant given by Eq. 8.70. The relation 8.74 is the same as for a dielectric layer placed in a homogeneous electric field.

Note that the assumption of a slow field increase, used in deriving Eq. 8.71, is optional. If the electric field is switched on, for example, instantaneously, the particles, which were previously at rest, begin to move in cycloids (Fig. 8.25*b*). This displaces the Larmor centers by the Larmor radius at $v_\perp = v_d = cE_p/H$, that is, by $\Delta x = c^2mE_p/ZeH^2$. Thus the displacement of the particles is independent of the nature of the field increase up to the final value.

8.8 PLASMA MOTION ACROSS MAGNETIC FIELD

In the presence of an external nonelectric force perpendicular to the magnetic field the electrons and ions drift in opposite directions. This drift leads to polarization of the plasma and the appearance of an electric field. It is easy to show that in this electric field particles of both signs drift in the direction of the external force. Since the drifts are unlimited, the polarization, and hence the electric field, increases with time. As a result the plasma as a whole moves in the direction of the force with an acceleration.

Let us consider the polarization of a plasma and its motion when the charged particles are subjected to constant forces: ions to a force F_i, and electrons to F_e. A homogeneous bounded plasma layer is easiest to analyze. Let the magnetic field be parallel to the $0z$ axis, and let the direction of the forces coincide with $0x$. These forces cause a drift of electrons and ions in the $0y$ direction. In accordance with Eq. 8.26 the components of their drift velocity are equal to

$$U_{di} = -\frac{cF_i}{eH}; \qquad u_{de} = \frac{cF_e}{eH}$$

The drift results in the appearance, on the plasma surface, of charges perpendicular to $0y$, whose surface density linearly increases with time according to the law $\partial P/\partial t = -en(u_{de} - u_{di})$.

The electric field in the plasma, which is caused by surface charges, satisfies the well-known electrostatic equation $E = -4\pi P/\epsilon$. Using it and taking into account Eq. 8.70 for ϵ, we obtain

$$\frac{\partial E}{\partial t} = \frac{4\pi en(u_{de} - u_{di})}{\epsilon} = \frac{4\pi cn(F_e + F_i)}{H(1 + 4\pi nm_i c^2/H^2)} \tag{8.75}$$

Under the effect of the polarization field the electrons and ions of the plasma drift in the direction of the force with a velocity $u = cE/H$. The equality 8.75 enables us to find the law of variation in this velocity:

$$\frac{\partial u}{\partial t} = \frac{F_e + F_i}{m_i(1 + H^2/4\pi nm_i c^2)} \tag{8.76}$$

Hence it follows that the motion of charged plasma particles in the direction of the acting force is uniformly accelerated. In a sufficiently dense plasma, where $e \approx 4\pi nm_i c^2/H^2 \gg 1$, the second term in the denominator of Eq. 8.76 can be neglected. The acceleration of the charged particles is then the same as in the absence of a magnetic field; it is equal to the ratio of the sum of the acting forces to the mass

$$g = \frac{F_e + F_i}{m_i} = n\frac{F_e + F_i}{\rho} \tag{8.77}$$

($\rho = nm_i$ is the mass density of the plasma). In particular, when the plasma motion under gravity is considered, the acceleration is equal to that of gravity.

The result obtained remains in force for a plasma with a nonuniform concentration. Suppose, for instance, that the charged particle density in the problem under study depends on the y coordinate. Then the electron and ion drift changes their density, as compared with the quasi-neutral density n_0, according to the law

$$\frac{\partial n_e}{\partial t} = -\operatorname{div}(n_0 \mathbf{u}_{de}); \qquad \frac{\partial n_i}{\partial t} = -\operatorname{div}(n_0 \mathbf{u}_{di})$$

Thus the change in space charge density is equal to

$$\frac{\partial \rho_e}{\partial t} = -e \operatorname{div}[n_0(\mathbf{u}_{di} - \mathbf{u}_{de})] \tag{8.78}$$

It is easy to see that this space charge induces an electric field (Eq. 8.75). Indeed, the field distribution for a medium with an electric constant is described by the equation div $\epsilon\mathbf{E} = +4\pi\rho_e$. Differentiating it with respect to time and substituting Eq. 8.78 for $\partial\rho_e/\partial t$ we obtain Eq. 8.75. Therefore the expression 8.76 for the rate of field increase, and hence for the acceleration of the drift in the electric field, remains valid in an inhomogeneous plasma.

The motion of a plasma in a magnetic field may be caused not only by external nonelectric forces, but also by forces produced by the inhomogeneous magnetic field. These are the centrifugal force due to the particle motion along the lines of force, and the diamagnetic force associated with the Larmor gyration of the particles. When the magnetic lines of force are planar the sum of these is equal to (see Eq. 8.48)

$$F_H = -\left(mv_\parallel^2 + \frac{1}{2}mv_\perp^2\right)\frac{|\mathrm{grad}_\perp H|}{H} \tag{8.79}$$

Averaging it over the velocities of charged particles of each type, we obtain, for an isotropic distribution,

$$\langle F_{H\alpha}\rangle = 2T_\alpha \frac{|\mathrm{grad}_\perp H|}{H} = \frac{2T_\alpha}{R} \tag{8.80}$$

Under the action of this force the charged plasma particles must move in the direction of decreasing magnetic field. In accordance with Eq. 8.77 their acceleration at $\epsilon \gg 1$ is equal to

$$g_H = \frac{F_e + F_i}{m_i} = \frac{2(T_e + T_i)}{m_i R} \tag{8.81}$$

In particular, in a toroidal magnetic field the drift of the charged particles in a direction perpendicular to grad H causes plasma polarization (Fig. 8.26). Under the effect of the electric field of polarization the charged particles are accelerated in the direction of the outer surface of the torus. The time of plasma displacement in this direction by the radius a is

$$\tau_\alpha = \sqrt{2a/g_H} = \sqrt{aRm_i/(T_e + T_i)} \tag{8.82}$$

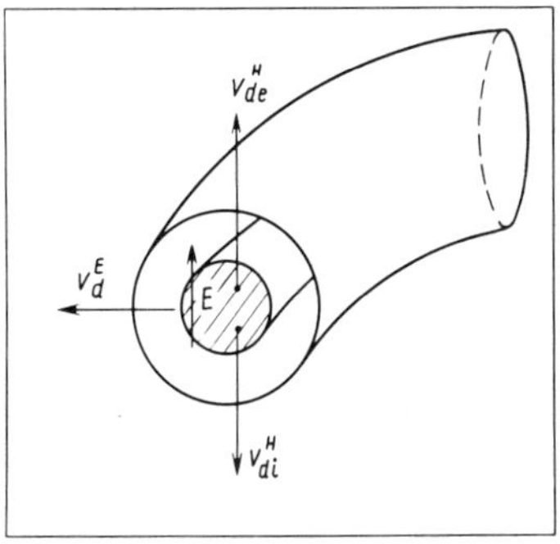

Fig. 8.26 Plasma polarization in a toroidal magnetic field.

Equation 8.81 enables us to discuss the plasma blob motion across the magnetic field. From this equation it follows, first of all, that in a homogeneous magnetic field $g_H = 0$ and that a plasma blob which initially had a directed velocity across the lines of force moves freely in the preassigned direction. The direction of motion remains the same because the charge separation and the associated electric field induced by the accelerating blob remain constant, since no energy sources can change the plasma polarization. This holds true, of course, in the absence of collisions leading to energy dissipation. Thus upon acceleration each plasma particle executes a drift in the electric field, which is carried over by the plasma blob.

The blob moving into the magnetic configuration across the lines of force is slowed down in the region of increasing field. The blob may either pass through the magnetic field region or be reflected, depending on the ratio between the initial transverse velocity and the maximum magnetic field. This happens in any case even if the magnetic field configuration is fundamentally capable of confining the plasma. The capture of the blob requires an additional mechanism for dissipating the energy of the directed motion inside the configuration. This can be achieved, for instance, by shooting two blobs at one another. Since their polarizations are opposite, the plasma polarization disappears on mutual penetration of the blobs, which then cease to move.

9

TRANSPORT PROCESSES IN MAGNETIC FIELD

9.1 DIRECTED VELOCITY AND HEAT FLUX OF CHARGED PARTICLES OF WEAKLY IONIZED PLASMAS IN MAGNETIC FIELD

In the preceding chapter we considered the charged plasma particle motion across a magnetic field for conditions where the collisions are insignificant. In most cases, however, it is necessary to consider the effect of collisions on transport in the plasma. We see below that this effect is substantial even at collision frequencies much below the cyclotron frequencies. We begin consideration of transport processes with a weakly ionized plasma in which the frequency of collisions of charged particles with neutrals greatly exceeds the frequency of their collisions with each other (see Eq. 7.1 and Fig. 7.1).

The directed motion of charged particles is described by the first-moment equation 6.62. For stationary or quasi-stationary conditions it is the force equilibrium equation, which can be written thus:

$$Z_\alpha e\mathbf{E} + \frac{Z_\alpha e}{c}[\mathbf{u}_\alpha \times \mathbf{H}] - \frac{\operatorname{grad}(nT_\alpha)}{n} + m\frac{\delta \mathbf{u}_\alpha}{\delta t} = 0 \tag{9.1}$$

This assumes that the plasma consists of electrons, single-charged ions, and neutral atoms (for electrons $\alpha = e$, $Z_e = -1$; for ions $\alpha = i$, $Z_i = 1$, $n_e = n_i = n$). The first term of this equation is the electric force, the second the Lorentz force, the third the pressure gradient per particle, and the fourth the force due to collisions. As shown in Chapter 7, Eq. 9.1 holds provided the characteristic time of plasma parameter variation greatly exceeds the intercollisional time. For a weakly ionized plasma the collision term of Eq. 9.1 is determined by collisions of charged particles with neutrals. If the collision frequency is velocity independent it is determined by the friction force (see Eq. 6.59):

$$m_\alpha \frac{\delta \mathbf{u}_\alpha}{\delta t} = -\mu_{\alpha a}\nu_{\alpha a}\mathbf{u}_\alpha$$

where $\nu_{\alpha a}$ is the transport frequency of α-type particle–atom collisions and $\mu_{\alpha a}$ is the reduced mass; it is assumed that the directed velocity of the charged particles $\mathbf{u}_\alpha$ is much higher than that of the neutrals. Substitution of the collision term changes Eq. 9.1 to

$$Ze\mathbf{E} + \frac{Ze}{c}[\mathbf{u}\times\mathbf{H}] - \frac{\text{grad}(nT)}{n} - \mu\nu\mathbf{u} = 0 \tag{9.2}$$

where we omitted, for brevity, the subscripts denoting the type of the particles. The expression 9.2 is a vector algebraic equation for $\mathbf{u}$. The projection of this equation onto the direction of the magnetic field results in the equality

$$Ze\mathbf{E}_\parallel - \frac{\text{grad}_\parallel(nT)}{n} - \mu\nu\mathbf{u}_\parallel = 0 \tag{9.3}$$

This projection lacks the Lorentz force. Therefore the solution of the equation for the longitudinal velocity component is the same as in the absence of a magnetic field:

$$\mathbf{u}_\parallel = \frac{Z_e E_\parallel}{\mu\nu} - \frac{\text{grad}_\parallel(nT)}{n\mu\nu} = \frac{Ze}{\mu\nu}E_\parallel - \frac{T}{\mu\nu}\left(\frac{\text{grad}_\parallel n}{n} + \frac{\text{grad}_\parallel T}{T}\right) \tag{9.4}$$

(cf. Eq. 7.7). Proceeding from this, we introduce the mobility and diffusion coefficients characterizing the motion in the direction of the field:

$$b_\parallel = \frac{e}{\mu\nu}; \qquad D_\parallel = D_\parallel^T = \frac{T}{\mu\nu} \tag{9.5}$$

The expressions for these quantities naturally coincide with the similar expressions 7.9 and 7.10.

The projection of Eq. 9.2 on a plane perpendicular to the field determines the projection of the velocity vector $\mathbf{u}_\perp$ on this plane. The resulting equation can be written as

$$\mu\nu\mathbf{u}_\perp - Zm\omega_H[\mathbf{u}_\perp\times\mathbf{h}] = Ze\mathbf{E}_\perp - \frac{\text{grad}_\perp(nT)}{n} \tag{9.6}$$

where $\mathbf{h} = \mathbf{H}/H$ is the unit vector in the direction of the field, and $\omega_H = eH/mc$ is the cyclotron frequency. To find the solution of Eq. 9.8 for $\mathbf{u}_\perp$, we multiply it vectorially by $\mathbf{h}$ and thus have

$$Zm\omega_H\mathbf{u}_\perp + \mu\nu[\mathbf{u}_\perp\times\mathbf{h}] = Ze[\mathbf{E}_\perp\times\mathbf{h}] + \frac{[\mathbf{h}\times\text{grad}(nT)]}{n} \tag{9.7}$$

Eliminating the vector product $[\mathbf{u}_\perp \times \mathbf{h}]$ from Eqs. 9.6 and 9.7, we find the transverse component of the directed velocity

$$\mathbf{u}_\perp = \frac{e[\mathbf{E}\times\mathbf{h}]}{m\omega_H(1+\mu^2\nu^2/m^2\omega_H^2)} + \frac{[\mathbf{h}\times \text{grad}(nT)]}{Zm\omega_H(1+\mu^2\nu^2/m^2\omega_H^2)n} + \frac{Ze\mu\nu\mathbf{E}_\perp}{m^2\omega_H^2+\mu^2\nu^2} - \frac{\mu\nu\,\text{grad}_\perp(nT)}{(m^2\omega_H^2+\mu^2\nu^2)n} \tag{9.8}$$

Expression 9.8 is the sum of four terms:

$$\mathbf{u}_\perp = \mathbf{u}_{dE} + \mathbf{u}_{dP} + \mathbf{u}_{tE} + \mathbf{u}_{tP} \tag{9.9}$$

The first two terms of this sum are the velocity of the drift perpendicular to the electric field and the pressure gradient. They can be written as

$$\mathbf{u}_{dE} = \frac{c[\mathbf{E}\times\mathbf{h}]}{H\left(1+\dfrac{\mu^2}{m^2}\dfrac{\nu^2}{\omega_H^2}\right)} = b_d[\mathbf{E}\times\mathbf{h}];$$

$$\mathbf{u}_{dp} = \frac{c[\mathbf{h}\times \text{grad}(nT)]}{Zen\,H\left(1+\dfrac{\mu^2}{m^2}\dfrac{\nu^2}{\omega_H^2}\right)} = D_d\left[\mathbf{h}\times\frac{\text{grad}\,p}{p}\right] \tag{9.10}$$

At a collision frequency much less than the cyclotron frequency ($\nu \ll \omega_H$) Eqs. 9.10 change to the relations 8.27 and 8.62 for the drift velocity in the collisionless regime. It is seen that the collisions reduce the drift velocity related to the electric field and the pressure gradient, and at $\nu \gg \omega_H$ this drift becomes insignificant. The last two terms of Eq. 9.8 can be written in a form similar to Eqs. 7.9 and 7.10 if we introduce transverse mobility and diffusion coefficients:

$$\mathbf{u}_{tE} = Zb_\perp\mathbf{E}_\perp; \qquad \mathbf{u}_{tp} = -\frac{D_\perp\,\text{grad}\,n}{n} - \frac{D_\perp^T\,\text{grad}\,T}{T} \tag{9.11}$$

where

$$b_\perp = \frac{e\mu\nu}{m^2\omega_H^2+\mu^2\nu^2}; \qquad D_\perp = D_\perp^T = \frac{\nu\mu T}{m^2\omega_H^2+\mu^2\nu^2} \tag{9.12}$$

The relationship between the diffusion and mobility coefficients is the same as without a magnetic field

$$\frac{b_\perp}{D_\perp} = \frac{e}{T}, \qquad D_\perp = D_\perp^T \tag{9.13}$$

As can be seen from Eqs. 9.5 and 9.12, the mobility and diffusion coefficients $b_\perp$ and $D_\perp$ differ from the corresponding coefficients in the absence of a magnetic field and from the longitudinal coefficients $b_\parallel$ and $D_\parallel$ by the factor $(1+m^2\omega_H^2/\mu^2\nu^2)^{-1}$. In weak magnetic fields (or at high collision frequencies), when $\omega_H/\nu \ll 1$, the transport coefficients are practically the same as without a magnetic field. This is natural because the inequality $\nu \gg \omega_H$ means that within the time between collisions the particle can only make a small fraction of a revolution along the Larmor circle; that is, the particle trajectory is close to a straight line. In other words, the effect of the magnetic field on the particle motion is insignificant. At $\omega_H \gg \nu$, on the contrary, the particle manages to make many revolutions between collisions. Therefore the nature of conduction and diffusion changes sharply. The mechanism of transport across a strong magnetic field will be considered in Section 9.2. At $\omega_H \gg \nu$ the coefficients $b_\perp$ and $D_\perp$ take the form:

$$b_\perp = \frac{e\mu\nu}{m^2\omega_H^2} = \frac{c^2\mu\nu}{eH^2}; \tag{9.14}$$

$$D_\perp = \frac{T\nu\mu}{m^2\omega_H^2} = \frac{c^2\mu T\nu}{e^2H^2}; \tag{9.15}$$

$$D_\perp^T = \frac{T\nu\mu}{m^2\omega_H^2} = \frac{c^2\mu T_\nu}{e^2H^2} \tag{9.16}$$

The ratio of these values to the respective longitudinal coefficients is

$$\frac{b_\perp}{b_\parallel} = \frac{D_\perp}{D_\parallel} = \frac{D_\perp^T}{D_\parallel^T} = \frac{\mu^2\nu^2}{m^2\omega_H^2} \tag{9.17}$$

At strong magnetic fields it may be much less than unity. The quantity $(\omega_H/\nu)^2$, which shows the decrease in transverse transport coefficients, is sometimes called *magnetization* of the charged particles. It can be seen that in a strong magnetic field the transport coefficients are proportional to the particle mass and the collision frequency. This means, in particular, that they are much larger for heavy ions than for light electrons. Recall that without a magnetic field the situation reverses: the transport coefficients are much larger for electrons than for ions and increase with reducing frequency.

The equations 9.4, 9.10, and 9.11 completely determine the value of the directed velocity. Taking into account the expressions for b and D, we can represent it thus:

$$\mathbf{u} = Zb_\parallel \mathbf{E}_\parallel + Z_{b_\perp}\mathbf{E}_\perp + b_d[\mathbf{E}\times\mathbf{h}] - D_\parallel \frac{\mathrm{grad}_\parallel p}{p}$$
$$-D_\perp \frac{\mathrm{grad}_\perp p}{p} + D_d \frac{[\mathbf{h}\times \mathrm{grad}\, p]}{p} \tag{9.18}$$

This sum is often written more concisely by using the tensor representation for mobility and the diffusion coefficients:

$$\mathbf{u} = Z\check{\mathbf{b}}\mathbf{E} - \frac{\check{\mathbf{D}}\,\text{grad}\,p}{p} \tag{9.19}$$

In a reference system where $\mathbf{H} \parallel 0z$ the tensors introduced have the following form:

$$\check{\mathbf{b}} = \begin{vmatrix} b_\perp & b_d & 0 \\ -b_d & b_\perp & 0 \\ 0 & 0 & b_\parallel \end{vmatrix}$$

$$= \begin{vmatrix} \dfrac{e\mu\nu}{m^2\omega_H^2 + \mu^2\nu^2} & \dfrac{c}{H(1+\mu^2\nu^2/m^2\omega_H^2)} & 0 \\ -\dfrac{c}{H(1+\mu^2\nu^2/m^2\omega_H^2)} & \dfrac{e\mu\nu}{m^2\omega_H^2+\mu^2\nu^2} & 0 \\ 0 & 0 & e/\mu\nu \end{vmatrix} \tag{9.20}$$

$$\check{\mathbf{D}} = \begin{vmatrix} D_\perp & D_d & 0 \\ -D_d & D_\perp & 0 \\ 0 & 0 & D_\parallel \end{vmatrix}$$

$$= \begin{vmatrix} \dfrac{\nu\mu T}{m^2\omega_H^2 + \mu^2\nu^2} & \dfrac{cT}{ZeH(1+\mu^2\nu^2/m^2\omega_H^2)} & 0 \\ -\dfrac{cT}{ZeH(1+\mu^2\nu^2/m^2\omega_H^2)} & \dfrac{\nu\mu T}{m^2\omega_H^2+\mu^2\nu^2} & 0 \\ 0 & 0 & T/\mu\nu \end{vmatrix} \tag{9.21}$$

The expressions for the directed velocity make it possible to obtain the density of the electric current in the plasma. In the general form, for a plasma consisting of electrons and single-charged ions we find from Eq. 9.19

$$\mathbf{j} = ne(\mathbf{u}_i - \mathbf{u}_e) = ne(\check{\mathbf{b}}_e + \check{\mathbf{b}}_i)\mathbf{E} - ne\left(\frac{\check{\mathbf{D}}_i\,\text{grad}\,p_i}{p_i} - \frac{\check{\mathbf{D}}_e\,\text{grad}\,p_e}{p_e}\right) \tag{9.22}$$

The first term describes the motion under the effect of the electric field. It can be expressed via the conductivity tensor:

$$\check{\boldsymbol{\sigma}} = ne(\check{\mathbf{b}}_e + \check{\mathbf{b}}_i) \tag{9.23}$$

For strong magnetic fields, where $\omega_{He} \gg \nu_{ea}$ and $\omega_{Hi} \gg \nu_{ia}$, the transverse conductivity component will be found using Eq. 9.14:

$$\mathbf{j}_\perp = \sigma_\perp \mathbf{E}_\perp;$$

$$\sigma_\perp = ne^2 \left(\frac{\mu_{ia}\nu_{ia}}{m_i^2 \omega_{Hi}^2} + \frac{\mu_{ea}\nu_{ea}}{m_e^2 \omega_{He}^2} \right) \approx \frac{ne^2 \nu_{ia}}{2m_i \omega_{Hi}^2} \tag{9.24}$$

(at $m_i = m_a$, $\mu_{ia} = m_i/2$). Note that the current across **E** and **H** is then zero, since the electron and ion drift velocities are equal.

The above expressions for the directed velocity were obtained for the case where the frequency of collisions of charged particles with neutrals is velocity independent. General expressions can be written in a form similar to Eqs. 9.10 through 9.12. They will contain some averaged collision frequencies and numerical factors of the order of unity. At near-Maxwellian velocity distributions the directed velocities can be obtained by substituting into Eq. 9.1 the expressions 6.58 for the collision term, which includes both the friction force and the thermal force:

$$\frac{m_\alpha \delta \mathbf{u}_\alpha}{\delta t} = \mathbf{R}_{\alpha a} + \mathbf{R}_{\alpha a}^T \tag{9.25}$$

The friction force

$$\mathbf{R}_{\alpha a} = -\mu_{\alpha a} \bar{\nu}_{\alpha a} \mathbf{u}_\alpha \tag{9.26}$$

includes the averaged collision frequency (see Eq. 7.18),

$$\bar{\nu}_{\alpha a} = \frac{1}{3} \sqrt{\frac{2}{\pi}} \int_0^\infty x^4 \nu_{\alpha a}^t \left(x \sqrt{\frac{T_{\alpha a}}{\mu_{\alpha a}}} \right) \exp\left(-\frac{x^2}{2} \right) dx \tag{9.27}$$

where $T_{\alpha a} = (m_a T_\alpha + m_\alpha T_a)/(m_a + m_\alpha)$. The averaging law (Eq. 7.18) is approximate (see Section 6.2), but it can be shown that its accuracy increases with $\omega_H/\bar{\nu}$. Its applicability at $\omega_H \gg \bar{\nu}$ for electrons is demonstrated in Section 9.2. The expression for the thermal force at $\omega_H \gg \bar{\nu}$ is

$$\mathbf{R}^T = -g_T \left(\frac{\bar{\nu}}{Z\omega_H} \right) [\mathbf{h} \times \text{grad } T] \tag{9.28}$$

where $g_T \approx (T/\nu)(\partial \bar{\nu}/\partial T)$ (Eq. 9.28 is obtained qualitatively in Section 9.3).

Substituting the collision term 9.25 with the friction force 9.26 and the thermal force 9.28 into the equation of motion 9.1, we obtain the former equation for the directed velocity components associated with the electric field and the density gradient. In particular, the transverse mobility $b_\perp$ and the diffusion coefficient $D_\perp$ are, as before, found from Eqs. 9.12 with the collision frequency 9.27. The thermal force changes the directed velocity due to the temperature gradient. The velocity of the drift caused by the thermal force at $\omega_H \gg \bar{\nu}$ is equal to

$$\mathbf{u}_\perp^{T*} = \frac{c}{ZeH}[\mathbf{R}^T \times \mathbf{h}] = -g_T \frac{\bar{\nu}}{m\omega_H^2}\left(\frac{\mu}{m}\right) \text{grad } T \tag{9.29}$$

Adding it to the velocity component u_{tp} (see Eq. 9.11), proportional to grad T,

$$\mathbf{u}_{tp}^T = -\frac{\mu T \bar{\nu}}{m^2 \omega_H^2} \frac{\text{grad } T}{T}$$

we obtain the thermal diffusion rate in the form

$$\mathbf{u}_\perp^T = -D_\perp^T \frac{\text{grad } T}{T} \tag{9.30}$$

where

$$D_\perp^T = \frac{\mu \nu T}{m^2 \omega_H^2}(1 + g_T)$$

To find the heat flux of charged particles we take advantage of Eq. 6.80. In the stationary case it has the form

$$\frac{5}{2}\frac{nT}{m} \text{grad } T - \frac{Ze}{mc}[\mathbf{q} \times \mathbf{H}] = \frac{\delta \mathbf{q}}{\delta t} \tag{9.31}$$

For electrons, at a velocity-independent collision frequency we obtain, by substituting the collision term 6.82 into Eq. 9.31,

$$\frac{5}{2}\frac{nT_e}{m_e} \text{grad } T_e + \frac{e}{m_e c}[\mathbf{q}_e \times \mathbf{H}] = -\nu_{ea}\mathbf{q}_e$$

This vector equation in $\mathbf{q}$ is similar in form to the vector equation 9.2. Solving it in the same way as was done on p. 281, we find

$$\begin{aligned}
\mathbf{q}_e &= \mathbf{q}_{e\|} + \mathbf{q}_{et} + \mathbf{q}_{ed}; \\
\mathbf{q}_{e\|} &= -\frac{5}{2}\frac{nT_e}{m_e \nu_{ea}} \text{grad}_\| \, T_e; \\
\mathbf{q}_{et} &= -\frac{5}{2}\frac{nT_e \nu_{ea}}{m_e(\omega_{He}^2 + \nu_{ea}^2)} \text{grad}_\perp \, T_e; \\
\mathbf{q}_{ed} &= -\frac{5}{2}\frac{cnT_e[\mathbf{h} \times \text{grad } T_e]}{eH(1 + \nu_{ea}^2/\omega_{He}^2)}
\end{aligned} \tag{9.32}$$

The first term is the longitudinal heat flux. It is the same as in the absence of a magnetic field. The component $\mathbf{q}_{ed}$ is the heat flux related to the diamagnetic flux perpendicular to grad T. At $\omega_H \gg \nu$ it is collision independent. Finally, the component $\mathbf{q}_{et}$ represents the heat flux in the direction of the transverse component of the temperature gradient. At $\omega_H \gg \nu$ it is proportional to the collision frequency:

$$\mathbf{q}_{et} = -\frac{5}{2}\frac{nT_e\nu_{ea}}{m_e\omega_H^2}\,\mathrm{grad}_\perp T_e \tag{9.33}$$

The expressions 9.32 and 9.33 make it possible, as before, to introduce the thermal- and temperature-conductivity coefficients. The longitudinal thermal conductivity is the same as without a magnetic field:

$$\mathcal{K}_\parallel = n\chi_\parallel = \frac{5}{2}\frac{nT_e}{m_e\nu_{ea}} \tag{9.34}$$

The transverse thermal- and temperature-conductivity coefficients can be written thus (as seen from Eq. 9.32):

$$\mathcal{K}_\perp = n\chi_\perp = \frac{5}{2}\frac{nT_e\nu_{ea}}{m_e(\omega_{He}^2 + \nu_{ea}^2)} \tag{9.35}$$

for $\omega_H \gg \nu$

$$\mathcal{K}_\perp = n\chi_\perp = \frac{5}{2}\frac{nT_e\nu_{ea}}{m_e\omega_{He}^2} \tag{9.36}$$

They are related to the corresponding components of the diffusion tensor by expressions similar to Eq. 7.23:

$$\chi_\perp = \frac{\mathcal{K}_\perp}{n} = \frac{5}{2}D_\perp \tag{9.37}$$

The expression for the ion heat flux in a weakly ionized plasma is more intricate, because this flux is usually associated with the heat flux of neutral particles. We do not consider it here.

9.2 TRANSVERSE COEFFICIENTS OF ELECTRON MOBILITY, DIFFUSION, AND THERMAL CONDUCTIVITY

As shown in Section 7.2, in order to obtain the transport coefficients of electrons, with an arbitrary velocity dependence of their collision frequency one can expand the distribution function in the degrees of anisotropy. Since for electrons in a weakly ionized plasma the directed velocity component is much below the random one, we can restrict ourselves to two components in this expansion (see Eq. 5.20):

$$f(\mathbf{v}) = f_0(v) + \frac{\mathbf{v}}{v}\mathbf{f}_1(v)$$

(f_0 is the isotropic and $\mathbf{f}_1$ the directed component of the distribution function). Here the directed velocity is related to the vector function f_1 by the expression (see Eq. 5.21)

$$\mathbf{u} = \int \mathbf{v} f(\mathbf{v})\, d^3v = \frac{4\pi}{3}\int_0^\infty v^3 \mathbf{f}_1\, dv \tag{9.38}$$

In a magnetic field the equation for the relationship between the directed component of the distribution function and the isotropic one can be written as (see Eq. 5.28)

$$-\frac{e\mathbf{E}}{m_e}\frac{\partial f_0}{\partial v} + v\,\frac{\mathrm{grad}(nf_0)}{n} - \frac{e}{m_e c}[\mathbf{H}\times\mathbf{f}_1] = -\mathbf{f}_1\nu_{ea}(v) \tag{9.39}$$

where $\nu_{ea}(v)$ is the summary frequency of electrons–atom collisions. The longitudinal component $\mathbf{f}_1$ is obtained from the projection of the equation onto the direction of the magnetic field:

$$f_{1\parallel} = \frac{eE_\parallel}{m_e\nu_{ea}}\frac{\partial f_0}{\partial v} - \frac{v}{\nu_{ea}}\frac{\mathrm{grad}_\parallel(nf_0)}{n} \tag{9.40}$$

As would be expected, this expression coincides with Eq. 7.31 for the directed component of the distribution function without a magnetic field. Accordingly, the expressions of the transport coefficients for the directed velocity also coincide. The projection of Eq. 9.39 onto a plane perpendicular to the magnetic field can be written as

$$\nu_{ea}\mathbf{f}_{1\perp} - \omega_{He}[\mathbf{h}\times\mathbf{f}_1] = \frac{e\mathbf{E}_\perp}{m_e}\frac{\partial f_0}{\partial v} - v\,\frac{\mathrm{grad}_\perp(nf_0)}{n} \tag{9.41}$$

The solution of this equation for $\mathbf{f}_{1\perp}$ is similar to that of Eq. 9.2 in Section 9.1. To find it we can use, along with Eq. 9.41, the equation obtained from it by vector multiplication by $\mathbf{h}$. As a result we find the two transverse components of the vector $\mathbf{f}_{1\perp}$: the one ($\mathbf{f}_{1t}$) parallel to the vector in the right-hand side of the equation, and the other ($\mathbf{f}_{1d}$) perpendicular to this vector:

$$\mathbf{f}_{1t} = \frac{e\mathbf{E}_\perp}{m_e}\frac{\nu_{ea}}{\omega_{He}^2+\nu_{ea}^2}\frac{\partial f_0}{\partial v} - \frac{v\nu_{ea}}{\omega_{He}^2+\nu_{ea}^2}\frac{\mathrm{grad}_\perp(nf_0)}{n}; \tag{9.42}$$

$$\mathbf{f}_{1d} = \frac{e}{m}\frac{\omega_{He}}{\omega_{He}^2+\nu_{ea}^2}\frac{\partial f_0}{\partial v}[\mathbf{h}\times\mathbf{E}] - \frac{v\omega_{He}}{\omega_{He}^2+\nu_{ea}^2}\frac{[\mathbf{h}\times\mathrm{grad}(nf_0)]}{n} \tag{9.43}$$

It is easy to see that $\mathbf{f}_{1d}$ represents the component of the directed velocity $\mathbf{u}_d$ describing the electron drift perpendicular to the electric field and the density and temperature gradients. At $\omega_H \gg \nu$ Eq. 9.43 ceases to depend on the collision frequency, and Eq. 9.38 leads to Eq. 8.62 for the directed drift velocity. At $\nu \gg \omega_H$ the expression 9.43 makes it possible to determine the decrease in drift velocity due to collisions.

The motion in the direction of the transverse component of the electric field and the transverse components of the density and temperature gradients is described by Eq. 9.42. Substituting it into Eq. 9.38, we obtain for the corresponding component of the directed velocity

$$\begin{aligned}\mathbf{u}_{et} &= \frac{4\pi}{3}\frac{e\mathbf{E}_\perp}{m_e}\int \frac{v^3 \nu_{ea}}{\omega_{He}^2 + \nu_{ea}^2}\frac{\partial f_0}{\partial v}\,dv \\ &\quad - \frac{4\pi}{3n}\,\mathrm{grad}\left[n\int \frac{v^4 \nu_{ea}}{\omega_{He}^2 + \nu_{ea}^2} f_0\,dv\right]\end{aligned} \tag{9.44}$$

Let us express $\mathbf{u}_{et}$ in terms of the mobility and diffusion coefficients:

$$\begin{aligned}\mathbf{u}_{et} &= -b_{e\perp}\mathbf{E}_\perp - \frac{1}{n}\,\mathrm{grad}_\perp(D_{e\perp} n) \\ &= -b_{e\perp}\mathbf{E}_\perp - D_{e\perp}\frac{\mathrm{grad}_\perp n}{n} - D_{e\perp}^T \frac{\mathrm{grad}_\perp T_e}{T_e}\end{aligned} \tag{9.45}$$

where

$$b_{e\perp} = -\frac{4\pi}{3}\frac{e}{m_e}\int \frac{v^3 \nu_{ea}}{\omega_{He}^2 + \nu_{ea}^2}\frac{\partial f_0}{\partial v}\,dv \tag{9.46}$$

$$D_{e\perp} = \frac{4\pi}{3}\int \frac{v^4 \nu_{ea}}{\omega_{He}^2 + \nu_{ea}^2} f_0\,dv \tag{9.47}$$

$$D_{e\perp}^T = T_e \frac{\partial D_{e\perp}}{\partial T_e} \tag{9.48}$$

When the collision frequency is velocity independent, Eqs. 9.46–9.48 change to Eqs. 9.12. With an arbitrary dependence $\nu_{ea}(v)$ and an arbitrary distribution function $f_0(v)$, they differ from Eqs. 9.12 by the numerical coefficients.

Let us particularize Eqs. 9.46–9.48 for a Maxwellian electron velocity distribution:

$$f_0(v) = \left(\frac{m_e}{2\pi T_e}\right)^{3/2} \exp\left(-\frac{m_e v^2}{2T_e}\right) \tag{9.49}$$

and for a cyclotron frequency greatly exceeding that of collisions. In this case the equations for the mobility and diffusion coefficients can be represented as

$$b_{e\perp} = \frac{e\bar{\nu}_{ea}}{m_e \omega_{He}^2} \tag{9.50}$$

$$D_{e\perp} = \frac{T_e \bar{\nu}_{ea}}{m_e \omega_{He}^2} \tag{9.51}$$

if we introduce the averaged collision frequency $\bar{\nu}_{ea}(T_e)$ (see Eq. 9.27). Then the thermal diffusion coefficient is equal to

$$D_{e\perp}^{T} = T_e \frac{\partial}{\partial T_e}\left(\frac{T_e \bar{\nu}_{ea}}{m_e \omega_{He}^2}\right) = \frac{T_e \bar{\nu}_{ea}}{m_e \omega_{He}^2}(1 + g_{T_e}) \tag{9.52}$$

where

$$g_{T_e} = \frac{T_e}{\bar{\nu}_{ea}} \frac{\partial \bar{\nu}_{ea}}{\partial T_e}$$

(cf. Eq. 9.30).

The electron heat flux in a magnetic field can be found from the directed component of the distribution function, as was done in Section 7.2. The general expression for the heat flux has the form

$$\mathbf{q} = \frac{4\pi}{6} n m_e \int \left(v^5 - \frac{5T_e}{m_e} v^3\right) \mathbf{f}_1 \, \mathrm{d}\mathbf{v} \tag{9.53}$$

Substituting $\mathbf{f}_1$ into this expression, we can easily find the components of the vector $\mathbf{q}$. Substituting $\mathbf{f}_{1\parallel}$ (see Eq. 9.40) yields the expression for the longitudinal heat flux $\mathbf{q}_\parallel$, which coincides with Eq. 7.45 for no magnetic field. The component $\mathbf{f}_{1d}$ is the heat flux vector component associated with the diamagnetic motion $\mathbf{q}_{ed}$. For a strong magnetic field $\omega_H \gg \nu$ it is independent of the collision frequency. The heat flux component parallel to the transverse component of the acting forces $\mathbf{q}_{et}$ can be obtained by substituting the component $\mathbf{f}_{1t}$ into Eq. 9.53:

$$\begin{aligned}
\mathbf{q}_{et} = &-\frac{4\pi}{6} ne\mathbf{E}_\perp \int_0^\infty \left(v^5 - \frac{5T_e}{m_e} v^3\right) \frac{\nu_{ea}}{\omega_{He}^2 + \nu_{ea}^2} \frac{df_0}{dv} \, dv \\
&-\frac{4\pi}{6} m_e \operatorname{grad}\left[n \int_0^\infty \left(v^6 - \frac{5T_e}{m_e} v^4\right) \frac{\nu_{ea}}{\omega_{He}^2 + \nu_{ea}^2} f_0 \, dv\right] \\
&-\frac{10\pi}{3} n \operatorname{grad} T_e \int_0^\infty v^4 \frac{\nu_{ea}}{\omega_{He}^2 + \nu_{ea}^2} f_0 \, dv
\end{aligned} \tag{9.54}$$

We transform this expression for $\omega_{He} \gg \nu_{ea}$ and a Maxwellian electron velocity distribution. Here the integrals in the first two terms are equivalent:

$$\begin{aligned}
J &= \frac{4\pi}{6} \frac{m_e}{T_e} \int_0^\infty \left(v^5 - 5\frac{T_e}{m_e} v^3\right) \nu_{ea} \frac{df_0}{dv} \, dv \\
&= -\frac{4\pi}{6} \left(\frac{m_e}{T_e}\right)^2 \int_0^\infty \left(v^6 - 5\frac{T_e}{m_e} v^4\right) \nu_{ea} f_0 \, dv \\
&= -\frac{1}{3\sqrt{2\pi}} \int_0^\infty (x^6 - 5x^4) \nu_{ea}\left(x\sqrt{\frac{T_e}{m_e}}\right) \exp\left(-\frac{x^2}{2}\right) dx
\end{aligned} \tag{9.55}$$

Integrating by parts and taking into account Eq. 9.27 for the average collision frequency, we transform Eq. 9.55 to

$$J = \frac{1}{3\sqrt{2\pi}} \int_0^\infty \frac{\partial \nu_{ea}}{\partial x} x^5 \exp\left(-\frac{x^2}{2}\right) dx = T_e \frac{\partial \bar{\nu}_{ea}}{\partial T_e} \tag{9.56}$$

Substituting Eq. 9.56 into Eq. 9.54, we obtain for $\omega_{He} \gg \nu_{ea}$

$$\mathbf{q}_{et} = -g_{T_e}\left[nT_e \frac{e\bar{\nu}_{ea}}{m_e\omega_{He}^2}\mathbf{E}_\perp + \operatorname{grad}_\perp\left(\frac{nT_e^2\bar{\nu}_{ea}}{m_e\omega_{He}^2}\right)\right] - \frac{5}{2}\frac{n\bar{\nu}_{ea}T_e}{m_e\omega_{He}^2}\operatorname{grad}_\perp T_e \tag{9.57}$$

or, with due allowance for Eqs. 9.45 and 9.50–9.52,

$$\mathbf{q}_{et} = -g_{T_e} nT_e \mathbf{u}_{et} - \left(\frac{5}{2} + g_{T_e}\right)\frac{nT_e}{m_e\omega_{He}^2}\bar{\nu}_{ea}\operatorname{grad}_\perp T_e \tag{9.58}$$

where $g_{T_e} = (T_e/\bar{\nu}_{ea})(\partial\bar{\nu}_{ea}/\partial T_e)$ is a coefficient determining the thermal diffusion. The first term in Eq. 9.58 represents the heat transfer related to the directed motion, and the second aids in finding the coefficient of transverse thermal conductivity. For an arbitrary dependence of ν_{ea} on v it is equal to

$$\mathcal{K}_{\perp e} = n\chi_{\perp e} = \left(\frac{5}{2} + g_{T_e}\right)\frac{nT_e}{m_e\omega_{He}^2}\bar{\nu}_{ea} \tag{9.59}$$

At $\nu_{ea} = \text{const}$, Eq. 9.59 changes to Eq. 9.36.

9.3 MECHANISM OF TRANSPORT OF CHARGED PARTICLES AND THEIR ENERGY ACROSS STRONG MAGNETIC FIELD

Let us consider the mechanism of transverse transport in a strong magnetic field where the cyclotron frequency of the charged particles greatly exceeds that of their collisions, $\omega_H \gg \nu$. As demonstrated in Chapter 8, without collisions the trajectory of charged particles in a plane perpendicular to the magnetic field is the sum of revolution along the Larmor circle with a drift across the electric field and the magnetic field gradient. The displacement of the Larmor centers in the direction of the forces acting in a plane perpendicular to the magnetic field and their random motion are due exclusively to the collisions, which lead to a sharp change in the velocities of the charged particles and hence to a jumpwise displacement of the Larmor centers. Therefore, when considering the mechanism of transport processes we must first of all determine the change in the position of the Larmor center as a result of collision.

To establish the relationship between the change in the velocity of a particle on collision and the displacement of the Larmor center we

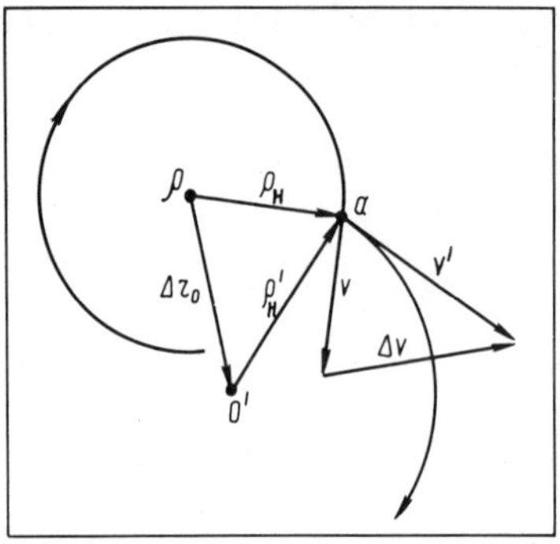

Fig. 9.1 Displacement of the Larmor center on collision.

introduce a vector $\boldsymbol{\rho}_H$ directed from the center of the Larmor circle to the site of the particle. It is clearly associated with the instantaneous particle gyration velocity $\mathbf{v}$ by the vector equation

$$\boldsymbol{\rho}_H = \frac{[\mathbf{h} \times \mathbf{v}]}{Z\omega_H} \tag{9.60}$$

(the direction of $\boldsymbol{\rho}_H$ is perpendicular to $\mathbf{v}$, and its length is equal to the Larmor radius). We assume that the collision occurs at a fixed point in space; this is true for a particle interaction radius much smaller than the Larmor. Then the displacement of the Larmor center $\Delta\mathbf{r}_0$ on collision is equal to the change in vector $\boldsymbol{\rho}_H$ (Fig. 9.1):

$$\Delta\mathbf{r}_0 = -\Delta\boldsymbol{\rho}_H = -\frac{[\mathbf{h} \times \Delta\mathbf{v}]}{Z\omega_H} \tag{9.61}$$

It can be seen that the displacement vector is perpendicular to the particle velocity change due to collision, and $\Delta\mathbf{r}_0$ depends, in each particular collision, on the change in v and in scattering angle. It is easy to ascertain with the aid of Eq. 9.61 that the displacement occurs predominantly in the direction of the collision, irrespective of the charge sign of the particle. For instance, if the collision took place to the right of the Larmor center it displaces to the right (Fig. 9.2); if to the left, then

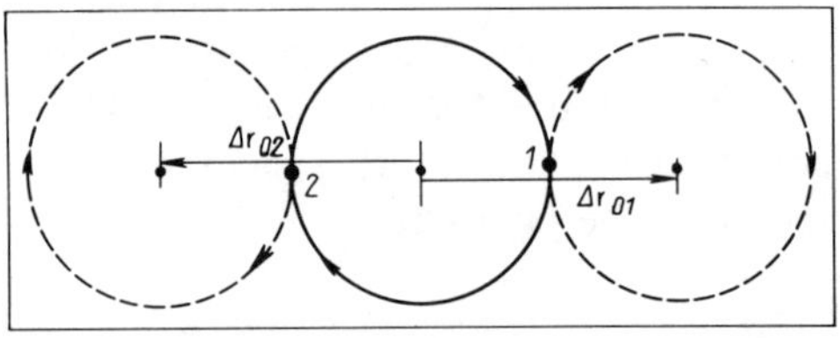

Fig. 9.2 Displacement occurs predominantly in the direction of the collision.

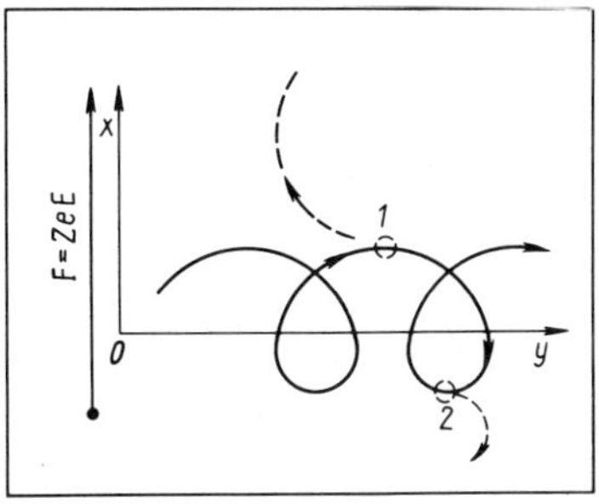

Fig. 9.3 Particle trajectory in the interval between collisions.

it displaces to the left. The maximum displacement corresponds to a head-on collision. On head-on collision, say, of an electron with an atom $|\Delta v| = 2v$ and $\Delta r_0 = 2v/\omega_H = 2\rho_H$.

Passing to transport processes, we begin with the motion under the effect of a transverse electric field. In the interval between collisions the particle trajectory in the crossed electric and magnetic fields is a trochoid, that is the sum of the Larmor gyration and the electric drift (Fig. 9.3). Owing to this superposition the curvature radius does not remain constant; it is at a maximum at points corresponding to the maximum particle energy (points 1 in Fig. 9.3) and at a minimum at points corresponding to the minimum energy (points 2). The displacement of the Larmor center as a result of collisions at points with a larger curvature radius are clearly larger than on collisions at points with a smaller curvature radius. Therefore, despite the equal collision probability at different points an average displacement in the direction of the force is observed.

Let us estimate this displacement by considering head-on electron–atom collisions. Let the electric field be directed along the $0x$ axis (Fig. 9.3). Then at point 1 the electron velocity component along the y axis can be treated as the sum of the gyration and drift velocities in the electric field, and at point 2, as the difference of these velocities:

$$v_{y1} = v_\omega + v_{dE};$$

$$v_{y2} = -(v_\omega - v_{dE})$$

A head-on collision results in reversal of these velocities. Thus we get

$$\Delta v_{y1} = -2(v_\omega + v_{dE}); \qquad \Delta v_{y2} = 2(v_\omega - v_{dE})$$

Since the drift velocity in the electric field and on collision remains unchanged, all the velocity changes are related to the rotation component. Hence the estimate of the average change in rotation velocity on head-on collision has the form

$$\overline{\Delta v}_\omega = \frac{1}{2}(\Delta v_{y1} + \Delta v_{y2}) = -2v_{dE}$$

A head-on collision corresponds to the maximum change in velocity. Assuming that the average change is equal to half the maximum, we find the estimate of the average electron displacement in the direction of the force on collision:

$$\langle \Delta x \rangle = \frac{1}{2}\frac{\overline{\Delta v_\omega}}{\omega_{He}} = -\frac{v_{dE}}{\omega_{He}} = -\frac{m_e c^2 E}{eH^2} \tag{9.62}$$

Summing over collisions per unit time, we obtain the estimate of the average electron velocity in the direction of the electric field:

$$u_{tE} = \langle \Delta x \rangle \nu_{ea} = -\frac{m_e c^2 \nu_{ea}}{eH^2} E \tag{9.63}$$

which corresponds to Eq. 9.14. A more precise determination of the directed velocity requires accurate averaging of the Larmor center displacements on collisions. Assuming, as before, that the electric field is directed along the $0x$ axis, we find from Eq. 9.61 the equation for the average rate of particle displacement in this direction:

$$u_{tE} = \frac{\langle \Delta x \rangle}{\Delta t} = \langle \Delta x \rangle_1 = \frac{\langle \Delta v_y \rangle_1}{\omega_H} \tag{9.64}$$

The symbol $\langle \ \rangle_1$ stands for collision-averaged changes in the corresponding values per unit time. The velocity change averaged over elastic collisions was found in Chapter 2. In accordance with Eqs. 2.16 and 2.55 we have

$$\langle \Delta v_{y\alpha} \rangle_1 = -\frac{\mu_{\alpha a}}{m_\alpha} \nu_{\alpha a}^t v_y \tag{9.65}$$

When averaging over the velocities we must bear in mind that the average particle velocity in the $0y$ direction is equal to the drift velocity $\langle v_y \rangle = u_{dE} = cE/H$. Therefore when the collision frequency is velocity independent the following equality holds:

$$\langle \Delta v_{y\alpha} \rangle_1 = -\frac{\nu_{\alpha a}^t u_{dE} \mu_{\alpha a}}{m_\alpha} \tag{9.66}$$

At a velocity-dependent collision frequency it also holds, but, as shown in Section 6.2, one must use the equation 6.56 for the average collision frequency. Substituting Eq. 9.66 into Eq. 9.64, we obtain the expression for the average velocity of motion in the direction of the electric field, which coincides with Eq. 9.11:

$$u_{\alpha tE} = \frac{\mu_{\alpha a}}{m_\alpha} \frac{\bar{\nu}_{\alpha a} u_{dE}}{\omega_{H\alpha}} = \frac{\mu_{\alpha a} Z_\alpha e E \bar{\nu}_{\alpha a}}{m_\alpha \omega_{H\alpha}^2} \tag{9.67}$$

Let us now consider the mechanism of charged particle diffusion across the magnetic field. The diffusion results from random jumps of Larmor centers under the effect of collisions. In the absence of an electric field such jumps reduce the average displacement to zero. Because of the density gradient, however, the particle flux from the region of higher density exceeds the opposite one. The difference in fluxes is, as always, the cause of diffusion. This difference may also be due to the temperature gradient, which results on the one hand in an asymmetry of the displacement amplitudes on collisions, since the Larmor radius depends on the temperature, and on the other hand, in different values of collision frequency when it is velocity dependent. The summary charged particle flux associated with these causes is thermal diffusion, of course.

It is easy to show that the diffusion motion of the charged particles of a weakly ionized plasma is determined by the mean square displacement of the Larmor centers. Let, for instance, the density and temperature gradients of the charged particles be directed along the $0x$ axis. We now find the flux across the plane with a coordinate x_0. The displacement of the particles on collisions depends on their velocity and the type of collision. Denote by $g(x, \xi)\, d\xi$ (here x is the initial coordinate of the particle) the fraction of the particles whose displacement along the $0x$ axis per unit time lies within the range from ξ to $\xi + d\xi$. Then the density of the flux associated with this group of particles is equal to

$$d\Gamma_x(\xi) = d\xi \int_{x_0-\xi}^{x_0} n(x) g(x, \xi)\, dx \tag{9.68}$$

The integral is taken within the range from $x_0 - \xi$ to x_0, since all the particles whose coordinates lie in this range will have crossed the area x_0 in unit time. Let us expand the integrand function in a Taylor series and restrict ourselves to the first two terms of the expansion:

$$n(x) g(x, \xi) = n(x_0) g(x_0, \xi) + (x - x_0) \frac{\partial (ng)}{\partial x}$$

Integrating this expression, we find

$$d\Gamma_x(\xi) = n(x_0) \xi g(x_0, \xi)\, d\xi - \frac{1}{2} \xi^2 \frac{\partial (ng)}{\partial x}\, d\xi \tag{9.69}$$

Summing the flux over all the possible displacements ξ and taking into account that

$$\int \xi g(x_0, \xi)\, d\xi = \langle \xi \rangle_1 = \langle \Delta x \rangle_1;$$

$$\int \xi^2 g(x_0, \xi)\, d\xi = \langle \xi^2 \rangle_1 = \langle (\Delta x)^2 \rangle_1$$

we obtain

$$\Gamma_x = n\langle \Delta x \rangle_1 - \frac{1}{2}\frac{\partial}{\partial x}[n\langle (\Delta x)^2 \rangle_1] \tag{9.70}$$

Without an electric field the first term vanishes, since the charged particle displacements on collisions are of a random nature, and the flux is defined by the second term, which can be written thus:

$$\Gamma_x = -\frac{\partial}{\partial x}(D_\perp n) \tag{9.71}$$

where the coefficient

$$D_\perp = \tfrac{1}{2}\langle (\Delta x)^2 \rangle_1 \tag{9.72}$$

is, by definition, the transverse diffusion coefficient. It can easily be estimated on the basis of the relationship between the diffusion coefficient and the mean square displacement. The order of magnitude of the displacement of charged particles on collisions with neutrals is, as noted above, equal to their Larmor radius. Therefore the diffusion coefficient, which is equal to the mean square displacement per unit time, is of the order of

$$D_\perp \approx \rho_H^2 \nu \approx \frac{v_\perp^2}{\omega_H^2}\nu \approx \frac{T\nu}{m\omega_H^2} \tag{9.73}$$

The transverse diffusion coefficient can be determined more precisely by accurate averaging in Eq. 9.72. Taking into account Eq. 9.61 (the relationship of the Larmor centers displacement with the velocity change on collisions), we get

$$D_\perp = \frac{1}{2}\langle (\Delta x)^2 \rangle_1 = \frac{1}{2}\frac{\langle (\Delta v_y)^2 \rangle_1}{\omega_H^2} \tag{9.74}$$

The velocity change Δv per collision is given by Eq. 2.15. Finding the value of $(\Delta v_y)^2$ with its aid, summing it over collisions per unit time, and averaging over the velocities, we obtain

$$\langle (\Delta v_{y\alpha})^2 \rangle_1 = \frac{2}{3}\left(\frac{\mu_{\alpha a}}{m_\alpha}\right)^2 \int (1 - \cos\vartheta) n_a \langle s_{\alpha a}^t \overline{v^3} \rangle\, d\Omega = \frac{2}{3}\left(\frac{\mu_{\alpha a}}{m_\alpha}\right)^2 \overline{\nu_{\alpha a}^t(v) v^2}$$

where $|\mathbf{v}| = |\mathbf{v} - \mathbf{v}_a|$ is the relative velocity on collisions, and $\nu^t v^2$ must be averaged over the relative velocities. For $\nu = \text{const}$ and a Maxwellian particle velocity distribution the averaging leads to the expression

$$\langle(\Delta v_{y\alpha})^2\rangle_1 = 2\frac{\mu_{\alpha a}}{m_\alpha^2}\nu_{\alpha a}^t T_\alpha \tag{9.75}$$

since $\bar{v}^2 = 3T_{\alpha a}/\mu_{\alpha a} \approx 3T_\alpha/\mu_{\alpha a}$, where $T_{\alpha a} = (m_a T_\alpha + m_\alpha T_a)/(m_a + m_\alpha)$ is the effective temperature, which defines the relative velocity distribution. With a velocity-dependent collision frequency, Eq. 9.75 obviously includes the quantity $\bar{\nu}^t$, averaged over the distribution of velocities with a weight of v^2, which is obtained from Eq. 9.27. Bearing this in mind and substituting Eq. 9.75 into Eq. 9.74, we find the expression for the transverse diffusion coefficient:

$$D_{\alpha\perp} = \frac{1}{2}\langle\overline{(\Delta x)_\alpha^2}\rangle_1 = \frac{\mu_{\alpha a}}{m_\alpha^2}\frac{\bar{\nu}_{\alpha a}T_\alpha}{\omega_{H\alpha}^2} \tag{9.76}$$

which is similar to Eq. 9.16. The coefficient obtained determines the diffusion caused by the density gradient.

The thermal diffusion is due to the spatial change of the mean square displacement $\langle(\Delta x)^2\rangle \sim \rho^2\nu$ because of its temperature dependence: $\rho_\alpha^2\nu_{\alpha a} \sim T_\alpha\nu_{\alpha a}(T_\alpha)$.

The thermal diffusion flux can be found from Eq. 9.70 at $n = \text{const}$:

$$\Gamma_{\alpha x}^T = -D_{\alpha\perp}^T\frac{n}{T_\alpha}\frac{\partial T_\alpha}{\partial x} = -\frac{\partial D_{\alpha\perp}}{\partial T_\alpha}\frac{\partial T_\alpha}{\partial x}n \tag{9.77}$$

whence we obtain the ratio for the coefficient of thermal diffusion, which coincides with Eq. 9.48:

$$D_{\alpha\perp}^T = T_\alpha\frac{\partial D_{\alpha\perp}}{\partial T_\alpha} = \frac{1}{2}T_\alpha\frac{\partial\langle\overline{(\Delta x)^2}\rangle_1}{\partial T_\alpha} \tag{9.78}$$

We can similarly find the heat flux across the magnetic field. By definition (see Section 6.1) it is an energy flux in a reference system where the directed velocity is equal to zero. In this system the energy is transferred because the particles that are moving from a region with a higher temperature to a cooler region carry more energy than the particles moving in the opposite direction. It is easy to estimate the resultant energy flux. We assume, as before, that the temperature gradient is parallel to the $0x$ axis. The choice of a reference system in which the directed velocity is zero means that in crossing any area, including one perpendicular to the temperature gradient, the clashing particle fluxes offset each other. The density of each of these fluxes is determined by the number of particles whose Larmor centers cross, as a result of collisions, a unit area in one of the directions in unit time. In the order of magnitude

$$\Gamma_{x+} = \Gamma_{x-} \approx n\nu\overline{|\Delta x|} \approx n\nu\rho_H \tag{9.79}$$

At the same time the energy carried by each particle moving toward the temperature gradient is less by a value of the order of $(\partial T/\partial x)|\Delta x| \approx \rho_H \partial T/\partial x$ than the energy carried by each particle moving in the opposite direction. As a result an uncompensated energy flux is formed:

$$q_x \approx -\frac{n\nu\rho_H^2 \partial T}{\partial x} \tag{9.80}$$

From Eq. 9.80 follows the estimate for the coefficient of transverse thermal conductivity:

$$\mathscr{K}_\perp \approx n\nu\rho_H^2 \approx \frac{nT\nu}{m\omega_H^2} \tag{9.81}$$

which corresponds to Eq. 9.36.

It is also easy to understand the origin of the other components of the heat flux. As noted above, the flux $\mathbf{q}_d$, proportional to $[\mathbf{h} \times \text{grad}\, T_e]$ (see Eq. 9.32), is associated with the diamagnetic electron flux; it results from summation of the Larmor trajectories in an inhomogeneous plasma. In a moving reference system, in which the heat flux is determined, the directed velocities must be zero; that is, the diamagnetic flux is compensated. But in the presence of a temperature gradient an uncompensated heat flux appears, since the energy of the particles moving in different directions is different (see the components v_y in Fig. 9.4.). It is

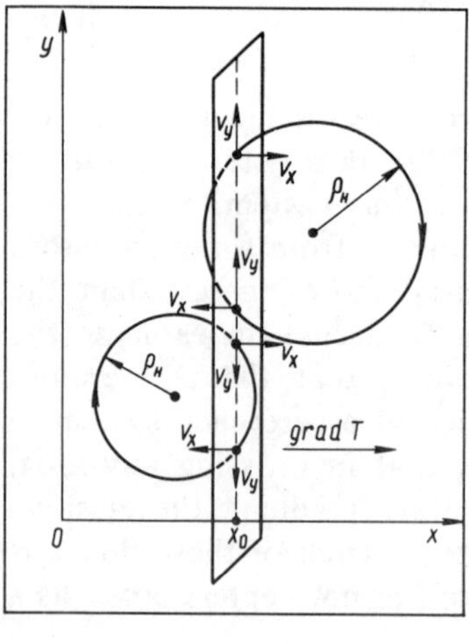

Fig. 9.4 Heat flux in the presence of a temperature gradient.

directed perpendicularly to the temperature gradient and is of the order of

$$q_d \approx n\langle v_\perp \rho_H\rangle \frac{\partial T}{\partial x} \approx \frac{n\langle v_\perp^2\rangle}{\omega_H}\frac{\partial T}{\partial x} \approx \frac{nT}{m\omega_H}\frac{\partial T}{\partial x} \tag{9.82}$$

which corresponds to the above result (see Eq. 9.32)].

At velocity-dependent collision frequencies one more component of the heat flux is possible ($\mathbf{q}_u$), which depends on the directed motion of the charged particles and is proportional to $[\mathbf{h}\times\mathbf{u}_\alpha]$ (see Section 9.2). This component arises only in the presence of external forces, which cause the appearance of the directed velocity. To understand its origin we must recall that in a system where the transverse directed velocity is zero the Lorentz force also vanishes. The external transverse forces acting on the charged particles are offset by their friction against the neutral gas, which has a velocity of $-\mathbf{u}_{\perp\alpha}$ in the adopted coordinate system. But at a velocity-dependent collision frequency it is only an average compensation: it does not mean compensation of forces for groups of particles with different thermal velocities. Therefore it can be said that the effective force acting on fast (hot) particles differs from that acting on slow (cold) particles. With a strong velocity dependence of the collision frequency, when $\Delta F = (dR/dv)\,\Delta v \approx R$, this difference is of the order of $\Delta F \approx R \approx m\nu(u_a - u_\alpha)$.

The velocities $\mathbf{u}_d = (c/eH)[\mathbf{F}\times\mathbf{h}]$ differ correspondingly. The difference in the drift velocities of the fast and slow particles leads to an energy transfer in a coordinate system with no particle transport. Estimation of the heat flux density yields

$$\mathbf{q}_u = nT_\alpha \Delta\mathbf{u}_F \approx \frac{cm_\alpha nT_\alpha\bar{\nu}_{\alpha a}}{eH}[\mathbf{h}\times(\mathbf{u}_a - \mathbf{u}_\alpha)] \tag{9.83}$$

We restrict ourselves here to order-of-magnitude estimates of the heat flux of the charged particles. It is readily and rigorously derived by considering the displacement of the charged particles on collisions, as was done in Section 7.3.

From the foregoing it follows that the mechanisms of charged particle and energy transport across a strong magnetic field differ substantially from the mechanism of longitudinal transport or of transport without a magnetic field. Indeed, longitudinal transport, as well as transport without a magnetic field, is due to the free motion of charged particles in intervals between collisions, and therefore the transfer coefficients reduce with increasing particle mass (particles velocity in the periods between collisions and their acceleration in the electric field decrease) and with increasing collision frequency (mean free path reduces). The transport across the magnetic field, on the contrary, is determined not by the displacement of the particles in the intercollisional period but by jumps of the Larmor centers at the moments of collision. Accordingly, the coefficients of transverse transport are proportional to the mass and inversely proportional to the square of the magnetic field strength, because the jumps are of the order of the Larmor radius, which increases with the mass and reduces with increasing magnetic field.

Let us now find the thermal force induced by the transverse temperature gradient. As in the absence of a magnetic field, the thermal force appears because at a velocity-dependent collision frequency the clashing particle fluxes experience different friction against the neutral gas. To estimate this difference we isolate a plane perpendicular to the temperature gradient and compare the momenta transported by the charged particles to this plane from regions with different temperatures. In the course of Larmor gyration each particle crosses the plane under consideration twice (see Fig. 9.4). During the two crossings the projections of the gyration velocity onto the direction grad T are opposite. Since in both phases collisions are equally probable, the average of the transported-momentum projection, which is parallel to the temperature gradient, is obviously equal to zero. Accordingly the projection of the thermal force on this direction is also zero. The projection of the thermal force on a direction perpendicular to the magnetic field and the temperature gradient can be estimated by summing the results of the collisions of the particles whose Larmor centers are to the right and to the left of the isolated plane. The velocity of the first group is directed parallel to $[\mathbf{h}\times \text{grad}\, T]$, and that of the second group, oppositely. In a coordinate system moving so that the summary momentum is zero the summary momentum transferred on collision is different from zero because the collision frequencies for these two groups of particles are different. Since the distance between their Larmor centers is of the order of the Larmor radius, the difference in collision frequencies is estimated by the equation $\delta\nu \approx \rho_H\, \partial\nu/\partial x$. This difference determines the average momentum transferred in unit time; that is, the thermal force

$$R^T \approx \mu v \delta\bar{\nu} \approx \mu \overline{v \rho_H} \frac{\partial \bar{\nu}}{\partial x} \approx \frac{\mu \bar{v}^2}{\omega_H} \frac{\partial \bar{\nu}}{\partial x} \approx \frac{T}{\omega_H} \frac{\partial \bar{\nu}}{\partial x} \tag{9.84}$$

Bearing in mind the direction of the momentum transferred on collisions, we can write Eq. 9.84 in vector form:

$$\mathbf{R}^T = -\frac{T}{Z\omega_H}[\mathbf{h}\times \text{grad}\, \bar{\nu}] = -\frac{\bar{\nu}}{Z\omega_H} g_T [\mathbf{h}\times \text{grad}\, T]; \qquad g_T = \frac{T}{\bar{\nu}} \frac{\partial \bar{\nu}}{\partial T} \tag{9.85}$$

The above estimate yields the expression for the thermal force with an accuracy to a numerical coefficient of the order of unity. A more accurate averaging over collisions and velocities also leads to Eq. 9.85, and the averaged collision frequency appearing in it is described by Eq. 9.27.

9.4 AMBIPOLAR DIFFUSION AND BALANCE OF CHARGED PARTICLES OF WEAKLY IONIZED PLASMAS IN MAGNETIC FIELD

With a magnetic field, as well as without it, separate electron and ion diffusion in a plasma is impossible because of the quasi-neutrality conditions. An electric field resulting from a relatively small charge separation equalizes the charged particle losses. In a three-component plasma consisting of electrons, ions, and neutral particles the equations of balance of the electron and ion components,

$$\frac{\partial n}{\partial t} + \operatorname{div}(n\mathbf{u}_i) = \frac{\delta n}{\delta t}; \qquad \frac{\partial n}{\partial t} + \operatorname{div}(n\mathbf{u}_e) = \frac{\delta n}{\delta t} \tag{9.86}$$

can be satisfied simultaneously only if the divergences of the directed fluxes,

$$\operatorname{div}(n\mathbf{u}_i) = \operatorname{div}(n\mathbf{u}_e) \tag{9.87}$$

are equal.

The solution of the system (9.86) in a magnetic field is generally much more complicated that outside it because of the anisotropy of the mobility and diffusion coefficients. These solutions coincide only for the unidimensional problem, when the plasma parameters depend on a single coordinate. For instance, when a homogeneous magnetic field is directed along the axis of the system and the plasma parameters depend only on the radius, the condition 9.87 takes the form

$$\frac{d}{dr}(rnu_{er}) = \frac{d}{dr}(rnu_{ir}) \tag{9.88}$$

Since the quantities n and u are finite as $r \to 0$, it follows from Eq. 9.88 that the radial components of the directed velocities are equal:

$$uer = u_{ir} \tag{9.89}$$

This equality describes the ambipolar diffusion. Substituting the general equation for the transverse directed velocity, we obtain (as without a magnetic field) the expressions for the strength of the electric field that ensures ambipolarity, and for the ambipolar-directed velocity. If there is no temperature gradient, they have a form similar to Eqs. 7.65 and 7.69:

$$\mathbf{E}_{A\perp} = \frac{D_{i\perp} - D_{e\perp}}{b_{i\perp} + b_{e\perp}} \frac{\mathrm{grad}_\perp n}{n} \tag{9.90}$$

$$\mathbf{u}_{At} = -D_{A\perp} \frac{\mathrm{grad}_\perp n}{n}; \qquad D_{A\perp} = \frac{D_{e\perp} b_{i\perp} + D_{i\perp} b_{e\perp}}{b_{i\perp} + b_{e\perp}} \tag{9.91}$$

Using Eqs. 9.12 and neglecting $m_e\nu_{ea}/m_i\nu_{ia}$ as compared with unity and with T_i/T_e, we get

$$\mathbf{E}_{A\perp} = -\frac{T_e}{e} \frac{1 - \dfrac{m_i}{\mu_{ia}} \dfrac{T_i}{T_e} \dfrac{\omega_{He}\omega_{Hi}}{\nu_{ia}\nu_{ea}}}{1 + \dfrac{m_i}{\mu_{ia}} \dfrac{\omega_{Hi}\omega_{He}}{\nu_{ia}\nu_{ea}}} \frac{\mathrm{grad}_\perp n}{n} \tag{9.92}$$

$$D_{A\perp} = \frac{T_e + T_i}{\mu_{ia}\nu_{ia}\left(1 + \dfrac{m_i}{\mu_{ia}} \dfrac{\omega_{Hi}\omega_{He}}{\nu_{ia}\nu_{ea}}\right)} \tag{9.93}$$

These relations make it possible to trace the transition from weak to strong magnetic fields. In weak magnetic fields, where $\omega_{He}\omega_{Hi} \ll \nu_{ea}\nu_{ia}$, the equations for $\mathbf{E}_{A\perp}$ and $D_{A\perp}$ are the same as without a magnetic field (cf. Eqs. 7.66 and 7.70). Owing to the faster removal of electrons the plasma then acquires a positive charge and the walls a negative one, the ambipolar electric field being of the order of $E_\perp \approx T_e/eL_\perp$. As the magnetic field increases, the electron diffusion and mobility coefficients reduce much faster than the ion coefficients. In a magnetic field with a strength at which $\omega_{He}\omega_{Hi}/\nu_{ea}\nu_{ia} = (T_e/T_i)\mu_{ia}/m_i$ the electron and ion diffusion coefficients are equal. Here the motion ambipolarity condition is fulfilled without an electric field. A further increase in magnetic field changes the ratio between the electron and ion diffusion coefficients—the ions begin to move across the field faster than the electrons. As a result the plasma charge becomes negative. In a strong magnetic field, where $\omega_{He}\omega_{Hi} \gg \nu_{ea}\nu_{ia}$, Eqs. 9.92 and 9.93 take the form

$$E_{A\perp} = \frac{T_i}{e} \frac{\mathrm{grad}_\perp n}{n} \tag{9.94}$$

$$D_{A\perp} = (T_e + T_i) \frac{\nu_{ea}}{m_e \omega_{He}^2} \tag{9.95}$$

The condition $\omega_{He}\omega_{Hi} \gg \nu_{ea}\nu_{ia}$ is sometimes called the *plasma magnetization condition.* When it is fulfilled the diffusion coefficient is proportional to the electron collision frequency and inversely proportional to the square of the magnetic field strength. It differs from the electron coefficient only by a factor of $1 + T_i/T_e$.

The effect of the temperature gradient on the ambipolar diffusion can be taken into account in the same way as in the absence of a magnetic field. In particular, when the collision frequency is velocity independent its inclusion leads to an expression for the directed velocity similar to Eq. 7.74:

$$\mathbf{u}_{A\perp} = -\frac{1}{n}\,\text{grad}(D_{A\perp} n) \tag{9.96}$$

where $D_{A\perp}$ is obtained from Eqs. 9.93 and 9.95.

Substituting the equations for the ambipolar directed velocity into the charged particle balance equation yields a diffusion equation similar to Eq. 7.77. At T_e, $T_i = \text{const}$, substituting Eq. 9.96 into Eq. 9.86, we obtain

$$\frac{\partial n}{\partial t} - D_{A\perp}\Delta n = \frac{\delta n}{\delta t} \tag{9.97}$$

The solution of the diffusion equation requires determining the boundary conditions for the density. As shown in Section 7.5, without a magnetic field the ratio of the boundary density to the density in the central part of the volume is of the order of

$$\frac{n_g}{n_0} \sim \frac{\lambda_i}{L}\sqrt{T_e/T_i} \tag{9.98}$$

Since usually $\lambda_i \ll L$, the density at the boundary is much lower than at the center, and the zero boundary conditions can be used with a sufficient accuracy. This conclusion is also true for the transverse diffusion of charged particles in a magnetic field. In a weak magnetic field, at $\omega_{Hi}\omega_{He} \ll \nu_{ea}\nu_{ia}$, the statements given in Section 7.5 still hold, since the magnetic field has almost no effect on the ion motion ($\omega_{Hi} \ll \nu_{ia}$), and the sign of the wall charge is negative, as in the absence of a magnetic field (see Eq. 9.92).

It is also easy to estimate the boundary density in the case of a strong magnetic field, when $\omega_{Hi}\omega_{He} \gg \nu_{ia}\nu_{ea}$. As demonstrated above the transverse diffusion coefficient for the ions is higher than for the electrons, and the walls of the plasma chamber, which are parallel to the magnetic field, become positively charged. Therefore the electrons that find themselves at the boundary of the wall layer must be attracted to the wall in the electric field of the layer. The value of the field in the layer can be found as in Section 7.5. The electron flux is easy to estimate by assuming that the size of the wall layer exceeds the Larmor radius. The motion of the electrons across the electric field at the layer boundary is due to collisions. Since in each collision their average displacement is of the order of the Larmor radius, the density of the flux from the plasma to the wall is approximately equal to

$$\Gamma_{eg} \approx n_g \rho_{He} \nu_{ea} \tag{9.99}$$

At the same time the ambipolar electron flux from the plasma is of the order of

$$\Gamma \approx D_{A\perp} |\text{grad}_\perp n| \approx \frac{(T_e + T_i)\nu_{ea}}{m_e \omega_{He}^2} \frac{n_0}{L_\perp} \tag{9.100}$$

where $L_\perp$ is the characteristic transverse dimension of the plasma. Equating this flux to the electron flux in the layer, we find the ratio of the charged particle densities at thc plasma boundary and at the center:

$$\frac{n_g}{n_0} \approx \frac{1}{\omega_{He} L_\perp} \sqrt{\frac{T_e}{m_e}} \approx \frac{\rho_{He}}{L_\perp} \tag{9.101}$$

In a strong magnetic field, usually $\rho_{He} \ll L_\perp$ and correspondingly $n_g \ll n_0$, and the boundary density can approximately be assumed equal to zero.

Thus for unidimensional diffusion across the magnetic field the diffusion equation 9.97 and the boundary conditions for the density are the same as in the absence of a magnetic field, except for the diffusion coefficients. Accordingly, the solutions of these equations are also the same. In particular, the solution of the equations for a stationary gas discharge in a long cylindrical chamber leads to a diffusion distribution described by a zero-order Bessel function (Eq. 7.94). The ratio between the average ionization frequency ν^i and the diffusion coefficient is determined by the condition of equality of the ionization and removal velocities (Eq. 7.96):

$$\nu^i = \frac{5.8 D_{A\perp}}{a^2} \tag{9.102}$$

The only difference is that in a strong magnetic field the diffusion coefficient and, accordingly, ν^i is much less than without a field, and as the field increases the diffusion coefficient drops abruptly ($D_{A\perp} \approx 1/H^2$).

The conclusions drawn in Section 7.8 remain true for the description of plasma decay in a magnetic field due to the transverse diffusion (when the chamber length greatly exceeds its radius). In particular, in the late stage of decay, when the diffusion distribution has been established, the law of time variation in concentration is exponential (Eq. 7.137). In this stage, for $\omega_{He}\omega_{Hi} \gg \nu_{ea}\nu_{ia}$ the decay time constant satisfies the equality

$$\tau_A = \frac{D_{A\perp}}{\Lambda_\perp^2} = \frac{11.6}{a^2} \frac{T_a \nu_{ea}}{m_e \omega_{He}^2} \tag{9.103}$$

Here we used Eq. 9.95 and took into account that in late stages of decay $T_e = T_i = T_a$.

We have considered the ambipolar diffusion and the charged particle balance in a plasma whose parameters depend on a single coordinate. As noted above, in a non-unidimensional case the solution of the problem becomes more complicated. Consider, for instance, plasma decay in a cylindrical chamber whose axis coincides with the direction of the magnetic field, assuming that the charged particle density depends both on the longitudinal coordinate z and the radius r. Here the electron and ion balance equation 9.86 takes the form

$$\frac{\partial n}{\partial t} + \frac{\partial}{\partial z}(nu_{\alpha z}) + \frac{1}{r}\frac{\partial}{\partial r}(rnu_{\alpha r}) = 0 \tag{9.104}$$

where the axial and radial components of the directed velocity are found from the relations

$$\begin{aligned} u_{z\alpha} &= Z_\alpha b_{\|\alpha} E_\| - D_{\|\alpha}\frac{1}{n}\frac{\partial n}{\partial z}; \\ u_{r\alpha} &= Z_\alpha b_{\perp\alpha} E_r - D_{\perp\alpha}\frac{1}{n}\frac{\partial n}{\partial r} \end{aligned} \tag{9.105}$$

The electron and ion densities are taken to be equal in view of the quasi-neutrality condition.

The solution of these equations need not correspond to the ambipolar diffusion regime. Such a regime can be realized however, on decay of a plasma in a chamber with dielectric walls. Then the ambipolar removal corresponds to the equality of the electron and ion velocities both along and across the magnetic field: $u_{ze} = u_{zi}$, $u_{re} = u_{ri}$. In accordance with the results of Section 7.4 and of this section they yield the expressions for the longitudinal and transverse components of the electric field:

$$\mathbf{E}_{Az} = -\frac{T_e}{e}\frac{\text{grad}_z\, n}{n}; \tag{9.106}$$

$$\mathbf{E}_{Ar} = -\frac{T_e}{e}\frac{1 - \dfrac{m_i}{\mu_{ia}}\dfrac{T_i}{T_e}\dfrac{\omega_{He}\omega_{Hi}}{\nu_{ea}\nu_{ia}}}{1 + \dfrac{m_i}{\mu_{ia}}\dfrac{\omega_{He}\omega_{Hi}}{\nu_{ia}\nu_{ea}}}\frac{\text{grad}_r\, n}{n} \tag{9.107}$$

It is seen that this field can be potential (i.e., curl $\mathbf{E} = 0$) only when

$$\frac{\partial^2(\ln n)}{\partial r\,\partial z} = 0 \tag{9.108}$$

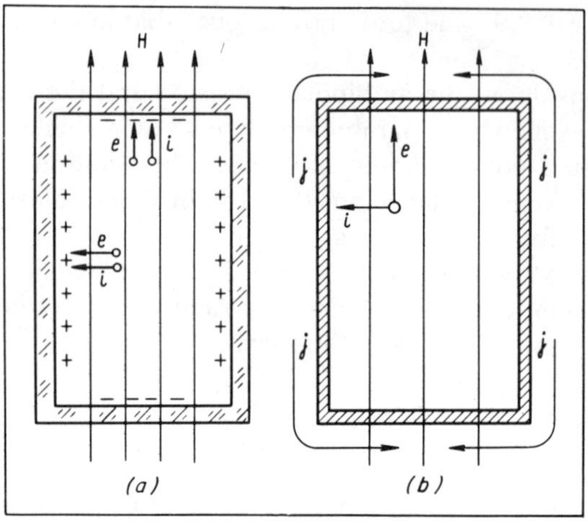

Fig. 9.5a,b Nonambipolar diffusion in the presence of conducting walls.

that is, if

$$n(z, r) = n_z(z)n_r(r) \tag{9.109}$$

From Eqs. 9.106 and 9.107 it follows that in a strong magnetic field the side and end walls of the chamber are oppositely charged (Fig. 9.5*a*). The end walls are charged negatively, since the electrons diffuse along the magnetic field much faster than the ions. The side walls, on the contrary, acquire a positive charge because the ion diffusion coefficient exceeds that of the electrons. Substituting the fields 9.106 and 9.107 into the decay equation 9.104 changes it to

$$\frac{\partial n}{\partial t} - D_{A\parallel}\frac{\partial^2 n}{\partial z^2} - D_{A\perp}\frac{1}{r}\frac{\partial}{\partial r}\left(r\frac{\partial n}{\partial r}\right) = 0 \tag{9.110}$$

where the coefficients $D_{A\parallel}$ and $D_{A\perp}$ can be found from Eqs. 7.70 and 9.93. The partial solution of the equation, which satisfies the boundary conditions, is

$$n = n_0 \exp\left(\frac{-t}{\tau}\right) J_0\left(\frac{r}{\Lambda_\perp}\right) \frac{\cos z}{\Lambda_\parallel} \tag{9.111}$$

where the lengths $\Lambda_\perp = a/2.405$ and $\Lambda_\parallel = d/\pi$ are chosen so that at the boundaries at $r = a$ and $z = \pm d/2$ (d being the chamber length) the density reduces to zero. The decay time constant in Eq. 9.111 is obtained from the relation

$$\tau = \left(D_{A\parallel}/\Lambda_\parallel^2 + \frac{D_{A\perp}}{\Lambda_\perp^2}\right)^{-1} \tag{9.112}$$

in accordance with which the effective velocity diffusion removal, characterized by the quantity $1/\tau$, is equal to the sum of $1/\tau_\parallel$ and $1/\tau_\perp$ corresponding to the longitudinal and transverse ambipolar particle transport. It can be shown that in the late stages of decay the overall solution of the diffusion equation for zero boundary conditions for n (Eqs 9.98 and 9.101) tends to the partial solution obtained.

It should, however, be noted that a nonambipolar diffusion regime, characterized by a passage of diffusion currents over the plasma, is often realized in a magnetic field. For instance, on decay of a plasma in a chamber with conducting walls the charge distribution depicted in Fig. 9.5*a* cannot be obtained. The currents flowing over the walls equalize the potential and change the distribution of the electric field in the plasma. Accordingly, currents appear in the plasma. In particular, one can imagine a situation in which the electrons diffuse largely along the magnetic field (their time of diffusion across the field greatly exceeds that of longitudinal diffusion $D_{e\parallel}/L_\parallel^2 \gg D_{e\perp}/L_\perp^2$), and the ions across the field (their time of longitudinal diffusion is large, $D_{i\parallel}/L_\parallel^2 \ll D_{i\perp}/L_\perp^2$). Then a current closing on the walls is induced in the plasma (Fig. 9.5*b*). In this case the decay time constant of the plasma is determined by the time of removal of the slower component, that is, $\tau_{e\parallel}$ or $\tau_{i\perp}$:

$$\frac{1}{\tau_{e\parallel}} \approx \frac{D_{e\parallel}}{L_\parallel^2}; \qquad \frac{1}{\tau_{i\perp}} \approx \frac{D_{i\perp}}{L_\perp^2} \tag{9.113}$$

It may be much less than the ambipolar diffusion time (Eq. 9.112). The diffusion acceleration due to the nonambipolar electron and ion transport to the conducting walls is usually called the *short-circuit effect.*

A similar effect can be observed when a cylindrical plasma chamber is placed at an angle to the magnetic field, during diffusion in an intricately shaped chamber, when metal bodies are placed inside the plasma, and so on. Thus analysis of the charged particle diffusion in a magnetic field requires a detailed study of the conditions of formation of the space-charge field and the current distribution in the plasma.

The electron and ion energy balance in a strong magnetic field is treated in the same way as in the absence of a magnetic field (see Sections 7.5 and 7.6). The differences consist in a decrease of the transverse coefficients of transport and a change in boundary conditions. Without embarking on detailed analysis of these differences we simply note that because of the abrupt decrease in the transverse thermal conductivity of the electron component in a strong magnetic field the energy losses by thermal conduction are negligibly small. Therefore the electron energy balance equation is local; that is, it is determined by the ratio of the Joule heating and the energy losses as a result of elastic and inelastic electron collisions.

9.5 DIRECTED MOTION OF CHARGED PARTICLES OF HIGHLY IONIZED PLASMAS ACROSS MAGNETIC FIELD

Let us consider the directed motion in a plasma where the frequency of collisions of charged particles with one another and the frequency of their collisions with neutral particles are comparable. The directed motion along the magnetic field is the same as in the absence of a field. Accordingly, for the longitudinal component of the directed velocity one can use the expressions obtained in Section 7.9. Let us find its transverse component.

We begin with a fully ionized plasma. The averaged transverse equations of the motion of electrons and ions of a fully ionized plasma in a magnetic field for stationary conditions can be written as

$$Z_\alpha e\mathbf{E} + \frac{Z_\alpha e}{c}[\mathbf{u}_\alpha + \mathbf{H}] - \frac{\operatorname{grad} p_\alpha}{n} + m_\alpha \frac{\delta \mathbf{u}_\alpha}{\delta t} = 0 \tag{9.114}$$

where $m_\alpha \delta \mathbf{u}_\alpha / \delta t$ is the collision term due to electron–ion collisions. Since the momentum is conserved in such collisons, we can write

$$\frac{m_e \delta \mathbf{u}_e}{\delta t} = -\frac{m_i \delta \mathbf{u}_i}{\delta t} \tag{9.115}$$

The expression for the collision term was derived in Section 7.9. In accordance with Eqs. 7.144 and 7.145,

$$\left(\frac{\delta \mathbf{u}_e}{\delta t}\right)_{ei} = -\bar{\nu}_{ei}(\mathbf{u}_e - \mathbf{u}_i) + \frac{3}{5}\bar{\nu}_{ei}\frac{\mathbf{q}_e}{nT_e} \tag{9.116}$$

where $\bar{\nu}_{ei} = (4\sqrt{2\pi}/3)(ne^4/m_e^{1/2}T_e^{3/2})L_e$ is the frequency of electron–ion collisions averaged over the Maxwellian distribution, and $\mathbf{q}_e$ is the electron heat flux vector. The transverse components of the heat flux vector are described by Eq. 9.32. For sufficiently large fields, where $\omega_{He} \gg \nu_{ei}, \nu_{ea}$, the drift component

$$\mathbf{q}_d = -\frac{5}{2m_e}\frac{nT_e}{\omega_{He}}[\mathbf{h} \times \operatorname{grad} T_e]$$

is the largest. Substituting it into Eq. 9.116 and omitting the other components, whose values are less than q_d by a factor of $\bar{\nu}_{ei}/\omega_{He}$ and $(\bar{\nu}_{ei}/\omega_{He})^2$, we get

$$\left(m_e \frac{\delta \mathbf{u}_e}{\delta t}\right)_{ei} = -\bar{\nu}_{ei} m_e (\mathbf{u}_e - \mathbf{u}_i) - \frac{3}{2} \cdot \frac{\bar{\nu}_{ei}}{\omega_{He}} [\mathbf{h} \times \text{grad } T_e] \tag{9.117}$$

The first term represents the friction force in electron–ion collisions:

$$\mathbf{R}_{ei} = -\mathbf{R}_{ie} = -m_e \bar{\nu}_{ei} (\mathbf{u}_e - \mathbf{u}_i) \tag{9.118}$$

and the second term, the thermal force:

$$\mathbf{R}_{ei}^T = -\mathbf{R}_{ie}^T = -\frac{3}{2} \frac{\bar{\nu}_{ei}}{\omega_{He}} [\mathbf{h} \times \text{grad } T_e] \tag{9.119}$$

It is easy to see that Eq. 9.119 coincides with Eq. 9.28 since for electron–ion collisions $\bar{\nu}_{ei} \sim 1/T_e^{3/2}$ and $g_{ei}^T = (T_e/\bar{\nu}_{ei})\, d\bar{\nu}_{ei}/dT_e = -\frac{3}{2}$.

Substituting the collision term 9.117 into Eq. 9.114 and taking into account Eq. 9.115, we obtain the transverse equations for the electron and ion motion in the form:

$$-e\mathbf{E}_\perp - \frac{1}{n} \text{grad}_\perp (nT_e) - \frac{e}{c} [\mathbf{u}_{e\perp} \times \mathbf{H}] - \frac{3}{2} \frac{\bar{\nu}_{ei}}{\omega_{He}} [\mathbf{h} \times \text{grad } T_e]$$
$$- m_e \bar{\nu}_{ei} (\mathbf{u}_{e\perp} - \mathbf{u}_{i\perp}) = 0; \tag{9.120}$$

$$e\mathbf{E}_\perp - \frac{1}{n} \text{grad}_\perp (nT_i) + \frac{e}{c} [\mathbf{u}_{i\perp} \times \mathbf{H}] + \frac{3}{2} \frac{\bar{\nu}_{ei}}{\omega_{He}}$$
$$\times [\mathbf{h} \times \text{grad } T_e] + m_e \bar{\nu}_{ei} (\mathbf{u}_{e\perp} - \mathbf{u}_{i\perp}) = 0 \tag{9.121}$$

Let us first consider the charged particle motion in a homogeneous, fully ionized plasma under the effect of an electric field. Assuming in this case grad $n = 0$ and grad $T_e = 0$, we get

$$-e\mathbf{E}_\perp - \frac{e}{c} [\mathbf{u}_{e\perp} \times \mathbf{H}] - m_e \bar{\nu}_{ei} (\mathbf{u}_{e\perp} - \mathbf{u}_{i\perp}) = 0;$$
$$e\mathbf{E}_\perp + \frac{e}{c} [\mathbf{u}_{i\perp} \times \mathbf{H}] + m_e \bar{\nu}_{ei} (\mathbf{u}_{e\perp} - \mathbf{u}_{i\perp}) = 0 \tag{9.122}$$

Adding these equations, we clearly see that they lead to the equality of the electron and ion velocities. The friction force reduces to zero, and the solution yields the electric drift velocity (Eq. 8.27) under no-collision conditions:

$$\mathbf{u}_{e\perp} = \mathbf{u}_{i\perp} = \frac{[\mathbf{E} \times \mathbf{H}]}{H^2} \tag{9.123}$$

The result means that in a homogeneous, fully ionized plasma a constant transverse electric field does not induce any current; that is, the transverse conductivity of such a plasma is zero. It is easy to understand this result. Without collisions, under the effect of the constant electric field the electrons and ions drift with the same velocity equal to Eq. 9.123. It is natural, therefore, that collisions between them do not cause friction. Following a different line of thought, one can adopt a reference system moving with the velocity of the drift. In it, the directed electron and ion velocities are equal to zero, and the electric field is nonexistent. Obviously, collisions cannot result in a directed motion here.

It should be emphasized once again, however, that the conclusion about the absence of a current is true only for a homogeneous plasma. In an inhomogeneous plasma collisions may induce a current. Imagine, for instance, that in a direction perpendicular to the electric and magnetic fields there is an obstacle impeding the motion of the charged particles, whose directed velocity is thus equal to zero. It follows from the general equation 9.114 that this is possible if there is a pressure gradient in the direction of the electric drift [**E** × **H**]. Then the projection of the Lorentz force onto the direction of the electric field will reduce to zero, and the equation of motion in the direction of the electric force (Eq. 9.122) are the same as in the absence of a magnetic field. Let, for instance, the electric field be directed along the 0x axis and the motion in the 0y direction be limited, so that $u_y = 0$. Then the projection of Eq. 9.122 on the 0x axis leads to the equality

$$eE = m_e \bar{\nu}_{ei}(u_{ix} - u_{ex})$$

whence we find the following expression for the current intensity:

$$j = ne(u_{ix} - u_{ex}) = \frac{ne^2}{m_e \bar{\nu}_{ei}} E \tag{9.124}$$

which is similar to Eq. 7.158 obtained in the absence of a magnetic field. This effect is called the *effect of restoration of the plasma transverse conduction* on suppression of the electric drift.

Let us now find the solution of Eqs. 9.120 and 9.121 in the general case, when there are temperature and density gradients along with the electric field. Adding up these equations, we obtain

$$\frac{e}{c}[(\mathbf{u}_{i\perp} - \mathbf{u}_{e\perp}) \times \mathrm{H}] = \frac{1}{n} \operatorname{grad}_\perp n(T_e + T_i)$$

This equality yields the difference of the transverse electron and ion velocities

$$\mathbf{u}_{i\perp} - \mathbf{u}_{e\perp} = \frac{c}{neH}[\mathbf{h} \times \operatorname{grad} n(T_e + T_i)] \tag{9.125}$$

Substituting it into Eq. 9.120, we obtain the vector equation

$$-e\mathbf{E}_\perp - \frac{1}{n}\,\text{grad}_\perp(nT_e) + \frac{\bar{\nu}_{ei}}{\omega_{He}}\frac{[\mathbf{h}\times\text{grad}\,n(T_e+T_i)]}{n} - \frac{3}{2}\frac{\bar{\nu}_{ei}}{\omega_{He}}[\mathbf{h}\times\text{grad}\,T_e]$$

$$= m_e\omega_{He}[\mathbf{u}_{e\perp}\times\mathbf{h}]$$

Multiplying it vectorially by **h**, we find the directed electron velocity:

$$\mathbf{u}_{e\perp} = \frac{c}{H}[\mathbf{E}\times\mathbf{h}] - \frac{c}{enH}[\mathbf{h}\times\text{grad}(nT_e)] - \frac{\bar{\nu}_{ei}(T_e+T_i)}{m_e\omega_{He}^2}$$
$$\times\frac{\text{grad}\,n}{n} + \frac{1}{2}\frac{\bar{\nu}_{ei}}{m_e\omega_{He}^2}\,\text{grad}\,T_e - \frac{\bar{\nu}_{ei}}{m_e\omega_{He}^2}\,\text{grad}\,T_i \tag{9.126}$$

and from Eq. 9.125, the directed ion velocity:

$$\mathbf{u}_{i\perp} = \frac{c}{H}[\mathbf{E}\times\mathbf{h}] + \frac{c}{enH}[\mathbf{h}\times\text{grad}(nT_i)] - \frac{\bar{\nu}_{ei}(T_e+T_i)}{m_e\omega_{He}^2}\frac{\text{grad}\,n}{n}$$
$$+\frac{1}{2}\frac{\bar{\nu}_{ei}}{m_e\omega_{He}^2}\,\text{grad}\,T_e - \frac{\bar{\nu}_{ei}}{m_e\omega_{He}^2}\,\text{grad}\,T_i \tag{9.127}$$

The first two terms of these equations yield the velocity of the electron and ion drift under the effect of the electric field, and the diamagnetic drift due to the pressure gradient. These components were considered above for a collisionless plasma (see Section 8.6). Therefore we only note that collisions of charged particles with each other do not affect them. The other terms are identical for the electrons and ions:

$$\mathbf{u}_{et} = \mathbf{u}_{it} = \frac{\bar{\nu}_{ei}(T_e+T_i)}{m_e\omega_{He}^2}\frac{\text{grad}\,n}{n} + \frac{\bar{\nu}_{ei}}{me\omega_{He}^2}\,\text{grad}\left(\frac{1}{2}T_e - T_i\right) \tag{9.128}$$

They give the diffusion of charged particles across the magnetic field. The first gives the diffusion due to the density gradient, and the second, the thermal diffusion. The coefficient of transverse diffusion in a magnetic field is equal to the factor at the first term:

$$D_{e\perp} = D_{i\perp} = \bar{\nu}_{ei}\,\frac{T_e+T_i}{m_e\omega_{He}^2} \tag{9.129}$$

It is seen from the equations that the diffusion in a fully ionized plasma is ambipolar (i.e., the electron and ion diffusion coefficients are equal) irrespective of the electric field strength.

It should be noted that the regime with different electron and ion temperatures can be maintained steadily only in the presence of external energy sources. In their absence collisions equalize the temperatures. Time change in the temperature of ions and hence in their diamagnetic drift velocity gives rise to an inertia force. The velocity of the ion inertial drift due to this force is of the same order of magnitude as the transverse diffusion velocity. The inertial ion drift leads to charge separation and to the appearance of an increasing electric field, which is necessary in this case for establishing an ambipolar regime of transverse charged particle motion.

Let us consider quantitatively the effects of ion and electron temperature equalization when there is no ion temperature gradient.* The change in ion temperature due to electron–ion collisions is described by the following equation (see Eq. 6.77):

$$\frac{\partial T_i}{\partial t} = 2\frac{m_e}{m_i}\bar{\nu}_{ei}(T_e - T_i) \tag{9.130}$$

This changes the velocity of the ion diamagnetic drift, which is proportional to the ion temperature (see Eq. 9.127):

$$\mathbf{u}_{idE} = \frac{c}{enH}[\mathbf{h}\times \mathrm{grad}(nT_i)] = \frac{cT_i}{eHn}[\mathbf{h}\times \mathrm{grad}\, n]$$

for grad $T_i = 0$. The resulting inertia force is equal to

$$\mathbf{F}_m = \frac{2\bar{\nu}_{ei}(T_e - T_i)}{\omega_{H_e} n}\mathbf{h}\times \mathrm{grad}\, n$$

for $(1/n)$ grad n = const. It causes inertial ion drift with a velocity of

$$\mathbf{u}_{im} = \frac{c[\mathbf{F}_m \times \mathbf{h}]}{eH} = -\frac{2\bar{\nu}_{ei}(T_e - T_i)}{m_e\omega_{H_e}^2}\frac{\mathrm{grad}\, n}{n} \tag{9.131}$$

It is seen that the velocity is parallel to that of the transverse diffusion under the action of the density gradient 9.128 and is of the same order of magnitude. For the electrons, the inertial drift velocity is m_i/m_e times less and can be neglected. Therefore the condition of transverse motion ambipolarity calls for compensation of the inertial ion drift due to the drift temperature change under the effect of the variable electric field. Such compensation evidently occurs when the change in electric strength is equal to that of the force associated with the pressure gradient

$$e\frac{\partial E}{\partial t} = \frac{\partial}{\partial t}\left[\frac{\mathrm{grad}(nT_i)}{n}\right] = \frac{\partial T_i}{\partial t}\frac{\mathrm{grad}\, n}{n}$$

Using Eq. 9.130, we obtain herefrom the equation for the ambipolar electric field in the case under study:

*We assume that the condition grad $T_i = 0$ is fulfilled in the course of change in T_i, bearing in mind that ion–ion collisions cause a relatively fast ion temperature equalization in the transverse direction (see Section 9.6).

$$\frac{\partial E_A}{\partial t} = 2\frac{m_e}{m_i}\bar{\nu}_{ei}(T_e - T_i)\frac{\text{grad } n}{n} \tag{9.132}$$

The transverse diffusion fluxes of charged particles in strong magnetic fields, where the cyclotron frequency greatly exceeds that of collisions, can also be determined by another method based on the summing of displacements of the Larmor centers of charged particles on collisions. This method enables one to analyze most easily the diffusion in a plasma of an arbitrary composition. Since collisions of different types are independent, the diffusion flux of the particles of a given type is obtained by summing the fluxes associated with all the types of collision:

$$\Gamma_\alpha = \sum_\beta \Gamma_{\alpha\beta} \tag{9.133}$$

Each of these partial fluxes can, in turn, be found using Eq. 9.70, which yields the relationship of the flux density with the particle displacement:

$$\Gamma_{\alpha\beta} = n_\alpha\langle\Delta x_{\alpha\beta}\rangle_1 - \frac{1}{2}\frac{\partial}{\partial x}[n_\alpha\langle(\Delta x_{\alpha\beta})^2\rangle_1] \tag{9.134}$$

where $\langle\Delta x_{\alpha\beta}\rangle_1$ is the average displacement, and $\langle(\Delta_{x\alpha\beta})^2\rangle_1$ is the mean square of the displacement of α-type particles in the direction of the density gradient due to their collisions with β-type particles per unit time. For simplicity we assume that there are no temperature gradients. The displacement of the Larmor center on a single collision is related to the change in velocity by the expression 9.61*:

$$\Delta x_\alpha = \frac{\Delta v_y}{\omega_{H\alpha}} = \frac{cm_\alpha \Delta v_y}{Z_\alpha eH}$$

The diffusion flux due to random displacements on collisions can be found, as in the case of a weakly ionized plasma, with the aid of Eq. 9.75 for the mean-square displacement:

$$\begin{aligned}\langle(\Delta x_{\alpha\beta})^2\rangle_1 &= \frac{\langle(\Delta v_{\alpha y})^2\rangle_1}{\omega^2_{H\alpha}} \\ &= 2\frac{\mu_{\alpha\beta}T_{\alpha\beta}}{m^2_\alpha\omega^2_{H\alpha}}\bar{\nu}_{\alpha\beta} \\ &= 2\frac{\mu_{\alpha\beta}c^2}{Z^2_\alpha e^2H^2}T_{\alpha\beta}\bar{\nu}_{\alpha\beta}\end{aligned} \tag{9.135}$$

where $T_{\alpha\beta} = (m_\alpha T_\beta + m_\beta T_\alpha)/(m_\alpha + m_\beta)$.

*Later in this section we assume $\omega_{H\alpha} = Z_\alpha eH/cm_\alpha$ so as to include the case of $|Z_\alpha| > 1$.

In a diffusion flux associated with charged particle collisions one must also take into account the effect of displacement asymmetry due to the density gradient of particles whose collisions induce diffusion. This flux component is related to the average displacement. Summing the displacements over collisions and averaging over the velocities of colliding particles in accordance with Eq. 9.65 yields

$$\langle \Delta x_{\alpha\beta} \rangle_1 = \frac{\langle \Delta v_{\alpha n} \rangle_1}{\omega_{H\alpha}} = -\frac{\mu_{\alpha\beta}}{m_\alpha \omega_{H\alpha}} \langle \nu^t_{\alpha\beta}(v_{\alpha y} - v_{\beta y}) \rangle \tag{9.136}$$

where the right-hand side is averaged over the velocities of α- and β-type particles. In averaging one must bear in mind that during collisions the Larmor center of β-type particles does not coincide with that of the α-type particle. The distance between them is determined by vectors ρ_α and ρ_β drawn from the centers to the collision point (Fig. 9.6)

$$\mathbf{r}_{\perp 0\beta} - \mathbf{r}_{\perp 0\alpha} = \boldsymbol{\rho}_\alpha - \boldsymbol{\rho}_\beta = \mathbf{h} \times \left(\frac{\mathbf{v}_\alpha}{\omega_{H\alpha}} - \frac{\mathbf{v}_\beta}{\omega_{H\beta}} \right);$$

$$x_{0\beta} - x_{0\alpha} = \frac{v_{\beta y}}{\omega_{H\beta}} - \frac{v_{\alpha y}}{\omega_{H\alpha}}$$

Accordingly, the density of charged β-type particles, appearing in the collision frequency $\nu_{\alpha\beta}$, must be taken at the point $x_{0\beta}$, which is spaced from $x_{0\alpha}$ by a distance depending on the particle velocity. Assuming the change in density over the length $x_{0\beta} - x_{0\alpha}$ to be small, we can use the following expansion for n_β:

$$n_\beta = n_\beta(x_{0\alpha}) + (x_{0\beta} - x_{0\alpha}) \frac{\partial n_\beta}{\partial x}$$

$$= n_\beta(x_{0\alpha}) + \left(\frac{v_{\beta y}}{\omega_{H\beta}} - \frac{v_{\alpha y}}{\omega_{H\alpha}} \right) \frac{\partial n_\beta}{\partial x}$$

Substituting it into Eq. 9.133 and considering that $\nu_{\alpha\beta} \approx n_\beta$, we find

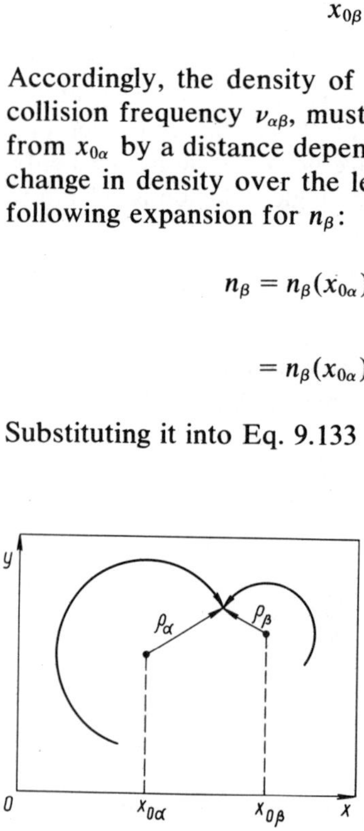

Fig. 9.6 Distance between Larmor centers during collisions.

$$\langle \Delta x_{\alpha\beta} \rangle_1 = -\frac{\mu_{\alpha\beta}}{m_\alpha \omega_{H\alpha}} \left\langle \left[\nu^t_{\alpha\beta}(v_{\alpha y} - v_{\beta y}) + \nu'^t_{\alpha\beta}(v_{\alpha y} - v_{\beta y}) \left(\frac{v_{\beta y}}{\omega_{H\beta}} - \frac{v_{\alpha y}}{\omega_{H\alpha}} \right) \frac{1}{n_\beta} \frac{\partial n_\beta}{\partial x} \right] \right\rangle_1$$

where $\nu'^t_{\alpha\beta}$ is found at the point $x_{0\alpha}$.

When averaging over the velocities we treat the velocity distribution of particles of both types as isotropic. Then the first term vanishes, and the equation takes the form

$$\langle \Delta x_{\alpha\beta} \rangle_1 = \frac{1}{3} \frac{\mu_{\alpha\beta}}{m_\alpha \omega_{H\alpha}} \bar{\nu}_{\alpha\beta} \left(\frac{\bar{v}^2_\beta}{\omega_{H\beta}} + \frac{\bar{v}^2_\alpha}{\omega_{H\alpha}} \right) \frac{1}{n_\beta} \frac{\partial n_\beta}{\partial x}$$

$$= \frac{\mu_{\alpha\beta} c^2}{Z_\alpha e^2 H^2} \bar{\nu}_{\alpha\beta} \left(\frac{T_\beta}{Z_\beta} + \frac{T_\alpha}{Z_\alpha} \right) \frac{1}{n_\beta} \frac{\partial n_\beta}{\partial x} \tag{9.137}$$

where $\bar{\nu}_{\alpha\beta}$ is the collision frequency averaged with a weight v^2. Substituting the obtained relations for $\langle \Delta x_{\alpha\beta} \rangle_1$ (Eq. 9.137) and $\langle (\Delta x_{\alpha\beta})^2 \rangle_1$ (Eq. 9.135) into Eq. 9.134, we get the general expression for the density of the diffusion flux of α-type particles due to their collisions with β-type particles under the effect of the density gradients of the colliding particles. In vector form this expression can be written as

$$\Gamma_{\alpha\beta} = -\frac{\mu_{\alpha\beta} c^2}{Z^2_\alpha e^2 H^2} \bar{\nu}_{\alpha\beta} \Big[T_{\alpha\beta} \operatorname{grad}_\perp n_\alpha + \left(T_{\alpha\beta} - T_\alpha - \frac{Z_\alpha}{Z_\beta} T_\beta \right) \frac{n_\alpha}{n_\beta} \operatorname{grad}_\perp n_\beta \Big] \tag{9.138}$$

For diffusion of electrons caused by their collisions with single-charged ions, substituting into Eq. 9.138

$$Z_\alpha = -1; \qquad Z_\beta = 1; \qquad n_\alpha = n_\beta = n; \qquad \mu_{\alpha\beta} = m_e;$$

$$T_{\alpha\beta} = T_\alpha = T_e; \qquad T_\beta = T_i; \qquad \nu_{\alpha\beta} = \nu_{ei}$$

we get

$$\Gamma_{ei} = -\frac{m_e c^2}{e^2 H^2} \bar{\nu}_{ei} \left(T_e \operatorname{grad}_\perp n_e + T_i \frac{n_e}{n_i} \operatorname{grad}_\perp n_i \right)$$

$$= -\frac{\bar{\nu}_{ei}}{m_e \omega^2_{H_e}} (T_e + T_i) \operatorname{grad}_\perp n \tag{9.139}$$

As would be expected, this expression coincides with Eq. 9.128 at $\operatorname{grad} T_e = \operatorname{grad} T_i = 0$.

Let us discuss in more detail the mechanism of diffusion of electrons colliding with ions under the effect of the density gradient. As we have noted more than once, diffusion in a strong magnetic field is induced by the displacement of the Larmor centers on collisions. For electrons these displacements are comparable with the Larmor radius. The usual diffusion flux due to the random displacement of Larmor centers on collisions is determined by the mean square of electron displacements:

$$\langle(\Delta x_e)^2\rangle_1 \approx \bar{\nu}_{ei}\bar{\rho}^2_{He} \approx \frac{2\bar{\nu}_{ei}T_e}{m_e\omega^2_{He}} \tag{9.140}$$

The intensity of this flux is

$$\Gamma'_{ei} = -\frac{1}{2}\frac{\partial}{\partial x}(n_e\langle(\Delta X_E)^2\rangle_1 \approx -\frac{\bar{\nu}_{ei}T_e}{m_e\omega^2_{He}}\left(\frac{\partial n_e}{\partial x} + \frac{n_e}{\bar{\nu}_{ei}}\frac{\partial \bar{\nu}_{ei}}{\partial x}\right)$$

$$\approx -\frac{\bar{\nu}_{ei}T_e}{m_e\omega^2_{He}}\left(\frac{\partial n_e}{\partial x} + \frac{n_e}{n_i}\frac{\partial n_i}{\partial x}\right) \tag{9.141}$$

The additional flux arises on inclusion of the asymmetry of Larmor center displacement due to the ion density gradient. Indeed, the displacement direction depends on which section of the trajectory the collision occured in. This illustrated in Fig. 9.2 for a head-on collision. The collisions to the right of the Larmor center result in rightward displacement, and to the left, leftward. Therefore, if the collision frequency on one side is higher than on the other (because of the higher ion density), an electron flux toward the higher collision frequency is induced. The difference in the collision frequencies on the opposite sides of the Larmor circle is of the order of

$$\delta\nu_{ei} \approx \frac{\partial \nu_{ei}}{\partial x}\rho_{He} \approx \rho_{He}\nu_{ei}\frac{1}{n_i}\frac{\partial n_i}{\partial x} \tag{9.142}$$

Since the average displacement on collisions is of the order of the Larmor radius, the flux due to this difference is equal to

$$\Gamma''_{ei} = n_e\langle\Delta x_e\delta\nu_{ei}\rangle_1 \approx \langle\nu_{ei}\rho^2_{He}\rangle_1\frac{n_e}{n_i}\frac{\partial n_i}{\partial x}$$

$$\approx \frac{T_e\bar{\nu}_{ei}}{m_e\omega^2_{He}}\frac{n_e}{n_i}\frac{\partial n_i}{\partial x} \tag{9.143}$$

Another cause for the appearance of the additional flux is the difference in the momenta transferred to the electron on collisions with ions moving in different directions (Fig. 9.7). It is easy to see that on head-on collision the additional change in electron velocity associated with the motion of the ion is $\delta v_e \approx 2v_i$. This change also causes a corresponding change in the displacement of the Larmor center of the electron by $\Delta x'_e \approx \delta v_e/\omega_{He} \approx v_i/\omega_{He}$. As can be seen from a comparison of collisions with different orientations of v_e and v_i (cases *a–d* in Fig. 9.7), the additional displacement $\Delta x'_e$ is directed away from the Larmor center of the ion. The difference in the frequency of collisions with ions arriving

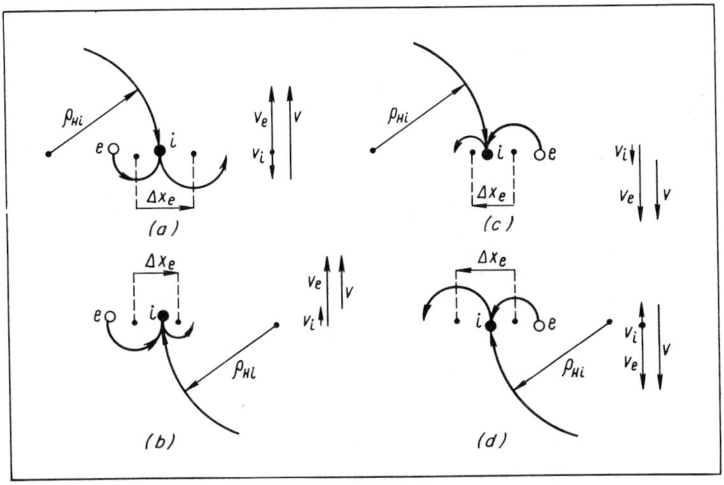

Fig. 9.7 Appearance of flux due to momenta transferred to electrons on collision with ions moving in different directions.

from different directions is of the order of

$$\delta\bar{\nu}_{ei} \approx \frac{\partial\bar{\nu}_{ei}}{\partial x}\rho_{Hi} \approx \overline{\rho_{Hi}\nu_{ei}}\frac{1}{n_i}\frac{\partial n_i}{\partial x} \tag{9.144}$$

The resulting flux is equal to

$$\Gamma'''_{ei} \approx -n_e\Delta x'_e\delta\nu_{ei} \approx -\frac{T_i\bar{\nu}_{ei}}{m_e\omega^2_{He}}\frac{n_e}{n_i}\frac{\partial n_i}{\partial x} \tag{9.145}$$

Adding up all the three components of the flux leads to Eq. 9.139; that is, it yields the correct estimate of the total electron flux.

From the general equation 9.138 it is possible to obtain the diffusion flux of ions due to their collisions with electrons. For single-charged ions the flux density is

$$\begin{aligned}\mathbf{\Gamma}_{ie} &= -\frac{m_e c^2}{e^2H^2}\bar{\nu}_{ei}\left[T_e \operatorname{grad}_\perp n + (2T_e - T_i)\frac{n_i}{n_e}\operatorname{grad}_\perp n_e\right]\\ &= -\frac{\bar{\nu}_{ei}(3T_e - T_i)}{m_e\omega^2_{He}}\operatorname{grad}_\perp n\end{aligned} \tag{9.146}$$

This equation differs from Eq. 9.139, although the diffusion flux of the ions due to their collisions with electrons must, according to Eq. 9.128, be equal to that of the electrons due to electron–ion collisions. The difference arises because without external energy sources the collisions (as noted above) equalize the electron and ion temperatures. This equalization is accompanied by an inertial ion drift. It is easy to see that the flux density Γ_{ie} is measured by the sum of the transverse diffusion velocity 9.128 and that of the inertial ion drift (Eq. 9.131) $\mathbf{\Gamma}_{ie} = n_i(\mathbf{u}_{it} + \mathbf{u}_{im})$. In the presence of external ion heating sources one must consider the effect of the change in the Larmor ion radius between collisions on the ion cross flux. In particular, at a time-invariant ion temperature the flux Γ_{ie} is equal to $\mathbf{\Gamma}_{ei}$ in accordance with Eq. 9.128. At equal ion and electron temperature $T_e = T_i$ the identity of the ion and electron transverse fluxes follows from Eq. 9.138. For single-charged ions, at $n_e = n_i$ we get

$$\mathbf{\Gamma}_{ie} = \mathbf{\Gamma}_{ei} = -\frac{2\bar{\nu}_{ei}T_e}{m_e\omega_{He}^2}\,\text{grad}_\perp\, n \tag{9.147}$$

Investigation shows that the ion and electron diffusion fluxes are equal despite the great difference in their Larmor radii. This result can be understood if we take into account that the displacement of the ion Larmor center on ion–electron collisions is much less than the Larmor radius. From Eq. 9.61 the displacement of the Larmor centers of the colliding particles is determined by the change in their momentum.

$$\Delta\mathbf{r}_{0\alpha} = \frac{[\Delta\mathbf{v}_\alpha \times \mathbf{h}]}{Z_\alpha\omega_{H\alpha}} = \frac{c}{Z_\alpha eH}[\Delta(m_\alpha\mathbf{v}_\alpha)\times\mathbf{h}]$$

Therefore, by the momentum conservation law for electron–ion collision $\Delta(m_e\mathbf{v}_e) = -\Delta(m_i\mathbf{v}_i)$ the value and direction of displacement of their Larmor centers on collision are the same $\Delta\mathbf{r}_{0e} = \Delta\mathbf{r}_{0i}$.

The pattern of displacement on electron–ion collision is schematized in Fig. 9.8 (for head-on collision). After such a collision the electron

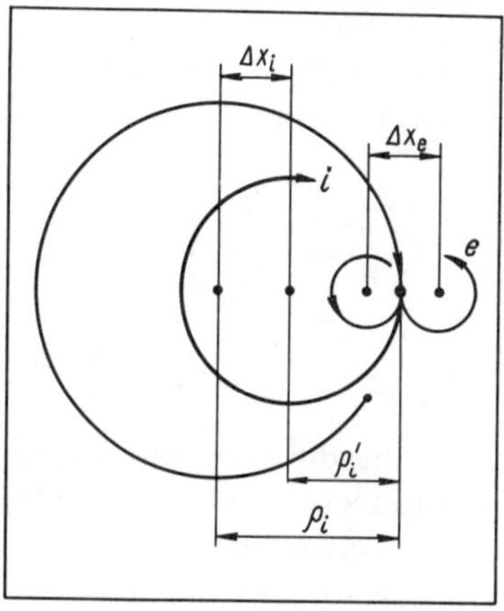

Fig. 9.8 Pattern of the displacement resulting from an electron-ion head-on collision.

reverses its motion, and its Larmor center shifts by a distance of $\Delta\rho_{He} \approx 2\rho_{He}$. The absolute value of the velocity of the ion reduces, and its Larmor radius changes accordingly by $2\rho_{He}$, which results in displacement of the Larmor center by the same distance and in the same direction as for the electron.

We now consider the effect of collisions of identical charged particles on the transverse diffusion. The general equation 9.138 indicates the absence of fluxes due to collisions of identical particles; substituting $Z_\alpha = Z_\beta$, $T_\alpha = T_\beta$, $n_\alpha = n_\beta$ yields $\mathbf{\Gamma}_{\alpha\alpha} = 0$. This means that the diffusion flux due to the mean-square displacement $\langle(\Delta x_{\alpha\alpha})^2\rangle$ is completely offset by the flux determining the average displacement $\langle \Delta x_{\alpha\alpha}\rangle$ and associated with the collision asymmetry in the presence of a density gradient. This result can be explained by considering displacements of identical particles on collisions. On head-on collision the particles simply change places (Fig. 9.9*a*). On arbitrary collision the displacements of the Larmor centers of the colliding particles are equal in magnitude and opposite in direction $\Delta\mathbf{r}_{01} = -\Delta\mathbf{r}_{02}$ in view of the momentum conservation law (Fig. 9.9*b*). In averaging, the two fluxes offset each other, provided the density gradient at the points $\mathbf{r}_{01}$ and $\mathbf{r}_{02}$ remains unchanged.

A more accurate calculation, taking into account the change in density gradient over a length of the order of the Larmor radius, suggests that incomplete compensation results in a flux proportional to the higher derivatives of the density. If the Larmor radius of the charged particles

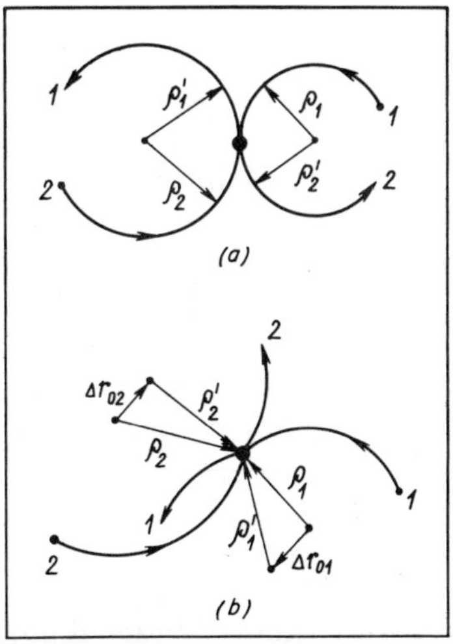

Fig. 9.9 Displacement of identical particles on collision. *a* head-on collisions, *b* arbitrary collisions.

is much less than the characteristic transverse lengths of density change, this flux is of the order of $\rho_{H_\alpha}^2/L_\perp^2$ and can usually be neglected. Then in a fully ionized plasma consisting of electrons and ions of the same type the transverse diffusion of electrons is entirely due to their collisions with ions, and the diffusion of ions is due to their collisions with electrons. The diffusion fluxes of the electrons and ions are equal; that is, the diffusion is ambipolar, irrespective of the electric field in the plasma.

If a plasma consisting largely of electrons and single-charged ions contains impurity ions whose charge exceeds unity, then a diffusion flux arises due to ion collisions. On such collisions the displacement of the Larmor center of the ions is of the order of the ion Larmor radius; that is, it greatly exceeds the displacement on ion–electron collisions. Therefore the diffusion of the ions of the main plasma due to ion–ion collisions may be substantial even at a low density of impurity ions. The corresponding diffusion flux can be found with the aid of Eq. 9.138. Assuming $T_\alpha = T_\beta = T_{\alpha\beta}$ and $Z_\alpha = 1$, we have

$$\boldsymbol{\Gamma}_{ip} = -\frac{\mu_{ip}c^2}{e^2H^2}\bar{\nu}_{ip}n_iT_i\left(\frac{1}{n_i}\operatorname{grad}_\perp n_i - \frac{1}{Z_pn_p}\operatorname{grad}_\perp n_p\right) \tag{9.148}$$

where p is used to denote the characteristics of the impurity ions. With equal relative density gradients the ratio of this flux to the ion flux due to ion–electron collisions is

$$\frac{\Gamma_{ip}}{\Gamma_{ie}} = \frac{Z_p - 1}{Z_p}\frac{\mu_{ip}}{m_e}\frac{\bar{\nu}_{ip}}{\nu_{ie}}\frac{n_iT_i}{n_e(T_e + T_i)} \tag{9.149}$$

and at $T_e \approx T_i$

$$\frac{\Gamma_{ip}}{\Gamma_{ie}} \approx \frac{n_p}{n}\sqrt{m_i/m_e} \tag{9.150}$$

In accordance with Eq. 9.138 the flux density of the impurity ions is equal to

$$\boldsymbol{\Gamma}_{pi} = \frac{\mu_{ip}c^2}{Z_p^2e^2H^2}\bar{\nu}_{pi}n_pT_i\left(Z_p\frac{1}{n_i}\operatorname{grad}_\perp n_i - \frac{1}{n_p}\operatorname{grad}_\perp n_p\right) \tag{9.151}$$

It is easy to see that this flux greatly exceeds the flux of impurity ions due to ion–electron collisions. With equal relative gradients we obtain, by comparing Eq. 9.148 and 9.151,

$$\boldsymbol{\Gamma}_{pi} = -\frac{1}{Z_p}\boldsymbol{\Gamma}_{ip} \tag{9.152}$$

It is significant that in this case the flux of impurity ions is directed not against the density gradient, as the ordinary diffusion flux, but toward increasing density. The opposite directions of the flux of the basic and impurity ions are associated with the ratio of the displacement of their Larmor centers on collision, since in view of the momentum conservation law the displacements of the Larmor centers of the colliding ions are oppositely directed. In the stationary state, when ions are produced in the central part of the plasma volume and their density reduces toward the periphery, the diffusion of ions of both types must also be directed toward the periphery. Therefore the stationary density distribution cannot be characterized by equal relative density gradients of both types of ions. Ions with a higher charge must concentrate in the central part of the volume more intensively than lower-charge ions. As is seen from Eq. 9.151, the following inequality must be satisfied for the flux of impurity ions to be directed toward decreasing density:

$$\text{grad}(\ln n_p) > Z_p \,\text{grad}(\ln n_i) \tag{9.153}$$

The condition 9.153 determines the much faster drop in impurity density toward the periphery as compared with the drop in the density of the main plasma component (faster than n^{Z_p}).

We have considered the transverse diffusion of charged particles of a fully ionized plasma due to collisions of charged particles with each other. When a plasma is partly ionized, the transverse motion may be affected both by collisions between charged particles and their collisions with neutral particles. In a strong magnetic field, where the cyclotron frequencies of the charged particles greatly exceed their collision frequencies, the effect of different types of collision on the transverse motion is additive. In accordance with Eq. 9.133 the total transverse flux of charged particles of each type in the direction of the density gradient and the electric field can be obtained by adding fluxes due to collisions of these particles to all the charged and neutral particles contained in the plasma. For a three-component plasma containing electrons, single-charged ions, and neutral particles the transverse electron and ion fluxes can be found by adding the expressions of Section 7.1 to Eqs. 9.139 and 9.146:

$$\mathbf{\Gamma}_e = \mathbf{\Gamma}_{ea} + \mathbf{\Gamma}_{ei} = -\frac{e\bar{\nu}_{ea} n}{m_e \omega_{He}^2}\mathbf{E}_\perp - \frac{\bar{\nu}_{ea} T_e}{m_e \omega_{He}^2}\text{grad}_\perp n - \frac{\bar{\nu}_{ei}(T_e + T_i)}{m_e \omega_{He}^2}\text{grad}_\perp n \tag{9.154}$$

$$\mathbf{\Gamma}_i = \mathbf{\Gamma}_{ia} + \mathbf{\Gamma}_{ie} = \frac{2e\bar{\nu}_{ia} n}{m_i \omega_{Hi}^2}\mathbf{E}_\perp - \frac{2\bar{\nu}_{ia} T_i}{m_i \omega_{Hi}^2}\text{grad}_\perp n - \frac{\bar{\nu}_{ei}(T_e + T_i)}{m_e \omega_{He}^2}\text{grad}_\perp n \tag{9.155}$$

where it is taken into account that at $m_i = m_a$, $\mu_{ia} = m_i/2$. In the ambipolar diffusion regime considered in Section 9.4 the electron and ion fluxes are equal. Equating them, we find the ambipolar strength of the electric field, which coincides with Eq. 9.94. In such a field the ambipolar transverse flux is the sum of the ambipolar flux (Eq. 9.96) due to collisions of charged particles with neutrals and the flux (Eq. 9.139) due to electron collisions:

$$\mathbf{\Gamma}_A = n\mathbf{u}_{A\perp} = -\frac{(\bar{\nu}_{ea} + \bar{\nu}_{ei})(T_e + T_i)}{m_e\omega_{He}^2}\,\mathrm{grad}_\perp\, n \tag{9.156}$$

In this expression the factor before $\mathrm{grad}_\perp\, n$ is the coefficient of ambipolar diffusion of a partly ionized plasma:

$$D_{A\perp} = \frac{(\bar{\nu}_{ea} + \bar{\nu}_{ei})}{m_e\omega_{He}^2}(T_e + T_i) \tag{9.157}$$

From the averaged equations of motion discussed above it follows that this equation holds when

$$\omega_{He}\omega_{Hi} \gg (\bar{\nu}_{ea} + \bar{\nu}_{ei})\bar{\nu}_{ia}$$

To conclude this section we note that the foregoing is true until the magnetic field begins to affect charged particle collisions. In very high magnetic fields, where the Larmor radius of the electrons is less than the Debye radius $\rho_{He} < r_D$ (and all the more so at $\rho_{Hi} < r_D$), one must take into account the effect of the magnetic field on electron–ion collisions. In this case the displacement of the Larmor center of the electrons in the course of collision is determined by its drift in the electric field of the ion. Analysis shows that the expressions for diffusion fluxes due to such a drift differ from those obtained above only by a numerical factor of the order of unity. The resulting change in the expression for the fluxes formally amounts to redetermination of the Coulomb logarithm in the equation for the effective collision frequency.

9.6 TRANSVERSE ENERGY TRANSPORT IN HIGHLY IONIZED PLASMAS

Passing to energy transport in a highly ionized plasma, we begin with determining the electron heat flux when the frequency of electron–ion collisions greatly exceeds the one for collisions of electrons with neutral atoms. The equations yielding the electron heat flux in such conditions can be obtained by substituting collision terms taking into account collisions of electrons with ions (Eq. 6.83) and with each other (Eq. 6.85) into the third-moment equations (Eq. 6.80). Then in the stationary case we obtain

$$\frac{5}{2}\frac{nT_e}{m_e}\operatorname{grad} T_e + \frac{e}{m_e c}[\mathbf{q}_e \times \mathbf{H}] - \frac{3}{2} nT_e\bar{\nu}_{ei}(\mathbf{u}_e - \mathbf{u}_i) + 1.87\bar{\nu}_{ei}\mathbf{q}_e = 0 \tag{9.158}$$

where $\bar{\nu}_{ei}$ is the averaged frequency of electron–ion collisions, and the coefficient 1.87 was obtained by summing the frequencies of electron–ion and electron–electron collisions (cf. Eq. 6.86). The projection of this equation on the direction of the magnetic field leads to the expression for the longitudinal heat flux $\mathbf{q}_{e\parallel}$, which coincides with Eq. 7.160 obtained in the absence of a magnetic field. The projection of the equation on a plane perpendicular to the magnetic field is a vector equation in $\mathbf{q}_{e\perp}$ similar to the one considered in Section 9.1. Its solution has the form

$$\mathbf{q}_{e\perp} = -\frac{1.87\bar{\nu}ei}{\omega_{He}^2 + (1.87\bar{\nu}_{ei})^2}\left[\frac{5}{2}\frac{nT_e}{m_e}\operatorname{grad}_\perp T_e - \frac{3}{2}nT_e\bar{\nu}_{ei}(\mathbf{u}_{e\perp} - \mathbf{u}_{i\perp})\right]$$
$$-\frac{\omega_{He}}{\omega_{He}^2 + (1.87\bar{\nu}_{ei})^2}\left\{\mathbf{h} \times \left[\frac{5}{2}\frac{nT_e}{m_e}\operatorname{grad}_\perp T_e - \frac{3}{2}nT_e\bar{\nu}_{ei}(\mathbf{u}_{e\perp} - \mathbf{u}_{i\perp})\right]\right\} \tag{9.159}$$

This expression is simplified in the case of large magnetic fields, that is, at $\omega_{He} \gg \nu_{ei}$. Neglecting the terms proportional to ν_{ei}^2 and substituting $(\mathbf{u}_{e\perp} - \mathbf{u}_{i\perp})$ from Eq. 9.125, we obtain the following:

$$\mathbf{q}_{e\perp} = \mathbf{q}_{et} + \mathbf{q}_{ed} + \mathbf{q}_{eu} \tag{9.160}$$

$$\mathbf{q}_{et} = -4{,}66\,\frac{\bar{\nu}_{ei}nT_e}{m_e\omega_{He}^2}\operatorname{grad}_\perp T_e \tag{9.161}$$

$$\mathbf{q}_{ed} = -\frac{5}{2}\frac{nT_e}{m_e\omega_{He}}[\mathbf{h} \times \operatorname{grad} T_e] \tag{9.162}$$

$$\mathbf{q}_{eu} = \frac{3}{2}\frac{nT_e\bar{\nu}_{ei}}{\omega_{He}}[\mathbf{h} \times (\mathbf{u}_e - \mathbf{u}_i)] \tag{9.163}$$
$$\approx \frac{3}{2}\frac{\bar{\nu}_{ei}T_e}{m_e\omega_{He}^2}\operatorname{grad}[n(T_e + T_i)]$$

where the component $\mathbf{q}_{et}$ is the heat flux in the direction of the temperature gradient, $\mathbf{q}_{ed}$ is the heat flux associated with the diamagnetic flux; and $\mathbf{q}_{eu}$ is the directed motion component, it appears, as in a weakly ionized plasma, because of the velocity dependence of the collision frequency. The factor of proportionality between the heat flux $\mathbf{q}_{et}$ and grad T_e is the electron thermal conductivity,

$$\mathcal{K}_{e\perp} = n\chi_{e\perp} = \frac{4.66\bar{\nu}_{ei}nT_e}{m_e\omega_{He}^2} \tag{9.164}$$

The numerical coefficient in Eq. 9.164 corresponds to an accurate calculation at $\omega_{He} \gg \nu_{ei}$. The origin of the transverse heat transfer due to thermal conduction is the same as for a weakly ionized plasma. Electron–ion collisions result in displacement of the electron by a distance of the order of its Larmor radius, and electrons with a higher and a lower temperature change places. Therefore, in the reference system where the directed velocity is zero the energy flux associated with electron–ion collisions is different from zero. Moreover, on electron–electron collisions electrons with different temperatures also change places at distances of the order of their Larmor radii. Naturally the heat flux transferred by electrons in the direction of the temperature gradient is described by an expression similar to Eq. 9.33, which was obtained for a weakly ionized plasma, where the flux is due to electron–atom collisions. The difference is in replacement of the electron–atom collision frequency ν_{ea} by the coulomb collision frequency ν_{ei} and the appearance of a numerical factor. The nature of the other components of the heat flux, one of which ($\mathbf{q}_{ed}$) is associated with the diamagnetic flux and the other ($\mathbf{q}_{eu}$) with the directed motion, was also discussed when considering heat transfer in a weakly ionized plasma; therefore we do not dwell on them here.

Let us now determine the transverse heat flux of ions for a plasma with a high degree of ionization, in which the frequency of ion–ion collisions greatly exceeds that of collisions with atoms. The collision term of the ion heat flux equation is associated with ion–ion collisions (ion–electron collisions can be neglected because of their weak effect on the ion motion and energy transport. Substituting the ion–ion collision term (Eq. 6.85) into the third-moment equation (Eq. 6.80), we obtain for stationary conditions

$$\frac{5}{2}\frac{nT_i}{m_i}\,\mathrm{grad}\;T_i - \frac{e}{m_i c}\left[\mathbf{q}_i \times \mathbf{H}\right] = -0.8\bar{\nu}_{ii}\mathbf{q}_i \qquad (9.165)$$

where the averaged frequency of ion–ion collisions is equal to

$$\bar{\nu}_{ii} = \frac{4}{3}\frac{\sqrt{\pi}}{(m_i)^{1/2}}\frac{e^4 n_i L_i}{T_i^{3/2}} \qquad (9.166)$$

Equation 9.165 is a vector equation with respect to the ion heat flux. The longitudinal flux component $\mathbf{q}_{i\parallel}$ defined by it is the same as in the absence of a magnetic field (Eq. 7.162). The transverse component $\mathbf{q}_{i\perp}$ is obtained by projecting the equation onto a plane perpendicular to the magnetic field. With an arbitrary ratio of $\bar{\nu}_{ii}$ and ω_{Hi} it has the form

$$\mathbf{q}_{i\perp} = -\frac{5}{2}\frac{nT_i}{m_i}\,\frac{0.8\bar{\nu}_{ii}\,\mathrm{grad}_\perp\,T_i - \omega_{Hi}(\mathbf{h}\times\mathrm{grad}\;T_i]}{\omega_{Hi}^2 + (0.8\bar{\nu}_{ii})^2} \qquad (9.167)$$

In strong magnetic fields, where $\omega_{Hi} \gg \bar{\nu}_{ii}$, the following relation holds:

$$\mathbf{q}_{i\perp} = -2\frac{nT_i\bar{\nu}_{ii}}{m_i\omega_{Hi}^2}\,\mathrm{grad}_\perp T_i + \frac{5}{2}\frac{nT_i}{m_i\omega_{Hi}}\,[\mathbf{h}\times \mathrm{grad}\, T_i] \tag{9.168}$$

The coefficient in the first term, determining the heat flux in the direction of the transverse temperature gradient, is the coefficient of transverse thermal conductivity of ions:

$$\mathcal{K}_{i\perp} = \chi_{i\perp} n = \frac{2nT_i\bar{\nu}_{ii}}{m_i\omega_{Hi}^2} \tag{9.169}$$

At an ion temperature comparable with that of the electrons it greatly exceeds the coefficient of thermal conductivity of electrons:

$$\frac{\mathcal{K}_{i\perp}}{\mathcal{K}_{e\perp}} = 0.6\sqrt{\frac{m_i}{m_e}}\left(\frac{T_e}{T_i}\right)^{1/2}\frac{L_i}{L_e} \approx \sqrt{\frac{m_i}{m_e}}\left(\frac{T_e}{T_i}\right)^{1/2} \tag{9.170}$$

This result is due to the association of the transverse thermal conductivity of ions with ion–ion collisions in which the displacement of the ion Larmor centers is of the order of the ion Larmor radius; that is, it greatly exceeds the displacement of the Larmor centers of electrons on collisions.

As noted in Section 9.5, collisions between like particles do not, on the average, lead to their displacement, but ions with different temperatures change places. Therefore a transverse heat flux of the order of

$$q_{i\perp} \sim \bar{\rho}_{Hi}^2\bar{\nu}_{ii} n T_i \frac{\partial T_i}{\partial x} \tag{9.171}$$

is possible. This estimate is similar to that of the electron heat flux in Section 9.3.

The expressions for the directed velocities and heat fluxes of charged particles enable us to write their energy balance equations. To do this, we must substitute the relations found into the second-moment equations (Eqs. 6.77 and 6.78), which were considered in Section 6.3. We do not give here the cumbersome general-case equations, but consider the summary energy balance equation for a fully ionized plasma, which results from adding up the electron and ion energy balance equations (Eqs. 6.77 and 6.78). For conditions where the energy losses are related to the transverse thermal conductivity and the plasma heating is determined by the longitudinal electric field, this equation has the form

$$\frac{3}{2}n\frac{\partial(T_i+T_e)}{\partial t} + \mathrm{div}(\mathcal{K}_{i\perp}\,\mathrm{grad}\, T_i) = \sigma_\parallel E_\parallel^2 \tag{9.172}$$

where $\mathcal{K}_{i\perp}$ is the transverse thermal conductivity of ions, described by

Eq. 9.169, and $\sigma_\parallel$ is the longitudinal conductivity defined by Eq. 7.153. Here we took into account that, in accordance with Eq. 9.170, $q_{i\perp} \gg q_{e\perp}$ and neglected the heat flux associated with the electron thermal conductivity, and also omitted the terms proportional to the directed velocity (it can be demonstrated that they are of the same order as div $\mathbf{q}_{e\perp}$).

To estimate the efficiency of the processes described by the energy balance equations we can introduce the characteristic times, as in Section 7.10. Let us find them for $T_e = T_i$ from the relation $\tau_p = T(\partial T/\partial t)_p^{-1}$, where p denotes the process at hand. Accordingly, using Eqs. 9.169 and 7.153 for $\mathcal{K}_{i\perp}$ and $\sigma_\parallel$, we obtain the characteristic times of the heat transfer and plasma heating by the electric field:

$$\tau_q \approx \frac{3nL_\perp^2}{\mathcal{K}_{i\perp}} \approx \frac{m_i\omega_{Hi}^2}{T_i\bar{\nu}_{ii}} L_\perp^2 \tag{9.173}$$

$$\tau_E = \frac{3nT_e}{\sigma_\parallel E_\parallel^2} = \frac{3}{2}\frac{m_e\nu_{ei}T_e}{e^2E_\parallel^2} \tag{9.174}$$

where $L_\perp = [(1/T)\,\mathrm{grad}_\perp\, T]^{-1}$ is the characteristic transverse dimension on which the temperature changes.

Equation 9.172 yields the change in the summary energy of the electrons and ions. Their temperature ratio can be found with the aid of the electron energy balance equation. If the heat losses due to the transverse electron thermal conductivity are neglected, it has the form:

$$\frac{\partial T_e}{\partial t} = \frac{2}{3n}\sigma_\parallel E^2 - \kappa_{ei}\bar{\nu}_{ei}(T_e - T_i) \tag{9.175}$$

where the first term on the right-hand side determines the heating of the electrons, and the second the transfer of the electron energy to ions on collisions. The electron energy balance equation was considered in Section 7.11. This consideration shows that the stationary solution of the equation exists only at moderate electric fields, at which $T_e - T_i < 0.5T_i$. At higher fields the electron energy "runs away," and other mechanisms of losses, in particular those associated with excitation of atoms (if any), must come into play, namely, radiation losses. Note that even with such mechanisms the stationary electron energy balance in a longitudinal electric field may prove unstable with respect to electron overheating (the so-called overheating instability). An accidental temperature rise in some flux tube increases the conductivity $\sigma_\parallel \sim 1/\nu_{ei} \sim T_e^{3/2}$. In this case the current increases, and so does the heating energy $\sigma_\parallel E^2$, which entails a further increase in conductivity. If the total current is limited, the current increase in the flux tube under consideration is due to the current decrease in another element of the cross section, which reduces the electron temperature and conductivity. As a result, the increase in perturbation leads to concentration of the current in one or several columns. The development of overheating instability is limited by the transverse electron thermal conductivity, which smears out the elevated-temperature region. Another possible limiting factor is increasing radiation losses with increasing electron temperature.

9.7 TRANSPORT IN TOROIDAL MAGNETIC CONFIGURATIONS

The description of transport processes in an inhomogeneous magnetic field is much more complicated than for a homogeneous one. The reason is primarily the drift of the charged particles due to the inhomogeneity, which leads to charge separation in the plasma. The resulting electric field may change the efficiency of transport of the charged particles and their energy across the magnetic field, and the nature of these changes depends on the particular type of magnetic configuration. Let us consider some peculiarities of transport processes in toroidal magnetic configurations. The principal cause for these peculiarities is the toroidal drift of charged particles.

We first discuss the transport of charged particles of a weakly ionized plasma in a toroidal magnetic field. We assume that the magnetic field is directed along the azimuth and decreases inversely as the radius $H = H_0R_0/R$. Assuming the minor radius of the torus to be much less than the major $r \ll R$, we can consider that the magnetic field gradient $|\text{grad}\, H|/H = 1/R_0$ is constant in the region containing the plasma. The toroidal inhomogeneity results in a drift of electrons and ions in a direction perpendicular to the magnetic field and the inhomogeneity, that is, in the direction of the torus axis. Introducing a local rectangular coordinate system in the torus cross section, we direct the $0x$ axis along the major radius; then the ion drift is directed along the $0y$ axis and the electron drift in the opposite direction (se Fig. 8.26). The components of the averaged velocity of the drift caused by the magnetic field gradient are equal, at $\omega_{He} \gg \nu_{ea}$ and $\omega_{Hi} \gg \nu_{ia}$, to (see Eq. 8.57)

$$u_{iy}^H = \frac{-2cT_i}{eHR}; \qquad u_{ey}^H = \frac{2cT_e}{eHR} \tag{9.176}$$

The rate of charge separation due to the drift is determined by the difference between the ion and electron velocities:

$$u_y^H = u_{iy}^H - u_{ey}^H = -2c\,\frac{T_i + T_e}{eHR} \tag{9.177}$$

The electric field induced by the charge separation causes a reverse motion. The velocities of the ions and electrons in the direction of the electric field in a weakly ionized plasma depend on their mobility. From Eqs. 9.11 and 9.14 we have

$$u_y^E = u_{iy}^E - u_{ey}^E = \left(\frac{e\mu_{ia}\nu_{ia}}{m_i^2\omega_{Hi}^2} + \frac{e\nu_{eu}}{m_e\omega_{He}^2}\right)E_y \approx \frac{e\nu_{ia}}{2m_i\omega_{Hi}^2}E_y$$

which takes into account that $\mu_{ia} = m_i/2$ at $m_i = m_a$. In the stationary state, the velocities u^H and u^E must offset each other. Equating them, we find the electric field strength

$$E_y = \frac{4H(T_e + T_i)}{m_i c \nu_{ia} R} \tag{9.178}$$

In this field the electrons and ions drift toward the outer surface of the torus with a velocity

$$u_{dx} = \frac{cE_y}{H} = \frac{4(T_e + T_i)}{m_i \nu_{ia} R} = \frac{2D_{A\|}}{R} \tag{9.179}$$

where $D_{A\|} = 2(T_e + T_i)/m_i\nu_{ia}$ is the longitudinal coefficient of ambipolar diffusion (see Eq. 7.70). Simultaneously the electrons and ions drift in the direction of the 0y axis with a velocity close to u_e^H (Eq. 9.176); since $u_e^H \ll u_d$ (Eq. 9.179), however, this drift component can be neglected.

Thus the toroidal inhomogeneity of the magnetic field in a weakly ionized plasma leads to a stationary motion of the charged particles in the direction of the major radius of the torus with the velocity 9.179. In a plasma enclosed in a toroidal chamber parallel to the magnetic field this motion is superposed on the ambipolar transverse diffusion. Summing the directed velocities due to the diffusion (Eq. 9.96) and the drift (Eq. 9.179), we get

$$\mathbf{u}_\perp = \mathbf{u}_{A\perp} + \mathbf{u}_d = -D_{A\perp}\frac{\mathrm{grad}_\perp n}{n} + \frac{2D_{A\|}}{R}\mathbf{x}_1 \tag{9.180}$$

where $\mathbf{x}_1$ is a unit vector in the direction of the major radius. It is easy to determine, with the aid of this equation, the charged particle balance in a stationary toroidal discharge. The stationary balance equation for the production of charged particles due to ionization in electron–atom collisions and for their losses through transverse transport has the form

$$\mathrm{div}(n\mathbf{u}_\perp) = \nu^i n \tag{9.181}$$

Substituting the directed velocity (Eq. 9.180), we obtain, at $T_e = \mathrm{const}$ and $T_i = \mathrm{const}$;

$$D_{A\perp}\Delta_\perp n - \frac{2D_{A\|}}{R}\frac{\partial n}{\partial x} + \nu^i n = 0 \tag{9.182}$$

The solution of this equation must satisfy the zero boundary conditions near the chamber walls. With a small torus curvature (at $R \gg a$) the effect of toroidality on $\Delta_\perp n$ can be neglected. Here the nonnegative solution of the equation has the form

$$n = n_0 \exp\left(\frac{D_{A\|}}{D_{A\perp}} \frac{x}{R}\right) J_0\left(\frac{r}{\Lambda}\right) \tag{9.183}$$

where $\Lambda = a/2.4$. This distribution differs from the diffusion distribution 7.94 by the exponential factor representing the displacement of the maximum in the direction of the toroidal drift. Substituting Eq. 9.183 into Eq. 9.182 results in the condition for particle balance in a stationary discharge:

$$\nu^i = \frac{D_{A\perp}}{\Lambda^2} + \left(\frac{1}{R}\right)^2 \frac{D_{A\|}^2}{D_{A\perp}} \tag{9.184}$$

The right-hand side of the equality characterizes the efficiency of removal of charged particles, which takes into account both the ordinary diffusion removal (the first term) and the transport acceleration due to the drift (the second term). Interestingly, with an increase in magnetic field, that is, with a decrease in $D_{A\perp}$, the efficiency of removal of charged particles varies nonmonotonically. It passes a minimum at $D_{A\perp} = D_{A\|}\Lambda/R$.

Let us now consider the transport of charged particles of a fully ionized plasma in a toroidal configuration. Note that stationary drift of a fully ionized plasma is impossible in a simple toroidal magnetic field, since the transverse electric field does not induce a current there and therefore cannot offset the charge separation due to the toroidal drift. We now briefly consider the behavior of a fully ionized plasma in the azimuthally symmetrical toroidal magnetic trap described in Section 8.5. Recall that the magnetic configuration used in such a trap is formed by the toroidal field H_θ and a poloidal field H_φ (see Fig. 8.1); usually, $H_\varphi \ll (r/R)H_\theta$. In our consideration we assume that $r \ll R$ and $H_\varphi \ll H_\theta$. As noted in Section 8.1, the poloidal field causes a rotational transformation producing a system of toroidal magnetic surfaces inserted in each other. Without collisions these surfaces must eliminate the consequences of the toroidal drift: the electric field induced during the drift is "short-circuited" as a result of electron motion along the lines of force. But in the presence of collisions the longitudinal conductivity of the plasma is finite, and the short-circuiting will be incomplete. The residual electric field leads to a drift of the charged plasma particles in the direction of the major radius of the torus; the drift increases the velocity of their transport due to the density and temperature gradients.

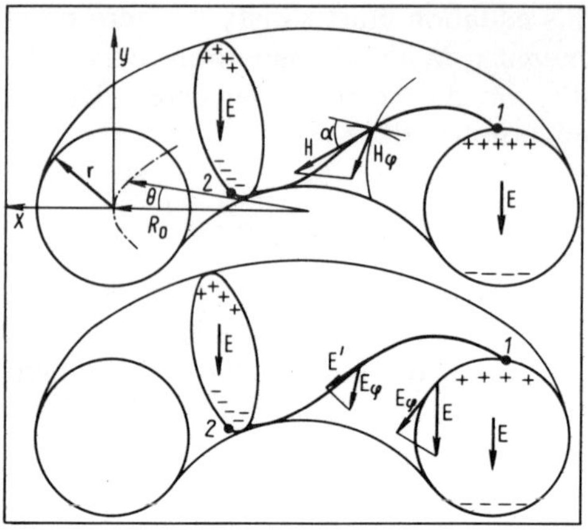

Fig. 9.10 Current induced along a helical line of force in a fully ionized plasma.

To estimate the effect of the toroidal drift let us consider the charge balance on some toroidal surface of the plasma with a radius r (Fig. 9.10). The charge separation rate due to the toroidal drift is defined by Eq. 9.177, as before. Substituting it into the continuity equation, we obtain the change in charge density due to the drift:

$$\left(\frac{\partial \rho}{\partial t}\right)_d = -\text{div}(en\mathbf{u}^H) = -\frac{\partial}{\partial y}(enu_y^H) = \frac{2c(T_i + T_e)}{HR}\frac{\partial n}{\partial y} \tag{9.185}$$

where the same local coordinate system as in the preceding problem is used, and the quantities T_i, T_e, and R are assumed to be constant throughout the plasma volume, as before.

The electric field induced by the charge separation leads to drift compensation. But, in contrast to a weakly ionized plasma, in a fully ionized plasma the field does not produce a transverse current. In the magnetic trap it induces a current along the helical lines of force (for instance, along the line 1–2 in Fig. 9.10). This current is determined by the longitudinal plasma conductivity (Eq. 7.153):

$$j = \sigma_{\|}E' \approx \frac{2ne^2}{m_e\bar{\nu}_{ei}}E' \tag{9.186}$$

where E' is the projection of the field onto the direction of the line of force:

$$E' = E_\varphi \sin\alpha = E \sin\varphi \sin\alpha \tag{9.187}$$

E_φ is the projection on the direction of φ; and $\sin\alpha = H_\varphi/H$ (see Fig. 9.10). The change in volume charge density due to the longitudinal current is determined by the current divergence. Assuming that the plasma parameters are constant on the magnetic surface (because of longitudinal mixing), we get

$$\left(\frac{\partial\rho}{\partial t}\right)_j = -\operatorname{div}\mathbf{j} = -\frac{1}{r}\frac{\partial j_\varphi}{\partial\varphi} = -\frac{1}{r}\frac{\partial}{\partial\varphi}(j\sin\alpha) \tag{9.188}$$

and , further, substituting Eq. 9.186 and Eq. 9.187:

$$\left(\frac{\partial\rho}{\partial t}\right)_j = -\frac{2ne^2}{m_e\bar{\nu}_{ei}r}\left(\frac{H_\varphi}{H}\right)^2 E\cos\varphi \tag{9.189}$$

In the stationary state the increase in volume charge due to the drift must be offset by its decrease caused by the current $(\partial\rho/\partial t)_d = -(\partial\rho/\partial t)_j$. We then obtain the field strength:

$$E = \frac{cm_e(T_e+T_i)\bar{\nu}_{ei}}{e^2H}\frac{r}{R}\left(\frac{H}{H_\varphi}\right)^2\frac{1}{n}\frac{\partial n}{\partial r} \tag{9.190}$$

[this takes into account that $(1/\cos\varphi)\partial n/\partial y = \partial n/\partial r$]. This field is directed along the $0y$ axis and induces a drift in the $0x$ direction (in the direction of the major radius). On one half of the surface being considered (on the outer side of the torus) the drift occurs toward increasing r, and on the other toward decreasing r. In the magnetic trap the toroidal drift does not, however, cause a substantial asymmetry in density distribution, since the longitudinal motion of the charged particles results in their intermixing within the magnetic surface. Therefore it is necessary to average the radial drift fluxes over the entire surface. The compensation of the drift on the different halves of the surface will be incomplete, since the outer surface is larger in size. The summary radial flux across the surface under consideration is obviously equal to

$$\Pi = \int_{(S)} nu_r\,dS = \int_0^{2\pi} n\frac{cE\sin\varphi}{H}2\pi(R - r\sin\varphi)r\,d\varphi \tag{9.191}$$

where $dS = 2\pi(R+\sin\varphi)r\,d\varphi$ is a torus surface element. Integration leads to the following equation for the average density of the radial flux:

$$\Gamma_r = \frac{\Pi}{2\pi R\cdot 2\pi r} = -n\frac{cE}{H}\frac{r}{R} \tag{9.192}$$

When integrating we included the inhomogeneity of the magnetic field:

$$H = \frac{H_0R_0}{R} = \frac{H_0R_0}{R_0 - r\sin\varphi}$$

Substituting Eq. 9.190 into Eq. 9.192, we have

$$\Gamma_r = -\frac{(T_e + T_i)\bar{\nu}_{ei}}{m_e\omega_{He}^2} q^2 \frac{\partial n}{\partial r} \tag{9.193}$$

where $q = (r/R)H/H_\varphi$.* This flux is directed toward decreasing density and has the same dependence on the plasma parameters as an ordinary diffusion flux with the diffusion coefficient 9.129; it differs only by the factor q^2. The summary flux can be characterized by the effective diffusion coefficient:

$$D_\perp = \frac{(T_e + T_i)\bar{\nu}_{ei}}{m_e\omega_{He}^2}(1 + q^2)\,\dagger \tag{9.194}$$

Since usually $q \gg 1$ (see Section 10.6), this coefficient greatly exceeds the coefficient of transverse diffusion across a homogeneous magnetic field.

The other transport coefficients can be derived in a similar way, with the toroidal drift taken into account. The thermal diffusion coefficient is then equal to the diffusion coefficient. The effective coefficient of ion thermal conductivity, which characterizes the transverse heat transfer, is equal to

$$\mathscr{K}_{i\perp} = \frac{2nT_i\bar{\nu}_{ii}}{m_i\omega_{Hi}^2}(1 + 1.6q^2) \tag{9.195}$$

As can be seen, the toroidal term, which determines the effect of the drift, exceeds the coefficient of thermal conductivity in a homogeneous field by a factor of $1.6q^2$.

At a low frequency of electron and ion collisions one more effect appears, which speeds up the transport processes. It is associated with the shape of the trajectory of the Larmor centers of the particles in a toroidal magnetic trap. As shown in Section 8.5, these trajectories depend on the ratio between the longitudinal and transverse velocity components. The particles for which this ratio is not too low are transit particles, and the projections of their trajectories in the toroidal cross section are nearly circular. The trapped particles, for which this ratio is low, have trajectories of the "banana" type. Changes in such trajectories on collisions may lead to transverse displacements of charged particles which greatly exceed their displacement in a homogeneous magnetic field, and hence to an increase in transport coefficients.

Collisions of trapped particles exert an especially strong effect on

*The coefficient q is sometimes called *the stability margin*, because it represents the plasma stability with respect to configuration disturbances (see Section 10.6).

†This expression is sometimes called the Pfirsch–Schlüter equation.

transport. Let us estimate their effect on the electron diffusion, for instance. The acceleration of diffusion due to collisions of trapped electrons can be estimated with the aid of Fig. 8.20. Suppose an electron has experienced a collision at point 1 such that it finds itself among the trapped particles. Moving along the banana trajectory, it may undergo another collision at point 2. As a rule, this collision makes it a transit particle, because the probability that the longitudinal velocity component will remain small on collision is low. Thus within the time between two collisions the particle may shift along the radius by the width of the banana trajectory. In accordance with the estimate of Section 8.5 this width is equal in its order of magnitude to

$$\Delta r_b = \frac{2u_e^H}{\bar{\nu}_{e\perp}} \frac{H}{H_\varphi} \sqrt{rR} \approx \frac{c\sqrt{m_e T_e}}{eH} \frac{H}{H_\varphi} \sqrt{\frac{r}{R}} \tag{9.196}$$

The effective diffusion coefficient is measured in terms of the mean square of the displacement, and for the mechanism under review can be estimated as follows:

$$D_{e\perp} \approx \eta \nu_e^b \langle (\Delta r_b)^2 \rangle = \eta \nu_e^b \frac{T_e}{m_e \omega_{He}^2} \frac{r}{R} \left(\frac{H}{H_\varphi}\right)^2 \tag{9.197}$$

where η is the fraction of trapped particles, and ν_e^b is the frequency of collisions of the trapped particles that change them to transit particles.

The fraction of trapped particles is found from the "mirror" ratio during the motion along a helical line of force $H_{\min}/H_{\max} = 1 - 2r/R$. This ratio yields the limiting angle corresponding to the change of trapped particles to transit ones:

$$\begin{gathered} \sin^2 \vartheta_0 = \frac{H_{\min}}{H_{\max}} = 1 - \frac{2r}{R}; \\ \frac{\pi}{2} - \vartheta_0 = \sqrt{2r/R} = \delta\vartheta \end{gathered} \tag{9.198}$$

For an isotropic velocity distribution the fraction of trapped particles can be expressed in terms of this angle:

$$\eta \approx \delta\vartheta \approx \sqrt{r/R} \tag{9.199}$$

In determining ν_e^b one must bear in mind that for particles to change from trapped to transit it is sufficient to change the direction of their velocity by a small angle $\delta\vartheta \approx \sqrt{r/R}$. The time of rotation of the velocity by this angle as a result of coulomb collisions leading largely to small deviations is much less than the average time between collisions of electrons with ions and with one another: $\tau_e = 1/(\bar{\nu}_{ee} + \bar{\nu}_{ei})$, which is defined as the time necessary for a substantial change in velocity. Considering that the effect of coulomb collisions is equivalent to diffusion in velocity space (see Section 3.3), we can obtain

$$\nu_e^b \approx \frac{\bar{\nu}_{ei} + \bar{\nu}_{ee}}{(\delta\vartheta)^2} \approx \bar{\nu}_{ei}\,\frac{R}{r} \tag{9.200}$$

Substituting the values of η and ν_e^b into Eq. 9.197, we get the following estimate to the diffusion coefficient:

$$D_{e\perp} \approx \frac{\bar{\nu}_{ei} T_e}{m_e \omega_{He}^2}\left(\frac{H}{H_\varphi}\right)^2 \sqrt{\frac{r}{R}} = \frac{\bar{\nu}_{ei} T_e}{m_e \omega_{He}^2}\, q^2 \left(\frac{R}{r}\right)^{3/2} \tag{9.201}$$

A more accurate calculation leads to an equation that differs from this one only by a factor of 1.07. As can be seen from a comparison of Eqs. 9.201 with 9.194 and 9.129 trapped particles increase considerably the diffusion coefficient, which exceeds $q^2(R/r)^{3/2}$ times the diffusion coefficient for a homogeneous field. Other transport coefficients increase similarly. The ion diffusion coefficient obtained with due allowance for ion–ion collisions may differ slightly from the electron coefficient. In the ambipolar regime, however, the diffusion coefficient practically coincides with Eqs. 9.201. The effective coefficient of transverse ion thermal conductivity with an allowance for the trapped particles is equal to*

$$\mathcal{K}_{i\perp} = 0.80\,\frac{n\bar{\nu}_{ii} T_i}{m_i \omega_{Hi}^2}\, q^2 \left(\frac{R}{r}\right)^{3/2} \tag{9.202}$$

and differs from the coefficient of transverse thermal conductivity in a homogeneous field (Eq. 9.169) by a factor of $0.4q^2(R/r)^{3/2}$.

Equations 9.196–9.202 were obtained under the assumption that the collision frequencies of charged particles are rather low. To apply them, it is necessary that the time the particles remain in the trapped group exceed the time of its traversal of the banana trajectory. The length of this trajectory is of the order of $L_b \approx rH/H_\varphi$; therefore the criterion amounts to the inequality $\nu^b < v_\parallel/L_b$, or, since for trapped particles $v_\parallel \leqslant v\sqrt{r/R}$, $\nu^b \approx \nu e/r$, we find

$$\lambda_\alpha > r\left(\frac{R}{r}\right)^{3/2} \frac{H}{H_\varphi} = qR\left(\frac{R}{r}\right)^{3/2} \tag{9.203}$$

where λ_α is the mean free path of electrons or ions $\lambda_e = \bar{v}_e/\bar{\nu}_{ei}$ and $\lambda_i = \bar{v}_i/\nu_{ii}$. With increasing collision frequency, when the condition 9.203

*Transfer coefficients due to trapped particles are sometimes called *Galeyev–Sagdeyev coefficients*.

is violated, the particles traverse only part of the banana trajectory while remaining in the trapped group. Accordingly, their displacement on collisions decreases. Thus despite the increase in collision frequency within a certain interval, the coefficients of diffusion and thermal conductivity remain approximately constant. This interval is determined by the condition

$$qR < \lambda_\alpha < qR\left(\frac{R}{r}\right)^{3/2} \tag{9.204}$$

At higher collision frequencies, when the mean free path exceeds the scope of the region 9.204, collisions reduce the mean free path to such an extent that the maximum displacement of the trajectory on collision is of the order of the Larmor radius. In this region the effective transfer coefficients are determined by collisions and toroidal drift in accordance with Eqs. 9.194 and 9.195.

The dependence of the diffusion coefficient in a toroidal magnetic trap on the collision frequency is illustrated in Fig. 9.11. The figure also shows the dependence of the diffusion coefficient $D_{\perp 0}$ on ν in a homogeneous field. In the graph, the frequency ν_1 is determined by the time the particle traverses the banana trajectory, and the frequency $\nu_2 = (R/r)^{3/2}\nu_1$. As noted above, at $\nu < \nu_1$ the principal contribution to the transport is made by displacements caused by trapped particle collisions; this region is called the *banana region.* In the region of $\nu > \nu_2$ the main effect on the transport is exerted by the toroidal drift. This is sometimes called the *hydrodynamic region,* because the diffusion in it can be described by averaged motion equations of the hydrodynamic type. The intermediate region $\nu_1 < \nu < \nu_2$, where the diffusion coefficient changes only slightly, is called the *plateau region.*

We note in conclusion that the above treatment of transport processes in a fully ionized plasma in a toroidal magnetic trap is based on idealized concepts. Actually an appreciable effect on the transport may be exerted

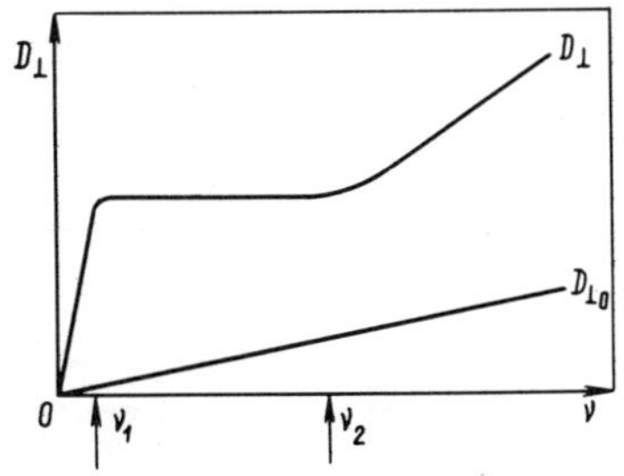

Fig. 9.11 Dependence of the diffusion coefficient in a toroidal magnetic trap on the collision frequency.

by factors, unaccounted for, associated with additional electric fields and with temperature perturbations caused by the diffusion of charged particles.

9.8 DRIFT INSTABILITIES AND ANOMALOUS DIFFUSION OF CHARGED PLASMA PARTICLES IN MAGNETIC FIELD

A plasma in a magnetic field is often unstable because of excitation of different kinds of oscillation. At sufficiently large amplitudes oscillations may change the plasma properties considerably. In particular, they increase the efficiency of transport of charged particles and their energy across the magnetic field. A detailed discussion of plasma instabilities and their effect on the transport processes is beyond the scope of this book. We consider here, by way of example, only the excitation of *drift waves*, whose source is the pressure gradient in a direction perpendicular to the magnetic field. Such waves usually propagate perpendicularly to the gradient and almost perpendicularly to the field; their phase velocity is close to the velocity of the diamagnetic drift (hence their name). Drift oscillations occupy a special place among the others since they can be excited when the only cause for plasma nonequilibrium is its inhomogeneity. Therefore the drift instability is sometimes called *universal.*

Let us first discuss qualitatively the development of drift oscillations under the effect of a density gradient perpendicular to the magnetic field (Fig. 9.12). We direct the z axis along the magnetic field and the x axis, along the unperturbed density gradient. The periodic density perturbation leading to a drift wave depends on the longitudinal and transverse coordinates perpendicular to the density gradient. We represent it in the form*

$$n^{(1)} = n_1 \cos(k_y y + k_z z) = \mathrm{Re}[n_1 \exp i(k_y y + k_z z)] \qquad (9.205)$$

and assume the perturbation amplitude to be small, $n_1 \ll n$.

With such a perturbation the more mobile electrons escape from the region of increased density along the magnetic field faster than the ions. As a result, excess ions remain in the blobs, while excess electrons collect in rarefied regions, producing a periodic potential field. At a small or moderate longitudinal wavelength the relationship between the density perturbations and the potential is given by the Boltzmann equation:

*As before, we use the complex form of writing harmonically varying quantities, implying their real part.

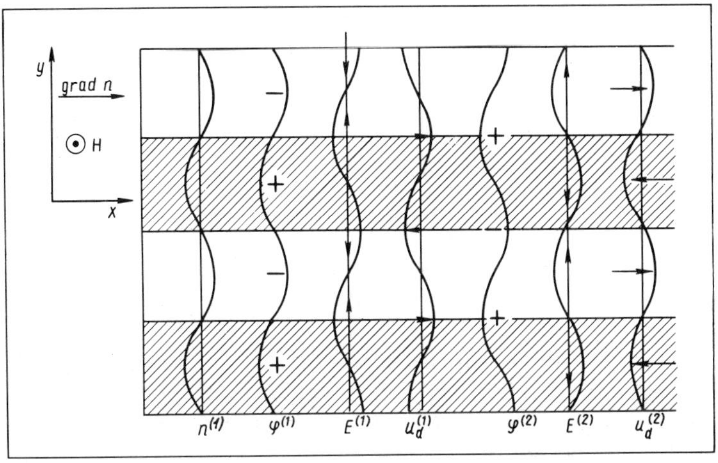

Fig. 9.12 Development of drift oscillations under the effect of a density gradient perpendicular to the magnetic field.

$$\frac{n^{(1)}}{n} = \exp\left(\frac{e\varphi^{(1)}}{T_e}\right) - 1 \approx \frac{e\varphi^{(1)}}{T_e} \tag{9.206}$$

(we assume that the perturbation is small and $e\varphi^{(1)} \ll T_e$). This potential gives the strength of the electric field directed along the y axis:

$$E_y^{(1)} = -\frac{\partial \varphi^{(1)}}{\partial y} = -\frac{ik_y T_e}{e}\frac{n^{(1)}}{n} \tag{9.207}$$

The electric field causes a drift in the direction of the x axis with a velocity

$$u_d^{(1)} = \frac{cE^{(1)}}{H} = -ick_y\left(\frac{T_e}{eH}\right)\left(\frac{n^{(1)}}{n}\right) \tag{9.208}$$

The drift velocity $u_d^{(1)}$ is shifted in phase by $\pi/2$ relative to the density perturbations (see Fig. 9.12). Therefore, as a result of the drift, one half of the blob receives charged particles from the region with a higher density, and the other half, from the region with a lower density. This produces a wave propagating across $\mathbf{u}_d^{(1)}$, that is, along the y axis.

The density change rate at each point associated with the drift is proportional to the unperturbed gradient and to $\mathbf{u}_d^{(1)}$:

$$\frac{\partial n^{(1)}}{\partial t} = -u_d^{(1)}\frac{dn}{dx} = \frac{ick_y T_e}{eH}\frac{n^{(1)}}{n}\frac{dn}{dx} = i\omega_d n^{(1)} \tag{9.209}$$

From Eq. 9.209 it follows that the perturbation $n^{(1)}$ and, correspondingly, the other associated quantities, fluctuate in time with a frequency

$$\omega_d = \frac{ck_y T_e}{eHn} \frac{dn}{dx} \tag{9.210}$$

which is called the *drift frequency*. The phase velocity of perturbation propagation in the direction of the y axis,

$$u_{py} = \frac{\omega_d}{k_y} = \frac{cT_e}{neH} \frac{dn}{dx} \tag{9.211}$$

is equal to the diamagnetic electron velocity related to the density gradient.

Let us now consider that in the drift wave the particles do not move together with the perturbation. The velocity of their drift $\mathbf{u}_d^{(1)}$ is perpendicular to the phase velocity $\mathbf{u}_p$. Therefore for all the particles the field of the moving wave is time variable. The variable field causes an inertial drift, whose velocity is proportional to the field derivative $dE^{(1)}/dt$ and to the particle mass. This drift is actually the cause of instability under certain conditions. With due regard to the concentration gradient the inertial drift of ions leads to an additional change in potential $\varphi^{(2)}$, which is shifted in phase by $\pi/2$ relative to $\varphi^{(1)}$ (see Fig. 9.12). The field $\mathbf{E}^{(2)}$, determined by it, induces an additional drift of charged particles in the x direction, whose velocity $u_d^{(2)}$ changes in phase with the density perturbation. In the blobs, the drift is directed against the density gradient, and in rarefied regions, along the gradient. As a result the charged particle density increases in the blobs and decreases in the rarefied regions. If dissipative processes do not offset this change, the amplitude of the drift wave increases with time; that is, the plasma is unstable to the wave excitation.

In the case just discussed the cause of instability is the inertial drift of ions. Under different conditions other factors may also be essential, such as the nonuniformity of the electron temperature, the disturbance of quasi-neutrality, and a longitudinal electric current.

Let us analyze quantitatively the drift instability when the longitudinal motion of the charged particles is limited by their collisions with neutrals, that is, when their mean free path is less than the longitudinal oscillation wavelength (here the instability is sometimes called the *drift-dissipative instability*). Under such conditions, consideration may be based on averaged transport equations. We assume, as before, that the source of instability is the charged particle density gradient. The electron temperature is assumed to be constant ($T_e = \text{const}$), and the ion temperature much lower than that of the electrons ($T_i \ll T_e$); the ambipolar electric field $E_A \sim T_i/L$ can be neglected. When analyzing the stability we represent the density and directed velocity of the charged particles as a sum of equilibrium values and small perturbations:

$$n_e = n_i = n^{(0)} + n^{(1)}(t, \mathbf{r}); \qquad \mathbf{u}_\alpha = \mathbf{u}_\alpha^{(0)} + \mathbf{u}_\alpha^{(1)}(t, \mathbf{r}) \tag{9.212}$$

As before, we assume the equilibrium density $n^{(0)}$ to depend on the x coordinate perpendicular to the magnetic field (it is supposedly directed along the z axis):

$$n^{(0)} = n^{(0)}(x); \qquad \frac{dn^{(0)}}{dx} = \kappa n^{(0)} \tag{9.213}$$

The perturbed values are sought in the form of waves propagating perpendicularly to the unperturbed density gradient:

$$\begin{aligned} n^{(1)}(t, \mathbf{r}) &= \mathrm{Re}\{n_1 \exp[\mathrm{i}(k_y y + k_z z - \omega t)]\}; \\ \mathbf{u}^{(1)}(t, \mathbf{r}) &= \mathrm{Re}\{\mathbf{u}_1 \exp[\mathrm{i}(k_y y + k_z z - \omega t)]\} \end{aligned} \tag{9.214}$$

and so on, where $\mathbf{k}$ is a wave vector with one component parallel to the magnetic field k_z and the other perpendicular to the field and the gradient k_y. The difference in directed electron and ion velocities results in charge separation and the appearance of an electric field. Since drift oscillations are low-frequency, the electric field accompanying them can be considered potential. Therefore, assuming that the potential $\varphi^{(1)}$ varies in space in the same way as the perturbations $n^{(1)}$ and $u^{(1)}$, we have

$$\mathbf{E}^{(1)} = -\mathrm{grad}\, \varphi^{(1)}; \qquad E_y^{(1)} = -\mathrm{i}k_y \varphi^{(1)}; \qquad E_z^{(1)} = -\mathrm{i}k_z \varphi^{(1)} \tag{9.215}$$

Using the expressions obtained in the preceding sections, we can find the charged particle velocity. For the electrons, the longitudinal component of the directed velocity is equal to

$$u_{ez} = -\frac{T_e}{m_e \nu_{ea}} \frac{1}{n} \frac{\partial n}{\partial z} - \frac{e}{m_e \nu_{ea}} E_z \tag{9.216}$$

Hence we obtain

$$u_{ez}^{(0)} = 0; \qquad u_{ez}^{(1)} = -\frac{\mathrm{i}k_z T_e}{m_e \nu_{ea}} \left(\frac{n^{(1)}}{n_0} - \frac{e\varphi^{(1)}}{T_e} \right) \tag{9.217}$$

Assuming that $\omega_{He} \gg \nu_{ea}$, we take into account, in the expression for the transverse component of the directed velocity (Eq. 9.8), only the electron drift related to the density gradient and the electric field. Then

$$u_{e\perp} = \frac{c}{H} \left[\mathbf{h} \times \left(\mathrm{grad}\, \varphi - \frac{T_e}{e} \mathrm{grad}\, n \right) \right] \tag{9.218}$$

Using the dependence of $n^{(0)}$ (Eq. 9.213) and $n^{(1)}$ $(\mathbf{r}, t)$ (Eq. 9.214) on the coordinates, we get

$$u_{ex}^{(0)} = 0; \qquad u_{ey}^{(0)} = -\frac{cT_e}{n^{(0)}eH}\frac{dn^{(0)}}{dx} = -\kappa\frac{cT_e}{eH};$$
$$u_{ex}^{(1)} = \mathrm{i}k_y\frac{c}{H}\varphi^{(1)} - \mathrm{i}k_y\frac{cT_e}{eH}\frac{n^{(1)}}{n^{(0)}} \tag{9.219}$$

These relations make it possible to find the change of the perturbation in the density of the electrons due to their directed motion:

$$\frac{\partial n^{(1)}}{\partial t} = -\mathrm{div}(n^{(1)}\mathbf{u}_e^{(0)}) - \mathrm{div}(n^{(0)}\mathbf{u}_e^{(1)}) \tag{9.220}$$

where we neglected the quadratic terms $\mathrm{div}(n^{(1)}\mathbf{u}^{(1)})$, assuming the perturbations $n^{(1)}$ and $\mathbf{u}^{(1)}$ to be small. Substituting the formulas for $u_{ex}^{(1)}$ and $u_{ez}^{(1)}$ into the last equation and allowing for Eq. 9.214, we obtain, after simple rearrangements,

$$\mathrm{i}\omega n^{(1)} = \frac{k_z^2 n^{(0)}T_e}{m_e\nu_{ea}}\left(\frac{n^{(1)}}{n^{(0)}} - \frac{e\varphi^{(1)}}{T_e}\right) + \mathrm{i}k_y\kappa\frac{c}{H}n^{(0)}\varphi^{(1)} \tag{9.221}$$

For the ions, the longitudinal velocity component can be neglected because of their large mass. The transverse component of the directed ion velocity is found by assuming $\omega_{Hi} \gg \nu_i$ and $T_i = 0$. In this case the ion drift is determined by the electric field. But, besides the main drift, we must consider the inertial drift of the ions in conformity with the foregoing. Taking it into account, we obtain, in accordance with Eqs. 8.27 and 8.33,

$$\mathbf{u}_{i\perp} = \frac{c}{H}[\mathbf{E}_\perp \times \mathbf{h}] + \frac{m_i c^2}{eH^2}\frac{\partial \mathbf{E}_\perp}{\partial t}$$
$$= \frac{c}{H}[\mathbf{h}\times \mathrm{grad}_\perp\,\varphi] - \frac{m_i c^2}{eH^2}\frac{\partial}{\partial t}(\mathrm{grad}_\perp\,\varphi) \tag{9.222}$$

Using Eq. 9.215, we find the components of the directed ion velocity:

$$u_{i\perp}^{(0)} = 0; \qquad u_{ix}^{(1)} = \mathrm{i}ky\frac{c}{H}\varphi^{(1)}; \qquad u_{iy}^{(1)} = -k_y\frac{c}{H}\frac{\omega}{\omega_{Hi}}\varphi^{(1)} \tag{9.223}$$

With their aid we can establish the change in ion density perturbation:

$$\frac{\partial n^{(1)}}{\partial t} = -\mathrm{div}(n^{(0)}\mathbf{u}_i^{(1)}) = -u_{ix}^{(1)}\kappa n^{(0)} - \mathrm{i}k_y n^{(0)}u_{iy}^{(1)} \tag{9.224}$$

[Here, as before, we neglect the quadratic terms $\mathrm{div}(n^{(1)}\mathbf{u}^{(1)})$]. Using Eq.

9.223 for $u_{ix}^{(1)}$ and $u_{iy}^{(1)}$ and taking into account the dependence of the perturabtions on the time and coordinates, we get

$$\mathrm{i}\omega n^{(1)} = \mathrm{i}k_y\kappa \frac{c}{H} n^{(0)}\varphi^{(1)} - \mathrm{i}k_y^2 \frac{\omega}{\omega_{Hi}} \frac{c}{H} n^{(0)}\varphi^{(1)} \tag{9.225}$$

The equations for the change in the density of electrons (Eq. 9.221) and ions (Eq. 9.225) can be rewritten:

$$\begin{gathered}\left(\omega + \mathrm{i}\frac{k_z^2 T_e}{m_e \nu_{ea}}\right)\frac{n^{(1)}}{n^{(0)}} = \left(\omega_d + \mathrm{i}\frac{k_z^2 T_e}{m_e \nu_{ea}}\right)\frac{e\varphi^{(1)}}{T_e};\\ \frac{n^{(1)}}{n^{(0)}} = \left(\frac{\omega_d}{\omega} - \frac{k_y^2 T_e}{m_i \omega_{Hi}^2}\right)\frac{e\varphi^{(1)}}{T_e}\end{gathered} \tag{9.226}$$

where $\omega_d = k_y\kappa c T_e/eH$ is the drift frequency (Eq. 9.210). Thus we obtain a set of linear equations for determining $n^{(1)}$ and $\varphi^{(1)}$. A nonzero solution of the set is possible if the determinant reduces to zero. This condition yields the so-called *dispersion equation*:

$$\omega^2 + \mathrm{i}\omega\left(\frac{k_z^2}{k_y^2}\frac{\omega_{He}\omega_{Hi}}{\nu_{ea}} + \frac{k_z^2 T_e}{m_e \nu_{ea}}\right) - \mathrm{i}\omega_d \frac{k_z^2}{k_y^2}\frac{\omega_{He}\omega_{Hi}}{\nu_{ea}} = 0 \tag{9.227}$$

In a strong magnetic field, where

$$\frac{k_y^2 T_e}{m_e \omega_{He}\omega_{Hi}} = \frac{k_y^2 T_e}{m_i \omega_{Hi}^2} \ll 1 \tag{9.228}$$

the second term in the parentheses can be neglected. Then the dispersion equation acquires the simple form:

$$\omega^2 + \mathrm{i}\omega\omega_s - \mathrm{i}\omega_d\omega_s = 0 \tag{9.229}$$

where $\omega_s = (k_z^2/k_y^2)\omega_{He}\omega_{Hi}/\nu_{ea}$. The solutions of the equation in the limiting cases are given by the following equalities:

$$\omega_1 = \omega_d + \frac{\mathrm{i}\omega_d^2}{\omega_s}; \qquad \omega_2 = -\mathrm{i}\omega_s \qquad \text{for } \omega_s \gg \omega_d \tag{9.230}$$

$$\omega_{1,2} = \pm(1+\mathrm{i})\sqrt{\omega_s\omega_d/2} \qquad \text{for } \omega_s \ll \omega_d \tag{9.231}$$

They define the complex frequency $\omega = \Omega + i\gamma$.

It is easy to see that of the two solutions the unstable one is that in which the imaginary part is positive ($\gamma > 0$). For this solution, the time variation in perturbation is characterized by the factor $\exp(-\mathrm{i}\omega t) = \exp(\gamma t)\exp(-\mathrm{i}\Omega t)$; that is, the perturbation increases exponentially. Here $\gamma = \mathrm{Im}(\omega)$ is the increment, which characterizes the rate of increase in perturbation. For unstable oscillations, in accordance with Eqs. 9.230 and 9.231,

$$\gamma = \frac{\omega_d^2}{\omega_s^2} \quad \text{for } \omega_s \gg \omega_d$$
$$\gamma = \sqrt{\frac{\omega_s \omega_d}{2}} \quad \text{for } \omega_s \ll \omega_d \tag{9.232}$$

It can be seen that with a change in $\omega_s = (k_z^2/k_y^2)\ \omega_{He}\omega_{Hi}/\nu_{ea}$ the increment passes a maximum. This maximum is achieved at $\omega_s \approx \omega_d$ and is, in its order of magnitude, $\gamma_{max} \approx \omega_d$.

It should be noted that we neglected the stabilizing effect of ion collisions. It can be shown that the stabilization associated with ion–atom collisions is achieved when the mean free path of the ions is of the order of, or less than, the characteristic transverse dimensions of the plasma. For a highly ionized plasma it is necessary to take into account the stabilization due to ion collisions as well. In an inhomogeneous magnetic field other mechanisms of oscillation stabilization are also essential; in particular, oscillations with a small k_z are stabilized by the shear.

The above discussion is a linear analysis of an instability in which the perturbations of the initial equilibrium state are assumed to be small. Such an analysis makes it possible to establish the conditions of instability, to find the frequency and configuration of the increasing oscillations as related to the plasma parameters and the nature of the perturbation. Obviously to establish the consequences of instability a nonlinear theory must be constructed covering the possible mechanisms of its limitation. A number of approaches to such a nonlinear theory are used, but it is still far from completion. Description of the contemporary concepts of the nonlinear stage in the development of an instability is beyond the scope of this book. We only note that an instability may develop by different paths, depending on the type of nonlinear dissipation. In one of the limiting cases the oscillation mode that corresponds to the optimum values of the wave vector components remains predominant, a near-periodic nonlinear wave being formed. In the other limiting case the instability results in the excitation of a broad spectrum of interacting oscillations with a wide range of values of the wave vector components. Under such conditions the plasma is usually called *turbulent*, since the interaction of the wave packets in it is similar to some extent to the interaction of eddies in a turbulent fluid. In both cases the oscillations give rise to particle and energy conduction across the magnetic field, which are not associated with collisions. These are called, respectively, the *anomalous*, (or *turbulent*) *diffusion and the thermal conduction.*

We now give a crude estimate of the coefficient of anomalous diffusion of charged particles during the development of the drift instability described above. Since the drift instability is caused by the density gradient, it naturally tends to smooth the density distribution, that is, conduction directed against the gradient. The density of the particle flux parallel to the density gradient is equal to the oscillation-averaged product of the density and the directed velocity of the particles, which is, for drift oscillations, practically equal to the velocity of their drift in the electric field.

As before, we represent the density and velocity as a sum of the unperturbed and perturbed quantities: $n = n^{(0)} + n^{(1)}$ and $\mathbf{u} = \mathbf{u}^{(0)} + \mathbf{u}^{(1)}$. Bearing in mind that $\langle n^{(1)} \rangle = 0$ and $\langle \mathbf{u}^{(1)} \rangle = 0$, we obtain

$$\Gamma_x^{(1)} = \langle n^{(1)} u_x^{(1)} \rangle = \frac{c}{H} \langle n^{(1)} E_y^{(1)} \rangle \tag{9.233}$$

where we assume, as before, that the unperturbed density gradient is directed along the x axis. In a harmonically varying field of a stationary drift wave the field $E_y^{(1)}$ is shifted in phase by 90° relative to the perturbed density $n^{(1)}$. Accordingly, the averaged current is zero for the following simple reason. A drift in a potential electric field leads to an incompressible motion at which $\operatorname{div} \mathbf{u} = \operatorname{div} c/H [\mathbf{h} \times \operatorname{grad} \varphi] = 0$. The transport of charged particles in such a motion is convective; it is related to the displacement of particles in the direction of the density gradient. With a periodically varying field the drift velocity changes sign periodically, and there is naturally no averaged current. But if we take into account the increase in wave amplitude with time, each successive period results in a larger displacement of the charged particles than the preceding one, and an averaged current appears. This effect is easy to estimate. The change in the density of the particles due to their incompressible motion is equal to

$$n^{(1)} = -\frac{\xi_x \, dn^{(0)}}{dx} = -\xi_x \kappa n^{(0)} \tag{9.234}$$

where $\dot{\xi}_x = u_x^{(1)}$. Taking this into consideration, we obtain

$$\Gamma_x^{(1)} = \langle n^{(1)} u_x^{(1)} \rangle = -\frac{\langle \xi_x \dot{\xi}_x \rangle \, dn^{(0)}}{dx} = -\frac{\gamma \langle \xi_x^2 \rangle \, dn^{(0)}}{dx} \tag{9.235}$$

where γ is the instability increment, and $\langle \xi_x^2 \rangle$ is the mean square displacement of the particles. Note that in an increasing drift wave the difference of the phase shift betwen $n^{(1)}$ and $u_x^{(1)} \sim E_y^{(1)}$ from 90° is due to the charge separation resulting from the inertial drift of the ions. (As mentioned above, it is precisely this separation that leads to an increase in oscillations.) Equation 9.235 enables us to introduce a diffusion coefficient describing the current against the gradient of the unperturbed density:

$$D_\perp = \gamma\langle\xi_x^2\rangle = \frac{\gamma}{\kappa^2}\left\langle\left(\frac{n^{(1)}}{n^0}\right)^2\right\rangle \tag{9.236}$$

The coefficient is seen to be proportional to the averaged square of the relative density perturbation. It is usually believed that the estimate obtained is also valid for large oscillation amplitudes, and even for the turbulent state. In the latter case we have a dynamic balance of exchanged energies between different modes of oscillation. Then the quantity $\gamma\langle(n^{(1)})^2\rangle$ appearing in Eq. 9.236 must be summed over the modes corresponding to the different values of the wave vector components.

With a developed turbulence, the oscillation amplitude of each mode may increase to such dimensions that the density gradient perturbation will be of the order of the initial gradient. For oscillations with a given value of k_y this yields $k_y n^{(1)} \sim \kappa n^{(0)}$. The transverse diffusion coefficient corresponding to such oscillations is equal to

$$D_\perp = \sum_k \frac{\gamma_k}{\kappa^2}\left\langle\left(\frac{n_k^{(1)}}{n^{(0)}}\right)^2\right\rangle \sim \sum_k \frac{\gamma_k}{k_y^2} \tag{9.237}$$

Among the oscillations with different values of k_z the greatest contribution to this sum is made by oscillations at which the maximum values of the increment $\gamma_k \approx \omega_d \approx k_y\kappa cT_e/eH$ are achieved. The least possible k_y must be of the order of the reciprocal characteristic dimension, that is, of the order of κ. Bearing this in mind, we obtain the estimate of the diffusion coefficient at the maximum oscillation amplitude

$$D_{\perp\max} = \frac{gcT_e}{eH} \tag{9.238}$$

where g is a factor of the order of unity. The coefficient with the numerical factor $g = 1/16$ is called the *Bohm diffusion coefficient* (after the physicist who suggested it for the first time from intuitive considerations). Equation 9.238 is often used for estimating the order of the coefficient of anomalous diffusion and thermal conductivity.

Let us derive the ratio of the Bohm coefficient of anomalous diffusion D_B, caused by oscillations, to the coefficient of transverse diffusion D_ν due to collisions (it is often called, rather inappropriately, *classical* as distinct from anomalous).* Using Eq. 9.95 or 9.157 for the coefficient of ambipolar collision diffusion, we get

$$\frac{D_B}{D_A} = \frac{g\omega_{He}}{\nu_{ea}} \tag{9.239}$$

where ν_{ea} is the frequency of electron–heavy particle collisions.

Note that in a strongly turbulent plasma the drift motion of the charged particles acquires a random character. The charged particles drifting in randomly changing fields undergo displacements whose magnitude and direction change randomly in space and time. Therefore their trajectories become quite similar to those of the particles whose displacements are caused by collisions. With this analogy in mind some physicists introduce effective collision frequencies characterizing anomalous transverse diffusion or thermal conductivity. To find these, they use equations for the transfer coefficients due to collisions and equate them to the anomalous coefficients. For the Bohm diffusion coefficient, for instance, we obtain

$$\frac{T_e \nu_{\text{eff}}}{m_e \omega_{He}^2} = \frac{gcT_e}{eH} \tag{9.240}$$

and $\nu_{\text{eff}} = g\omega_{He}$. At $\omega_{He} \gg \nu_{ea}$ the effective collision frequency, which determines the anomalous diffusion, greatly exceeds the actual electron–heavy particle collision frequency. This means that the anomalous diffusion coefficient greatly exceeds the transverse diffusion coefficient due to collisions.

*Diffusion in a toroidal magnetic field due to collisions of trapped particles is sometimes called *neoclassical*.

10

PLASMA CONFINEMENT BY MAGNETIC FIELD

10.1 MAGNETOHYDRODYNAMIC EQUATIONS

As shown in Chapter 9, a strong magnetic field sharply reduces the transverse coefficients of transport of charged particles. Therefore a fully ionized plasma, containing charged particles only, can long be confined by a magnetic field away from the material walls. This confinement is used for thermal insulation and heating of the plasma in investigations on controlled fusion.

Analysis of plasma confinement by a magnetic field must obviously be based on a simultaneous (self-consistent) solution of the equations describing the behavior of a plasma in a magnetic field and those defining the magnetic field in a plasma. The set of these equations is rather complex, in the general case, and various simplifications are used for its solution. It is often convenient to describe a plasma as a unified, neutral but conducting fluid that can be subjected to the effect of a magnetic field. This description is called *magnetohydrodynamic*. When it is used, short-term and small-scale processes leading to perturbation in quasi-neutrality are not considered. The characteristic times of the processes in question must clearly be longer than the microscopic times—the reciprocal plasma frequency $1/\omega_p$ and the periods of revolution of the particles in the magnetic field $1/\omega_{He}$ and $1/\omega_{Hi}$. The distances thus must greatly exceed the Debye and Larmor radii of the particles.

Let us derive the magnetohydrodynamic equations for a fully ionized two-component plasma. We introduce the mass density of the plasma:

$$\rho = m_i n_i + m_e n_e = (m_i + m_e)n \approx m_i n \tag{10.1}$$

and the center-of-mass velocity (hydrodynamic velocity):

$$\mathbf{u} = \frac{n}{\rho}(m_i\mathbf{u}_i + m_e\mathbf{u}_e) \approx \mathbf{u}_i + \frac{m_e}{m_i}\mathbf{u}_e \tag{10.2}$$

Since for drift motions in a sufficiently strong magnetic field $\mathbf{u}_e \approx \mathbf{u}_i$ and in a longitudinal electric field $u_e/u_i \approx (m_i/m_e)^{1/2}$ it can be approximately assumed that $\mathbf{u} \approx \mathbf{u}_i$.

Multiplying the zero-moment equation for the electrons and ions by m_e and m_i, respectively, and adding them, we obtain the plasma mass conservation equation, which determines the density variation:

$$\frac{\partial \rho}{\partial t} + \operatorname{div} \rho\mathbf{u} = 0 \tag{10.3}$$

The hydrodynamic equation of plasma motion as a whole can be found from the motion equations for the components, which have the following form in accordance with Eq. 6.61:

$$\begin{aligned} nm_e\left[\frac{\partial \mathbf{u}_e}{\partial t} + (\mathbf{u}_e \operatorname{grad})\mathbf{u}_e\right] &= -en\mathbf{E} - \frac{en}{c}[\mathbf{u}_e \times \mathbf{H}] \\ &\quad -\operatorname{grad} p_e + n\mathbf{R}_{ei} + n\mathbf{R}_{ei}^T; \\ nm_i\left[\frac{\partial \mathbf{u}_i}{\partial t} + (\mathbf{u}_i \operatorname{grad})\mathbf{u}_i\right] &= en\mathbf{E} + \frac{en}{c}[\mathbf{u}_i \times \mathbf{H}] \\ &\quad -\operatorname{grad} p_i + n\mathbf{R}_{ie} + n\mathbf{R}_{ie}^T \end{aligned} \tag{10.4}$$

Adding these equations and considering that $\mathbf{R}_{ei} = -\mathbf{R}_{ie}$ and $\mathbf{R}_{ei}^T = -\mathbf{R}_{ie}^T$, we obtain the equation for the average velocity $\mathbf{u}$:

$$\rho\left(\frac{d\mathbf{u}}{dt}\right) = \left(\frac{1}{c}\right)[\mathbf{j} \times \mathbf{H}] - \operatorname{grad} p \tag{10.5}$$

where $p = p_i + p_e$ is the summary pressure, $\mathbf{j} = en(\mathbf{u}_i - \mathbf{u}_e)$ is the current density; and the total derivative $d\mathbf{u}/dt = \partial\mathbf{u}/dt + (\mathbf{u}\operatorname{grad})\mathbf{u}$ is the so-called hydrodynamic acceleration.

The hydrodynamic equation of motion obtained has a clear physical meaning. It describes the acceleration of a unit volume of plasma due to the sum of the forces acting on the unit volume: the electrodynamic force associated with current–magnetic field interaction and the summary-pressure gradient. Since the plasma as a whole is neutral, the electric field does not affect it and does not appear on the right-hand side of Eq. 10.5. For slow processes, when the change in directed velocity, as determined by the left-hand side of the equation of motion (Eq. 10.5), is small, the equation amounts to the condition of equilibrium of the forces acting on the plasma, and it takes the form

$$\frac{1}{c}[\mathbf{j}\times\mathbf{H}] = \operatorname{grad} p \tag{10.6}$$

The equation for the current in the plasma can also be obtained from the equations of motion of the electron and ion components (Eq. 10.4). Multiplying these equations by e/m and subtracting one from the other, we obtain

$$\begin{aligned}\frac{\partial \mathbf{j}}{\partial t} - \frac{\mathbf{j}}{n}\frac{\partial n}{\partial t} &= ne^2\left(\frac{1}{m_e}+\frac{1}{m_i}\right)\mathbf{E} + \frac{ne^2}{c}\left[\left(\frac{\mathbf{u}_i}{m_i}+\frac{\mathbf{u}_e}{m_e}\right)\times\mathbf{H}\right] \\ &\quad - e\left(\frac{\operatorname{grad} p_i}{m_i} - \frac{\operatorname{grad} p_e}{m_e}\right) - ne(\mathbf{R}_{ei}+\mathbf{R}_{ei}^T)\left(\frac{1}{m_e}+\frac{1}{m_i}\right)\end{aligned} \tag{10.7}$$

where the small quadratic terms $(\mathbf{u}_e \operatorname{grad})\mathbf{u}_e$ and $(\mathbf{u}_i \operatorname{grad})\mathbf{u}_i$ are omitted. Neglecting the terms of the order of m_e/m_i and taking into account that $\mathbf{j} = ne(\mathbf{u}_i - \mathbf{u}_e) \approx ne(\mathbf{u} - \mathbf{u}_e)$, we transform Eq. 10.7 to

$$\begin{aligned}\frac{\partial \mathbf{j}}{\partial t} - \frac{\mathbf{j}}{n}\frac{\partial n}{\partial t} &= \frac{ne^2}{m_e}\mathbf{E} + \frac{ne^2}{m_e c}[\mathbf{u}\times\mathbf{H}] - \frac{e}{m_e c}[\mathbf{j}\times\mathbf{H}] \\ &\quad + \frac{e}{m_e}\operatorname{grad} p_e - \frac{ne}{m_e}(\mathbf{R}_{ei}+\mathbf{R}_{ei}^T)\end{aligned} \tag{10.8}$$

The friction and thermal force in a strong magnetic field appearing in this equation can be found from Eqs. 7.146–7.148, 9.118, and 9.119. In accordance with them,

$$\begin{aligned}\mathbf{R}_{ei} &= -0.51 m_e \bar{\nu}_{ei}(\mathbf{u}_{e\|} - \mathbf{u}_{i\|}) - m_e\bar{\nu}_{ei}(\mathbf{u}_{e\perp} - \mathbf{u}_{i\perp}) \\ &= \frac{m_e\bar{\nu}_{ei}}{en}(0.51\mathbf{j}_\| + \mathbf{j}_\perp);\end{aligned} \tag{10.9}$$

$$\mathbf{R}_{ei}^T = -0.71 \operatorname{grad}_\| T_e - \frac{3}{2}\frac{\bar{\nu}_{ei}}{\omega_{He}}[\mathbf{h}\times\operatorname{grad}_\perp T_e]$$

where $\bar{\nu}_{ei}$ is the averaged electron–ion collision frequency (Eq. 7.145). Equation 10.8 is greatly simplified for slow processes in which one can neglect its left-hand side and use Eq. 10.6 yielding the plasma equilibrium condition. Substituting the expression for $\mathbf{R}_{ei}$ into Eq. 10.8 and using Eq. 10.6 to find $[\mathbf{j}\times\mathbf{H}]$, we reduce this equation to

$$\frac{\mathbf{j}_\|}{\sigma_\|} + \frac{\mathbf{j}_\perp}{\sigma_t} = \mathbf{E} + \frac{1}{c}[\mathbf{u}\times\mathbf{H}] - \frac{1}{en}\operatorname{grad} p_i - \frac{1}{e}\mathbf{R}_{ei}^T \tag{10.10}$$

where $\sigma_\| \approx 2ne^2/m_e\bar{\nu}_{ei}$ and $\sigma_t = ne^2/m_e\bar{\nu}_{ei}$ [$\sigma_\|$ is the longitudinal conductivity of the plasma (Eq. 7.153) and σ_i is sometimes called the transverse conductivity]. The quantities $\sigma_\parallel$ and σ_t differ only by a factor of 2. This difference is often neglected, and Eq. 10.10 is written as

$$\mathbf{j} = \sigma \left\{ \mathbf{E} + \frac{1}{c} [\mathbf{u} \times \mathbf{H}] - \frac{1}{en} \operatorname{grad} p_i - \frac{1}{e} \mathbf{R}_{ei}^T \right\} \tag{10.11}$$

Equations 10.10 and 10.11, which yield the relationship between the current and strength of the electric field in the plasma, are called the *generalized Ohm's law.* Note that it includes not only the electric and magnetic fields, but also the parameters of the plasma components—the ion–pressure and electron–temperature gradients. The equation of the generalized Ohm's law (Eq. 10.10) can be simplified for the commonly encountered case, where the electron and ion density and temperature gradients have the same direction. Since the summary-pressure gradient is perpendicular to the magnetic field in accordance with Eq. 10.6, all the partial gradients are also perpendicular to the field. Under these conditions it is convenient to write the longitudinal and transverse Ohm's laws separately. Projecting the vector equation (Eq. 10.10) onto the direction of the magnetic field and taking into consideration that grad p, grad $T_e \perp \mathbf{H}$, we find the usual equation of Ohm's law for the longitudinal component of the current:

$$\mathbf{j}_\parallel = \sigma_\parallel \mathbf{E}_\parallel \tag{10.12}$$

Projecting Eq. 10.10 onto a direction perpendicular to the magnetic field, we get

$$\mathbf{j}_\perp = \sigma_t \left\{ \mathbf{E}_\perp + \frac{1}{c} [\mathbf{u} \times \mathbf{H}] + \frac{3\nu_{ei}}{2e\omega_{He}} [\mathbf{h} \times \operatorname{grad} T_e] - \frac{1}{ne} \operatorname{grad} p_i \right\} \tag{10.13}$$

(here we used Eq. 10.9). Representing the electric field in a reference system moving together with the plasma ($\mathbf{E}_\perp^* = \mathbf{E}_\perp + (1/c)[\mathbf{u} \times \mathbf{H}]$) as a sum of the components parallel and perpendicular to the current $\mathbf{E}_{\perp j}^*$ and $\mathbf{E}_{\perp g}^*$, and taking into account that $\mathbf{j} \perp \operatorname{grad} p$ in accordance with Eq. 10.6, we divide this equation into two expressions, the first of which can be called the transverse Ohm's law:

$$\mathbf{j}_\perp = \sigma_t \left\{ \mathbf{E}_{\perp j}^* + \frac{3\nu_{ei}}{2e\omega_{He}} [\mathbf{h} \times \operatorname{grad} T_e] \right\} \tag{10.14}$$

and the second describes the field component perpendicular to the current (the so-called *Hall field*):

$$\mathbf{E}_{\perp g}^* = \frac{1}{ne} \operatorname{grad} p_i \tag{10.15}$$

This field is necessary for maintaining the quasi-neutrality of the plasma.

The force of interaction of the current with the field $(1/c)[\mathbf{j} \times \mathbf{H}]$ acts on the electron component and produces an electron–gas pressure gradient, which, in a quasi-neutral plasma, gives rise to an ion pressure gradient, the latter being offset by the electric field 10.15. In this case, as seen from Eqs. 10.12 and 10.14, without a temperature gradient Ohm's law takes a form similar to the conventional one. Then the transverse Ohm's law includes the field in a coordinate system moving together with the plasma, and the transverse conductivity differs slightly from the longitudinal one.

Let us now derive the expressions for the magnetic field strength. To this end we use the Maxwell equations:

$$\text{rot}\,\mathbf{H} = \frac{4\pi}{c}\mathbf{j} + \frac{1}{c}\frac{\partial \mathbf{E}}{\partial t};$$
$$\text{rot}\,\mathbf{E} = -\frac{1}{c}\frac{\partial \mathbf{H}}{\partial t} \tag{10.16}$$

Assuming the processes in the plasma to be rather slow, we neglect the displacement current $(1/4\pi)\partial\mathbf{E}/\partial t$, considering it small compared with the conduction current. It is easy to ascertain that such a neglect is admissible when the characteristic times of plasma parameters variation greatly exceed the intercollisional time $\tau \gg 1/\nu_{ei}$. The equation for the magnetic field takes the form

$$\text{curl}\,\mathbf{H} = \frac{4\pi}{c}\mathbf{j} \tag{10.17}$$

We transform it, applying the curl operator to the right- and left-hand sides and taking into consideration that curl (curl $\mathbf{H}$) = grad(div $\mathbf{H}$) $-\Delta\mathbf{H} = -\Delta\mathbf{H}$, since div $\mathbf{H} = 0$. As a result we get

$$\Delta\mathbf{H} + \frac{4\pi}{c}\,\text{curl}\,\mathbf{j} = 0 \tag{10.18}$$

The expression for the current density (Eq. 10.10) must be substituted into this equation. In the general case the equation is extremely cumbersome. It is greatly simplified if we use the approximate form of Ohm's law (Eq. 10.11) for the case where the temperature gradient is negligibly small. Then we may neglect the thermal force and assume the plasma conductivity $\sigma \approx ne^2/m_e\nu_{ei} \sim T_e^{3/2}$ to be constant. As a result Eq. 10.18 transforms to

$$\Delta\mathbf{H} = -\frac{4\pi}{c}\,\sigma\,\text{curl}\left\{\mathbf{E} + \frac{1}{c}[\mathbf{u} \times \mathbf{H}]\right\} \tag{10.19}$$

Substituting curl $\mathbf{E}$ from the second Maxwell equation into it, we obtain the simplified equation for the magnetic field strength:

$$\frac{\partial \mathbf{H}}{\partial t} = \frac{c^2}{4\pi\sigma} \Delta \mathbf{H} + \text{curl}[\mathbf{u} \times \mathbf{H}] \tag{10.20}$$

Note that we eliminated the electric field strength and the current.

Equations 10.5 and 10.6 include the plasma pressure $p = n(T_e + T_i)$. To find it we use the energy balance equations. For the electron and ion components of a fully ionized plasma these equations take the form (see Eqs. 6.78 and 6.79)

$$\begin{aligned} &\frac{\partial}{\partial t}\left(\frac{3}{2} nT_e\right) + \text{div}\left(\frac{3}{2} nT_e \mathbf{u}_e\right) + nT_e \text{ div } \mathbf{u}_e = -\text{div } \mathbf{q}_e - p_{ei} + P_e; \\ &\frac{\partial}{\partial t}\left(\frac{3}{2} nT_i\right) + \text{div}\left(\frac{3}{2} nT_i \mathbf{u}_i\right) + nT_i \text{ div } \mathbf{u}_i = -\text{div } \mathbf{q}_i - P_{ie} + P_i, \end{aligned} \tag{10.21}$$

where the term $P_{ei} = -P_{ie}$ describes the energy exchange between electrons and ions as a result of elastic collisions, and the quantities P_e and P_i represent the energy acquired and lost by electrons and ions due to inelastic collisions and on interaction with external energy sources. Adding the two equations, we get

$$\begin{aligned} &\frac{\partial}{\partial t}\left(\frac{3}{2} p\right) + \text{div}\left(\frac{3}{2} p\mathbf{u}\right) - \text{div}\left(\frac{3}{2}\frac{p_e}{en}\mathbf{j}\right) + p \text{ div } \mathbf{u} - p_e \text{ div} \frac{\mathbf{j}}{en} \\ &\quad = -\text{div } \mathbf{q} + P \end{aligned} \tag{10.22}$$

Here $P = P_e + P_i$ and $\mathbf{q} = \mathbf{q}_e + \mathbf{q}_i$. Since in accordance with Eq. 10.17 $\mathbf{j} = (c/4\pi)\text{curl } \mathbf{H}$ and $\text{div } \mathbf{j} = 0$, the equation takes the form

$$\begin{aligned} &\frac{\partial}{\partial t}\left(\frac{3}{2} p\right) + \frac{3}{2}(\mathbf{u} \text{ grad})p + \frac{5}{2} p \text{ div } \mathbf{u} \\ &\quad = \frac{3}{2}\frac{\mathbf{j}}{e} \text{ grad } T_e - \mathbf{j}\frac{T_e}{en} \text{ grad } n - \text{div } \mathbf{q} + P \end{aligned} \tag{10.23}$$

It is seen that the plasma energy balance equation generally contains not only the summary characteristics, but also the parameters of the electron component. These parameters and the current drop out of the equation only in particular cases; for instance, if the plasma is homogeneous, the current is perpendicular to grad T_e, grad n, or P offsets the other terms on the right-hand side of Eq. 10.23. In the last case all the sources of release and loss of heat are compensated. Using the continuity equation, in accordance with which $\rho \text{ div } \mathbf{u} = -\partial\rho/\partial t - \mathbf{u} \text{ grad } \rho =$

$-d\rho/dt$, we obtain

$$\frac{d}{dt}\left(\frac{3}{2}p\right)-\frac{5}{2}\frac{p}{\rho}\frac{d\rho}{dt}=0$$

or

$$\frac{d}{dt}\left(\frac{p}{\rho^\gamma}\right)=0, \qquad p\sim\rho^\gamma \tag{10.24}$$

where $\gamma=\frac{5}{3}$ is the adiabatic index for a monatomic gas. As would be expected, in the absence of exchange with the environment a plasma can be regarded as an ideal gas. Equation 10.24 supplies, in this particular case, the missing link between p and ρ. The temperature then remains arbitrary. It depends on the initial conditions, that is, on the total energy accumulated in the plasma. This reasoning holds good for sufficiently fast processes, which leave no time for energy exchange with the environment to be accomplished.

The equations for the plasma density (Eq. 10.3), for the average velocity (Eq. 10.5), for the magnetic field and current intensity (Eqs. 10.20 and 10.17), and for the pressure (Eq. 10.24) form a complete set of magnetohydrodynamics equations. Together, they have the form

$$\begin{aligned}&\frac{\partial\rho}{\partial t}=-\operatorname{div}\rho\mathbf{u}; \qquad \rho\frac{d\mathbf{u}}{dt}=-\frac{1}{4\pi}[\mathbf{H}\times\operatorname{rot}\mathbf{H}]-\operatorname{grad}p;\\ &\frac{\partial\mathbf{H}}{\partial t}=\frac{c^2}{4\pi\sigma}\Delta\mathbf{H}+\operatorname{rot}[\mathbf{u}\times\mathbf{H}]; \qquad \frac{dp}{dt}=\gamma\frac{p}{\rho}\frac{d\rho}{dt}\end{aligned} \tag{10.25}$$

In deriving the set 10.25 the pressure of the charged particles of each type was assumed isotropic. For a plasma with a low collision frequency in intricate magnetic configurations this assumption may fail. For such conditions it is customary to introduce a diagonal pressure tensor in which the longitudinal and transverse components are different. A magnetohydrodynamic description of a plasma is possible in this case, too. Changes in the equations amount to the difference in the longitudinal and transverse components of the pressure gradient in the equation of motion and to a change in the pressure equation. The description method that takes into account the pressure anisotropy is called *anisotropic magnetic hydrodynamics.**

10.2 PLASMA EQUILIBRIUM IN MAGNETIC FIELD

The first problem we face in considering plasma confinement in a magnetic field is establishing the conditions under which an equilibrium

*Sometimes it is called Chew–Goldberger–Low hydrodynamics.

sets in, that is, when the electrodynamic forces acting on each element of the plasma volume offset the pressure gradient.

The expression for the electrodynamic force $\mathbf{G}_H$, which appears on interaction of the current flowing in the plasma with the magnetic field, can be obtained from Eq. 10.17:

$$\begin{aligned}\mathbf{G}_H &= \frac{1}{c}[\mathbf{j}\times\mathbf{H}] = -\frac{1}{4\pi}[\mathbf{H}\times\operatorname{curl}\mathbf{H}] \\ &= -\frac{1}{8\pi}\operatorname{grad}(H^2) + \frac{1}{4\pi}(\mathbf{H}\operatorname{grad})\mathbf{H}\end{aligned} \tag{10.26}$$

(here we use the well-known vector equality for $[\mathbf{H}\times\operatorname{curl}\mathbf{H}]$). It is seen that this force can be represented as the gradient of the Maxwellian strength tensor:

$$\mathbf{G}_H = -\operatorname{grad}\check{p}_M; \qquad G_{Hk} = -\sum_l \frac{\partial p_{Mkl}}{\partial x_l} \tag{10.27}$$

where $p_{Mkl} = (H^2/8\pi)\delta_{kl} - H_kH_l/4\pi$.

In a coordinate system with the z axis directed along the magnetic field this tensor is diagonal:

$$\check{p}_M = \begin{pmatrix} H^2/8\pi & 0 & 0 \\ 0 & H^2/8\pi & 0 \\ 0 & 0 & -H^2/8\pi \end{pmatrix} \tag{10.28}$$

Another form of expression for $\mathbf{G}_H$ can be obtained by introducing the curvature radius of the lines of force $\mathbf{R}$ (see Section 8.1). Taking into account Eq. 8.9, we transform the second term of Eq. 10.26:

$$\begin{aligned}(\mathbf{H}\operatorname{grad})\mathbf{H} &= H(\mathbf{h}\operatorname{grad})\mathbf{h}H = H^2\operatorname{grad}_\parallel\mathbf{h} + \frac{1}{2}\operatorname{grad}_\parallel H^2 \\ &= -\frac{H^2}{R^2}\mathbf{R} + \frac{1}{2}\operatorname{grad}_\parallel H^2\end{aligned} \tag{10.29}$$

where, as before, $\mathbf{h} = \mathbf{H}/H$; and $\operatorname{grad}_\parallel = (\mathbf{h}\operatorname{grad})$ is the projection of the gradient on the direction of the magnetic field. Substituting Eq. 10.29 into Eq. 10.26, we get

$$\mathbf{G} = -\operatorname{grad}_\perp\frac{H^2}{8\pi} - \frac{H^2}{4\pi R^2}\mathbf{R} \tag{10.30}$$

The first term in Eq. 10.30 is the transverse gradient of the magnetic pressure $p_H = H^2/8\pi$ introduced in Section 8.6. It corresponds to the transverse diagonal elements of the strength tensor 10.28 and defines the electrodynamic force acting on the conducting medium when the lines of force are straight ($R\to\infty$, $\mathbf{H}\perp\operatorname{grad}H$). The action of this force can be described as mutual "elbowing" of lines of force. The second term represents the force directed toward the center of curvature of the lines of force, and it is called the *magnetic field tension.* It is formally obtained by ascribing the properties of a taut string to the lines of force. The equivalent tension force acting on the lines of force that cross a unit area is equal to $H^2/4\pi$ as seen from Eq. 10.30. Substituting the force 10.30 into the equilibrium equation 10.6, we reduce it to the form

$$\operatorname{grad} p + \operatorname{grad}_\perp \frac{H^2}{8\pi} + \frac{H^2}{4\pi R^2}\mathbf{R} = 0 \tag{10.31}$$

If the lines of force are straight ($R \to \infty$), the equation reduces to the constancy of the sum of the kinetic and magnetic pressures in a plane perpendicular to the magnetic field:

$$p + \frac{H^2}{8\pi} = \text{const} \tag{10.32}$$

Since Eq. 10.32 must be fulfilled throughout the plasma volume, it describes the reduction in magnetic field from the plasma boundary toward the region of maximum kinetic pressure. In particular, the relationship between the magnetic field outside the plasma H_e and the magnetic field in the maximum-pressure region H_0 takes the form

$$p_0 + \frac{H_0^2}{8\pi} = \frac{H_e^2}{8\pi} \tag{10.33}$$

This relationship is obviously true for a plasma with a clear-cut boundary as well. In that case it describes the boundary equilibrium for any configuration of the magnetic field, since with a finite R the last term in Eq. 10.31 can be neglected as compared with the first.

The equality 10.33 shows that the maximum plasma pressure that can be contained by the magnetic field is equal to the magnetic pressure outside the plasma $p_{\max} = H_e^2/8\pi$. When describing the magnetic confinement it is customary to introduce the coefficient β representing the ratio of the pressure of the confined plasma to the maximum pressure possible:

$$\beta = \frac{p}{p_{\max}} = \frac{8\pi p}{H_e^2} \tag{10.34}$$

This coefficient describes the efficiency of utilization of the magnetic field for plasma confinement.

While the plasma configuration detached from the walls is confined, a sharp plasma boundary at which there is a magnetic field potential difference cannot exist indefinitely. It will be smeared out as a result of the diffusion due to collisions. Within the frame of magnetohydrodynamic description the mutual mixing of the magnetic field and the plasma must manifest itself in a change of the magnetic field strength, which is described by Eq. 10.20. For a fixed conducting medium the equation takes the form

$$\frac{\partial \mathbf{H}}{\partial t} = \frac{c^2}{4\pi\sigma} \Delta \mathbf{H} \tag{10.35}$$

This equation for each field component is the much-discussed diffusion equation with a diffusion coefficient:

$$D_H \approx \frac{c^2}{4\pi\sigma} \tag{10.36}$$

Hence it follows that the smearing out of a sharp jump can be described as the *magnetic field diffusion*, which tends to straighten out the field on both sides of the boundary. Such straightening is caused by the damping of the inductance currents due to the finite conductivity. Equation 10.35 makes it possible to estimate the characteristic time during which the magnetic field penetrates to a depth L (the field diffusion time):

$$\Delta t_H \approx \frac{L^2}{D_H} = \frac{4\pi\sigma L^2}{c^2} \tag{10.37}$$

Accordingly, the depth of field penetration into the medium within the time τ is equal to

$$\Delta s_H \approx \sqrt{D_H \tau} = \sqrt{c^2\tau/4\pi\sigma} \tag{10.38}$$

The quantity Δt_H is called the *skin time* and the quantity Δs_H the *skin-layer thickness.* From the above relations it is seen that the diffusion of the magnetic field into the plasma can be neglected if the process in question is so fast that the field penetration depth Δs_H for the characteristic time τ is much less than the characteristic size of the plasma L, that is, when the following inequality holds:

$$\tau \ll \frac{4\pi\sigma L^2}{c^2} = \frac{4\pi n e^2}{m_e c^2 \bar{\nu}_{ei}} L^2 \tag{10.39}$$

In describing the change of the magnetic field in a plasma we cannot usually regard it as a fixed conducting medium. Therefore in Eq. 10.20 we should, along with the diffusion term, take into account the term representing the magnetic field change associated with the directed motion. In the stationary state (at $\partial H/\partial t = 0$) these terms must offset each other:

$$\frac{c^2}{4\pi\sigma} \Delta \mathbf{H} + \operatorname{curl}[\mathbf{u} \times \mathbf{H}] = \operatorname{curl}\left\{\frac{c^2}{4\pi\sigma} \operatorname{curl} \mathbf{H} - [\mathbf{u} \times \mathbf{H}]\right\} = 0 \tag{10.40}$$

Hence it is easy to find the transverse velocity $u_\perp$ corresponding to the stationary state:

$$\mathbf{u}_\perp = \frac{c^2}{4\pi\sigma H^2}[\mathbf{H}\times \text{curl}\,\mathbf{H}] = -\frac{c^2}{\sigma H^2}\mathbf{G}_H \tag{10.41}$$

In this expression the term describing the motion perpendicular to the density and the magnetic field gradient is omitted. Substituting the expression for G_H (Eq. 10.30) and allowing for the equilibrium equation 10.31, we find

$$\mathbf{u}_\perp = -\frac{c^2\,\text{grad}_\perp p}{\sigma H^2} = \frac{c^2(T_e+T_i)}{\sigma H^2}\,\text{grad}_\perp n \tag{10.42}$$

Using the equation for σ (see Eq. 10.10), we finally obtain

$$\mathbf{u}_\perp = -\frac{m_e c^2(T_e+T_i)\bar{\nu}_{ei}}{e^2H^2}\frac{\text{grad}_\perp n}{n} = -D_{e\perp}\frac{\text{grad}_\perp n}{n} \tag{10.43}$$

This equation coincides with Eq. 9.128, which defines the diffusion rate of a fully ionized plasma. Thus the stationary magnetic field in a plasma with a finite collision frequency can be maintained only in the presence of a stationary diffusion flux across the field.

As noted above, with sufficiently small times of variation in plasma parameters at which Eq. 10.39 is fulfilled the magnetic field diffusion can be neglected. Then the magnetic field equation 10.20 takes the form

$$\frac{\partial \mathbf{H}}{\partial t} = \text{curl}[\mathbf{u}\times\mathbf{H}] \tag{10.44}$$

Transition from Eq. 10.20 to Eq. 10.44 corresponds to the limit $\sigma\to\infty$ or $\nu_{ei}\to 0$. In magnetohydrodynamics it is sometimes called the *ideal-conductivity limit.* Let us consider the nature of the plasma motion, which is described by Eq. 10.44. To do this we transform Eq. 10.44 with the aid of the relation

$$\text{curl}[\mathbf{u}\times\mathbf{H}] = (\mathbf{H}\,\text{grad})\mathbf{u} - (\mathbf{u}\,\text{grad})\mathbf{H} - \mathbf{H}\,\text{div}\,\mathbf{u} + \mathbf{u}\,\text{div}\,\mathbf{H} \tag{10.45}$$

Substituting Eq. 10.45 into Eq. 10.44 and bearing in mind that $\text{div}\,\mathbf{H}=0$, we obtain

$$\frac{d\mathbf{H}}{dt} = (\mathbf{H}\,\text{grad})\mathbf{u} - \mathbf{H}\,\text{div}\,\mathbf{u} \tag{10.46}$$

where we introduced the operator $d/dt = \partial/\partial t + (\mathbf{u}\,\text{grad})$. Let us compare the change in field strength described by this equation with the change in plasma density that follows from the continuity equation:

$$\frac{d\rho}{dt} = \frac{\partial \rho}{\partial t} + (\mathbf{u}\ \text{grad})\rho = -\rho\ \text{div}\ \mathbf{u} \tag{10.47}$$

Canceling div **u** from Eqs. 10.46 and 10.47, we get

$$\frac{d\mathbf{H}}{dt} - \frac{\mathbf{H}}{\rho}\frac{d\rho}{dt} = (\mathbf{H}\ \text{grad})\mathbf{u}$$

or (10.48)

$$\frac{d}{dt}\left(\frac{\mathbf{H}}{\rho}\right) = \left(\frac{\mathbf{H}}{\rho}\ grad\right)\mathbf{u}$$

Hence it follows that if the directed velocity is constant along **H**, that is, if it changes only on transition from one line of force to another, then during the plasma motion the ratio of the magnetic field strength to the density of the substance does not change with time:

$$\frac{d(\mathbf{H}/\rho)}{dt} = 0 \tag{10.49}$$

Since the density of the lines of force is proportional to the magnetic field strength, this equality is interpreted as *"freezing" of the lines of force into the plasma.* Freezing into the plasma implies that on any transverse displacements and deformations of an arbitrary closed circuit associated with the plasma the magnetic flux piercing the surface bounded by this contour remains unchanged. Then the plasma behaves as a superconductor. Naturally, the term "freezing" should not be taken literally. The condition $H/\rho = \text{const}$ does not forbid mutual transverse motions of matter and the magnetic field, for instance, a plasma drift, across the density gradient.

The freezing of the magnetic field substantially affects the plasma equilibrium conditions on fast changes in environment. For instance, a fast change in external magnetic field disturbs the equilibrium, since this field does not penetrate into the plasma. As a result the plasma boundaries begin to change their position; under the effect of the excess magnetic pressure the plasma volume contracts with an increase in magnetic field or expands under the influence of the kinetic pressure with a reduction in external field. Then the plasma density changes, and so does the magnetic field in the plasma. Thus a new equilibrium state is established, in which the kinetic pressure is equal to the magnetic pressure differential. The parameters of this new state depend on the specifics of the process.

10.3 STABILITY OF PLASMA CONFINEMENT BY MAGNETIC FIELD

In Section 10.2 we considered the conditions for equilibrium of a plasma confined by a magnetic field. But the equilibrium state may prove unstable with respect to some small perturbations in plasma configuration if the resultant force, which appears on disturbance of the equilibrium, promotes the growth of the perturbation. In this case it is impossible to maintain the plasma equilibrium for a long time, since accidental small fluctuations must increase with time. In confining a plasma by some method or other the most difficult task is to maintain its stability.

An analysis of plasma stability with respect to small perturbations may be based on the solution of a set of equations describing the plasma behavior. In describing configurational instabilities it is customary to use the equations of magnetohydrodynamics (therefore these instabilities are also called *magnetohydrodynamic*). In such description, the configuration perturbation is assigned by determining the displacement of a plasma volume element $\xi(\mathbf{r})$. Assuming this displacement to be small compared with the characteristic dimension on which plasma parameters vary and neglecting the terms quadratic with respect to ξ, we can obtain the linear differential equation

$$\frac{\rho d^2\xi}{dt^2} = \mathscr{F}(\xi) = -\check{\mathbf{K}}\xi \tag{10.50}$$

where $\check{\mathbf{K}}$ is a differential operator, which includes the derivatives with respect to the coordinates. This equation is similar to the small-oscillation equation. The quantity $\mathscr{F}(\xi)$ can be interpreted as a quasi-elastic force, and the operator $\check{\mathbf{K}}$ plays the part of the elasticity coefficient. The solution of Eq. 10.50 with given boundary and initial conditions is found by the method of variables separation. The time dependence of each of the particular solutions can be written in the complex form

$$\xi = \xi_0(\mathbf{r}) \exp(i\omega t) \tag{10.51}$$

where, as before, the real displacements are represented by the real part of the complex expression. The coordinate dependence of the particular solutions is described by equations obtained by substituting Eq. 10.51 into Eq. 10.50:

$$\check{\mathbf{K}}\xi = \omega^2\rho\xi \tag{10.52}$$

with given boundary conditions, this equation is known to have a spectrum of eigenfunctions $\xi_n(r)$ and a spectrum of eigenvalues $K_n = \omega_n^2\rho$. If the quantity ω_n, which is determined by the eigenvalue K_n, has a negative imaginary part, the displacement of ξ_n increases exponentially with time $[\xi_n \sim \exp(-\mathrm{Im}(\omega_n)t)]$. This means that the perturbation described by the function $\xi_n(\mathbf{r})$ increases the instability. The increment of this instability is equal to $\gamma_n = -\mathrm{Im}(\omega_n)$.

Let us find the linearized displacement equation for conditions where the plasma instability can be described by equations of ideal magnetohydrodynamics (i.e., when the plasma conductivity can be assumed infinite in Eq. 10.20). This is usually permissible in considering the most hazardous, rapidly developing instabilities, since the characteristic times of their increase are much less than the skin time (see Eq. 10.39). Then the magnetohydrodynamics equations (Eqs. 10.25) take the form

$$\frac{\partial \rho}{\partial t} = -\mathrm{div}(\rho \mathbf{u}); \ \frac{\rho d\mathbf{u}}{dt} = -\mathrm{grad}\, p - \frac{1}{4\pi}[\mathbf{H} \times \mathrm{curl}\, \mathbf{H}];$$
$$\frac{\partial \mathbf{H}}{\partial t} = \mathrm{curl}[\mathbf{u} \times \mathbf{H}]; \ \frac{dp}{dt} = \gamma \frac{p}{\rho}\frac{d\rho}{dt} \tag{10.53}$$

where $d/dt = \partial/\partial t + (\mathbf{u}\,\mathrm{grad})$ and $\gamma = \frac{5}{3}$ is the adiabatic index. In the unperturbed state the equilibrium condition must hold; that is, the sum of the forces acting on the plasma must be equal to zero. Accordingly, the directed velocity must be constant. It can be assumed equal to zero without the loss of generality. (This only means the choice of a definite reference system.)

To obtain the equations describing the perturbations we represent the density, pressure, and magnetic field as a sum of equilibrium values and small perturbations:

$$\rho = \rho^{(0)} + \rho^{(1)}; \ p = p^{(0)} + p^{(1)}; \ \mathbf{H} = \mathbf{H}^{(0)} + \mathbf{H}^{(1)} \tag{10.54}$$

Substituting these sums into Eq. 10.53 and neglecting the terms quadratic with respect to the perturbations (i.e., the terms containing squares or products of small values), we obtain a set of equations for the perturbed quantities:

$$\frac{\partial \rho^{(1)}}{\partial t} + \mathrm{div}(\rho^{(0)}\mathbf{u}) = 0;$$
$$\frac{\rho^{(0)}\partial \mathbf{u}}{\partial t} = -\mathrm{grad}\, p^{(1)} - \frac{1}{4\pi}[\mathbf{H}^{(1)} \times \mathrm{curl}\, \mathbf{H}^{(0)}] - \frac{1}{4\pi}[\mathbf{H}^{(0)} \times \mathrm{curl}\, \mathbf{H}^{(1)}];$$
$$\frac{\partial \mathbf{H}^{(1)}}{\partial t} = \mathrm{curl}[\mathbf{u} \times \mathbf{H}^{(0)}]; \tag{10.55}$$

$$\frac{\partial p^{(1)}}{\partial t} = -(\mathbf{u}\,\mathrm{grad})p^{(0)} - \gamma p^{(0)}\,\mathrm{div}\,\mathbf{u}$$

which takes into account that the directed velocity $\mathbf{u}$ is a perturbation, since $\mathbf{u}^{(0)} = 0$. In place of $\mathbf{u}$ we now introduce the displacement, assuming $\mathbf{u} = \partial\boldsymbol{\xi}/\partial t$. Then the equations for the disturbances in ρ, H, and p can be integrated with respect to time:

$$\rho^{(1)} = -\text{div}(\rho^{(0)}\boldsymbol{\xi});\ p^{(1)} = -\boldsymbol{\xi}\ \text{grad}\ p^{(0)} - \gamma p^{(0)}\ \text{div}\ \boldsymbol{\xi}$$
$$\mathbf{H}^{(1)} = \text{curl}[\boldsymbol{\xi} \times \mathbf{H}^{(0)}]. \tag{10.56}$$

Substituting these equations into those for $\mathbf{u}$, we have

$$\frac{\rho\partial^2\boldsymbol{\xi}}{\partial t^2} = \text{grad}[(\boldsymbol{\xi}\ \text{grad})p + \gamma p\ \text{div}\ \boldsymbol{\xi}] + \frac{1}{4\pi}[\text{curl}\ \mathbf{H} \times \text{curl}[\boldsymbol{\xi} \times \mathbf{H}]]$$
$$-\frac{1}{4\pi}[\mathbf{H} \times \text{curl}(\text{curl}[\boldsymbol{\xi} \times \mathbf{H}])] \tag{10.57}$$

where the superscripts $^{(0)}$ at the values characterizing the equilibrium state are omitted for brevity. The second-order differential equation 10.57 is actually the displacement equation, which was written symbolically in the form of Eq. 10.50. The right-hand side of the equation represents a quasi-elastic force $\mathcal{F}(\boldsymbol{\xi})$ arising on small displacements; it determines the operator $\check{\mathbf{K}}$. The displacement equation 10.57 can be supplemented by the boundary conditions. One of them is obtained from the requirement of the constancy of the sum of the kinetic and magnetic pressures $p + H^2/8\pi$ at the plasma–vacuum interface. The other is found from the conditions of continuity of the normal component of the magnetic field at the boundary; it must reduce to zero, because in a plasma with an infinite conductivity the magnetic field must be parallel to the boundary.

It is seen that the displacement equation is in general quite complicated, and its solution can be found only for simple plasma configurations. It is possible, however, to investigate the stability of a confined plasma without solving the equation. For such analysis it is customary to use the variational principle, which is called energetic. This principle is similar to the condition of the minimum potential energy for stable mechanical systems. In the case at hand it is also possible to introduce the effective potential energy. Since the displacement equation 10.50 contains a force depending linearly on the displacement $\mathcal{F}(\boldsymbol{\xi}) = -\check{\mathbf{K}}\boldsymbol{\xi}$, the change in the potential energy of each element on displacement is clearly defined by the integral of the product of this force by the displacement $\int_0^{\xi} \mathcal{F}\, d\boldsymbol{\xi} = \frac{1}{2}\boldsymbol{\xi}\mathcal{F}$. Summing this quantity over the entire volume yields

$$\delta W = \frac{1}{2}\int \boldsymbol{\xi}\mathcal{F}(\boldsymbol{\xi})\, dV = -\frac{1}{2}\int \boldsymbol{\xi}\check{\mathbf{K}}\boldsymbol{\xi}\, dV \tag{10.58}$$

If $\delta W > 0$, that is, a displacement $\boldsymbol{\xi}(\mathbf{r})$ results in an increase of the potential energy, then the system is stable; otherwise it is unstable with respect to the perturbation discussed. Substituting the force $\mathcal{F}$ from Eq. 10.57 into the integral of Eq. 10.58 and taking into consideration the boundary conditions, one can (after some transformations) reduce it to the following general form:

$$\delta W = \frac{1}{2}\int_{V_i} \left\{ \gamma p (\operatorname{div} \boldsymbol{\xi})^2 + \frac{1}{4\pi} (\operatorname{curl}[\boldsymbol{\xi} \times \mathbf{H}])^2 + \boldsymbol{\xi} \operatorname{grad} p \operatorname{div} \boldsymbol{\xi} \right.$$
$$\left. - \frac{1}{4\pi} [\boldsymbol{\xi} \times \operatorname{curl} \mathbf{H}] \operatorname{curl}[\boldsymbol{\xi} \times \mathbf{H}] \right\} dV + \frac{1}{8\pi}\int_{(V_e)} H^2\, dV \qquad (10.59)$$
$$- \frac{1}{2}\int_S \left[\frac{\partial p}{\partial \eta} + \frac{1}{8\pi}\left(\frac{\partial H_i^2}{\partial \eta} - \frac{\partial H_e^2}{\partial \eta} \right) \right] \xi_\eta^2\, dS$$

where the first integral is taken over the plasma volume, the second over the external (vacuum) volume, and the third over the surface bounding the plasma; the derivatives $\partial/\partial\eta$ in the last integral are taken along the normal to the surface; and ξ_η is the projection of the displacement $\boldsymbol{\xi}$ on the normal. Precisely this expression for the potential energy is generally used in analyzing stability with the aid of the energy principle. Equation 10.52 $\check{\mathbf{K}}\boldsymbol{\xi} = \omega^2 \rho \boldsymbol{\xi}$ can be obtained from the following variational principle in view of the self-conjugation of the operator $\check{\mathbf{K}}$:

$$\delta(\omega^2) = 0, \qquad \omega^2 = \frac{\int \boldsymbol{\xi}\check{\mathbf{K}}\boldsymbol{\xi}\, dV}{\int \rho \xi^2\, dV} = \frac{2\delta W}{\int \rho \xi^2\, dV} \qquad (10.60)$$

This principle makes it possible to find the frequencies of natural oscillations or the instability increments without solving the equations. It also follows from the above expression that in this case the quantity ω^2 is always real. At $\delta W > 0$ the frequency ω is real, and the displacement oscillations do not increase (they damp out if due account is taken of the dissipation). At $\delta W < 0$ the quantity ω^2 is negative; then there is a solution corresponding to an exponential increase in perturbation with the increment $\gamma = |\omega|$.

The linearized displacement equation 10.57 and the energy principle stemming from it describe the linear stage of instability in which the displacement is much less than the characteristic size. In order to determine the consequences of the magnetohydrodynamic instability it is necessary to solve the nonlinear problem. The solution can be found only for some simple applications. It is sometimes useful to apply the phenomenological approach based on an analogy with conventional hydrodynamics. We cannot dwell on this at length and only indicate that rapidly developing instabilities usually result in a sharp deterioration of plasma confinement. The velocity of plasma escape across the field may be close to the thermal velocity of ions.

With the aid of the displacement equations and the energy principle it has been possible to analyze the conditions for the appearance of the basic magnetohydrodynamic instabilities limiting the plasma confinement in the magnetic traps. Many monographs and surveys are dedicated to this analysis (see Section 4 of the Bibliography). A systematic exposition of the plasma stability problems is beyond the scope of this book. Below, we consider only some characteristic instabilities, focusing our attention on the qualitative discussion of the processes leading to the growth of perturbations.

10.4 STABILITY OF PLASMA BOUNDARY IN MAGNETIC FIELD

Let us consider quantitatively the comparatively simple problem of stability of a plane plasma boundary in a field of a constant force, the gravitational force in particular. Let a constant-density plasma fill the half-space $x > 0$ (Fig. 10.1), the force $\mathbf{G}$, directed toward the boundary, acting on a unit plasma volume. The magnetic field inside the plasma, $\mathbf{H}_i$, and outside it, $\mathbf{H}_e$, is assumed homogeneous and parallel to the boundary ($\mathbf{H}_i = \text{const}$ and $\mathbf{H}_e = \text{const}$). These fields are not equal and need not be parallel. The conditions of equilibrium in the plasma volume then reduce to the equality of the external force and the pressure gradient:

$$\mathbf{G} = \text{grad}\, p, \qquad G = -\frac{dp}{dx} \tag{10.61}$$

The pressure at the plasma boundary must be offset by the magnetic pressure drop:

$$p_g = \frac{H_{eg}^2}{8\pi} - \frac{H_{ig}^2}{8\pi} \tag{10.62}$$

Let us investigate the stability of equilibrium with respect to small perturbations. Since the system is homogeneous with regard to the y and

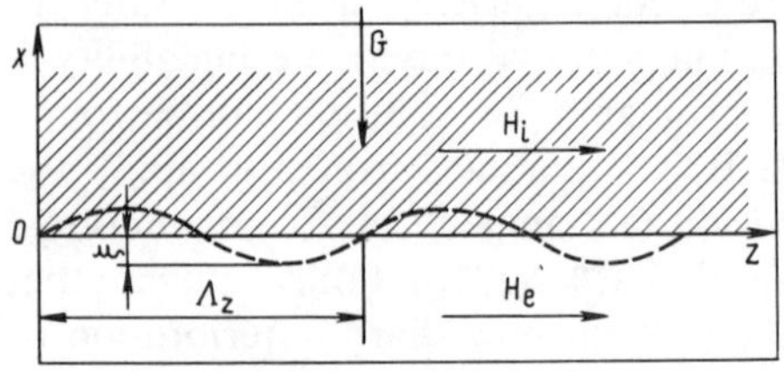

Fig. 10.1 Plane plasma boundary in a gravitational field.

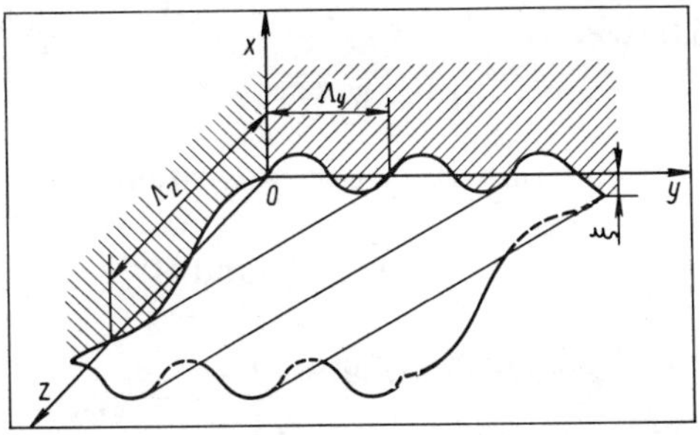

Fig. 10.2 Wave vector appearing in a perturbed plasma surface.

z coordinates, we seek all perturbations (displacements field and pressure perturbations) in the form of periodic functions of these coordinates:

$$\begin{aligned}\boldsymbol{\xi}(x, y, z, t) &= \mathrm{Re}[\boldsymbol{\xi}(x)\exp[\mathrm{i}(k_y y + k_z z - \omega t)]];\\ \mathbf{H}(x, y, z, t) &= \mathrm{Re}[\mathbf{H}(x)\exp(\mathrm{i}(k_y y + k_z z - \omega t)]]\end{aligned} \tag{10.63}$$

and so on. The wave vector $\mathbf{k}(k_y, k_z)$ appearing in them determines the orientation and magnitude of the perturbations; the wavelength $\Lambda = 2\pi/k$ characterizes their width; and the quantities $\Lambda_y = 2\pi/k_y$ and $\Lambda_z = 2\pi/k_z$ characterize the dimensions in the y and z directions (see Figs. 10.1 and 10.2; the latter depicts part of the perturbed surface).

We use the perturbation equations 10.56 and 10.57. For simplicity we consider such perturbations on which the plasma density remains unchanged; that is, we use the incompressible fluid approximation. It can be shown that it enables us to correctly determine the stability boundaries and the characteristics of perturbations near the boundary. Therefore we assume

$$\rho^{(1)} = \mathrm{div}(\rho^{(0)}\boldsymbol{\xi}) = 0 \tag{10.64}$$

or, since $\rho^{(0)} = \mathrm{const}$,

$$\mathrm{div}\,\boldsymbol{\xi} = 0 \tag{10.65}$$

Equation 10.56 for a perturbed magnetic field in a plasma takes the following form, due allowance being made for the constancy of the unperturbed field:

$$\begin{aligned}\mathbf{H}_i^{(1)} = \mathrm{rot}[\boldsymbol{\xi} \times \mathbf{H}_i^{(0)}] &= \mathbf{H}_i^{(0)}\,\mathrm{div}\,\boldsymbol{\xi} + (\mathbf{H}_i^{(0)}\,\mathrm{grad})\boldsymbol{\xi}\\ &= (\mathbf{H}_i^{(0)}\,\mathrm{grad})\boldsymbol{\xi}\end{aligned}$$

Taking into account the dependence of the displacement on the coordinates (Eq. 10.63), we get

$$\mathbf{H}_i^{(1)} = \mathrm{i}(\mathbf{k}\mathbf{H}_i^{(0)})\boldsymbol{\xi} \tag{10.66}$$

The displacement equation 10.57 can be written thus:

$$\frac{\rho d^2\boldsymbol{\xi}}{dt^2} = -\mathrm{grad}\,\mathcal{P}^{(1)} + \frac{1}{4\pi}(\mathbf{H}_i^{(0)}\,\mathrm{grad})\mathbf{H}_i^{(1)} \tag{10.67}$$

where we introduced the summary pressure

$$\mathcal{P} = p + \frac{H^2}{8\pi} \tag{10.68}$$

and $\mathcal{P}^{(1)}$ represents its perturbation. Bearing in mind the dependence of the perturbation on the coordinates and time (Eq. 10.63) and substituting the field equation 10.66 into Eq. 10.67, we find

$$\boldsymbol{\xi} = \frac{\mathrm{grad}\,\mathcal{P}^{(1)}}{\rho\omega^2 - \dfrac{1}{4\pi}(\mathbf{k}\mathbf{H}_i^{(0)})^2} \tag{10.69}$$

Substituting this displacement into Eq. 10.65, we obtain the Laplace equation for $\mathcal{P}^{(1)}$:

$$\Delta\mathcal{P}^{(1)} = \frac{\partial^2\mathcal{P}^{(1)}}{\partial x^2} - k^2\mathcal{P}^{(1)} = 0 \tag{10.70}$$

Its solution of the form 10.63, which vanishes as $x \to \infty$, is

$$\mathcal{P}^{(1)} = C_i\rho\,\exp(-kx)\exp[\mathrm{i}(\mathbf{kr} - \omega t)] \tag{10.71}$$

where $k = (k_y^2 + k_z^2)^{1/2}$, and C_i is a constant. Substituting this solution into Eq. 10.69 yields the displacement, and substituting it into Eq. 10.66, the perturbation of the magnetic field in the internal region (at $x > 0$).

We now find the field perturbation in the external region at $x < 0$. Since there are no currents in this vacuum region, that is, curl $\mathbf{H}_e = 0$, we introduce a scalar magnetic potential, which determines the magnetic field perturbation:

$$\mathbf{H}_e^{(1)} = \mathrm{grad}\,\Psi \tag{10.72}$$

It must satisfy the Laplace equation,

$$\mathrm{div}\,\mathbf{H}_e^{(1)} = \Delta\Psi = 0 \tag{10.73}$$

The solution of an equation of the type 10.63, which vanishes as $x \to -\infty$, has the form (cf. Eq. 10.71)

$$\Psi = C_e\exp(kx)\exp[\mathrm{i}(\mathbf{kr} - \omega t)] \tag{10.74}$$

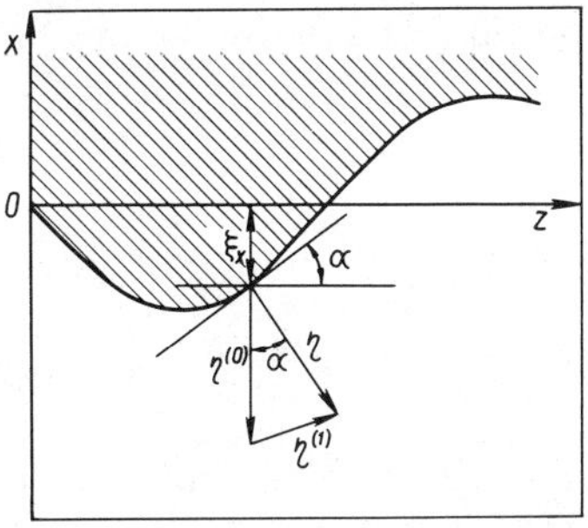

Fig. 10.3 Matching conditions at a perturbed boundary.

The equalities 10.72 and 10.74 thus yield the field perturbation in the external region.

The solutions for the external and internal regions must now be matched at the plasma boundary at $x = 0$. The boundary condition for the magnetic field reduces to the vanishing of the normal field component at the perturbed boundary. Considering the smallness of the perturbation, we can represent it as follows:

$$(\eta \mathbf{H}_e) = \boldsymbol{\eta}^{(1)}\mathbf{H}_e^{(0)} + \boldsymbol{\eta}^{(0)}\mathbf{H}_e^{(1)} = 0 \tag{10.75}$$

where $\boldsymbol{\eta} = \boldsymbol{\eta}^{(0)} + \boldsymbol{\eta}^{(1)}$ is a unit vector of the normal to the surface.

The correction $\boldsymbol{\eta}^{(1)}$ can be related to the displacement ξ_x. Using Fig. 10.3, we find $\eta_z^{(1)} = 2\sin(\alpha/2) \approx \tan\alpha \approx \partial\xi_x/\partial z$, or, in vector form,

$$\boldsymbol{\eta}^{(1)} = -\text{grad}\,\xi_x \tag{10.76}$$

Bearing in mind the last relation, we transform the boundary condition 10.75:

$$H_{ex}^{(1)} = (\mathbf{H}_e^{(0)}\,\text{grad})\xi_x \tag{10.77}$$

Substituting $H_{ex}^{(1)} = \partial\Psi/\partial x$ (see Eqs. 10.72 and 10.74) and ξ_x (see Eqs. 10.69 and 10.71) into Eq. 10.77, we get the ratio between C_e and C_i:

$$C_e = -i\frac{(\mathbf{kH}_e)}{\omega^2 - (\mathbf{kH}_i)^2/4\pi\rho}C_i \tag{10.78}$$

The second boundary condition is determined by the continuity of the summary pressure $\mathcal{P}$ on crossing the boundary. The pressure at the displaced boundary can be found from

$$\mathcal{P}_g + \xi\,\text{grad}\,\mathcal{P}_g = \mathcal{P}_g^{(0)} + \mathcal{P}_g^{(1)} + \xi\,\text{grad}\,\mathcal{P}_g^{(0)}$$

where the quantities $\mathcal{P}_g$, $\mathcal{P}_g^{(0)}$, and $\mathcal{P}_g^{(1)}$ are determined at the unperturbed boundary, and the small quadratic term $\xi\,\text{grad}\,\mathcal{P}_g^{(1)}$ is omitted. This relation makes it possible to write the condition of continuity of the summary pressure as

$$\mathcal{P}_{ge}^{(1)} + \xi \text{ grad } \mathcal{P}_{ge}^{(0)} = \mathcal{P}_{gi}^{(1)} + \xi \text{ grad } \mathcal{P}_{gi}^{(0)}$$

or

$$\mathcal{P}_{gi}^{(1)} - \mathcal{P}_{ge}^{(1)} = -\xi \text{ grad } \mathcal{P}_{gi}^{(0)} = \xi_x G \tag{10.79}$$

where we included the equilibrium conditions (Eqs. 10.61 and 10.62). The quantities $\mathcal{P}_{gi}^{(1)}$ and ξ_x at the boundary (at $x = 0$) satisfy Eqs. 10.71 and 10.69, respectively. The quantity $\mathcal{P}_{ge}^{(1)}$ and the external field are related by

$$\mathcal{P}_{ge}^{(1)} = \left(\frac{H_e^2}{8\pi}\right)^{(1)} = \frac{\mathbf{H}_e\mathbf{H}_e^{(1)}}{4\pi} = \frac{1}{4\pi}(\mathbf{H}_e \text{ grad})\Psi \tag{10.80}$$

Taking into account Eqs. 10.80 and 10.74, we find, with the aid of Eq. 10.79, the second relation for the coefficients C_i and C_e:

$$\rho C_i - \frac{\mathrm{i}}{4\pi}(\mathbf{kH}_e)C_e = -\frac{kC_i}{\omega^2 - (\mathbf{kH}_i)^2/4\pi\rho}G \tag{10.81}$$

Eliminating the constants C_i and C_e from Eqs. 10.78 and 10.81, we obtain the dispersion ratio for ω:

$$\rho\omega^2 = -kG + \frac{(\mathbf{kH}_e)^2}{4\pi} + \frac{(\mathbf{kH}_i)^2}{4\pi} \tag{10.82}$$

Equation 10.82 enables one to find the conditions of stability of the plasma boundary, the oscillation frequency of the perturbation, and the instability increment. As noted above, the stability region corresponds to the real values of ω, that is, to the positive right-hand side of Eq. 10.82. Therefore the absolute stability of the plasma boundary is only possible if the external force is nonexistent or is directed from the boundary into the plasma. With an outward-directed force the possibility of instabilities with different $\mathbf{k}$ depends on the relationship between the negative term of Eq. 10.82 and the positive terms describing this stabilizing factor. When the magnetic fields outside and inside are parallel ($\mathbf{H}_e \parallel \mathbf{H}_i$), the positive terms reduce to zero at $\mathbf{k} \perp \mathbf{H}$. Therefore the plasma boundary is unstable with respect to perturbations extending along the magnetic field. This is the so-called *flute instability*. In accordance with Eq. 10.82 its increment is equal to

$$\gamma = \mathrm{i}\omega = \sqrt{kG/\rho} \tag{10.83}$$

It increases with k, that is, with decreasing size of the flutes* ($\Lambda = 2\pi/k$). If the external magnetic field is not parallel to the internal, the stabilizing terms of Eq. 10.82 are different from zero. Hence it follows that the crossing of the magnetic fields (shear) is a stabilizing factor. With crossing, however, the dispersion relation 10.82 again points to the existence of instabilities with a sufficiently small k, since the stabilizing term is proportional to k^2. The value of k in a system of finite dimensions cannot be infinitely small. The maximum perturbation wavelength is usually equal to the doubled length of the system, that is,

$$k_{\min} = \frac{\pi}{L} \tag{10.84}$$

This condition corresponds to "fixed ends," for instance, to the vanishing of the displacement at the plasma boundary. Assuming $k > k_{\min}$, we obtain from Eq. 10.82 the boundary stability condition when $\mathbf{H}_i \perp \mathbf{H}_e$:

$$\frac{H_i^2}{4\pi} > \left(\frac{G}{k}\right)_{\max} = \frac{GL}{\pi} \tag{10.85}$$

Let us now discuss the physical mechanism of instability of the plasma boundary in the field of an external force for the limiting cases of a high and a low plasma pressure. At a high pressure, when $\beta = 8\pi p/H_e^2 = 1$, the magnetic field inside the plasma is equal to zero. The formation of troughs extending along the magnetic field is, in this case, accompanied by plasma interchange as indicated by the arrows in Fig. 10.4. Owing to the force **G** the plasma pressure is higher than the confining magnetic pressure on the convex areas of the surface, and lower on the concave areas. As a result, the perturbation grows. The instability is similar to the *Rayleigh–Taylor instability*, which appears

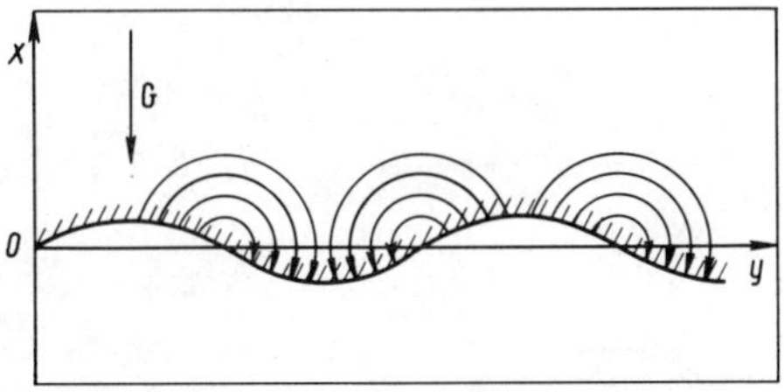

Fig. 10.4 Rayleigh–Taylor instability.

*Note that with very small sizes, that is, at large k, there come into play stabilizing factors associated with a finite collision frequency (transverse diffusion in particular) which are not accounted for.

when a heavy fluid is placed over a light one; the role of the heavy fluid is played by the plasma, and the role of the light one, by the magnetic field. Proceeding from the qualitative picture it is easy to estimate the increment of the flute instability. The increase in plasma pressure in the flute is obviously determined by the work of the force G. We can write approximately $\Delta p \approx \xi G$. This additional pressure acts on a unit surface of the plasma, causing its accelerated motion. If the condition of incompressibility of matter is fulfilled, the displacement of a surface element must be accompanied by an opposite displacement of the other element. It is easy to see that the depth to which the perturbation extends has the scale $\Lambda \sim 1/k$ (see Fig. 10.4). Therefore the equation of motion of a unit surface of the flute can be written approximately as

$$\frac{\rho}{k}\frac{d^2\xi}{dt^2} = \Delta p = \xi G \tag{10.86}$$

The equation yields an exponential increase in perturbation with increment 10.83. The decrease in increment with increasing wavelength in this case is associated with the increase in the size of the perturbation region, which increases the mass of the substance involved in the motion.

The above estimate yields the increment of a flute extending along the magnetic field. With a different orientation, the perturbation increase is impeded by the distortion of the magnetic field. Since the field cannot penetrate into a plasma with an infinite conductivity, perturbations oriented at an angle to the field bend the lines of force (Fig. 10.5). The bending gives rise to a tension force

$$G_R = \frac{H^2}{4\pi R} \approx \frac{H^2}{4\pi}\left|\frac{d^2\xi}{dz^2}\right| = \frac{H^2}{4\pi} k_z^2 \xi \tag{10.87}$$

which retards the perturbation (here the z axis is directed along the

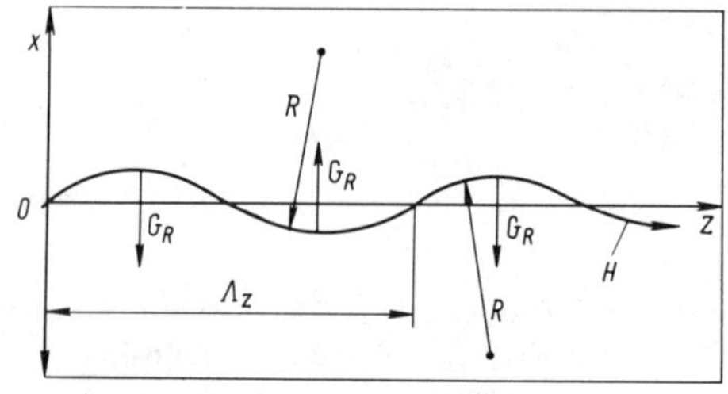

Fig. 10.5 Perturbations oriented at an angle to the field bend the lines of force.

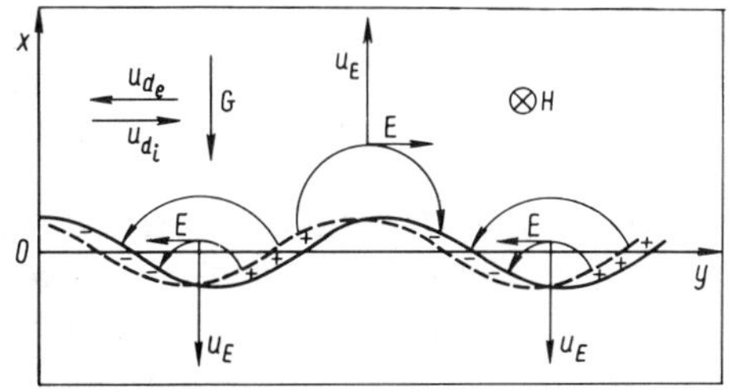

Fig. 10.6 Drift in the presence of flute perturbations.

unperturbed magnetic field). At sufficiently large k_z, the tension force is so large that the instability cannot develop at all.

Let us now consider the microscopic picture of the flute instability at low plasma pressures: $\beta = 8\pi p/H^e \ll 1$ at which the magnetic field inside the plasma practically coincides with the external one. Under the effect of the force **G** the charged plasma particles drift across the magnetic field and the force. In the absence of perturbations such a drift is parallel to the plasma boundary and does not affect the density. In the presence of flute perturbations the drift results in accumulation of ions on one side of the flute and electrons on the other (Fig. 10.6). The electric field of polarization causes a drift of the plasma as a whole in the direction of increasing perturbation. This drift actually determines the rate of increase of instability. It is easy to estimate this rate by considering the motion of individual charged particles in the perturbation region. The drift velocity of the electrons and ions depends on the forces acting on them. In accordance with Eq. 8.26,

$$u_{di} = \frac{cF_i}{eH}; \qquad u_{de} = \frac{cF_e}{eH} \tag{10.88}$$

$\mathbf{u}_{di}$ being directed toward $[\mathbf{F} \times \mathbf{h}]$, that is, toward the y axis, and $\mathbf{u}_{de}$ in the opposite direction. The electron and ion motion leads to accumulation of charges on the flute sides. The rate of charge accumulation on a unit surface is approximately equal to

$$\frac{\partial P}{\partial t} \approx en\xi \frac{u_{di} + u_{de}}{\Lambda} \approx \frac{ck\xi G}{H} \tag{10.89}$$

where $\Lambda \approx 1/k$ characterizes the width of the perturbation in the y direction, and $G = n(F_e + F_i)$. The surface density of the charge P at the flute boundaries sets up an electric field in the plasma, which is of the order of

$$E_y \approx \frac{4\pi P}{\epsilon} \tag{10.90}$$

where $\epsilon = 1 + 4\pi\rho c^2/H^2 \approx 4\pi\rho c^2/H^2$ is the effective dielectric constant of the plasma (see Eq. 8.70). The electric field causes a joint drift of the electrons and ions parallel to the acting force, with a velocity of

$$u_x = \frac{d\xi}{dt} = \frac{cE_y}{H} = \frac{4\pi cP}{\epsilon H} \approx \frac{HP}{c\rho} \tag{10.91}$$

Differentiating this equation with respect to time and including Eq. 10.89, we obtain the displacement equation:

$$\frac{d^2\xi}{dt^2} \approx \frac{H}{c\rho}\frac{\partial P}{\partial t} \approx \frac{kG}{\rho}\xi \tag{10.92}$$

from which follows the exponential increase in displacement with the increment 10.83. Note that the decrease in increment with increasing wavelength $\Lambda \sim 1/k$, which follows from Eq. 10.92, is due to the diminishing charge accumulation on the surface of a large flute (see Eq. 10.89).

We have considered the flute instability of the plasma boundary under the effect of a constant force. In the plasma of cosmic objects this may be the gravitational force (in this case the instability is called *gravitational*). In a laboratory plasma the gravitational force does not play an important part, but there may be other sources of flute instability. The most widespread among them is the inhomogeneity of the magnetic field. In an inhomogeneous field the plasma is subjected to a force directed against the field gradient. This force, which is due to the diamagnetic properties of the plasma, results in an instability that pushes the plasma out into the region of the weak field.

With a high plasma pressure, corresponding to $\beta_{\max} = 8\pi p/H_e^2 = 1$ (there is no magnetic field inside the plasma), the inhomogeneity of the external field creates an effective force at the boundary, which is equal to the magnetic pressure gradient:

$$\mathbf{G} = -\text{grad}\left(\frac{H^2}{8\pi}\right) \tag{10.93}$$

If this force is directed along the normal to the plasma surface, that is, if the magnetic field decreases in the direction of the normal, an instability may arise. The condition of increasing perturbation is obtained by substituting Eq. 10.93 into Eq. 10.82. It can be represented as the inequality

$$\frac{\partial}{\partial \eta}\left(\frac{H^2}{8\pi}\right) > \frac{1}{4\pi}\frac{k_z^2 H^2}{k} \tag{10.94}$$

which means that the magnetic pressure gradient in the direction of the normal (η) exceeds the stabilizing force of tension of the lines of force. The inequality 10.94 shows that not only flute perturbations with $k_z = 0$, but also perturbations with a finite k_z, are unstable. From Eq. 10.94 we obtain the following limitations:

$$k_z < \sqrt{k/L}; \qquad \Lambda_z > \sqrt{2\pi L \Lambda} \tag{10.95}$$

where $L = |[(1/H)\partial H/\partial \eta]^{-1}|$ is the characteristic length on which the magnetic field decays; $\Lambda = 2\pi/k$; and $\Lambda_z = 2\pi/k_z$. Unstable perturbations at $\Lambda \ll L$ resemble "tongues" extending along the magnetic field. But their length in the direction of the field Λ_z may be much less than the characteristic size of the plasma (of the order of L). Therefore the instability of the plasma boundary is local and is determined by the local direction of the magnetic field gradient. It can develop on those boundary portions where the magnetic field decays in the direction of the normal. From the foregoing it follows that in order to ensure the stability of the plasma boundary at $\beta = 1$ it is necessary that the magnetic field increase away from the plasma boundary at all points of the surface; in other words, that the plasma be placed in the region of the magnetic field minimum.

At low plasma pressures, when $\beta = 8\pi p/H^2 \ll 1$, the energy accumulated in the plasma is much less than that of the magnetic field. Therefore instabilities caused by inhomogeneity cannot lead to a substantial perturbation in an equilibrium field (the energy expenditures on such perturbations cannot be offset by redistributing the plasma in space). Accordingly, only flute perturbations extending along the field can be unstable. The effective force causing instabilities in an inhomogeneous field is the sum of the forces acting on the individual charged particles (see Section 8.4): the centrifugal force $F_R = mv_\parallel^2/R$ and the diamagnetic force $F_\mu = (mv_\perp^2/2H)|\operatorname{grad}_\perp H| = mv_\perp^2/2R$ (for the case where the lines of force are plane curves). Summing these forces for the electrons and ions in a unit volume and averaging over the velocities, we obtain

$$\mathbf{G} = n(\langle F_e \rangle + \langle F_i \rangle) \approx 2n\frac{T_e + T_i}{R} = \frac{2p}{R} \tag{10.96}$$

where we put $m\langle v_{\alpha\parallel}^2 \rangle = m\langle v_{\alpha\perp}^2 \rangle/2 = T_\alpha$. The direction of the force $\mathbf{G}$, as well as that of $\mathbf{F}_R$ and $\mathbf{F}_\mu$, coincides with the direction of the curvature of the lines of force. If the force is everywhere directed inward from the boundary, the plasma is stable. If the force is directed outward, it promotes the instability. Note that in contrast to the limit $\beta = 1$ the criterion of the flute instability at $\beta \ll 1$ is not local. The development of the instability is determined by the magnitude and direction of the effective force averaged over the length of the line of force.

The development of the flute instability in a plasma with $\beta < 1$ can be ascribed to interchange of the flux tubes filled with plasma. The magnetic field strength and the shape of the lines of force remain constant, but the plasma volume and pressure in the tubes change. The energy accumulated in the plasma–magnetic field system changes accordingly. Analyzing the change in energy, we can find the conditions under which instabilities appear, which are caused by the interchange of the flux tubes for different magnetic field configurations (this instability is often called the *interchange instability*).

The volume of the flux tube is equal to (see Eq. 8.7)

$$V = \int_{(l)} S\, dl = \Phi \int_{(l)} \frac{dl}{H} \tag{10.97}$$

where $\Phi = HS$ is the magnetic flux piercing the tube cross section; and $\int_l dl/H$ is the so-called specific volume. (The integral is taken over the entire length of the flux tube.*) The plasma energy accumulated in this volume is measured by the product of the volume and the plasma pressure $W = pV$. The flux tube with the plasma tends to shift so that it could expand. If it is surrounded by "vacant" tubes, an interchange that increases the volume of the tube with the plasma reduces the internal energy. If two adjacent flux tubes are filled with plasma, but the pressure in them is lower than in the isolated one, this interchange is also energetically advantageous. Conversely, a flux tube surrounded by a plasma with a higher pressure tends to shift where its volume is smaller, because then the energy accumulated in the entire system decreases: an increase in energy $\delta(Vp_1)$ owing to the compression of an isolated tube cannot offset the decrease in energy $\delta(Vp_2)$ of the adjacent tube, which has taken its place. The displacement of a flux tube with the plasma resembles the motion of a liquid drop with a density proportional to p_1 suspended in another liquid with a density proportional to p_2 and located in an effective gravitational field, whose potential increases with decreasing tube volume: $U = -\int dl/H$. The drop either sinks or floats up, depending on the ratio between p_1 and p_2. This analogy shows that an interchange of flux tubes with a plasma in a nonuniform magnetic field is

*The integral is determined rigorously only for configurations with closed lines of force. For open configurations it can approximately be assumed that the integration limits bound the region enclosing the plasma.

similar to the convection of an inhomogeneous fluid in the gravitational field. Therefore the interchange instability is also called the *convective instability*.

Let us formulate the criterion of the interchange instability at $\beta \ll 1$. In a state of equilibrium on a constant-pressure surface ($p = \text{const}$) the energy must also be constant: $pV = p\Phi \int dl/H = \text{const}$. Therefore the equilibrium pressure p can be regarded as a function of $U = -\int dl/H$. In analyzing the stability of this equilibrium we assume that some flux tube containing plasma displaces by a small distance, pushing the other tubes aside. If the plasma in the tube does not exchange energy with the remaining plasma during displacement, that is, if the pressure in the tube varies adiabatically ($p \sim V^{-\gamma}$), the pressure variation can be related to that of the effective potential U:

$$\delta p = -\frac{\gamma p \delta V}{V} = -\frac{\gamma p \delta U}{U} \tag{10.98}$$

(γ is the adiabatic index). At the same time the pressure in the tubes surrounding the displaced one is equal to

$$p(U + \delta U) = p + \frac{dp}{dU}\delta U; \qquad \Delta p = \frac{dp}{dU}\delta U \tag{10.99}$$

By comparing the pressure in the displaced tube with the ambient pressure we can determine whether or not the displacement will increase. So that it does increase at $\delta U < 0$ (i.e., with expansion of the plasma), the pressure of the surrounding plasma must be less than that in the tube, $\Delta p < \delta p$, and at $\delta U > 0$ the reverse relation must obtain, $\Delta p > \delta p$. Thus the stability condition can be written as follows:

$$\frac{dp}{dU} < \frac{\gamma p}{|U|}, \qquad U = -\int \frac{dl}{H} \tag{10.100}$$

(Kadomtsev's criterion).* This criterion can be adjusted if the magnetic field configuration is known. For instance, if the field is induced by a linear current ($H \sim 1/r$ and $|U| = \int_0^{2\pi} r d\varphi/H \sim r^2$), the stability criterion amounts to the condition

$$\left|\frac{dp}{dr}\right| < \frac{2\gamma p}{r} \tag{10.101}$$

which yields the maximum pressure gradient when the pressure in-

*Strictly speaking, this criterion is applicable to configurations with closed lines of force, for which the integral appearing in U is uniquely determined; for open magnetic traps it is necessary to consider the effect of the pressure anisotropy.

creases with U. A plasma with a sharp boundary is stable only when the pressure decreases with increasing $U = -\int dl/H$, that is, if the plasma is placed in the region of the minimum of U. In magnetic configurations where the lengths of the lines of force are approximately equal, the minimum of U corresponds to that of the magnetic field. In more intricate configurations, particularly in those with closed lines of force, the minimum of $U = -\oint dl/H$ may correspond to the minimum of the magnetic field strength averaged over the line of force (the so-called *average field minimum*).

Thus to ensure plasma stability with respect to flute perturbations it is necessary to place the plasma in the region of the minimum, or at least of the average minimum, of the magnetic field. This considerably limits the configuration of the field of magnetic traps suitable for plasma confinement. As shown in Section 8.1, the magnetic field in a current-free space decreases in the direction of the curvature radius of the lines of force. Therefore magnetic traps with convex lines of force (i.e., with lines of force bulging outward) cannot ensure a stable plasma confinement, since the field falls of in them from the center outward. This refers, in particular, to the simplest magnetic traps with mirrors (see Section 8.5, Fig. 8.16). To ensure stability, it is best to use a configuration in which the field increases from center outward, that is, where the lines of force are concave. If the lines of force have both concave and convex sections, the stability is determined by the variation in $U = -\int dl/H$ on transition from the center to the periphery of the plasma.

10.5 EQUILIBRIUM AND STABILITY OF CURRENT-CARRYING PLASMA COLUMN

A plasma can be confined not only by an external magnetic field, but also by a field induced by currents passing through the plasma. The simplest system in which such confinement can be accomplished is a linear electric discharge. At a high-current discharge the magnetic field induced by it is sufficient to ensure the stability of a plasma that does not touch the walls. This method of confinement was widely investigated in the early experiments on controlled fusion. It was named the *pinch effect*; a linear discharge confined by the current field is called a *linear* or *zeta pinch.*

We first consider equilibrium conditions for a cylindrically symmetrical plasma column with a current passing along it (Fig. 10.7). We assume that the current density and the plasma parameters depend exclusively on the radius: $j = j_z(r)$, $p = p(r)$. Obviously, the magnetic

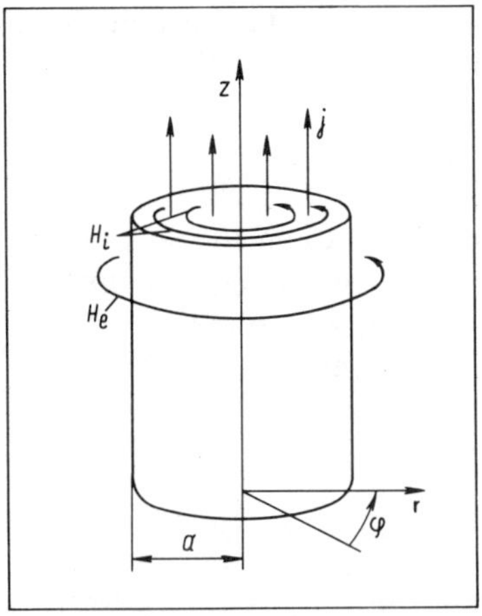

Fig. 10.7 Cylindrical plasma column with a current passing along it.

field has only the azimuthal component $H = H_\varphi(r)$. The relationship between the current density and the field is given by the equations

$$\mathbf{j} = \frac{c}{4\pi}\,\text{curl}\,\mathbf{H}; \qquad j = \frac{c}{4\pi r}\frac{d}{dr}(Hr) \tag{10.102}$$

The total current can be obtained by integrating Eq. 10.102 over the cross section of the plasma column:

$$I = 2\pi \int_0^a jr\,dr = \frac{ca}{2}H(a)$$

where a is the column radius. This equality describes the magnetic field at the plasma boundary

$$H(a) = \frac{2I}{ca} \tag{10.103}$$

Outside the current-carrying plasma column the magnetic field decreases inversely as the radius (this follows from Eq. 10.102 at $j = 0$). Taking this into account, we obtain the well-known relation for the magnetic field of a linear current:

$$H_e(r) = \frac{2I}{cr} \tag{10.104}$$

The equation of plasma equilibrium in a magnetic field of a current (Eq. 10.6), grad $p = (1/c)[\mathbf{j} \times \mathbf{H}]$, can be written as

$$\frac{dp}{dr} = -\frac{H}{4\pi r}\frac{d}{dr}(Hr) \tag{10.105}$$

The right-hand side of the equation represents the current-field interaction force. It can be written in a form similar to Eq. 10.30:

$$\frac{H}{4\pi r}\frac{d}{dr}(Hr) = \frac{d}{dr}\left(\frac{H^2}{8\pi}\right) + \frac{H^2}{4\pi r} \tag{10.106}$$

The first term yields the magnetic pressure gradient and the second, the force associated with the tension of the lines of force. The integral condition of equilibrium is found by multiplying Eq. 10.105 by r^2 and integrating over the radius from 0 to a. Then we get

$$-\int_0^a \frac{dp}{dr} r^2\, dr = \frac{1}{8\pi} a^2 H^2(a) \tag{10.107}$$

Integrating by parts on the left-hand side, we find

$$\int_0^a \frac{dp}{dr} r^2\, dr = -2\int_0^a p(r) r\, dr = -\langle p\rangle a^2 \tag{10.108}$$

where $\langle p\rangle = \int_0^a p(r) r\, dr / \int_0^a r\, dr$ is the pressure averaged over the plasma cross section. Substituting it into Eq. 10.107, we obtain the equilibrium condition in the form

$$\langle p\rangle = \frac{1}{8\pi} H^2(a) \tag{10.109}$$

The right-hand side of the equality can be expressed in terms of the total current (Eq. 10.103). It then yields the relationship between the current and the plasma pressure:

$$I^2 = 2\pi a^2 c^2 \langle p\rangle \tag{10.110}$$

When the temperature of the charged particles is constant throughout the cross section, the pressure is proportional to the concentration $p = n(T_e + T_i)$ and Eq. 10.110 can be represented in a different form:

$$I^2 = 2c^2(T_e + T_i)N \tag{10.111}$$

where $N = \pi a^2\langle n\rangle$ is the linear density of the plasma (number of charged particles per unit length of column).

The relations 10.109–10.111 are the integral conditions of equilibrium for a current-carrying plasma column. They are independent of the radial current distribution. Under actual conditions the current distribution is determined by the method of formation of the plasma and by its conductivity. In pulse discharges at high conductivity, when the pulse duration is much less than the skin time (Eq. 10.39), the current flows over the plasma surface (so-called *skin pinch*). Then the plasma pressure inside the column must be constant, since $j = H = 0$, and the equilibrium condition 10.109 reduces to the equality of the plasma pressure to the magnetic pressure at the boundary $p = H^2(a)/8\pi$. Conversely, when the current density is constant throughout the plasma cross section, Eq. 10.102 describes the linear increase in magnetic field inside a plasma with a radius

$$H_i(r) = \frac{2\pi j}{c} r \tag{10.112}$$

Here, in accordance with Eq. 10.105, the pressure distribution is parabolic:

$$p = p_0\left(1 - \frac{r^2}{a^2}\right) \tag{10.113}$$

where

$$p_0 = \frac{\pi a^2 j^2}{c^2} = \frac{I^2}{\pi a^2 c^2} \tag{10.114}$$

The conditions of equilibrium of a current-carrying plasma column can readily be extended to the case where there is a longitudinal magnetic field $H_z(r)$ inside and outside the plasma. Then the equilibrium equation takes the form

$$\frac{dp}{dr} = -\frac{1}{c}[\mathbf{H} \times \operatorname{curl} \mathbf{H}]_r = -\frac{H_\varphi}{4\pi r}\frac{d}{dr}(H_\varphi r) - \frac{1}{4\pi} H_z \frac{dH_z}{dr} \tag{10.115}$$

It can be written thus:

$$\frac{d}{dr}\left(p + \frac{H_z^2}{8\pi}\right) = -\frac{H_\varphi}{4\pi r}\frac{d}{dr}(H_\varphi r) \tag{10.116}$$

Multiplying Eq. 10.116 by r^2 and integrating from 0 to a, we obtain, instead of Eq. 10.109,

$$\begin{aligned} \langle p \rangle &= \frac{1}{8\pi} H_\varphi^2(a) - \frac{\langle H_{zi}^2 \rangle - H_{ze}^2}{8\pi} \\ &= \frac{I^2}{2\pi c^2 a^2} - \frac{\langle H_{zi}^2 \rangle - H_{ze}^2}{8\pi} \end{aligned} \tag{10.117}$$

where the quantity $\langle H_{zi}^2 \rangle$, as well as $\langle p \rangle$, is averaged over the plasma cross section. This expression takes into account that at the plasma column boundary (at $r = a$) $H_{zi} = H_{ze}$.

The relations obtained yield the equilibrium conditions for a current-carrying plasma column. It should be borne in mind, however, that the passage of a current through a plasma is accompanied by the release of Joule heat. Until this release is offset by losses (due to radiation, thermal conduction, etc.), the temperature, and hence the plasma pressure, must increase, and therefore the equilibrium current (Eq. 10.111) must also increase with time.

Let us now consider the stability of equilibrium of a current-carrying plasma column when the current is skinned, that is, when it flows over the column surface. In this case there is no magnetic field inside the plasma ($\beta = 1$) and, in accordance with the analysis in the preceding section, the stability of the plasma-vacuum interface is determined by the direction of variation in the strength of the external magnetic field. In conformity with Eq. 10.104 the magnetic field decreases with the radius $H \sim 1/r$; therefore $\partial H/\partial \eta = \partial H/\partial r < 0$, and the plasma boundary must be unstable to various types of perturbation—in the first place, to perturbations along the field.

Let us first discuss qualitatively the development of some typical instabilities of a current-carrying plasma column. Since the magnetic field of the current is directed along the azimuth (see Fig. 10.7), a flute perturbation along the field must be azimuthally symmetrical. It may represent a local contraction of the column (sausage instability), a local expansion, or a periodic modulation of the column thickness. The cause for the instability is clear enough. A local change in column radius evidently changes the magnetic field strength at the boundary (Fig. 10.8*a*); namely, it increases in the neck region and weakens in the expansion region (if the current persists, the field is inversely proportional to the radius). The magnetic pressure changes accordingly, whereas the gas-kinetic pressure of the plasma remains the same. Therefore forces appear, which increase the perturbation. This in-

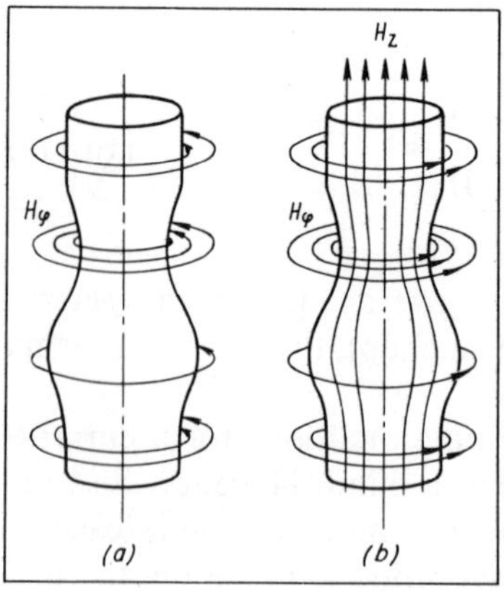

Fig. 10.8 Sausage instability of a plasma column.

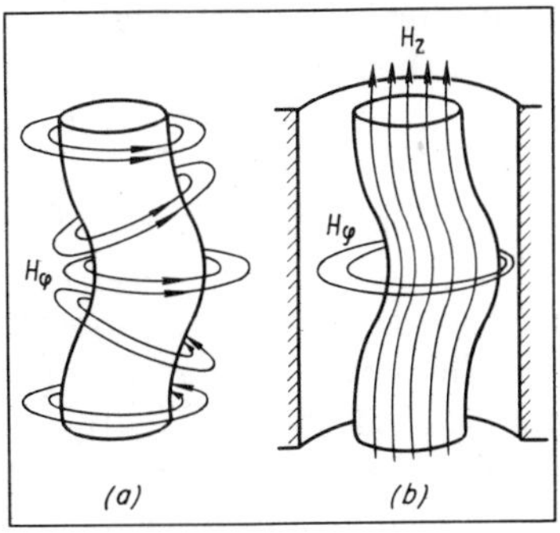

Fig. 10.9 Instability related to the bending of a plasma column.

stability may result in a complete discontinuity of the current at the neck.

Another type of instability is related to the bending of the plasma column (Fig. 10.9*a*). Bending decreases the magnetic field on the side where the generation of the column surface is convex, and increases the field on the other side. This causes a difference in magnetic pressures, which tends to increase the perturbation. Helical perturbations, which are more complicated, are also unstable.

Some of the instabilities can be stabilized by adding a homogeneous magnetic field directed inside the plasma. Then the boundary equilibrium condition reduces to the equality of the sum of the kinetic pressure and the pressures of the internal magnetic field to the external-field pressure:

$$p+\frac{H_{zi}^2}{8\pi}=\frac{H_{\varphi e}^2}{8\pi} \tag{10.118}$$

It is easy to see that the longitudinal field, which is frozen into the plasma, may stabilize sausage perturbations. Indeed, necking bends the boundary, thus producing the corresponding curvature in the lines of force of the longitudinal field and changing its strength in the volume (Fig. 10.8*b*). The tension of the lines of force due to the bending primarily interferes with the development of short neckings, whose length is less than, or comparable with, the radius. With a large length, the tension effect is relatively small, and the main role in stabilization is played by the change in internal magnetic pressure. The magnetic flux frozen into the plasma column is constant: $\pi a^2 H_z = \text{const}$ and $H_z \sim 1/a^2$. The external field is determined by the current and $H_\varphi \sim 1/a$. To suppress the instability it is necessary that the internal magnetic pressure change faster than the external pressure $|\partial(H_z^2/8\pi)/\partial a| > |\partial(H_\rho^2/8\pi)/\partial a|$. Therefore the stability condition reduces to the inequality $H_z^2 > \frac{1}{2}H_\varphi^2$, which yields the minimum value of the longitudinal field necessary for stabilization. The longitudinal field also results (because of the tensile forces) in suppression of bending-type instabilities, whose characteristic length is comparable with the column radius or smaller than it. Long bends are not stabilized by a longitudinal field, however. To stabilize them, use is made of a thick-walled metal casing surrounding a current-carrying plasma (Fig. 10.9*b*). Since the perturbation life is much shorter than the time of penetration of the field into the metal, we can assume that the flux H_φ, which embraces the column and is enclosed between plasma and casing wall, remains unchanged. This increases the magnetic pressure on the external side of the bend (where the gap is smaller) and promotes the suppression of the instability.

A quantitative analysis of the instability of a current-carrying skinned plasma column can be conducted in the same way as in Section 10.4 for a plane boundary. For generality we assumed that both inside (at $r < a$) and outside the plasma (at $r > a$) the azimuthal field $H_\varphi(r)$ and the constant longitudinal field H_z are at equilibrium. Outside the plasma the magnetic field region is bounded by the casing ($r = b$). The main difference from the problem considered in Section 10.4 consists in replacing the plane geometry by a cylindrical one. Therefore the problem must be solved in the cylindrical system of coordinates. Since the equilibrium conditions are independent of the azimuthal and longitudinal coordinates φ and z, the dependence of the perturbations on these coordinates may be assumed exponential. In particular, for the boundary displacement we have

$$\xi_r(r, \varphi, z, t) = \mathrm{Re}[\xi(r) \exp[i(k_z z + k_\varphi a\varphi - \omega t)]] \qquad (10.119)$$

Since a complete traversal along the azimuth brings us back to the same point, the dependence of ξ_r on φ must have a period multiple of 2π: $\xi_r(\varphi) = \xi_r(\varphi + 2\pi)$, that is, $k_\varphi a = m$, where m is a whole number called the *mode of oscillation*. A perturbation with $m = 0$, which is independent of the azimuth, is a periodic sequence of neckings and expansions of the column $\xi \sim \cos k_z z$. A perturbation with $m = 1 \xi \sim \cos(k_z z + \varphi)$ at each cross section represents the column displacement in the direction of $\varphi = -k_z z$. On the whole, such displacements form a screw-type perturbation. Perturbations with larger m have a structure of a multi mode screw. The view of a perturbed column at different values of m is given in Fig. 10.10.

The solution of linearized equations of magnetohydrodynamics with boundary conditions for a perturbation of the type 10.118 leads to a dispersion equation similar to Eq. 10.82 obtained for a plane boundary in Section 10.4. It can be written as

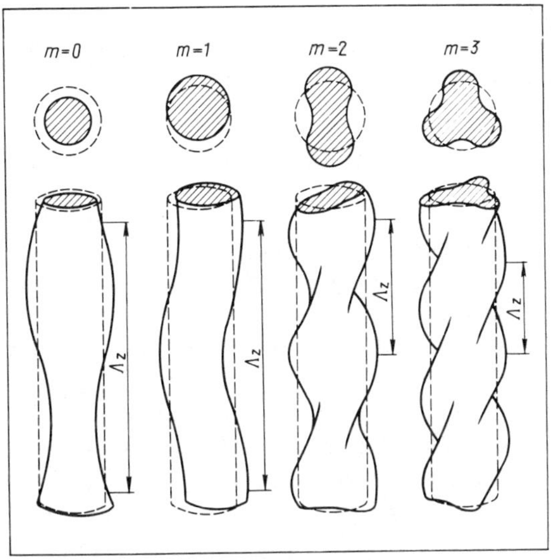

Fig. 10.10 Perturbed column for different values of the mode of oscillation.

$$\rho\omega^2 = -\alpha_1 k_z G + \frac{(\mathbf{kH}_{ig})^2}{4\pi} + \frac{\alpha_2(\mathbf{kH}_{eg})^2}{4\pi} \tag{10.120}$$

where $G = |(d/dr)(H_\varphi^2/8\pi)| = H_\varphi^2(a)/4\pi a$ is an effective force acting on the plasma boundary and equal to the magnetic pressure gradient. It is assumed that the vector $\mathbf{k}$ has a longitudinal projection k_z and an azimuthal k_φ. The coefficients α_1 and α_2, which distinguish Eq. 10.120 from Eq. 10.82, are equal to

$$\alpha_1 = \frac{I'_m(k_z a)}{I_m(k_z a)};$$

$$\alpha_2 = \alpha_1 \frac{K_m(k_z a) I'_m(k_z b) - K'_m(k_z b) I_m(k_z a)}{K'_m(k_z b) I'_m(k_z a) - K'_m(k_z a) I'_m(k_z b)}$$

where I_m and K_m are modified m-order Bessel functions, and I'_m and K'_m are their derivatives. These coefficients take into account the cylindrical character of the geometry and the casing effect. For short waves, when $kb \gg ka \gg 1$, the boundary is practically plane, and the casing effect is small; here $\alpha_1 \sim \alpha_2 \approx 1$. In the absence of a longitudinal magnetic field [$H_i = 0$, $H_{eg} = H_\varphi(a)$, and $(\mathbf{kH}_e) = k_\varphi H_\varphi = (m/a)H_\varphi$] Eq. 10.120 takes the form

$$\omega^2 = \left(\frac{H_{\varphi a}^2}{4\pi\rho}\right)\left(-\frac{\alpha_1 k_z}{a} + \frac{\alpha_2 m^2}{a^2}\right) \tag{10.121}$$

where we introduce the notation $H_\varphi(a) = H_{\varphi a}$. It follows that at small wavelengths perturbations with $m < \sqrt{k_z a}$ are unstable. The perturbation increment is of the order of

$$\gamma \approx \left(\frac{H_{\varphi a}^2 k_z}{4\pi\rho a}\right)^{1/2} \approx \left(\frac{2pk_z}{\rho a}\right)^{1/2} \tag{10.122}$$

[This takes into account the equilibrium conditions (Eq. 10.118)]. It is seen that at $(k_z a) \approx 1$ the instability life $\tau \approx 1/\gamma \approx (1/a)\sqrt{p/\rho} \approx (1/a)\sqrt{T/m_i}$ is of the order of the time of ion displacement by the length of a.

Equation 10.120 makes it possible to analyze quantitatively the conditions of stabilization of various types of instability by the longitudinal magnetic field and the casing. When the stabilization is ensured by the internal longitudinal field; that is, $H_i = H_{zi} \neq 0$ and $H_{ze} = 0$, the dispersion equation takes the form

$$\omega^2 = \frac{k_z^2 H_{\varphi a}^2}{4\pi\rho}\left(-\frac{\alpha_1}{k_z a} + \frac{\alpha_2 m^2}{k_z^2 a^2} + \frac{H_{zi}^2}{H_{\varphi a}^2}\right) \tag{10.123}$$

It can be seen that the equation acquires an additional stabilizing term proportional to H_{zi}^2. Note that its value is limited, since $H_{zi} < H_{\varphi a}$ in conformity with the equilibrium condition. The condition of stabilization of perturbations with $m = 0$ (sausage instabilities) is determined by the relationship between the first and the third terms in Eq. 10.123. This condition amounts to the inequality

$$\frac{H_{zi}^2}{H_{\varphi a}^2} > \left(\frac{\alpha_1}{k_z a}\right)_{\max} = \left[\frac{I_0'(k_z a)}{I_0(k_z a) k_z a}\right]_{\max} = \frac{1}{2}$$

The stabilization of perturbations with $m \neq 0$ is described by the second and third terms of Eq. 10.123. It can be shown that it requires the use of a casing with a radius $b < 5a$.

Suppose there is a strong longitudinal field both inside and outside the plasma column $H_{zi} = H_{ze} \gg H_\varphi$. Here, perturbations with a wavelength of the order of the radius or less cannot increase, since their development will result in large changes of energy associated with the bending of the lines of force of the longitudinal field H_{zi}. The greatest hazard is offered by screw-type instabilities with a longitudinal wavelength similar to the pitch of the line of force at the plasma boundary $\Lambda_z \approx (H_z/H_{\varphi a})2\pi a$. Such a perturbation, which is parallel to the magnetic field, minimally distorts the shape of the lines of force. A considerable influence on the screw-type instability is exerted by crossing lines of force (shear). The lines cross because the azimuthal field falls off with the radius, while the longitudinal field is constant. As a result, the slope of the lines of force

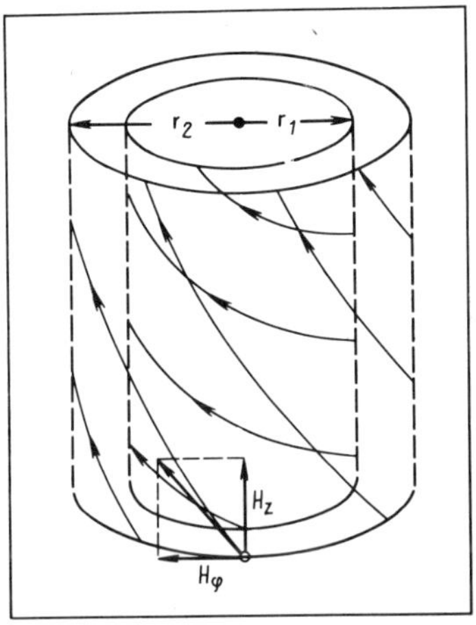

Fig. 10.11 Linear stage of the screw-type instability.

with respect to the column axis diminishes with the radius (Fig. 10.11). If the slope of the perturbation at the column boundary relative to the axis is less than that of the line of force, then the angle between the perturbation and the line of force decreases with increasing perturbation, whose development is thus facilitated. If, however, the perturbation has the shape of a screw with a pitch less than that of the line of force at the boundary, the angle between the magnetic field and the perturbation increases with an increase in r, and therefore its development is impeded. It is clear from the above that perturbations may be considered unstable if

$$\Lambda_z \geqslant \frac{2\pi a H_z}{H_\varphi} \tag{10.124}$$

A quantitative consideration of the stability of the plasma cylinder boundary in a strong longitudinal magnetic field can be carried out using Eq. 10.120. Putting $H_{zi} = H_{ze}$, we get

$$\rho\omega^2 = -\frac{\alpha_1 k_z H_{\varphi a}^2}{4\pi a} + \frac{k_z^2 H_z^2}{4\pi} + \alpha_2 \frac{(k_z H_z + k_\varphi H_{\varphi a})^2}{4\pi} \tag{10.125}$$

Since $H_z \gg H_{\varphi a}$, only long-wave perturbations with $k_z a \ll 1$ can obviously be unstable (otherwise the second and third terms on the right-hand side would exceed the first). An asymptotic representation of the Bessel function here leads to $\alpha_1 \approx m/k_z a$, $\alpha_2 \approx 1$, and the dispersion equation takes the form

$$\rho\omega^2 = -\frac{mH_{\varphi a}^2}{4\pi a^2} + \frac{k_z^2 H_z^2}{4\pi} + \frac{[k_z H_z + (m/a) H_\varphi a]^2}{4\pi} \tag{10.126}$$

It is easy to see that the right-hand side of the equation cannot become negative at $m \geqslant 2$. Therefore only perturbations with $m = 1$ can be unstable. For these we get, at $k_z < 0$,

$$\omega^2 = \left(\frac{k_z^2 H_z^2}{2\pi\rho}\right)\left(1 - \frac{H\varphi\alpha}{H_z|k_z a|}\right) \tag{10.127}$$

It is seen that the plasma boundary is unstable with respect to a screw-type instability with $m = 1$, provided

$$|k_z a| < \frac{H_{\varphi a}}{H_z} \tag{10.128}$$

The maximum increment of the screw-type instability is equal to

$$\gamma_{\max} = \frac{1}{a}\sqrt{H_{\varphi a}^2/8\pi\rho} = \frac{1}{a}\sqrt{p/\rho} \tag{10.129}$$

The condition for a screw-type instability corresponds to Eq. 10.124, which was obtained previously from qualitative considerations. It determines the minimum perturbation wavelength. On the other hand, the wavelength is usually limited to the length of the plasma column l. The limitation is imposed by the conditions at the ends (mostly by the condition of the freezing of the lines of force into the metal electrodes). Therefore, in a plasma column of finite length and with a sufficiently large longitudinal field a screw-type instability cannot develop. In accordance with Eq. 10.124 the stability criterion can be written thus:

$$H_z > \frac{H_\varphi l}{2\pi a} \tag{10.130}$$

It is called the *Shafranov–Kruskal criterion.*

We have considered the linear stage of the screw-type instability. Its development must result in the twisting of the current-carrying plasma column into a helix. Then the helical lines of force outside the column contract and become straight lines (Fig. 10.12a, b). The further development is associated with attraction of the currents in adjacent turns, which must transform the helical column into a hollow cylinder. In this state the magnetic field energy is at a minimum. Thus the screw-type instability is due to the tensile forces, which tend to straighten and shorten the magnetic lines of force.

The discussion of the stability of the plasma column in this section was based on the skinned-current model. Analysis of stability for a plasma cylinder with a distributed current is much more complicated; therefore we do not dwell on it. The stability criterion here is generalization of the criterion of stability to convective perturbations (Eq.

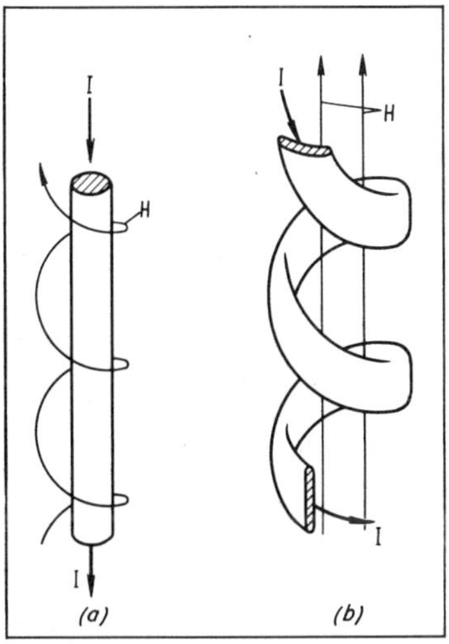

Fig. 10.12 Development of screw-type instability.

10.100) for a wider class of perturbations. They are largely determined by the shear in the plasma volume. For instance, for perturbations with a large m (which can be regarded as local) a sufficient stability criterion (the *Sydem criterion*) can be represented as

$$-8\pi r \frac{dp}{dr} > \frac{H_z^2}{4} \frac{r}{\psi} \frac{d\psi}{dr} \tag{10.131}$$

where $\psi = H_\varphi / rH_z$ gives the pitch of the line of force $s = 2\pi/\psi$, and $d\psi/dr$ characterizes the shear (see also Eq. 8.12).

10.6 EQUILIBRIUM AND STABILITY OF TOROIDAL PLASMA COLUMN

In Chapter 8 we considered the structure of the magnetic field of toroidal magnetic traps formed by superposition of a toroidal (H_θ) and a poloidal (H_φ) magnetic field. The simplest method for setting such traps is used in Tokamak installations, which have gained wide recognition in investigations on controlled fusion. A toroidal magnetic field is induced there by an external solenoid, and a poloidal field by the plasma current; $H_\theta \gg H_\varphi$. In this section we briefly consider the equilibrium and magnetohydrodynamic stability of a toroidal plasma column under such conditions.

Let us first look into the integral equilibrium conditions for a toroidal plasma turn with a current I in the toroidal field (Fig. 10.13), assuming that the condition $H_\varphi \ll H_\theta$ is met and that the minor torus radius is

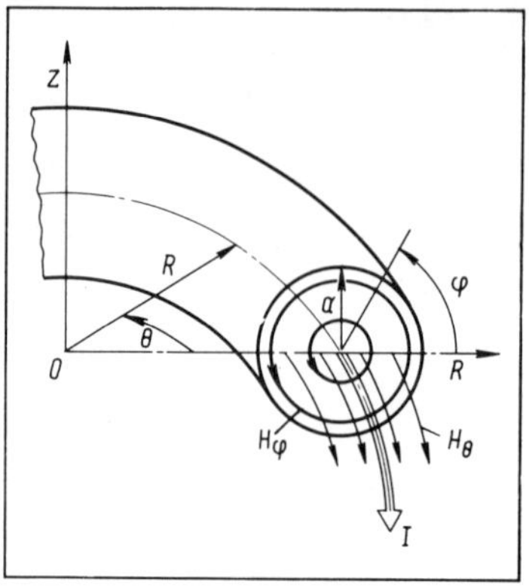

Fig. 10.13 Toroidal plasma in a toroidal magnetic field.

much less than the major ($a \ll R$). The equilibrium equation 10.6 can be expanded in a series with respect to the small ratio a/R. In the zero approximation ($a/R = 0$) the toroidal plasma column becomes cylindrical, and the plasma equilibrium condition becomes (see Eq. 10.117)

$$\langle p\rangle = \frac{H_\varphi^2(a)}{8\pi} + \frac{H_{\Theta e}^2 - \langle H_{\Theta i}^2\rangle}{8\pi} = \frac{I^2}{2\pi a^2 c^2} + \frac{H_{\Theta e}^2 - \langle H_{\Theta i}^2\rangle}{8\pi} \qquad (10.132)$$

This equality is called the condition of equilibrium "with respect to the minor torus radius." In the next approximation one must take into account the forces proportional to a/R, which are due to the toroidality of the plasma column. There are several sources of forces acting toward an increase in the major torus radius and hence in the volume of the plasma column.

These are the gas-kinetic pressure, which tends to expand the plasma volume, the pressure from the side of the inhomogeneous toroidal magnetic field (due account being taken of the tension of the lines of force), and the electrodynamic interaction ("elbowing") of the elements of the current-carrying turn.

To determine the influence of the first two effects, we consider an element of the volume bounded by the toroidal surface and coinciding with the boundary of the plasma column and two planes passing through the major torus axis $0Z$, so that the azimuthal angle $\delta\theta$ between them is small.

To calculate the forces associated with the gas-kinetic pressure we introduce a local system of coordinates xy (Fig. 10.14a). Because of the toroidal symmetry the pressure obviously depends only on these coordinates. The pressure of a plasma enclosed in a small volume element

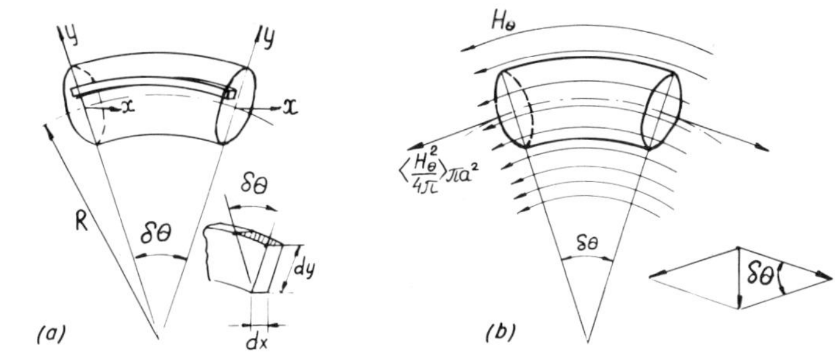

Fig. 10.14 *a* Forces associated with gas-kinetic pressure. *b* Forces associated with the magnetic field.

with a cross section $dx\,dy$ and a length $(R + y)\delta\theta$ creates a radial force due to the inequality of the areas of the surfaces bounding this volume element. It is evident from the figure that the surface external to the torus axis exceeds the internal by $dx\,dy\,\delta\theta$. Hence an uncompensated force arises, which is directed along R and equal to $p(x, y)\delta\theta\,dx\,dy$. Integration over the cross section gives the magnitude of the radial force acting on the segment of the toroidal column with an angular dimension $\delta\theta$, which is equal to $\langle p \rangle \pi a^2 \delta\theta$.

To find the radial force acting on the entire plasma column, the expression for the volume element must be multiplied by $2\pi/\delta\theta$. For the force associated with the gas-kinetic pressure we find

$$\mathcal{F}_p = \langle p \rangle 2\pi^2 a^2 \tag{10.133}$$

The force acting from the side of the magnetic field H_θ can be computed as follows. With no plasma present, the space element under consideration (as well as its immediate environment) contains nothing but the field $H_{\theta e}$ induced by external sources. Thus the forces acting on the lateral surface from inside and outside the torus are mutually compensated. The magnetic field enclosed in the toroidal surface sets up a lateral pressure $H_{\theta e}^2/8\pi$. This pressure, as well as the gas-kinetic, gives rise to a force directed toward increasing R and equal to $\langle H_{\theta e}^2/8\pi \rangle \pi a^2 \delta\theta$. The effect of the tension of the lines of force can be represented as the result of the tensile stresses applied to each of the toroidal surfaces and equal to $\langle H_{\theta e}^2/4\pi \rangle \pi a^2$. Therefore their resultant is equal to $\langle H_{\theta e}^2/4\pi \rangle \pi a^2 \delta\theta$ and directed toward decreasing R (Fig. 10.14*b*). Thus the magnetic field enclosed in the toroidal surface acts with a summary force directed toward the center of curvature of the lines of force and equal to $\langle H_{\theta e}^2/8\pi \rangle \pi a^2 \delta\theta$. The resultant of the forces involved in the pressure of the magnetic field on the lateral (toroidal) surface from the outside clearly has the same magnitude, but acts in the opposite direction.

If the toroidal volume in question encloses a plasma column, the force equilibrium is disturbed. The pressure from the external magnetic field remains unchanged, but the field inside the plasma ($H_{\theta i}$) diminishes (see Eq. 10.132). This gives rise to an uncompensated force $[\langle(H_{\theta e}^2/8\pi) - (H_{\theta i}^2/8\pi)\rangle]\,\pi a^2 \delta\theta$ acting on the isolated element of the plasma column away from the center.

For the force associated with the toroidal magnetic field,

$$\mathcal{F}_H = \frac{\langle H_{\theta e}^2\rangle - \langle H_{\theta i}^2\rangle}{8\pi} 2\pi^2 a^2 \tag{10.134}$$

The force of electrodynamic interaction of the elements of the current-carrying turn is obtained from the relation

$$\mathcal{F}_L = \frac{I^2}{2}\frac{\partial L}{\partial R} \tag{10.135}$$

For inductance, we use the well-known equation

$$L = \frac{4\pi R}{c^2}\left[\ln\frac{8R}{a} - 2 + \frac{l_i}{2}\right] \tag{10.136}$$

where $l_i = \langle H_{\varphi i}^2\rangle / H_{\varphi i}^2(a)$ characterizes the internal inductance of the distributed current per unit length of the column. Substituting Eq. 10.136 into Eq. 10.135, we get

$$\mathcal{F}_L = \frac{2\pi I^2}{c^2}\left(\ln\frac{8R}{a} - 1 + \frac{l_i}{2}\right) \tag{10.137}$$

Summing up Eqs. 10.133, 10.134, and 10.137, we obtain the tensile centrifugal force:

$$\begin{aligned}\mathcal{F}_R &= 2\pi^2 a^2 \left\{\langle p\rangle + \frac{H_{\Theta e}^2 - \langle H_{\Theta i}^2\rangle}{8\pi} + \frac{I^2}{\pi c^2 a^2}\left(\ln\frac{8R}{a} - 1 + \frac{l_i}{2}\right)\right\} \\ &= \frac{2\pi I^2}{c^2}\left(\ln\frac{8R}{a} - \frac{3}{2} + \frac{l_i}{2} + \frac{2\pi a^2 c^2 \langle p\rangle}{I^2}\right)\end{aligned} \tag{10.138}$$

where we used the condition of equilibrium with respect to the minor radius (Eq. 10.132). Compensation of this force is possible only in the presence of an additional magnetic field directed along the torus axis H_Z (see Fig. 10.13). Interaction of such a field with the current, if the directions of the field and current are appropriate, sets up a centripetal force. With a homogeneous field H_Z the force summed over the torus volume is equal to

$$\mathcal{F}_R^* = \frac{1}{c}\int j_\Theta H_Z\,dV = \frac{2\pi R}{c} I H_Z \qquad (10.139)$$

Equating the forces $\mathcal{F}_R$ and $\mathcal{F}_R^*$, we get the value of the field H_Z necessary for equilibrium:

$$H_Z = \frac{I}{cR}\left(\ln\frac{8R}{a} - \frac{3}{2} + \frac{l_i}{2} + \frac{2\pi a^2 c^2}{I^2}\langle p\rangle\right) \qquad (10.140)$$

This equality is called the condition of equilibrium "with respect to the major torus radius."

To maintain equilibrium on variation in plasma parameters, such as current, pressure, and their volume distribution, one must change the field H_Z in a rather complex way in conformity with Eq. 10.140. There is, however, a method for automatic maintenance of the equilibrium value of the field by placing the plasma column in a toroidal metal casing. With a sufficiently large conductivity and thickness of the casing and with the time of penetration of the field through it exceeding the equilibrium time, the casing behaves as a superconductor. The magnetic field external to the plasma, in particular the field of the current, is concentrated in the space between the plasma cylinder and the casing. At the same time, eddy currents are induced in the casing; they are directed oppositely to the plasma current and prevent the penetration of the field into the casing. If the plasma cylinder is equidistant from the chamber walls, the surface density of the eddy currents is such that the forces of repulsion between these currents and the plasma current offset each other. If the plasma cylinder shifts toward the outer wall of the casing, the current density on this wall is higher than on the opposite one. Then a field H_Z is induced inside the chamber, which sets up a force directed inside; that is, the casing "repulses" the plasma column. Thus inside the metal casing the current-carrying toroidal plasma cylinder may be at equilibrium provided it is displaced toward the outer wall of the casing. The necessary bias is achieved automatically under the effect of the tensile forces.

It is easy to obtain the expression for the equilibrium bias when the toroidality is small, that is, when the plasma radius a and the casing radius b are much less than the major torus radius R. Let the axis of the current-carrying cylinder be displaced by a distance Δ relative to the chamber axis (Fig. 10.15). Then the effect of the casing at $b \gg a$ is equivalent to the appearance of an "image" current at a distance $d = b^2/\Delta$ from the axis (the equation is strictly valid for a linear current in a cylinder). The field induced by the image current in the cylinder

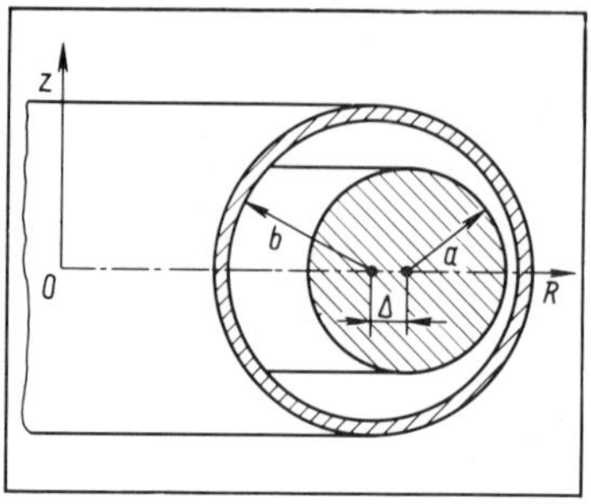

Fig. 10.15 Axis of a current-carrying plasma cylinder displaced relative to the chamber axis.

region is equal to

$$H_Z \approx \frac{2I}{cd} = \frac{2I\Delta}{cb^2} \tag{10.141}$$

It can be considered homogeneous, because $d \gg a$. This field must ensure equilibrium with respect to the major radius. In determining the equilibrium conditions one must also remember that the inductance of a current-carrying turn in the casing differs from Eq. 10.136, being equal to

$$L = 4\pi R \left(\ln \frac{b}{a} + \frac{1}{2} l_i\right) \tag{10.142}$$

Substituting the field (Eq. 10.141) into the equilibrium condition (Eq. 10.140) and allowing for Eq. 10.142, we get the expression for the equilibrium bias, which holds for $b \gg a$:

$$\Delta = \frac{b^2}{2R}\left(\ln \frac{b}{a} + \frac{l_i - 1}{2} + \frac{2\pi a^2 c^2}{I^2} \langle p \rangle\right) \tag{10.143}$$

A more accurate calculation, free from the assumption of the smallness of a/b, leads to the equality

$$\Delta = \frac{b^2}{2R}\left[\ln \frac{b}{a} + \left(1 - \frac{a^2}{b^2}\right)\left(\frac{l_i - 1}{2} + \frac{2\pi a^2 c^2}{I^2} \langle p \rangle\right)\right] \tag{10.144}$$

which is valid for $\Delta \ll b$.

These equations represent integral equilibrium conditions. To be able to correlate the magnetic field and pressure distributions one must take into account the toroidal structure of the magnetic field. As shown in Section 8.1, this structure is determined by the set of toroidal magnetic surfaces inserted in each other (see Fig. 8.3). With small toroidality the meridian cross sections of the surfaces are close to circles whose centers are displaced with respect to each other. As the radius of the

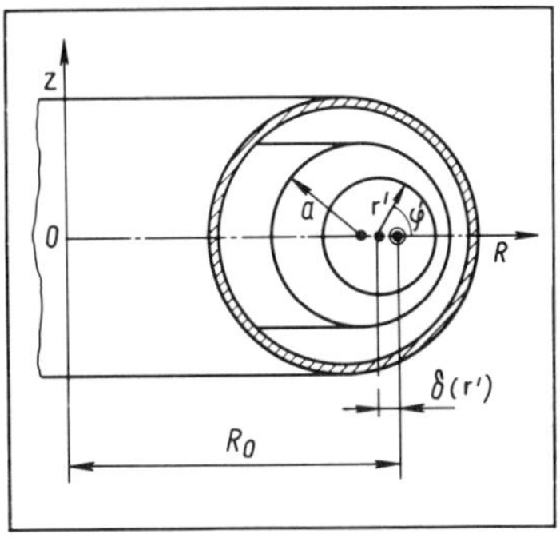

Fig. 10.16 Magnetic structure of a toroidal plasma cylinder.

circles diminishes, they shift toward the major torus radius. This shift is caused by an increase in the tensile forces, associated with the plasma pressure, on transition from the peripheral magnetic surface to the central (to the magnetic axis). In describing the magnetic structure of a toroidal plasma cylinder it is customary to introduce toroidal coordinates: the magnetic surface radius r', the azimuthal angle in the meridian cross section of the torus, φ, and the equatorial angle θ (Fig. 10.16). When the torus is "straightened," these coordinates obviously change to cylindrical ones ($r' \to r$, $\varphi' \to \varphi$, $R\theta \to z$).

When toroidality is small, $a \ll R$, $r' \ll R$, it is easy to determine the magnetic field distribution. To a first approximation with respect to a/R the toroidal (longitudinal) magnetic field is equal to

$$H_\Theta = \frac{H_0}{R} R_0 = H_0 \left[1 - \left(\frac{r'}{R_0}\right) \cos \varphi'\right] \qquad (10.145)$$

where R_0 is the magnetic axis radius. The second term characterizes the toroidal inhomogeneity of the field.

The inhomogeneity of the magnetic field of the current (of the poloidal field) leads to concentration of the magnetic surfaces toward the major radius. As a first approximation with respect to a/R this field can be represented thus:

$$H_\varphi = H_{\varphi 0}(r) \left[1 + \left(\frac{r'}{R_0}\right) \zeta(r') \cos \varphi'\right] \qquad (10.146)$$

The parameter $\zeta(r')$ is related uniquely with the displacement of the center of the magnetic surfaces $\delta(r')$ relative to the magnetic axis

$$\frac{d\delta}{dr'} = \frac{r'}{R}(\zeta + 1); \qquad \delta(r') = \int_0^{r'} \frac{r}{R}(\zeta + 1)\, dr \qquad (10.147)$$

where we put $\delta(0) = 0$.

Substituting the fields 10.145 and 10.146 into the general equilibrium equation (Eq. 10.6), written in the form

$$\operatorname{grad} p = -\frac{1}{4\pi}[\mathbf{H} \times \operatorname{curl} \mathbf{H}]$$

we can find the relationship of the equilibrium distributions of the magnetic field and pressure. With small toroidality the solution of the equation is found from the parameter a/R by the method of successive approximations. Without going into the calculations, we give their results. In the zero approximation (for $a/R \to 0$) the equilibrium equation amounts to Eq. 10.115 and corresponds to straightening the torus into a cylinder. In the next approximation (in which the terms proportional to r'/R and a/R remain) the equilibrium equation makes it possible to find the parameter $\zeta(r')$ appearing in Eq. 10.147:

$$\zeta(r') = \frac{8\pi[\langle p(r')\rangle - p(r')]}{H_\varphi^2(r')} + \frac{l_i(r')}{2} - 1 \tag{10.148}$$

where $\langle p(r')\rangle$ stands for the pressure averaged over the cross section area bounded by a magnetic surface of radius r', and $H_\varphi(r')$ is the field of the current on the surface r'; $l_i(r') = \langle H_\varphi^2(r')\rangle / H_\varphi^2(r')$. [The values of $p(r')$ and $H_\varphi(r')$ are taken in the zero approximation, neglecting the toroidal corrections.] This equality yields the poloidal field distribution (Eq. 10.146) and the relative displacement of the magnetic surfaces (Eq. 10.147).

In particular, for a parabolic pressure distribution and a current density constant throughout the cross section $[p(r') = p_0(l - r'^2/a^2); j = j_0]$ we obtain

$$\zeta(r') = \frac{8\pi\langle p(a)\rangle}{H_\varphi^2(a)} - \tfrac{3}{4}$$

and in accordance with Eq. 10.147,

$$\delta(r') = \frac{r'^2}{2R}\left[\frac{\pi c^2 a^2 p_0}{I^2} + \frac{1}{4}\right] = \frac{r'^2}{2R}\left[\frac{8\pi\langle p(a)\rangle}{H_\varphi^2(a)} + \frac{1}{4}\right] \tag{10.149}$$

It is seen that the shift between the magnetic axis ($r' = 0$) and the magnetic surface centers increases with the plasma pressure. The maximum shift, characterizing the position of the boundary of the magnetic surface with $r' = a$, is determined by the ratio of the average pressure to the magnetic pressure of the poloidal field $\beta_I = 8\pi\langle p\rangle / H_\varphi^2(a)$. In conformity with Eq. 10.149 it is equal to

$$\delta(a) = \frac{a^2}{2R}(\beta_I + \tfrac{1}{4}) \tag{10.150}$$

In estimating the limiting pressure that can be maintained in the toroidal field we can assume that the maximum displacement must be less than the radius. Then we have

$$\beta_I = \frac{8\pi\langle p\rangle}{H_\varphi^2(a)} < \frac{2R}{a} \tag{10.151}$$

More accurate computations, taking into account higher powers of a/R, show that as β_I approaches the limit (Eq. 10.151), the magnetic surfaces deform; their cross sections lose their circular shape and are pressed against the casing. The plasma pressure gradient on its outer surface must then increase abruptly.

Let us now discuss briefly the problem of stability of a toroidal plasma cylinder with respect to small perturbations in its configuration. For regimes in which the current is skinned, that is, when it flows over the column surface, the stability conditions with small toroidality correspond to those in the straight current-carrying plasma column considered in Section 10.5. If the toroidal (longitudinal) field greatly exceeds the poloidal one (the field of the current), a screw-type instability with $m = 1$ is hazardous. The condition of its stabilization in the torus can be obtained from Eq. 10.130, due consideration being given to the fact that the maximum longitudinal wavelength of the perturbation in the torus is equal to its perimeter $l_{\max} = 2\pi R$. Substituting this length into Eq. 10.130, we obtain the Shafranov–Kruskal criterion for a toroidal plasma cylinder

$$H_\Theta > \frac{R}{a} H_\varphi(a) = \frac{2RI}{ca^2} \tag{10.152}$$

which describes the region of stability with respect to a screw-type perturbation. The ratio of the magnetic field to the limiting field (Eq. 10.152) is called the *stability margin*

$$q(a) = \frac{aH_\Theta}{RH_\varphi(a)} \tag{10.153}$$

Analysis of a screw-type instability with a current distributed over the cross section is much more complicated. It turns out that in this case, at $q(a) > 1$, perturbations of the type of a multistart screw with $m > 1$ are possible. Their stabilization can be achieved with the aid of a shear due to a change in the direction of the field H_φ because of a change in radius. For a smooth current distribution $j(r')$ stabilization is attained by increasing the stability margin $q(a)$. The stability condition can be written approximately in the form of the inequality

$$q(r') > \frac{j(r')}{\langle j(r')\rangle} \tag{10.154}$$

which must hold at all r'. Here $j(r')$ is the current density on a magnetic surface of radius r' and $\langle j(r')\rangle$ is the current density averaged over the area bounded by the cross section of this magnetic surface. In particular, the following condition must hold:

$$q(0) = \frac{cH_\Theta}{2\pi R_j(0)} > 1 \qquad (10.155)$$

This is usually a sufficient condition for stability.

A plasma column with a distributed current may also have a convective (interchange) instability, which was considered in Section 10.4. It can be stabilized by placing the plasma in the region of the average minimum of the field, or more precisely, in the region of the minimum of the effective potential $U = -\int (dl/H)$ (Eq. 10.100). For a toroidal plasma with open lines of force the integration limits in the equation for U are not established. Therefore it is best to revert to the expression for the specific volume of the flux tube (Eq. 8.7), according to which

$$U = -\int \frac{dl}{H} = -\frac{\delta V}{\delta \Theta} \qquad (10.156)$$

where δV is the volume of the tube enclosing a magnetic flux $\delta\phi$. Considering the toroidal structure of the field, it is convenient to adopt, for δV, the volume bounded by adjacent magnetic surfaces. Here the quantity $\delta\phi$ is the flux of the longitudinal magnetic field H_θ across the section of this volume. Proceeding from this concept it is easy to find the effective potential U, using Eqs. 10.145 and 10.150, which yield the distribution of the longitudinal field and the position of the magnetic surfaces. Calculation leads to the expression

$$U = -\frac{2\pi R}{H_\Theta}\left[1 - \frac{r'^2}{R^2}(\beta_I + 1)\right] \qquad (10.157)$$

It follows that U increases with r', that is, the region of $r' \approx 0$ adjoining the magnetic axis is the region of the minimum of U (the region of the average minimum of the field). This result is easy to understand. On the peripheral magnetic surfaces the lines of force alternately pass both the regions where the field is lower than on the magnetic axis, $R > R_0$, and the regions where the field is higher, $R < R_0$. But owing to the inward displacement of the surface centers with increasing r' (see Fig. 10.16) the length of the external areas ($R > R_0$) is less than that of the internal ones, and therefore the average magnetic field increases with transition from the magnetic axis to the periphery.

The existence of the average minimum of the longitudinal field promotes stabilization of the convective instability. The existence of the average minimum of the field alone is not sufficient, however, for achieving stability. In a more rigorous study one must give due consideration to the effect exerted on the stability of distribution not only by the poloidal field, but also by the longitudinal. An investigation shows that convective instabilities do not develop if Eq. 10.155 holds. Thus Eq. 10.155 can be considered a sufficient condition for magnetohydrodynamic stability of a current-carrying toroidal plasma cylinder in a strong magnetic field.

BIBLIOGRAPHY

1 GENERAL

Alfvén, H., and Fälthammar, C. Q., *Cosmical Electrodynamics*, Clarendon Press, 1963.

Boyd, T., and Sanderson, J., *Plasma Dynamics*, Nelson, 1969.

Cap, F., *Handbook on Plasma Instabilities*, Vol. 1, Academic Press, 1976.

Chen, F. F., *Introduction to Plasma Physics*, Plenum, 1974.

Drummond, J. E., *Plasma Physics*, McGraw-Hill, 1961.

Frank-Kamenetsky, D. A., *Lektsii po fizike plazmy* [Lectures on Plasma Physics], Atomizdat, 1964.

Krall, N. A., and Trivelpiece, A. W., *Principles of Plasma Physics*, McGraw-Hill, 1973.

Longmire, C. L., *Elementary Plasma Physics*, Wiley-Interscience, 1963.

Smirnov, B. M., *Fizika slaboionizirovannogo gaza* [Physics of Weakly Ionized Gas], Nauka, 1972.

Spitzer, L., *Physics of Fully Ionized Gases*, Wiley, 1962.

Thompson, W. B., *An Introduction to Plasma Physics*, Pergamon, 1962.

Trubnikov, B. A., *Vvedeniye v teoriyu plazmy* [Introduction to Plasma Theory (Lectures)], Parts 1,2, MIFI Press, 1969.

2 COLLISIONS IN PLASMAS

Brown, S. C., *Basic Data on Plasma Physics*, MIT Press, 1967.

Hasted, J. B., *Physics of Atomic Collisions*, Butterworths, 1964.

McDaniel, E. W., *Collisions Phenomena in Ionized Gases*, Wiley, 1964.

Massey, H. S. W., Burhop, E. H. S., and Gilbods, H. R., *Electronic and Ionic Impact Phenomena*, Clarendon, Vol. 1, 1969; Vol. 2, 1969; Vol. 3, 1971; Vol. 4, 1973.

Sivuchin, D. V., Coulomb Collisions in Fully Ionized Plasma, in *Voprosy teorii plazmy* [Problems of Plasma Theory], Issue 4, Atomizdat, 1964.

Smirnov, B. M., *Atomnye stolknoveniya i elementarnye protsessy v plazme* [Atomic Collisions and Elementary Processes in Plasma], Atomizdat, 1968.

Smirnov, B. M., *Iony i vozbuzhdënnye atomy v plazme* [Ions and Excited Atoms in Plasma], Atomizdat, 1974.

Trubnikov, B. A., Particles Collisions in the Fully Ionized Plasma, in *Reviews of Plasma Physics*, Vol. 1, Consultants Bureau, 1965.*

3 KINETIC THEORY AND TRANSPORT PROCESSES IN PLASMAS

Balescu, R., *Statistical Mechanics of Charged Particles*, Academic Press, 1963.

Braginsky, S. I., Transport Phenomena in Plasma, in *Reviews of Plasma Physics*, Vol. 1, Consultants Bureau, 1965.*

Chapman, S., and Cowling, T. G., *The Mathematical Theory of Nonuniform Gases*, Cambridge University, 1958.

Echer, G., *Theory of Fully Ionized Plasmas*, Academic Press, 1972.

Granovsky, V. L., *Elektricheskii tok v gaze* [Electric Current in Gas], Vol. 1, Gostekhizdat, 1952; Vol. 2, Nauka, 1973.

Gurevich, A. V. and Shvartsburg, A. B., *Nelineinaya teoriya rasprostraneniya radiovoln v ionosfere* [Nonlinear Theory of Radiowave Propagation in Ionosphere], Nauka, 1972.

Huxley, L. G. H., and Crompton, R. W., *The Diffusion and Drift of Electrons in Gases*, Wiley, 1974.

Jancel, R., and Kahan, T., *Electrodynamics of Plasmas*, Wiley, 1966.

Klimontovich, Yu. L., *The Statistical Theory of Non-Equilibrium Processes in a Plasma*, Pergamon, 1967.*

Krall, N. A., and Trivelpiece, A. W., *Principles of Plasma Physics*, McGraw-Hill, 1973.

McDaniel, E. W., and Mason, E. A., *The Mobility and Diffusion of Ions in Gases*, Wiley, 1973.

MacDonald, A. D., *Microwave Breakdown in Gases*, Wiley, 1969.

Mitchner, H., and Kruger, C. H., *Partially Ionized Gases*, Wiley, 1973.

Montgomery, D. C., and Tidman, D. A., *Plasma Kinetic Theory*, McGraw-Hill, 1964.

Raizer, N. P., *Lazernaya iskra i rasprostranenie razryadov* (Laser Spark and Charge Propagation], Nauka, 1974.

Shkarofsky, I. P., Yohnston, T. W., and Bachynski, M. P., *The Particle Kinetics of Plasma*, Addison-Wesley, 1966.

Silin, V. P., *Vvedeniye v kineticheskuyu teoriyu gazov* [Introduction to Kinetic Theory of Gases], Nauka, 1971.

Smirnov, B. M., *Fizika slaboionizirovannogo gaza* [Physics of Weakly Ionized Gas], Nauka, 1972.

Van Kampon, N. G., and Felderhaf, B. U., *Theoretical Methods in Plasma Physics*, Wiley, 1967.

Vedenov, A. A., Thermodynamics of Plasma, in *Reviews of Plasma Physics*, Vol. 1, Consultants Bureau, 1965.*

*Hereafter the asterisk denotes translation from the Russian.

4 PLASMAS IN MAGNETIC FIELD

Alfvén, H., and Fälthammar, C. Q., *Cosmical Electrodynamics*, Clarendon Press, 1963.

Artsimovich, L. A., *Upravlyaemye termoyadernye reaktsii* [Controlled Thermonuclear Reactions], Fizmatgiz, 1963.

Artsimovich, L. A., *Configuration de Plasma Fermees*, Universite de France, 1968.*

Artsimovich, L. A., *Zamknutye plazmennye konfiguratsii* [Closed Plasma Configurations], Nauka, 1969.

Braginsky, S. I., Transport Phenomena in Plasma, in *Reviews of Plasma Physics*, Vol. 1, Consultants Bureau, 1965.*

Cowling, T. G., *Megnetohydrodynamics*, Wiley-Interscience, 1957.

Galeyev, A. A., and Sagdeyev, R. Z., Neoclassical Diffusion Theory, in *Voprosy teorii plazmy* [Problems of Plasma Theory], Issue 7, Atomizdat, 1973.

Glasstone, S., and Lovberg, R. H., *Controlled Thermonuclear Reactions*, Van Nostrand, 1960.

Kadomtsev, B. B., Hydromagnetic Stability of Plasma, in *Reviews of Plasma Physics*, Vol. 2, Consultants Bureau, 1966.*

Kadomtsev, B. B., and Pogutse, O. P., Turbulent Processes in Toroidal Systems, in *Voprosy teorii plazmy* [Problems of Plasma Theory], Issue 5, Atomizdat, 1967.

Lehnert, B., *Dynamics of Charged Particles*, North-Holland, 1964.

Longmire, C. L., *Elementary Plasma Physics*, Wiley-Interscience, 1963.

Lukianov, S. Yu., *Goryachaya plazma i upravlyaemyi yadernyi sintez* [Hot Plasma and Controlled Fusion], Nauka, 1975.

Morosov, A. I., and Solovjev, L. S., The Structure of Magnetic Field, in *Reviews of Plasma Physics*, Vol. 2, Consultants Bureau, 1966.*

Morosov, A. I., and Solovjev, L. S., Charged Particles Motion in Electromagnetic Fields, in *Reviews of Plasma Physics*, Vol. 2, Consultants Bureau, 1966.*

Rose, D. J., and Clark, M., *Plasmas and Controlled Fusion*, MIT and Wiley, 1961.

Schmidt, G., *Physics of High Temperature Plasmas*, Academic Press, 1966.

Schafranov, V. D., Plasma Equilibrium in a Magnetic Field, in *Reviews of Plasma Physics*, Vol. 2, Consultants Bureau, 1966.

Simon, A., *An Introduction to Thermonuclear Research*, Pergamon, 1960.

Sivuchin, D. V., The Drift Theory of Charged Particles Motion in Electromagnetic Fields, in *Reviews of Plasma Physics*, Vol. 1, Consultants Bureau, 1965.*

Soloviev, L. S., Hydromagnetic Stability of Closed Plasma Configurations, in *Voprosy teorii plazmy* [Problems of Plasma Theory], Issue 6, Atomizdat, 1972.

Soloviev, L. S., and Shafranov, V. D., Closed Magnetic Configurations, in *Voprosy teorii plazmy* [Problems of Plasma Theory], Issue 5, Atomizdat, 1967.

Spitzer, L., *Physics of Fully Ionized Gases*, Wiley, 1962.

Trubnikov, B. A., *Vvedeniye v teoriyu plazmy* [Introduction to Plasma Theory) (Lectures)], Parts 1, 2, MIFI Press, 1969.

5 OSCILLATIONS AND WAVES IN PLASMAS

Achiezer, A. I., Achiezer, I. A., Polovin, R. V., Sitenko, A. G., and Stepanov, K. N., *Collective Oscillations in a Plasma*, Pergamon, 1967.*

Achiezer, A. I., Achiezer, I. A., Polovin, R. V., Sitenko, A. G., and Stepanov, K. N., *Elektrodinamika plazmy* [Plasma Electrodynamics], Nauka, 1974.

Allis, W. P., Buchsbaum, S. J., and Bers, A., *Waves in Anisotropic Plasmas*, MIT, 1963.

Bekefi, G., *Radiation Processes in Plasmas*, Wiley, 1966.

Brundstatler, J. J., *An Introduction to Waves, Rays, and Radiation in Plasma Media*, McGraw-Hill, 1963.

Budden, K. G., *Radio Waves in the Ionosphere*, Cambridge University, 1961.

Cap, F., *Handbook on Plasma Instabilities*, Vols. 1, 2, Academic Press, 1976.

Davidson, R. S., *Methods in Nonlinear Plasma Theory*, Academic Press, 1971.

Denisse, J. F., and Delcroix, J. L., *Plasma Waves*, Wiley-Interscience, 1963.

Fried, B., and Conte, S., *The Plasma Dispersion Function*, Academic Press, 1961.

Galeyev, A. A., and Sagdeyev, R. Z., Nonlinear Plasma Theory, in *Voprosy teorii plazmy* [Problems of Plasma Theory], Issue 7, Atomizdat, 1973.

Ginsberg, V. L., *The Properties of Electromagnetic Waves in Plasmas*, Pergamon, 1970.*

Ginsberg, V. L., and Ruchadze, A. A., Waves in Magnetoactive Plasmas, in *Handbuch der Physik*, Vol. 49, 1972, p. 395.*

Golant, V. E., *Sverkhvysokochastotnye metody issledovaniya plazmy* [Ultrahigh-Frequency Methods of Plasma Investigation], Nauka, 1968.

Hasegawa, A., *Plasma Instabilities and Nonlinear Effects*, Springer, 1975.

Heald, M. A., and Wharton, C. B., *Plasma Diagnostics with Microwaves*, Wiley, 1965.

Kadomtsev, B. B., *Plasma Turbulence*, Academic Press, 1965.*

Kadomtsev, B. B., *Kollektivnye yavleniya v plazme* [Collective Phenomena in Plasma], Nauka, 1976.

Krall, N. A., and Trivelpiece, A. W., *Principles of Plasma Physics*, McGraw-Hill, 1973.

Michailovskii, A. B., *Theory of Plasma Instabilities*, Vols. 1, 2, Consultants Bureau, 1974.

Shafranov, V. D., Electromagnetic Waves in Plasma, in *Reviews of Plasma Physics*, Vol. 3, Consultants Bureau, 1967.*

Silin, V. P., *Parametricheskoe vozdeistvie izlucheniya bolśhoi moshchnosti* [Parametric Effect of High-Power Radiation], Nauka, 1973.

Silin, V. P., and Rukhadze, A. A., *Electromagnitnye svoistva plazmy i plazmopodobnykh sred* [Electromagnetic Properties of Plasma and Plasma-Like Media], Atomizdat, 1961.

Sitenko, A. G., *Fluktuatsii i nelineinye vzaimodeistviya voln v plazme* [Fluctuations and Nonlinear Wave Interactions in Plasmas], Naukova Dumka, 1977.

Stix, T.H., *The Theory of Plasma Waves*, McGraw-Hill, 1962.

Tidman, D. A., and Krall, N. A., *Shock Waves in Collisionless Plasma*, Wiley, 1971.

Tsytovich, V. N., *Teoriya turbulentnoi plazmy* [Theory of Turbulent Plasmas], Atomizdat, 1971.

Vandenplas, P., *Electron Waves and Resonances in Bounded Plasma*, Wiley, 1968.

Vedenov, A. A., An Introduction to the Theory of Weakly Turbulent Plasmas, in *Reviews of Plasma Physics*, Vol. 3, Consultants Bureau, 1967.*

AUTHOR INDEX

SUBJECT INDEX